NSTAR
2005

editors

Simon Capstick

Volker Crede

Paul Eugenio

Florida State University

Proceedings of the Workshop on the Physics of Excited Nucleons

NSTAR 2005

Florida State University, Tallahassee, USA   12 – 15 October 2005

World Scientific

NEW JERSEY • LONDON • SINGAPORE • BEIJING • SHANGHAI • HONG KONG • TAIPEI • CHENNAI

Published by

World Scientific Publishing Co. Pte. Ltd.
5 Toh Tuck Link, Singapore 596224
USA office: 27 Warren Street, Suite 401-402, Hackensack, NJ 07601
UK office: 57 Shelton Street, Covent Garden, London WC2H 9HE

British Library Cataloguing-in-Publication Data
A catalogue record for this book is available from the British Library.

NSTAR 2005
Proceedings of the Workshop on the Physics of Excited Nucleons

ISBN 1-86094-657-7

Printed in Singapore by B & JO Enterprise

The 10th workshop on *The Physics of Excited Nucleons* was hosted by the Florida State University in Tallahassee, Florida, October 12-15, 2005. This workshop was the latest of a series of successful conferences, which started at Florida State University in 1994, followed by Jefferson Lab (1995), the INT in Seattle (1996), George Washington University (1997), ECT* in Trento (1998), Jefferson Lab (2000), University of Mainz (2001), University of Pittsburgh (2002), and the LPSC in Grenoble (2004). A summary of the Baryon Resonance Analysis Group (BRAG) premeeting, held on October 11, is also included in this volume. The workshop was attended by about 90 scientists from about 15 countries.

The goal of the meeeting was to bring together theoretical and experimental experts on all areas of physics relevant to baryon spectroscopy. The participants presented new experimental results in their talks, from facilities such as BES, BNL, ELSA, GRAAL, JLab, MAMI, and LEPS located all around the world. This unprecedented quality of the data produced by experiments at these facilities is enabling rapid advances in the field. Special emphasis was made on experimental programs designed to measure polarization observables. The participants shared their latest findings in plenary focus sessions on topics such as coupled-channel analysis, strange particle production, doubly-strange baryons (Cascades), polarization experiments and observables, and recent developments in the theoretical description of the spectrum and properties of baryons using lattice QCD, chiral symmetry restoration, coupled-channel unitarized chiral models, and the large N_c approach. A similar emphasis was made in 36 detailed and interesting talks given in two parallel sessions. Each focus session ended with a critical discussion of important issues involving that sessions' speakers and the rest of the participants.

The valuable contributions to the success of the workshop of all of the speakers is gratefully acknowledged. We wish to thank all of our institutional sponsors: the Department of Physics, the College of Arts and Sciences, and the Office of Research at the Florida State University; Thomas Jefferson National Accelerator Facility and the US Department of Energy; the International Society of Technical Environmental Professionals (INSTEP); and the Tallahassee Convention and Visitors Bureau. We are grateful for the advice provided by the International Advisory Committee and the invaluable help of the Organizing Committee. We would like to give special thanks to: Loreen Kollar, for helping make this meeting happen; Eva Crowdis, who tirelessly resolved problems ranging from financial to culinary; Lorena Barahona, Sandy Heath, Nick Nguyen, and Son Nguyen for their hard work with logistics and shuttle services; Ken Ford for his invaluable help with graphics and photographs; Wlodzimierz Blaszczyk and David Caussyn for their expert help with our computing services; and our graduate students Lucasz Blasczyk, Shifeng Chen, Charles Hanretty, Mica Lyczek-Way, and Blake Sharin, who collected talks, prepared presentations, and provided shuttle services to and from the hotel, along with countless other duties. Without the efforts of every one of these people this meeting would not have been successful.

Tallahassee, March, 2006

Simon Capstick, Volker Credé, Paul Eugenio

NSTAR 2005

Workshop on the Physics of Excited Nucleons

October 12-15, 2005

Florida State University, Tallahassee - USA

ORGANIZATION

LOCAL ORGANIZING COMMITTEE

S. Capstick, V. Credé, and P. Eugenio
Florida State University, USA

ORGANIZING COMMITTEE

V. Burkert (JLab), D. Carman (Ohio), S. Dytman (Pittsburgh), M. Manley (Kent State), W. Roberts (JLab/Old Dominion/DOE), S. Schadmand (Jülich), H. Schmieden (Bonn), R. Schumacher (CMU), E. Swanson (Pittsburgh), U. Thoma (Giessen), A. Thomas (JLab), L. Tiator (Mainz)

INTERNATIONAL ADVISORY COMMITTEE

R. Beck (Bonn), C. Bennhold (GWU), B. Briscoe (GWU), J.-P. Chen (JLab), M. Giannini (Genova), L.-Y. Glozman (Graz), R. Gothe (South Carolina), K. Hicks (Ohio), F. Klein (Catholic), S. Krewald (COSY Jülich), B. Krusche (Basel), T.-S.H. Lee (Argonne), V. Metag (Giessen), C. Meyer (CMU), V. Mokeev (Moscow State), T. Nakano (Osaka), B. Nefkens (UCLA), E. Oset (Valencia), M. Ripani (Genova), D.-O. Riska (Helsinki), D. Richards (JLab), M. Sadler (ACU), B. Saghai (Saclay), A. Sandorfi (BNL), S. Strauch (South Carolina), P. Stoler (RPI), T. Walcher (Mainz), D. P. Weygand (JLab), B. S. Zou (Beijing)

CONTENTS

Session on Pentaquarks and Exotics, Recent Experimental Results, GDH Sum Rule

Focus Session on Polarization

MODELS FOR EXTRACTING N^* PARAMETERS FROM MESON-BARYON REACTIONS

T.-S. H. LEE

Physics Division, Argonne National Laboratory Argonne, IL 60439, USA

Models for extracting the nucleon resonance parameters from the data of meson-baryon reactions are reviewed. The development of a dynamical coupled-channel model with $\pi\pi N$ unitarity is briefly reported.

1. Introduction

In the past few years, very extensive data of electromagnetic meson production reactions have been accumulated at several electron and photon facilities. We now have a great opportunity to learn about the properties of nucleon resonances (N^*) from these data. The advance in this direction is an important step toward understanding the non-perturbative QCD. For example, the extracted N-N^* transition form factors could shed lights on the dynamical origins of the confinement of constituent quarks and the meson cloud associated with baryons. Furthermore, it is important to resolve the long-standing "missing" resonance problem.

To make progress, we need to perform amplitude analyses of the data to extract the N^* parameters. More importantly, we need to develop dynamical reaction models to interpret the extracted N^* parameters in terms of QCD. At the present time, the achievable goal is to test the predictions from various QCD-based hadron models such as the well-developed constituent quark model[1] and the covariant model based on Dyson-Schwinger Equations[2]. In the near future, we hope to understand the N^* parameters in terms of Lattice QCD (LQCD).

In the Δ region, both the amplitude analyses and reaction models have been well developed. We find that these two efforts are complementary. For example, all amplitude analyses[3] gave the result that the N-Δ M1 transition strength is $G_M(0) = 3.18 \pm 0.04$. This value is about 40 % larger than the constituent quark model predictions $G_M(0) = \frac{4}{3}\sqrt{\frac{m_N}{2m_\Delta}}G_p(0) \sim 2.30$.

This problem is resolved by developing dynamical models[4,5] within which one can show that the discrepancy is due to the pion cloud which is not included in the constituent quark model calculations. The meson cloud effect on the N-Δ transitions has been further revealed in the study of pion electroproduction reactions. Some of the recent developments are :

- Sixteen reponse functions of $p(\vec{e}, e\vec{p})$ at $Q^2 = 1$ (GeV/c)^2 have been measured[6] at JLab. These data allow for the first time an almost model independent amplitude analysis.
- LQCD calculations of the $\gamma N \to \Delta$ form factors are becoming available[7].
- High precision data for exploring the meson cloud effects in the low Q^2 region have been obtained at JLab and Mainz.
- Double polarization data of $\vec{d}(\vec{\gamma}, \pi N)N$ have been obtained at LEGS of BNL for improving our understanding of the pion photo-production amplitudes on the neutron target.

In the higher mass N^* region, the situation is much more complicated because of many open channels. Any reliable analysis of the meson production data in this higher energy region must be based on a coupled-channel approach. The main objective of this contribution is to report the development in this direction. In section 2, we review most of the models of meson production reactions and also assess the recent coupled-channel analyses. In section 3, the development of a dynamical coupled-channel model with $\pi\pi N$ unitarity will be reported. A summary is given in section 4.

2. Models of Meson Production Reactions

Most of the models for meson production reactions can be derived by considering the following coupled-channel equations

$$T_{a,b}(E) = V_{a,b} + \sum_c V_{a,c} g_c(E) T_{c,b}(E) \,, \tag{1}$$

where $a, b, c = \gamma N,\ \pi N,\ \eta N,\ \omega N,\ KY,\ \pi\Delta,\ \rho N\ \sigma N,\ \cdots$. The interaction term is defined by $V_{a,b} =< a|V|b >$ with

$$V = v^{bg} + v^R \,. \tag{2}$$

Here v^{bg} represents the non-resonant(background) mechanisms such as the tree diagrams illustrated in Figs. 1(a)-(d), and v^R describes the N^* excitation shown in Fig. 1(e). Schematically, the resonant term can be written

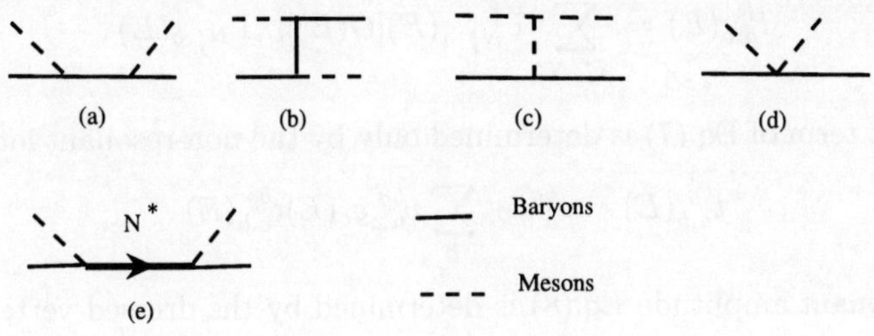

Fig. 1. Tree-diagram mechanisms of meson-baryon interactions.

as

$$v^R(E) = \sum_{N_i^*} \frac{\Gamma_i^\dagger \Gamma_i}{E - M_i^0}\,, \tag{3}$$

where Γ_i defines the decay of the i-th N^* state into meson-baryon states, and M_i^0 is a mass parameter related to the resonance position.

The meson-baryon propagator in Eq.(1) is

$$g_c(E) = < c \mid g(E) \mid c >\,,$$

with

$$\begin{aligned} g(E) &= \frac{1}{E - H_0 + i\epsilon} \\ &= g^P(E) - i\pi\delta(E - H_0)\,, \end{aligned} \tag{4}$$

where H_0 is the free Hamiltonian and

$$g^P(E) = \frac{P}{E - H_0}\,. \tag{5}$$

Here P denotes taking the principal-value part of any integration over the propagator. If $g(E)$ is replaced by $g^P(E)$ and $T_{a,b}(E)$ by $K_{a,b}(E)$, Eq.(1) then defines the K-matrix which is related to T-matrix by

$$T_{a,b}(E) = K_{a,b}(E) - \sum_c T_{a,c}(E)[i\pi\delta(E - H_0)]_c K_{c,b}(E)\,. \tag{6}$$

By using the two potential formulation, one can cast Eq.(1) into

$$T_{a,b}(E) = t^{bg}_{a,b}(E) + t^{R}_{a,b}(E) \tag{7}$$

with

$$t^{R}_{a,b}(E) = \sum_{N^*_i, N^*_j} \bar{\Gamma}^{\dagger}_{N^*_i,a}(E)[G(E)]_{i,j}\bar{\Gamma}_{N^*_j,b}(E)\,. \tag{8}$$

The first term of Eq.(7) is determined only by the non-resonant interaction

$$t^{bg}_{a,b}(E) = v^{bg}_{a,b} + \sum_{c} v^{bg}_{a,c} g_c(E) t^{bg}_{c,b}(E)\,. \tag{9}$$

The resonant amplitude Eq.(8) is determined by the dressed vertex

$$\bar{\Gamma}_{N^*,a}(E) = \Gamma_{N^*,a} + \sum_{b} \Gamma_{N^*,b} g_b(E) t^{bg}_{b,a}(E)\,, \tag{10}$$

and the dressed propagator

$$[G(E)^{-1}]_{i,j}(E) = (E - M^0_i)\delta_{i,j} - \Sigma_{i,j}(E)\,. \tag{11}$$

Here the mass-shift term is

$$\Sigma_{i,j}(E) = \sum_{a} \Gamma^{\dagger}_{N^*,a} g_a(E) \bar{\Gamma}_{N^*_j,a}(E)\,. \tag{12}$$

Note that the meson-baryon propagator $g_a(E)$ for channels including an unstable particle, such as $\pi\Delta$, ρN and σN, must be modified to include a width. In the Hamiltonian formulation[4], this amounts to the following replacement

$$g_a(E) \to < a \mid \frac{1}{E - H_0 - < a|\Sigma_V(E)|a >} \mid a >, \tag{13}$$

where the mass shift operator is

$$\Sigma_V(E) = \sum_{i} \Gamma^{\dagger}_{V}(i) \frac{P_{\pi\pi N}}{E - H_0 + i\epsilon} \Gamma_V(i)\,, \tag{14}$$

with

$$\Gamma_V = \Gamma_{\Delta,\pi N} + h_{\rho,\pi\pi} + h_{\sigma,\pi\pi}\,. \tag{15}$$

Clearly, Γ_V describes the decay of ρ, σ or Δ in the unstable two-particle channels.

With the above equations, we now can derive in the next few subsections most of the recent models for analyzing the data of meson production reactions.

2.1. *Unitary Isobar Model*

In recent years, the most widely used model is the Unitary Isobar Model (UIM) developed[8] by the Mainz group. This model, called MAID, is based on the on-shell relation Eq.(6). By including only one hadronic channel, πN (or ηN), Eq.(6) leads to

$$T_{\pi N,\gamma N} = e^{i\delta_{\pi N}} cos\delta_{\pi N} K_{\pi N,\gamma N} , \tag{16}$$

where $\delta_{\pi N}$ is the pion-nucleon scattering phase shift. By further assuming that $K \to V = v^{bg} + v^R$, Eq.(16) can be cast into the following form

$$T_{\pi N,\gamma N}(UIM) = e^{\delta_{\pi N}} cos\delta_{\pi N}[v^{bg}_{\pi N,\gamma N}] + \sum_{N^*_i} T^{N^*_i}_{\pi N,\gamma N}(E) . \tag{17}$$

The non-resonant term v^{bg} in Eq.(17) is calculated from the standard Born terms but with an energy-dependent mixture of pseudo-vector (PV) and pseudo-scalar (PS) πNN coupling. For the resonant term in Eq.(17), MAID uses the following Walker's parameterization[10]

$$T^{N^*_i}_{\pi N,\gamma N}(E) = f^i_{\pi N}(E) \frac{\Gamma_{tot} M_i e^{i\Phi}}{M_i^2 - E^2 - iM_i\Gamma^{tot}} f^i_{\gamma N}(E) \bar{A}^i , \tag{18}$$

where $f^i_{\pi N}(E)$ and $f^i_{\gamma N}(E)$ are the form factors describing the decays of N^*, Γ^{tot} is the total decay width, $\bar{A}^i$ is the $\gamma N \to N^*$ excitation strength. The phase Φ is required by the unitarity condition and is determined by an assumption that relates the phase of the total photoproduction amplitude to the πN scattering phase shift.

The UIM developed by the JLab-Yerevan collaboration[9] is similar to MAID. The main difference is that this model uses the Regge parameterization to define the amplitudes at high energies.

Both MAID and JLab-Yeveran UIM have been applied extensively to analyze the data of π and η production reactions. However, their results must be further examined since the important two-pion production channels are not treated explicitly in thses two models. Attempt is being made to improve the JLab-Yeveran approach, as reported by V. Mokeev and I.G. Aznauryan in this proceeding.

2.2. *VPI-GWU Model*

The VPI-GWU model[11] (SAID) can be derived from Eq.(6) by considering three channels: γN, πN, and $\pi\Delta$. The solution of the resulting 3×3 matrix equation leads to

$$T_{\gamma N,\pi N}(SAID) = A_I(1 + iT_{\pi N,\pi N}) + A_R T_{\pi N,\pi N} , \tag{19}$$

where

$$A_I = K_{\gamma N,\pi N} - \frac{K_{\gamma N,\pi\Delta}K_{\pi N,\pi N}}{K_{\pi N,\pi\Delta}}, \qquad (20)$$

$$A_R = \frac{K_{\gamma N,\pi\Delta}}{K_{\pi N,\pi\Delta}}. \qquad (21)$$

In actual analyses, they simply parameterize A_I and A_R as

$$A_I = [v^{bg}_{\gamma N,\pi N}] + \sum_{n=0}^{M} \bar{p}_n z Q_{l_\alpha+n}(z), \qquad (22)$$

$$A_R = \frac{m_\pi}{k_0}(\frac{q_0}{k_0})^{l_\alpha} \sum_{n=0}^{N} p_n(\frac{E_\pi}{m_\pi})^n, \qquad (23)$$

where k_0 and q_0 are the on-shell momenta for pion and photon respectively, $z = \sqrt{k_0^2 + 4m_\pi^2}/k_0$, $Q_L(z)$ is the legendre polynomial of second kind, $E_\pi = E_\gamma - m_\pi(1 + m_\pi/(2m_N))$, and p_n and $\bar{p}_n$ are free parameters. SAID calculates $v^{bg}_{\gamma N,\pi N}$ of Eq.(22) from the standard PS Born term and ρ and ω exchanges. The πN amplitude $T_{\pi N,\pi N}$ needed to evaluate Eq.(19) is also available in SAID.

Once the parameters $\bar{p}_n$ and p_n in Eqs.(22)-(23) are determined, the N^* parameters are then extracted by fitting the resulting amplitude $T_{\gamma N,\pi N}$ at energies near the resonance position to a Breit-Wigner parameterization(similar to Eq.(18)).

Very extensive data of pion photoproduction have been analyzed by SAID. The extension of SAID to also analyze pion electroproduction data is being pursued. Similar to the UIM models described in the previous subsection, the results from SAID must also be examined because it also does not treat the important two-pion production channels explicitly. Furthermore, their parameterizations Eqs.(22)-(23) need to be justified or improved theoretically.

2.3. *Giessen and KVI Models*

The coupled-channel models developed by the Giessen group[12] and the KVI group[13] can be obtained from Eq.(6) by taking the approximation $K = V$. This leads to a matrix equation involving only the on-shell matrix elements of V

$$T_{a,b}(E) \to \sum_c [(1 + iV(E))^{-1}]_{a,c} V_{c,b}(E). \qquad (24)$$

The interaction $V = v^{bg} + v^R$ is calculated from tree diagrams such as those illustrated in Fig.1. The Giessen group has recently completed an analysis with γN, πN, $2\pi N$, ηN, and ωN channels. They find strong evidence for the N^* $D_{15}(1675)$ in $\pi N \to \omega N$ and $F_{15}(1680)$ in $\gamma N \to \omega N$. The KVI group has focused on the hyperon production reactions, as reported by O. Scholten in this proceeding. While the development of these two K-matrix coupled-channel models is an important step forward, their treatments of the important two-pion channels still need improvements.

2.4. *KSU Model*

To derive the Kent State University (KSU) model[14], we first note that the non-resonant amplitude t^{bg}, defined by a *hermitian* v^{bg} in Eq.(9), specifies a S-matrix with the following properties

$$S^{bg}_{a,b}(E) = \delta_{a,b} - 2\pi i\delta(E - H_0)t^{bg}_{a,b}(E) \tag{25}$$

$$= \sum_c \omega^{(+)T}_{a,c}(E)\omega^{(+)}_{c,b}(E)\,, \tag{26}$$

where the non-resonant scattering operator is

$$\omega^{(+)}_{a,c}(E) = \delta_{a,c} + g_a(E)t^{bg}_{a,c}(E)\,. \tag{27}$$

With some derivations, the S-matrix for the scattering T-matrix defined by Eqs.(7)-(14) can then be cast into following $distorted - wave$ form

$$S_{a,b}(E) = \sum_{c,d} \omega^{(+)T}_{a,c}(E)R_{c,d}(E)\omega^{(+)}_{c,b}(E)\,, \tag{28}$$

with

$$R_{c,d}(E) = \delta_{c,d} + 2iT^R_{c,d}(E)\,. \tag{29}$$

(30)

Here we have defined

$$T^R_{c,d}(E) = \sum_{i,j} \Gamma^\dagger_{N^*_i,c}(E)[G(E)]_{i,j}\Gamma_{N^*_j,d}(E)\,. \tag{31}$$

The above set of equations is identical to that used in the KSU model of Ref.[14]. In practice, the KSU model makes the separable parameterizations $T^R \sim [x_1\Gamma_1/2/(E - M_1 - i\Gamma_1/2)] \cdots [x_n\Gamma_n/2/(E - M_n - i\Gamma_n/2)]$ and $\omega^{(+)} = exp(iX_1\Delta_1)\cdots exp(iX_n\Delta_n)$.

In recent years, the KSU model has been applied mainly for extracting the Λ^* and Σ^* resonances from the $S = -1$ meson-baryon amplitudes. It has also been used to analyze the data of $K^-p \rightarrow neutrals$ ($\bar{K}^0 n$, $\pi^0 n$, $\pi^0\Sigma^0 \cdots$) from the Crystal Ball Collaboration. To make further progress in using this model to extract N^* parameters, more theoretical input must be implemented to improve their separable parameterizations of $\omega^{(+)}$ and T^R.

2.5. *The CMB Model*

In 1970's a unitary multi-channel isobar model with analyticity was developed[15] by the Carnegie-Mellon Berkeley(CMB) collaboration for analyzing the πN data. The CMB model can be derived by assuming that the non-resonant potential v^{bg} is also of the separable form of v^R of Eq.(3)

$$v^{bg}_{a,b} = \frac{\Gamma^\dagger_{L,a}\Gamma_{L,b}}{E - M_L} + \frac{\Gamma^\dagger_{H,a}\Gamma_{H,b}}{E - M_H} \,. \tag{32}$$

The resulting coupled-channel equations are identical to Eqs.(7)-(11), except that $t^{bg}_{a,b} = 0$ and the sum over N^*_i is now extended to include these two distant poles L and H.

By changing the integration variables and adding a subtraction term, Eq.(12) leads to CMB's dispersion relations

$$\Sigma_{i,j}(s) = \sum_c \gamma_{i,c}\Phi_c(s)\gamma_{j,c} \,, \tag{33}$$

$$Re[\Phi_c(s)] = Re[\Phi_c(s_0)] + \frac{s - s_{th,c}}{\pi}\int_{s_{th}}^{\infty} \frac{Im[\Phi_c(s')]}{(s'-s)(s'-s_0)} ds' \,. \tag{34}$$

Thus CMB model is analytic in structure which marks its difference with all K-matrix models described above.

The CMB model has been revived in recent years by the Zagreb group[16] and a Pittsburgh-ANL collaboration[17] to extract the N^* parameters from fitting the available empirical πN reaction amplitudes. However, the extension of the CMB model to also analyze the data of electromagnetic meson production reactions remains to be pursued.

2.6. *Dynamical Models*

The dynamical models of meson-baryon reactions are the models which account for the off-shell scattering dynamics through the use of the integral equation Eq.(1) or its equivalence Eqs.(7)-(12). The off-shell dynamics is closely related to the meson-baryon scattering wavefuntions in the short-range region where we want to map out the structure of N^*. Thus the development of dynamical models is an important step to interpret the extracted N^* parameters.

In recent years, the predictions from the dynamical models of Sato and Lee (SL)[4] and the Dubna-Mainz-Taiwan (DMT) collaboration[5] are most often used to analyze the data in the Δ region. The SL model can be derived from Eqs.(7)-(12) by keeping only one resonance $N^* = \Delta$ and two channels $a, b = \pi N, \gamma N$. In solving exactly the resulting equations the non-resonant interactions $v^{bg}_{\pi N,\pi N}$ and $v^{bg}_{\pi N,\gamma N}$ are derived from the standard PV Born terms and ρ and ω exchanges by using an unitary transformation method.

The DMT model also only includes πN and γN channels. They however depart from the exact formulation based on Eq.(1) or Eqs.(7)-(12) by using the Walker's parameterization Eq.(18) to describe the resonant amplitude. Accordingly, their definition of the non-resonant amplitude also differs from that defined by Eq.(9) : $t^{bg}_{c,b}$ in the right-hand side of Eq.(9) is replaced by the full amplitude $T_{c,b}$. Furthermore, they follow MAID to calculate the non-resonant interaction $v^{bg}_{\pi N,\gamma N}$ from an energy-dependent mixture of PS and PV Born terms.

Extensive data of pion photoproduction and electroproduction in the Δ region can be described by both the SL and DMT models. However, they have significant differences in the extracted electric E2 and Coulomb C2 form factors of the $\gamma N \rightarrow \Delta$ transition. Both models show very large pion cloud effects on the $\gamma N \rightarrow \Delta$ transition form factors in the low Q^2 region. New data from JLab and Mainz will further test their predictions.

The Ohio model[18] has also succeeded in describing the data in the Δ region. Despite some differences in treating the gauge invariance problem and the Δ excitation amplitude, its dynamical content is similar to that of SL and DMT models.

Eqs.(7)-(12) are used in a 2-N^* and 3-channels (πN, ηN, and $\pi\Delta$) study[19] of πN scattering in S_{11} partial wave. This work illustrated the extent to which the quark-quark interactions in the constituent quark model can be determined directly by the πN reaction data. Eqs.(7)-(12) have also been used to show that the coupled-channel effects due to πN channel are very large in ω photoproduction[20] and K photoproduction[21]. All N^* identified by the Particle Data Group are included in these two dynamical calculations.

The coupled-channel study of both the πN scattering and $\gamma N \rightarrow \pi N$ in S_{11} channel by Chen et al[22] includes πN, ηN, and γN channels. Their πN scattering calculation is performed by using Eq.(1). In their $\gamma N \rightarrow \pi N$

calculation, they neglect the $\gamma N \to \eta N \to \pi N$ coupled-channel effect, and follow the procedure of the DMT model to evaluate the resonant term in terms of the Walker's parameterization (Eq.(18)). They find that four N^* are needed to fit the empirical amplitudes in S_{11} channel up to $W = 2$ GeV. This approach is being extended to also fit the $\pi N \to \pi N$ and $\gamma N \to \pi N$ amplitudes in all partial waves.

A coupled-channel calculation based on Eq.(1) has been carried out by the Jülich group[24] for πN scattering. They are able to describe the πN phase shifts up to $W = 1.9$ GeV by including πN, ηN, $\pi\Delta$, ρN and σN channels and 5 N^* resonances : $P_{33}(1232)$, $S_{11}(1535)$, $S_{11}(1530)$, $S_{11}(1650)$ and $D_{13}(1520)$. They find that the Roper resonance $P_{11}(1440)$ is completely due to the meson-exchange coupled-channel effects.

A coupled channel calculation based on Eq.(1) for both the πN scattering and $\gamma N \to \pi N$ up to $W = 1.5$ GeV has been reported by Fuda and Alarbi[23]. They include πN, γN, ηN, and $\pi\Delta$ channels and 4 N^* resonances : $P_{33}(1232)$, $P_{11}(1440)$, $S_{11}(1535)$, and $D_{13}(1520)$. They adjust the parameters of their model to fit the empirical multipole amplitudes in low partial waves.

Much simpler coupled-channel calculations have been performed by using separable interactions. In the model of Gross and Surya[25], such separable interactions are from simplifying the meson-exchange mechanisms in Figs 1.(a)-(c) as a contact term like Fig. 1(d). They include only πN and γN channels and 3 resonances: $P_{33}(1232)$, $P_{11}(1440)$ and $D_{13}(1520)$, and restrict their investigation up to $W < 1.5$ GeV. To account for the inelasticities in P_{11} and D_{13}, the $N^* \to \pi\Delta$ coupling is introduced in these two partial waves. The inelasticities in other partial waves are neglected.

A similar separable simplification is also used in the chiral coupled-channel models[26,27] for strange particle production. There the separable interactions are directly deduced from the SU(3) effective chiral lagrangians. They are able to fit the total cross section data for various strange particle production reaction channels without introducing N^* resonance states. In recent years, this model has been further extended by Lutz and Kolomeitsev[28] to also fit the πN scattering data. It will be interesting to explore the consequencies of their model in descrbing the electromagnetic meson production reactions.

3. Dynamical Coupled-channel model with $\pi\pi N$ unitarity

All of the models described in section 2 do not account for all of the effects due to the $\pi\pi N$ channels which contribute about 1/2 of the πN and γN

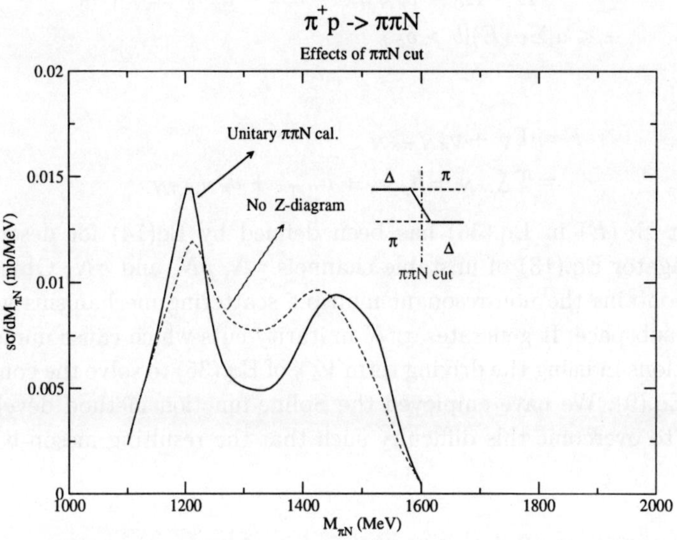

Fig. 2. Invariant mass distribution $d\sigma/dM_{\pi^+p}$ of $\pi^+p \to \pi\pi N$ at $W = 1880$ MeV. The results are from the dynamical coupled-channel model of Ref.29.

total cross sections in the higher mass N^* region. Consequently, the N^* parameters extracted from these models could have uncertainties due to the violation of $\pi\pi N$ unitarity condition. One straightforward way to improve the situation is to extend the Hamiltonian formulation of Ref.[4] to include (a) $\rho \to \pi\pi$ and $\sigma \to \pi\pi$ decay mechanisms as specified in Γ_V of Eq.(15), (b) $v_{\pi\pi}$ for non-resonant $\pi\pi$ interactions, (c) $v_{MN,\pi\pi N}$ for non-resonant $MN \to \pi\pi N$ transitions with $MN = \gamma N$ or πN, and (d) $v_{\pi\pi N,\pi\pi N}$ for non-resonant $\pi\pi N \to \pi\pi N$ interactions. Such a dynamical coupled-channel model has been developed recently by Matsuyama, Sato, and Lee[29] (MSL). The coupled-channel equations from this model can also be cast into the form of Eqs.(7)-(14) except that the driving term of Eq.(9) is replaced by

$$v^{bg}_{a,b} \to \hat{V}_{a,b} = v^{bg}_{a,b} + Z_{a,b}(E)\,, \tag{35}$$

where

$$Z_{a,b}(E) = < a \mid F \frac{P_{\pi\pi N}}{E - H_0 - v_{\pi N,\pi N} - v_{\pi\pi} - v_{\pi\pi N,\pi\pi N} + i\epsilon} F^\dagger \mid b > \\ - < a|\Sigma_V(E)|b > \delta_{a,b} \tag{36}$$

with

$$F = \Gamma_V + v_{\pi N,\pi\pi N} \\ = \Gamma_{\Delta,\pi N} + h_{\rho,\pi\pi} + h_{\sigma,\pi\pi} + v_{\pi N,\pi\pi N} \,. \tag{37}$$

Note that $\Sigma_V(E)$ in Eq.(36) has been defined by Eq(14) for describing the propagator Eq.(13) of unstable channels $\pi\Delta$, ρN, and σN. Obviously $Z_{a,b}(E)$ contains the non-resonant multiple scattering mechanisms within the $\pi\pi N$ subspace. It generates $\pi\pi N$ unitarity cuts which cause numerical complications in using the driving term $\hat{V}_{a,b}$ of Eq.(35) to solve the coupled-channel Eq.(9). We have employed the Spline-function method developed in Ref.[30] to overcome this difficulty such that the resulting meson-baryon

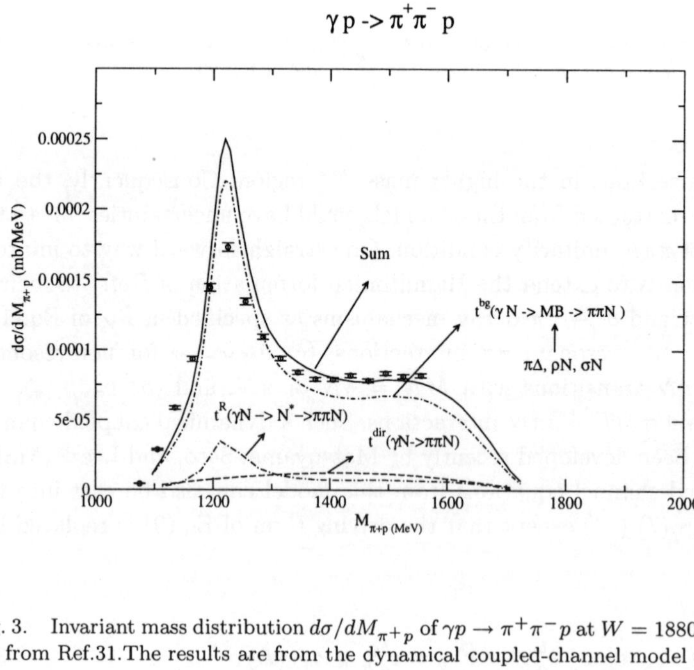

Fig. 3. Invariant mass distribution $d\sigma/dM_{\pi^+p}$ of $\gamma p \to \pi^+\pi^-p$ at $W = 1880$ MeV. Data are from Ref.31.The results are from the dynamical coupled-channel model of Ref.29.

amplitudes can be directly used to calculate the two-pion production cross sections. It is well known that the usual contour-rotation method can not be used easily to calculate two-pion production cross sections.

Since two-pion production accounts for about 1/2 of the γN total cross sections in the N^* region, it is necessary to explore whether our model can describe this rather complex process. This has been the focus of our first calculations which include γN, πN, ηN, $\pi\pi N$ ($\pi\Delta$, ρN, σN) channels and 23 four-star and three-star N^* states listed by the Particle Data Group. The two-pion production amplitudes within our model can be calculated from the following expression

$$T_{\pi\pi N,MN}(E) = t^{dir}_{\pi\pi N,MN}(E) + t^{bg}_{\pi\pi N,MN}(E) + t^{R}_{\pi\pi N,MN}(E), \quad (38)$$

where $MN = \pi N$ or γN, and

$$t^{dir}_{\pi\pi N,MN}(E) = \Omega^{(-)\dagger}_{\pi\pi N}(E) v_{\pi\pi N,MN},$$

$$t^{bg}_{\pi\pi N,MN}(E) = \sum_{c=\pi\Delta,\rho N,\sigma N} \Omega^{(-)\dagger}_{\pi\pi N}(E) < \pi\pi N|F^\dagger_V|c > g_c(E) t^{bg}_{c,MN}(E),$$

$$t^{R}_{\pi\pi N,MN}(E) = \sum_{c=\pi\Delta,\rho N,\sigma N} \Omega^{(-)\dagger}_{\pi\pi N}(E) < \pi\pi N|F^\dagger_V|c > g_c(E) t^{R}_{c,MN}(E),$$

In the above equations $t^R_{c,MN}(E)$ and $t^{bg}_{c,MN}(E)$ are calculated from Eq.(8) and Eq.(9), F_V has been defined by Eq.(37), and the $\pi\pi N$ scattering operator is defined by

$$\Omega^{(-)\dagger}_{\pi\pi N}(E) = 1 + t_{\pi\pi N,\pi\pi N}(E)\frac{1}{E - H_0 + i\epsilon}, \quad (39)$$

where $t_{\pi\pi N,\pi\pi N}(E)$ is the $\pi\pi N$ scattering amplitude calculated only from the non-resonant interactions $v_{\pi N,\pi N}$, $v_{\pi\pi}$ and $v_{\pi\pi N,\pi\pi N}$.

The details of our calculations will be given in Ref.[29]. Here we only present two illustrative results. The first result is to show the importance of the $\pi\pi N$ interaction term $Z_{a,b}(E)$ which contains the effects due to the $\pi\pi N$ cuts. We see in Fig.2 that the predicted invariant mass distribution $d\sigma/dM_{\pi N}$ of $\pi N \to \pi\pi N$ can be significantly different if the term $Z_{a,b}(E)$ is neglected in the calculation. Fig.3 shows that our results for $\gamma p \to \pi^+\pi^- p$ are close to the preliminary data from Jlab[31]. The contributions from three different mechanisms (as indicated in Fig.3) are also shown there.

4. Summary

We have reviewed most of the available models for extracting the N^* parameters from the data of meson-baryon reactions. A recently developed dynamical coupled-channel model with $\pi\pi N$ unitary is briefly reported. The first results from this model will soon be published[29] as a step to develop a program for extracting the N^* parameters from the very extensive data from JLab and other laboratories.

Acknowledgments

This work is support in part by U.S. Department of Energy, Office of Nuclear Physics, under Contract No. W-31-109-ENG-38.

References

1. See the review by Simon Capstick and W. Roberts, Prog.Part.Nucl.Phys. **45**, S241 (2000)
2. See the review by P. Maris and C.D. Roberts, Int.J.Mod.Phys. **E12** 297(2003)
3. R. Arndt et al., Proccedings of the workshop on the Physics of Excited Nucleons, eds. D. Drechsel adnL. Tiator, World Scientific (2001)
4. T. Sato and T.-S. H. Lee, Phys. Rev. **C 54**, 2660 (1996); Phys. Rev. **C63**, 055201 (2001).
5. Kamalov and S.N. Yang, Phys. Rev. Lett. **83**, 4494 (1999); S.S. Kamalov, S. N. Yang, D. Drechsel, O. Hanstein, and L. Tiator, Phys, Rev, **C64**, 032201(R) (2001),
6. J. Kelly et al., nucl-ex/0509004 (2004).
7. C. Alexandrou et al., Phys. Rev. **C69**, 114506 (2004); Phys. Rev. Lett. **94**, 021601 (2005)
8. D. Drechsel, O. Hanstein, S.S. Kamalov, and L. Tiator, Nucl. Phys. **A645**, 145 (1999)
9. I. G. Aznauryan, Phys. Rev. **C68**, 065204 (2003)
10. R.L. Walker, Phys. Rev. **182**, 1729 (1969)
11. R.A. Arndt, I.I. Strakovsky, R.L. Workman, Int. J. Mod. Phys. **A18**, 449 (2003)
12. V. Shklyar, H. Lenske, U. Mosel, G. Penner, Phys. Rev. **C71**, 055206 (2005)
13. A. Usov and O. Scholten, Phys. Rev. **c72**, 025205 (2005)
14. D. M. Manley, Int. J. of Mod. Phys., **A18**, 441 (2003)
15. R.E. Cutkosky, C.P. Forsyth, R.E. Hendrick, and R.L. Kelly, Phys. Rev. **D20**, 2839 (1979).
16. M. Batinic, I. Slaus, and A. Svarc, Phys. Rev. **C51**, 2310 (1995)
17. T.P. Vrana, S.A. Dytman, and T.-S. H. Lee, Phys. Rept. **328**, 181 (2000).
18. G.L. Cala, V. Pascalutsa, J.A. Tjon, and L.E. Wright, Phys. Rev. **C70**, 032201 (2004)

19. Yoshimoto, T. Sato, M. Arima, and T.-S. H. Lee, Phys. Rev. **C61** 065203 (2000).
20. Y. Oh and T.-S. H. Lee, Phys. Rev. **C66**, 045201 (2002).
21. W.-T. Chiang, F. Tabakin, T.-S. H. Lee and B. Saghai, Phys. Lett. **B517**, 101 (2001); B. Julia-Diaz, B. Saghai, T.-S. H. Lee and F. Tabakin, in preparations (2005).
22. G.-Y. Chen, S.S. Kamalov, S. N. Yang, D. Drechsel, and L. Tiator, Nucl. Phys. **A723**, 447 (2003).
23. M.G. Fuda and H. Alharbi, Phys. Rev. **C68**, 064002 (2003)
24. O. Krehl, C. Hanhart, S. Krewald, and J. Speth, Phys. Rev. **C62**, 025270 (2000)
25. F. Gross and Y.Surya, Phys. Rev. **C47**, 703 (1993); Y.Surya and F. Gross, Phys. Rev. **C 53**, 2422 (1996).
26. N. Kaiser, T. Waas, and W. Weise, Nucl. Phys. **A612**, 297 (1997).
27. E. Oset and A. Ramos, Nucl. Phys. **A635**, 99 (1998).
28. M.F. Lutz and E.E. Kolomeitsev, Nucl. Phys. **A700**, 193 (2002)
29. A. Matsuyama, T. Sato, and T.-S. H. Lee, in preparation (2005)
30. A. Matsuyama and T.-S. H. lee, Phys. Rev. **C34**, 1900 (1986)
31. M.Bellis, et. al. (CLAS Collaboration) Proceedings of NSTAR04 Workshop. March 24-27 2004,Grenoble, World Scientific, ed by J.-P.Bocquet, V.Kuznetsvov, D.Rebreyend, 139.

MAID ANALYSIS TECHNIQUES*

LOTHAR TIATOR[1] AND SABIT KAMALOV[1,2]

[1] *Institut für Kernphysik, Universität Mainz, D-55099 Mainz, Germany*
[2] *JINR Dubna, 141980 Moscow Region, Russia*
E-mail: tiator@kph.uni-mainz.de, kamalov@kph.uni-mainz.de

MAID is a unitary isobar model for a partial wave analysis of pion photo- and electroproduction in the resonance region. It is fitted to the world data and can give predictions for multipoles, amplitudes, cross sections and polarization observables in the energy range from pion threshold up to $W = 2$ GeV and photon virtualities $Q^2 < 5$ GeV^2. Using more recent experimental results from Mainz, Bates, Bonn and JLab for Q^2 up to 4.0 GeV^2, the Q^2 dependence of the helicity couplings $A_{1/2}, A_{3/2}, S_{1/2}$ has been extracted for a series of four star resonances. We compare single-Q^2 analyses with a superglobal fit in a new parametrization of Maid2005. Besides the (pion) MAID, at Mainz we maintain a collection of online programs for partial wave analysis of η, η' and kaon photo- and electroproduction which are all based on similar footings with field theoretical background and baryon excitations in Breit-Wigner form.

1. Introduction

In 1998 the first version of MAID (MAID98) was developed and implemented on the web for an easy access for the whole community. MAID98 was based on a unitary isobar model constructed with a limited set of the seven most important nucleon resonances for pion photoproduction in Breit-Wigner form and a non-resonant background with Born terms and t-channel vector meson exchange contributions.[1] The model was unitarized for each partial wave up to the 2π threshold, the region where Watson's theorem is strictly valid. The model was later extended to eight resonances and the unitarization procedure was modified in accordance with the dynamical model (MAID2000). But yet the model was not fitted to the full world data base of pion photo- and electroproduction, and some background parameters were adjusted to multipoles from the SAID partial wave analysis.

*This work is supported by Deutsche Forschungsgemeinschaft (SFB443).

Results for finite Q^2 were mere predictions or extrapolations from photoproduction, where only the magnetic form factor of the Delta excitation was obtained from experimental results.

Since 2003 MAID has become a partial wave analysis program, where all parameters are fitted to experimental observables as cross sections and polarization asymmetries from pion photo- and electroproduction.

Besides the mostly developed Maid program for pion photo- and electroproduction on the nucleon, at Mainz we have developed and collected a series of programs for photo- and electroproduction of π, η, K and η', covering the whole pseudoscalar meson nonet. The programs are available in the internet for online calculations under the URL http://www.kph.uni-mainz.de/MAID. In detail we can report on the following status:

- **MAID** is the main part of our project and describes the reaction $e + N \to e' + N + \pi$ in the kinematical range of 1.073 GeV $< W <$ 2 GeV and $Q^2 <$ 5 GeV^2, see Refs.[1,2]. The current version is $Maid2005$.
- **DMT** (Dubna-Mainz-Taipei) is a dynamical model based on a previous work of the Taipei group[3]. It also describes the reaction $e + N \to e' + N + \pi$ in the same kinematical range as Maid, see Ref. [4,10]. The current version is $DMT2001$.
- **KAON-MAID** is an isobar model developed by Mart and Bennhold. It describes the reaction $e + N \to e' + \{\Lambda, \Sigma\} + K$ and is applicable in the kinematical range of 1.609 GeV $< W <$ 2.2 GeV and $Q^2 <$ 2.2 GeV^2, see Ref.[5,6]. The current version is $KaonMaid2000$.
- **ETA-MAID** is an isobar model for the reaction $e+N \to e'+N+\eta$. It exists in two versions, an older version ($EtaMaid2001$) that can give predictions for photo- and electroproduction of η from proton and neutron. It can be used in the kinematical range of 1.486 GeV $< W <$ 2.0 GeV and $Q^2 <$ 5 GeV^2, see Ref.[7].
  The more recent version ($EtaMaid2003$) incorporates in addition an option to choose Regge tails for the t-channel vector meson exchange. It can only be applied to photoproduction on the proton in the kinematical range of 1.486 GeV $< W < 3.5\ GeV$ and $Q^2 <$ 5 GeV^2, see Ref.[8].
- **ETA'-MAID** is an isobar model with t-channel ω, ρ Regge trajectories for the reaction $\gamma + p \to p + \eta'$ in the kinematical range of 1.896 GeV $< W <$ 3.5 GeV and $Q^2 = 0$, see Ref.[8]. The current

version is *EtaprimeMaid2003*.

- **DR-MAID** is a dispersion theoretical analysis of $e+N \to e'+N+\pi$ and is still in progress. It is based on fixed-t dispersion relations and uses as an input the imaginary parts of the MAID amplitudes, see Ref.[9]. The current version *DrMaid2004* is not yet available on the web.

2. The dynamical approach to meson electroproduction

In the dynamical approach to pion photo- and electroproduction[3,10], the t-matrix is expressed as

$$t_{\gamma\pi}(E) = v_{\gamma\pi} + v_{\gamma\pi}\, g_0(E)\, t_{\pi N}(E)\,, \tag{1}$$

where $v_{\gamma\pi}$ is the transition potential operator for $\gamma^* N \to \pi N$, and $t_{\pi N}$ and g_0 denote the πN t-matrix and free propagator, respectively, with $E \equiv W$ the total energy in the CM frame. A multipole decomposition of Eq. (1) gives the physical amplitude[4]

$$\begin{aligned} t_{\gamma\pi}^{(\alpha)}(q_E, k; E + i\epsilon) &= \exp\left(i\delta^{(\alpha)}\right) \cos\delta^{(\alpha)} \times [v_{\gamma\pi}^{(\alpha)}(q_E, k) \\ &+ P \int_0^\infty dq' \frac{q'^2 R_{\pi N}^{(\alpha)}(q_E, q'; E)\, v_{\gamma\pi}^{(\alpha)}(q', k)}{E - E_{\pi N}(q')}]\,, \end{aligned} \tag{2}$$

where $\delta^{(\alpha)}$ and $R_{\pi N}^{(\alpha)}$ are the πN scattering phase shift and reaction matrix in channel α, respectively; q_E is the pion on-shell momentum and $k = |\mathbf{k}|$ is the photon momentum. The multipole amplitude in Eq. (2) manifestly satisfies the Watson theorem and shows that the γ, π multipoles depend on the half-off-shell behavior of the πN interaction.

In a resonant channel the transition potential $v_{\gamma\pi}$ consists of two terms

$$v_{\gamma\pi}(E) = v_{\gamma\pi}^B + v_{\gamma\pi}^R(E), \tag{3}$$

where $v_{\gamma\pi}^B$ is the background transition potential and $v_{\gamma\pi}^R(E)$ corresponds to the contribution of the bare resonance excitation. The resulting t-matrix can be decomposed into two terms

$$t_{\gamma\pi}(E) = t_{\gamma\pi}^B(E) + t_{\gamma\pi}^R(E)\,. \tag{4}$$

The background potential $v_{\gamma\pi}^{B,\alpha}(W, Q^2)$ is described by Born terms obtained with an energy dependent mixing of pseudovector-pseudoscalar πNN coupling and t-channel vector meson exchanges. The mixing parameters and coupling constants were determined from an analysis of nonresonant multipoles in the appropriate energy regions. In the new version of MAID, the S, P, D and F waves of the background contributions are unitarized in accordance with the K-matrix approximation,

$$t_{\gamma\pi}^{B,\alpha}(\text{MAID}) = \exp\left(i\delta^{(\alpha)}\right)\cos\delta^{(\alpha)} v_{\gamma\pi}^{B,\alpha}(W,Q^2). \tag{5}$$

From Eqs. (2) and (5), one finds that the difference between the background terms of MAID and of the dynamical model is that off-shell rescattering contributions (principal value integral) are not included in MAID, therefore, after re-fitting the data, they are implicitly contained in the resonance part leading to dressed resonances.

Following Ref.[1], we assume a Breit-Wigner form for the resonance contribution $\mathcal{A}_\alpha^R(W,Q^2)$ to the total multipole amplitude,

$$\mathcal{A}_\alpha^R(W,Q^2) = \bar{\mathcal{A}}_\alpha^R(Q^2)\,\frac{f_{\gamma R}(W)\Gamma_R\,M_R\,f_{\pi R}(W)}{M_R^2 - W^2 - iM_R\Gamma_R}\,e^{i\phi}, \tag{6}$$

where $f_{\pi R}$ is the usual Breit-Wigner factor describing the decay of a resonance R with total width $\Gamma_R(W)$ and physical mass M_R. The expressions for $f_{\gamma R}$, $f_{\pi R}$ and Γ_R are given in Ref.[1]. The phase $\phi(W)$ in Eq.(6) is introduced to adjust the phase of the total multipole to equal the corresponding πN phase shift $\delta^{(\alpha)}$. While in the original version of MAID only the 7 most important nucleon resonances were included with mostly only transverse e.m. couplings, in our new version all four star resonances below $W = 2$ GeV are included. These are $P_{33}(1232)$, $P_{11}(1440)$, $D_{13}(1520)$, $S_{11}(1535)$, $S_{31}(1620)$, $S_{11}(1650)$, $D_{15}(1675)$, $F_{15}(1680)$, $D_{33}(1700)$, $P_{13}(1720)$, $F_{35}(1905)$, $P_{31}(1910)$ and $F_{37}(1950)$.

The resonance couplings $\bar{\mathcal{A}}_\alpha^R(Q^2)$ are independent of the total energy and depend only on Q^2. They can be taken as constants in a single-Q^2 analysis, e.g. in photoproduction, where $Q^2 = 0$ but also at any fixed Q^2, where enough data with W and θ variation is available. Alternatively they can also be parametrized as functions of Q^2 in an ansatz like

$$\bar{\mathcal{A}}_\alpha(Q^2) = \bar{\mathcal{A}}_\alpha(0)(1+\beta_1 Q^2 + \beta_2 Q^4 + \cdots)\,e^{-\gamma Q^2}\,. \tag{7}$$

With such an ansatz it is possible to determine the parameters $\bar{\mathcal{A}}_\alpha(0)$ from a fit to the world database of photoproduction, while the parameters β_i and γ can be obtained from a combined fitting of all electroproduction data at different Q^2. The latter procedure we call the 'superglobal fit'. In MAID the photon couplings $\bar{\mathcal{A}}_\alpha$ are direct input parameters. They are directly related to the helicity couplings $A_{1/2}, A_{3/2}$ and $S_{1/2}$ of nucleon resonance excitation. For further details see Ref.[11].

3. Data analysis

The unitary isobar model MAID was used to analyze the world data of pion photo- and electroproduction. In a first step we have fitted the background parameters of MAID and the transverse normalization constants $\bar{\mathcal{A}}_{\alpha}^{R}(0)$ for the nucleon resonance excitation. The latter ones give rise to the helicity couplings shown in Tables 1 and 2. Most of the couplings are in good agreement with PDG and the GW/SAID analysis. It is very typical for such a global analysis, where about 15000 data points are fitted to a small set of 10-20 parameters, that the fit errors appear unrealistically small. Such errors only reflect the statistical uncertainty of the experimental errors, but the model uncertainty can be ten times larger. Therefore we do not report these fit errors which are very similar as in the GW02 fits. The only reliable error estimate can be obtained by comparing different analyses like SAID, MAID and coupled channels analyses.

In Fig. 1 we give a comparison between MAID and SAID for three important multipoles, $E_{0+}(S_{11}), M_{1-}(P_{11})$ and $E_{2-}(D_{13})$. For both analyses we show the global (energy dependent) curves together with the local (single energy) fits, where only data in energy bins of 10-20 MeV are fitted. Such a comparison demonstrates the fluctuations due to a limited data base, especially in the case of the Roper multipole M_{1-}. It also shows systematic differences between the MAID and SAID analyses in the real parts of E_{0+} and E_{2-}. Because of correlations between these amplitudes, these differences cannot be resolved with our current data base. Because of isospin 1/2, they can, however, lead to sizeable differences in the γ, π^{+} channel, where the data base is still quite limited.

In a second step we have fitted recent differential cross section data on $p(e, e'p)\pi^0$ from Mainz[14], Bates[15], Bonn[16] and JLab[17–19]. These data cover a Q^2 range from $0.1 \cdots 4.0$ GeV^2 and an energy range 1.1 GeV $< W <$ 2.0 GeV. In a first attempt we have fitted each data set at a constant Q^2 value separately. This is similar to a partial wave analysis of pion photoproduction and only requires additional longitudinal couplings for all the resonances. The Q^2 evolution of the background is described with nucleon Sachs form factors in the case of the $s-$ and $u-$ channel nucleon pole terms. At the e.m. vertices of the π pole and seagull terms we apply the pion and axial form factors, respectively, while a standard dipole form factor is used for the vector meson exchange. Furthermore, as mentioned above, we have introduced a Q^2 evolution of the transition form factors of the nucleon to N^* and Δ resonances and have parameterized each of the transverse $A_{1/2}$

Table 1. Proton helicity amplitudes at $Q^2 = 0$ of the major nucleon resonances. The results from our own analyses with Maid2003 and the current Maid2005 version are compared to the Particle Data Tables[12] and the GW/SAID[13] analysis. Numbers are given in units of 10^{-3} GeV$^{-1/2}$.

		PDG	GW02	MD03	MD05
$P_{33}(1232)$	$A_{1/2}$	-135±6	-129±1	-140	-137
	$A_{3/2}$	-255±8	-243±1	-265	-260
$P_{11}(1440)$	$A_{1/2}$	-65±4	-67±4	-77	-61
$D_{13}(1520)$	$A_{1/2}$	-24 ±9	-24 ±4	-30	-27
	$A_{3/2}$	166 ±5	135 ±4	166	161
$S_{11}(1535)$	$A_{1/2}$	90±30	30±4	73	66
$S_{31}(1620)$	$A_{1/2}$	27±11	-13±4	71	66
$S_{11}(1650)$	$A_{1/2}$	53±16	74±4	32	33
$D_{15}(1675)$	$A_{1/2}$	19 ±8	33 ±4	23	15
	$A_{3/2}$	15 ±9	9 ±4	24	22
$F_{15}(1680)$	$A_{1/2}$	-15 ±6	-13 ±4	-25	-25
	$A_{3/2}$	133 ±12	129 ±4	134	134
$P_{13}(1720)$	$A_{1/2}$	18 ±30		55	73
	$A_{3/2}$	-19 ±20		-32	-11

Table 2. Neutron helicity amplitudes at $Q^2 = 0$ of the major nucleon resonances. Notation as in Table 1.

		PDG	GW02	MD03	MD05
$P_{11}(1440)$	$A_{1/2}$	40±10	47±5	52	54
$D_{13}(1520)$	$A_{1/2}$	-59 ±9	-67 ±4	-85	-77
	$A_{3/2}$	-139 ±11	-112 ±3	-148	-154
$S_{11}(1535)$	$A_{1/2}$	-46±27	-16±5	-42	-51
$S_{11}(1650)$	$A_{1/2}$	-15±21	-28±4	27	9
$D_{15}(1675)$	$A_{1/2}$	-43 ±12	-50 ±4	-61	-62
	$A_{3/2}$	-58 ±13	-71 ±5	-74	-84
$F_{15}(1680)$	$A_{1/2}$	29 ±10	29 ±6	25	28
	$A_{3/2}$	-33 ±9	-58 ±9	-35	-38
$P_{13}(1720)$	$A_{1/2}$	1 ±15		17	-3
	$A_{3/2}$	-29 ±61		-75	-31

and $A_{3/2}$ and longitudinal $S_{1/2}$ helicity couplings. In a combined fit with all electroproduction data from the world data base of GWU/SAID[13] and the data of our single-Q^2 fit we obtained a Q^2 dependent solution (superglobal fit).

In Fig. 2 we show our results for the $\Delta(1232)$, the $D_{13}(1520)$ and the $F_{15}(1680)$ resonances. Our superglobal fit agrees very well with our single-Q^2 fits, except in the case of the Δ resonance for the 2 lowest points of $S_{1/2}$ from our analysis of the Hall B data. Whether this is an indication

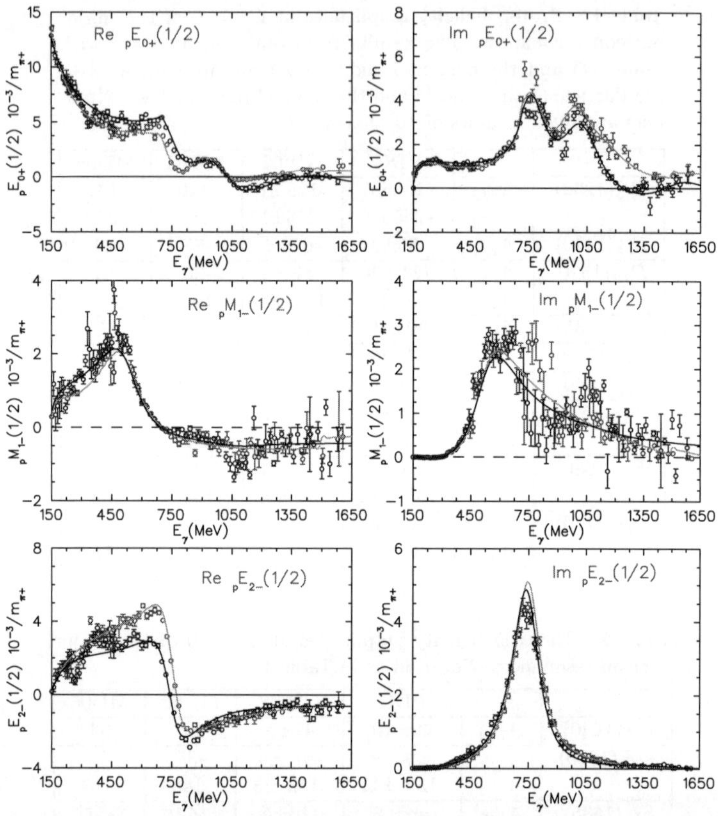

Fig. 1. Comparison of MAID and SAID multipoles. The black lines and points show the Maid2005 global and single-energy solutions, respectively. The SAID solutions are shown in lighter (gray) colours.

for a different Q^2 dependence has still to be investigated. Generally, all our single-Q^2 points are shown with statistical errors from χ^2 minimization only. A much bigger error has to be considered for model dependence.

In Fig. 3 we show our results for the helicity amplitudes of the Roper resonance $P_{11}(1440)$ and the $S_{11}(1535)$ in the region of $Q^2 < 1$ GeV^2. In addition to our own singe-Q^2 analysis we also compare to the analysis of Aznauryan and Burkert[20] who used both an isobar model similar to Maid and an analysis based on fixed-t dispersion relation. In general we get a good agreement with the results of Ref.[20]. Only for the longitudinal excitation of the S_{11} resonance one may observe a different tendency of the Q^2 dependence, however, in this case the statistical fluctuations of our

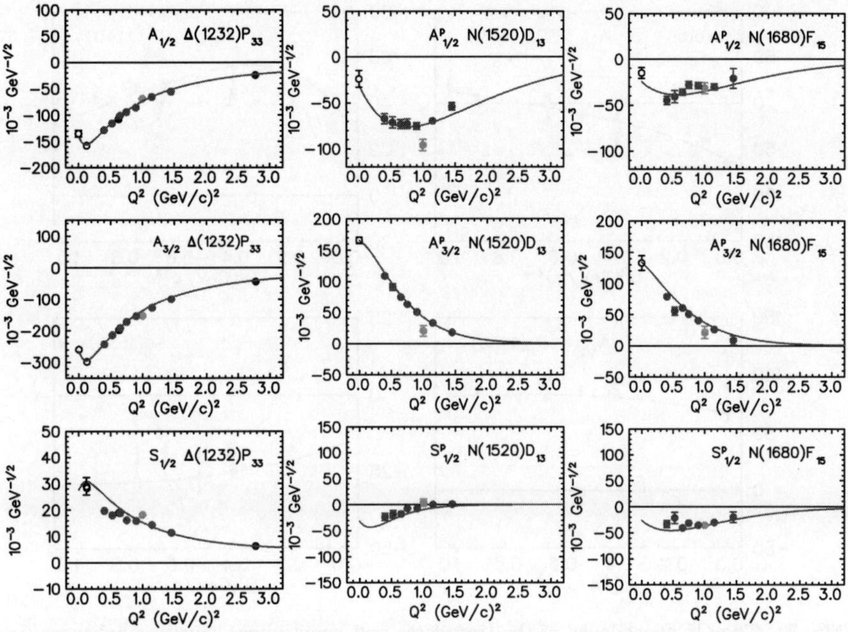

Fig. 2. The Q^2 dependence of the transverse ($A_{1/2}, A_{3/2}$) and longitudinal ($S_{1/2}$) helicity couplings for the $P_{33}(1232)$, $D_{13}(1520)$ and $F_{15}(1680)$ resonance excitation. The solid curves show our superglobal fit. The data points at finite Q^2 are obtained from our single-Q^2 analysis using the data from MAMI and Bates for $Q^2 = 0.1\text{GeV}^2$, from ELSA for 0.6 GeV^2, JLAB(Hall A) for 1.0 GeV^2, JLab(Hall C) for 2.8 GeV^2 and JLab(Hall B) for the remaining points. At the photon point ($Q^2 = 0$) we show our result from Table 1 obtained from the world data base.

analysis is quite large.

Furthermore we also have some empirical results for the partial waves that are not shown here, but most of them come out of the fit with rather large errors bars in the single-Q^2 analysis. This gives us less confidence also for our superglobal fit. The reason for it is mainly that we have fewer data points to analyze at higher energies.

4. Conclusions

Using the world data base of pion photo- and electroproduction and recent data from Mainz, Bonn, Bates and JLab we have made a first attempt to extract all longitudinal and transverse helicity amplitudes of nucleon resonance excitation for four star resonances below $W = 2$ GeV. For this purpose we have extended our unitary isobar model MAID

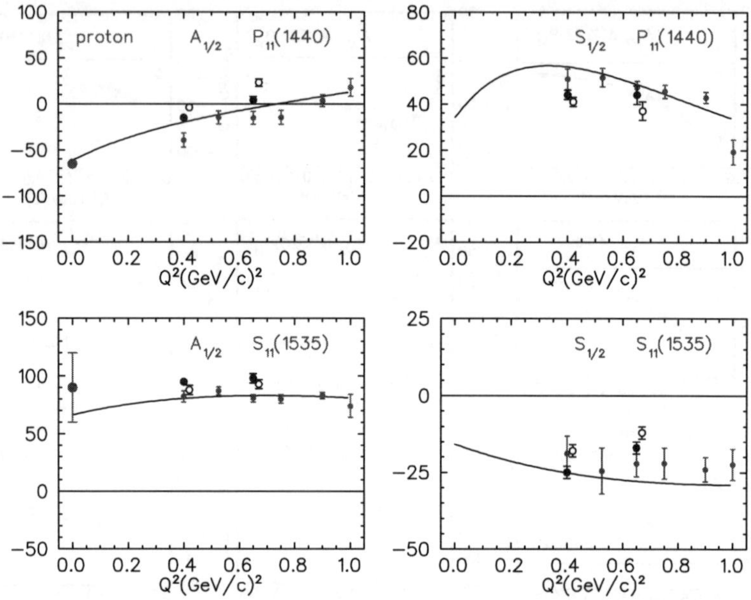

Fig. 3. The Q^2 dependence of the transverse and longitudinal helicity amplitudes for the $P_{11}(1440)$ and the $S_{11}(1535)$ resonance excitation of the proton. The solid lines are the superglobal Maid2005 solutions. The solid red (gray) points are our single-Q^2 fits to the exp. data from CLAS/JLab[18], the solid and open blue circles show the isobar and dispersion analysis of Aznauryan [20] using a similar data set.

and have parameterized the Q^2 dependence of the transition amplitudes. Comparisons between single-Q^2 fits and a Q^2 dependent superglobal fit give us confidence in the determination of the helicity couplings of the $P_{33}(1232), P_{11}(1440), S_{11}(1535), D_{13}(1520)$ and the $F_{15}(1680)$ resonances, even though the model uncertainty of these amplitudes can be as large as 50% for the longitudinal amplitudes of the D_{13} and F_{15}.

For other resonances the situation is more uncertain. However, this only reflects the fact that precise data in a large kinematical range are absolutely necessary. In some cases double polarization experiments are very helpful as has already been shown in pion photoproduction. Furthermore, without charged pion electroproduction, some ambiguities between partial waves that differ only in isospin as S_{11} and S_{31} cannot be resolved without additional assumptions. While all electroproduction results discussed here are only for the proton target, we have also started an analysis for the neutron, where much less data are available from the world data base and no new data has been analyzed in recent years. Since we can very well rely on isospin symmetry, only the electromagnetic couplings of the neutron resonances with isospin 1/2 have to be determined. We have obtained a global solution for the neutron which is implemented in MAID2005. However, for most resonances this is still highly uncertain. So it will be a challenge for the experiment to investigate also the neutron resonances in the near future.

5. Acknowledgements

We wish to thank Cole Smith for having access on recent experimental data. This work was supported in part by the Deutsche Forschungsgemeinschaft (SFB443).

References

1. D. Drechsel, O. Hanstein, S.S. Kamalov and L. Tiator, Nucl. Phys. **A645** (1999) 145; http://www.kph.uni-mainz.de/MAID/.
2. D. Drechsel, S.S. Kamalov and L. Tiator, Maid2005, to be published.
3. S.N. Yang, J. Phys. G **11** (1985) L205.
4. S. Kamalov, S.N. Yang, D. Drechsel, O. Hanstein, L. Tiator, Phys. Rev. C **64** (2001) 032201.
5. F.X. Lee, T. Mart, C. Bennhold, H. Haberzettl, L.E. Wright, Nucl. Phys. **A695** (2001) 237.
6. C. Bennhold, H. Haberzettl and T. Mart, Proc. of 2nd ICTP Int. Conf. on Perspectives in Hadronic Physics, Trieste 1999, p. 328, nucl-th/9909022.
7. W.-T. Chiang, S.N. Yang, L. Tiator and D. Drechsel, Nucl. Phys. **A700** (2002) 429.
8. W.-T. Chiang, S.N. Yang, L. Tiator, M. Vanderhaeghen and D. Drechsel, Phys. Rev. C **68** (2003) 045202.
9. S.S. Kamalov, L. Tiator, D. Drechsel, R.A. Arndt, C. Bennhold, I.I. Strakovsky and R.L. Workman, Phys. Rev. C **66** (2000) 065206.
10. S.S. Kamalov and S.N. Yang, Phys. Rev. Lett. **83** (1999) 4494.
11. L. Tiator, D. Drechsel, S. Kamalov, M.M. Giannini, E. Santopinto, and A. Vassallo, EPJ A **19** (2004) 55.
12. S. Eidelman *et al.* (Particle Data Group), Phys. Lett. B **592** (2004) 1.
13. R.A. Arndt, W.J. Briscoe, I.I.Strakovsky, and R.L. Workman, Phys. Rev. C **66** (2002) 055213; http://gwdac.phys.gwu.edu/.
14. Th. Pospischil *et al.*, Phys. Rev. Lett. **86** (2001) 2959.
15. C. Mertz *et al.*, Phys. Rev. Lett. **86** (2001) 2963.
16. T. Bantes, PhD thesis Bonn 2003, BONN-IR-2003-08.
17. G. Laveissiere *et al.*, Phys. Rev. C **69** (2004) 045203.
18. K. Joo *et al.*, Phys. Rev. Lett. **88** (2002) 122001-1.
19. V.V. Frolov *et al.*, Phys. Rev. Lett. **82** (1999) 45.
20. I. Aznauryan, V. Burkert, H. Egiyan, K. Joo, R. Minehart and L.C. Smith, Phys. Rev. C **71** (2005) 015201.

MESON PRODUCTION ON THE NUCLEON IN THE GIESSEN K-MATRIX APPROACH *

H. LENSKE, V. SHKLYAR, U. MOSEL

Institut für Theoretische Physik, Universität Giessen
Heinrich-Buff-Ring 16
D-35392 Giessen, Germany
E-mail: horst.lenske@physik.uni-giessen.de

Meson production on the proton is described in a coupled channels K-matrix approach. Both hadronic and photonic scenarios are taken into account on the same theoretical footing. At tree-level the Born amplitudes are obtained from an effective Lagrangian with phenomenologically adjustable coupling constants. The spectral distributions, resonances and their width and the background contributions are obtained within the same approach, thus accounting properly for interference effects. Applications to omega- and associated strangeness production on the proton are discussed.

1. Introduction

The nucleon as an entity of strongly interacting constituents is playing the gateway to low-energy QCD as realized in hadrons. Investigating its structure and dynamics by various probes and measuring a variety of observables will complete the much wanted data base on spectral properties of resonances, their excitations and decays, including information about branching ratios into the various meson-nucleon channels. An equally important motivation is the search for *missing resonances* trying to bridge the gap between the number of excited states of the nucleon predicted by quark models and the – at least until now – much fewer resonances seen in pion- or photon-induced reactions. In any respect, research on the nucleon structure will necessarily include the need for a good understanding of reaction dynamics. For that aim a realistic description is necessary, accounting properly for the interplay of various production channels, the interference among resonant and non-resonant parts of the scattering amplitudes. Ac-

*Work partially supported by FZ Jülich

cepting this as a guide line, it is obvious that single channel descriptions in general will fail.

Coupled channel approaches are meeting these goals. They are able to account for the various aspects of meson-nucleon and photon-nucleon interactions and the *cross talk* between the various dynamical sectors. Only with such a more involve description we will be able to distinguish between the dynamical and the intrinsic, QCD related properties of spectral structures observed in cross sections. The Giessen model formulated a few years ago in [1,2] and extended subsequently in [3–5] is in line with these requirements. The model uses a unitary coupled-channel effective Lagrangian approach. It has been successfully applied in the analysis of pion- and photon-induced reactions in the energy region up to 2 GeV. The resonance couplings are simultaneously constrained by available experimental data from all open channels. While our previous analyses [3,4] have been restricted to resonances with spin $J \leq \frac{3}{2}$ we have extended the description quite recently to higher spin states [6], thus enlarging the model space and increasing the predictive power of the calculations. We now include essentially all channels contributing significantly to the cross sections in the energy region up to 2 GeV. The Giessen model is briefly summmarized in sect. 2. We then discuss recent applications to ω-meson production in sect. 3 and to the associated strangeness production in sect. 4. The report closes with a short summary and outlook in sect. 5.

2. The Giessen Model: Coupled Channels K-Matrix Description of Meson Production

The Giessen model describes meson production on the nucleon on the basis of an effective Lagrangian. At the energy scales considered here the appropriate degrees of freedom are the nucleons and hyperons from the basic SU(3) flavor octet and their excited states and, on the meson side, the pseudoscalar and vector meson octet states, supplemented by the photon and the electromagnetic coupling of the hadrons [3]. The principal structure of the model at tree-level is depicted in Fig.1. Here, we only briefly discuss the K-matrix part of the approach. The starting point is the decomposition of a Green function into a principal value and pole part given by a Dirac delta-function:

$$G_{bs} \equiv \frac{P}{H-\omega} + i\pi\delta(H-\omega) \quad . \tag{1}$$

With the propagators cast into this form the Bethe-Salpeter equation can be represented by a set of two coupled equations

$$\mathcal{K} = \mathcal{V} + \mathcal{V}\frac{P}{H-\omega}\mathcal{K} \sim \mathcal{V} \tag{2}$$

$$\mathcal{M} = \mathcal{K} + i\mathcal{K}\delta(H-\omega)\mathcal{M} \tag{3}$$

where the first equation defines the K-matrix, corresponding for a Hermitian Born amplitude $\mathcal{V} = \mathcal{V}^\dagger$ to the real part of the scattering amplitude. The solution of the above equation for a multi-channel problem is a matrix containing the scattering amplitudes from channels α to channels β

$$\mathcal{M}_{\alpha\beta} \equiv \left[\frac{\mathcal{K}}{1-i\mathcal{K}}\right]_{\alpha\beta} \sim \left[\frac{\mathcal{V}}{1-i\mathcal{V}}\right]_{\alpha\beta} \tag{4}$$

where α and β denote any of the photoproduction or hadronic production channels. The validity and quality of the K-matrix approach has been tested positively by various groups, e.g. [7].

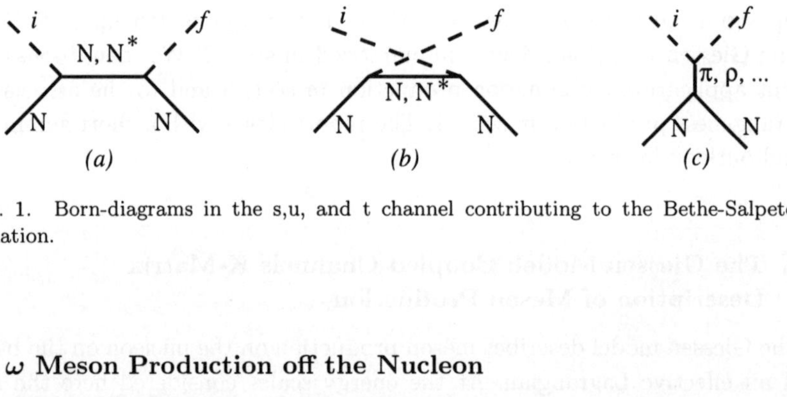

Fig. 1. Born-diagrams in the s,u, and t channel contributing to the Bethe-Salpeter equation.

3. ω Meson Production off the Nucleon

In this section our primary interest is the ω meson production in πp and γp reactions, as discussed in detail in [8]. Most of the theoretical studies of this reaction are based on a rather simplified single channel effective Lagrangian approach, e.g. [11,12]. But there is agreement on the importance of the t-channel π_0-exchange contributions, which were studied by Friman and Soyeur [13]. However, the claims by the various models on the contributions of different resonances to the ωN final state are controversial [8]. Compared to our previous findings [3,4] we observe significant changes by inclusion of spin-$\frac{5}{2}$ resonance contributions. To provide an additional constraint on the resonance couplings to ωN we also included the recent data on the spin density matrix obtained by the SAPHIR group [10].

3.1. *Hadronic Production: $\pi N \to \omega N$*

All experimental data on the ω-meson production in the πN scattering have been measured before 1980 and therefore have rather poor statistics. In total, there are 115 data points which includes differential and total cross sections data. The inclusion of spin-$\frac{5}{2}$ resonance contributions affects the ωN final state considerably. The main contributions close to the threshold come from the P_{13} and D_{15} partial waves. The resonance part of the production amplitude is dominated by the $D_{15}(1675)$ state [8]. Overall, the inclusion of spin-$\frac{5}{2}$ resonances shifts strength to the P_{13} and D_{15} partial waves.

We also find strong contributions from the P_{13} partial wave to the $\pi N \to \omega N$ reaction what has been already reported in [3]. The strength in this partial wave is shifted to the lower energies and becomes more pronounced at the reaction threshold. A peaking behavior seen in the P_{13} partial cross section is due to the interference pattern between P_{13} resonances and background contributions to the ωN channel. Hence, this is a representative example that the collaboration of resonance and background features can produce structures in cross sections which are easily misinterpreted as a resonance. Since the major contributions to the $\pi N \to \omega N$ reaction come from the P_{13} and D_{15} waves, it is interesting to look at the πN inelasticity for these partial waves which are found in [8]. They lead to the conclusion that probably inelasticities from other channels, e.g. $(3\pi N)$, should also be included.

3.2. *Photoproduction: $\gamma N \to \omega N$*

The differential ω meson photoproduction cross sections are presented in Fig.2. With the $\frac{5}{2}$ components included we obtain $\chi^2_{\gamma\omega}$=4.5 which significantly improves our previous result ($\chi^2_{\gamma\omega}$=6.25) [3,4]. The strong π^0 exchange lead to a peaking behavior of the calculated differential cross sections at forward angles which are clearly visible in the SAPHIR measurements [10] above 1.783 GeV and the theoretical cross sections, both displayed in Fig. 2. More detailed information of the production mechanism is obtained from observables measuring the spin degree of freedom of the ω meson. In the Gottfried-Jackson frame, where the initial photon and exchange particle are in their rest frame, and z-axis is in the direction of the incoming photon momentum, the calculation gives $\rho^{GJ}_{00} = 0$. The experimental value of ρ^{GJ}_{00} for forward directions, where the π^0 exchange dominates, was measured by SAPHIR and found to be in the range of $\rho^{GJ}_{00} = 0.2 \cdots 0.3$. Thus,

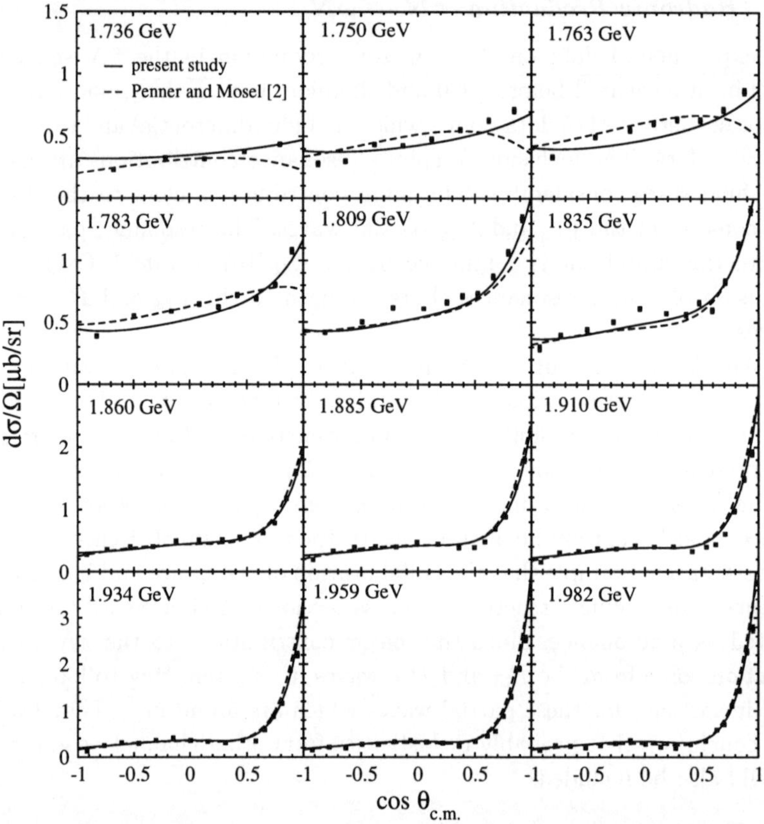

Fig. 2. $\gamma N \to \omega N$ differential cross sections in comparison with the SAPHIR data [10] and our previous results from [4].

the nonzero matrix element testifies that even in this kinematical region other mechanisms (rescattering effects, interference with resonances) must be important.

Beside the π^0 exchange the largest contributions to ω meson photoproduction comes from the subthreshold spin-$\frac{5}{2}$ resonances: $D_{15}(1675)$ and $F_{15}(1680)$. Since the π^0 exchange above 1.8 GeV strongly influences the $\gamma N \to \omega N$ reaction a consistent identification of individual resonance contributions from only the partial wave decomposition is difficult. The $P_{13}(1900)$, and $F_{15}(2000)$, and $D_{13}(1950)$ states which lie above the reaction threshold hardly influence the reaction due to their small couplings to ωN. Despite of the small relative contribution from the D_{15} and F_{15}

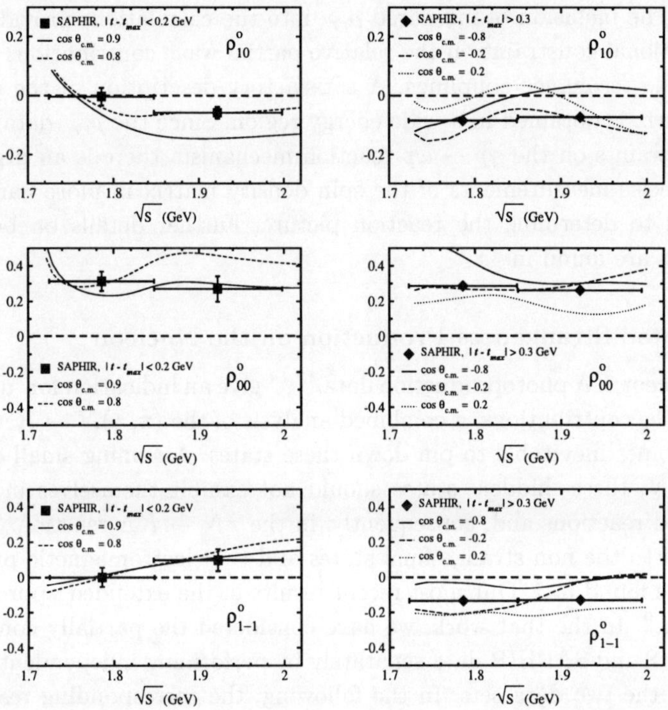

Fig. 3. Spin density matrix elements in the helicity frame compared to the SAPHIR measurements [10].

waves to the ω photoproduction the cross sections are strongly affected by spin-$\frac{5}{2}$ states because of the destructive interference pattern between the π^0 exchange and these resonance contributions.

While $F_{15}(1680)$ plays only a minor role in the $\pi N \to \omega N$ reaction the contribution from this state becomes more pronounced in the ω meson photoproduction because of its large $A^p_{\frac{3}{2}}$ helicity amplitude. The importance of the $F_{15}(1680)$ resonance to the ω meson photoproduction was also found by Titov and Lee [12] and by Zhao [15]. However, in contrast to [12] where also a large effect from $D_{13}(1520)$ was observed we do not find any visible contribution from this state. In fact, as discussed in [8] a strong contribution found in the D_{13} partial wave, resembling a resonance structure, comes in fact from non-resonant π^0 exchange.

The spin density matrix elements $\rho_{rr'}$ extracted from the SAPHIR data [10] are an outcome of the averages over rather wide energy and angle regions, see Fig. 3. The inclusion of measured $\rho_{rr'}$ into the calculations provides a strong additional constraint on the relative partial wave contributions and finally on the resonance couplings. A satisfactory description of the spin density matrix is obtained in a wide energy region. Since the $\rho_{rr'}$ data put strong constraints on the $\gamma p \to \omega p$ reaction mechanism there is an urgent need for precise measurements of the spin density matrix in more narrow energy bins to determine the reaction picture. Further details on beam asymmetries are found in [8].

4. Associated Strangeness Production on the Nucleon

Since the recent $K\Lambda$ photoproduction data [17,18] give an indication for 'missing' resonance contributions, a combined analysis of the $(\pi, \gamma)N \to K\Lambda$ reactions becomes inevitable to pin down these states. Assuming small couplings to πN, these 'hidden' states should not exhibit themselves in the pion-induced reactions and, consequently, in the $\pi N \to K\Lambda$ reaction. The decay ratios to the non-strange final states and the electromagnetic properties can be found in [8]. Our most recent results in the extended approach are given in [9]. In the that work, we have considered the partially contradicting CLAS and SAPHIR data separately by performing independent fits to either of the two data sets. In the following, the corresponding results are denoted by the indices C and S, respectively.

4.1. *Hadronic Strangeness Production: $\pi N \to K\Lambda$*

The S- and C- calculations differ in their description of the non-resonance couplings to $K\Lambda$. As a consequence, different background strengths are obtained for the S_{11}, P_{11}, and P_{13} partial waves while leaving the $P_{13}(1720)$ and $P_{13}(1900)$ resonance couplings almost unchanged [9]. Comparing the S- and C-parameter sets, the largest difference in the resonance parameters is observed for the $P_{11}(1710)$ state. This resonance is found to be almost completely of inelastic origin with a small branching ratio to πN [9]. This state gives only a minor contribution to the reaction and the observed difference in the P_{11} partial wave between S- and C-results is due to the Born term and the t-channel exchange contributions.

The calculated differential cross sections corresponding to the S- and C-coupling sets are found in [9]. Both results show a good agreement with the experimental data in the whole energy region. A difference between the two solutions is only found at forward and backward scattering angles. This is due to the fact that the CLAS photoproduction cross sections rise at backward angles which is not observed by the SAPHIR group (see discussion below). At other scattering angles the S and C results are very similar. The differences between S- and C-calculations are more pronounced for the Λ-polarization. Again, the main effect is seen at the backward angles where the polarization changes its sign in the C-calculations. Unfortunately, the quality of the data does not allow to determine the reaction mechanism further.

4.2. *Photoproduction of Strangeness: $\gamma N \to K\Lambda$*

The older SAPHIR measurements [16] show a resonance-like peak in the total photoproduction cross section around 1.9 GeV. The more recent SAPHIR [17] and CLAS [18] data confirm the previous findings. However, the interpretation of these data is controversial leaving open questions whether in these measurements contributions from presently unknown resonances are observed or if they can be explained by already established reaction mechanisms.

Guided by the results of [4] we have performed a new coupled-channel study of this reaction using separately the CLAS and SAPHIR measurements as two independent input sets. The main difference between the CLAS and SAPHIR data is seen at backward and forward directions, Fig. 4. Both measurements show two peaks but disagree in the absolute values of the corresponding differential cross sections. Also, the second bump in the CLAS data is shifted to the lower energy 1.8 GeV for the scattering angles corresponding to $\cos\theta$=0.35 and $\cos\theta$=0.55.

Similar to $\pi N \to K\Lambda$ the major difference between the S and C solutions is the treatment of the non-resonant contributions. As seen in Fig. 4, both calculations show two peak structures in the differential cross sections at 1.7 and 1.9 GeV. In both cases the first bump at 1.67 GeV is produced by the $S_{11}(1650)$ resonance. The relative contributions to the second peak at 1.9 GeV are different in the C and S solutions. In the C-calculations this structure is described by the S_{11} partial wave. At higher energies the S_{11} channel is dominated by the non-resonant reaction mechanisms and there is no need to include a third S_{11} resonance, as done e.g. in [20]. The P_{13} partial wave is entirely driven by the $P_{13}(1720)$ and $P_{13}(1900)$ resonance contributions. Switching off these resonance couplings to $K\Lambda$ leads to an almost vanishing P_{13} partial wave. In the S-calculations no peaking behavior is found in the S_{11} partial wave at 1.95 GeV. However, the non-resonant effects in the S_{11} channel are still important. The role of the P_{13} resonances are slightly enhanced in the S-calculations. The effect from the $P_{11}(1710)$

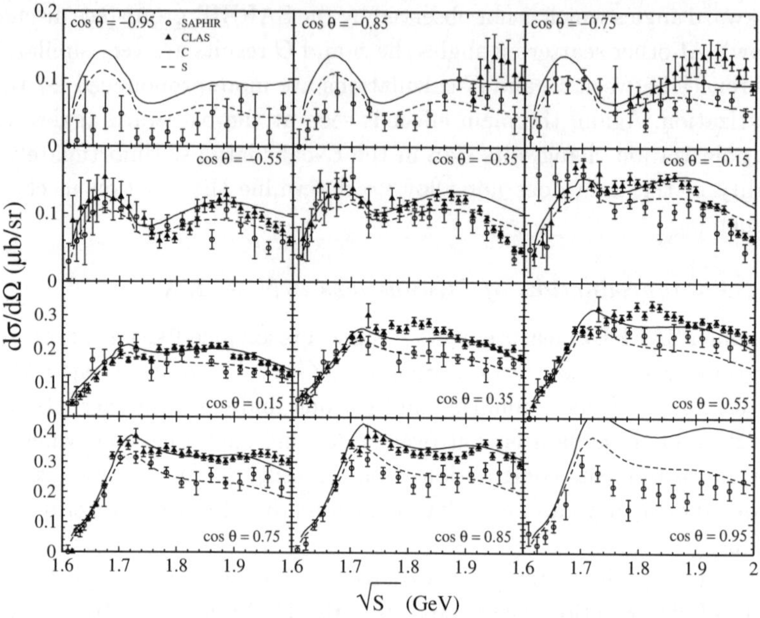

Fig. 4. Comparison of the differential cross sections for the reaction $\gamma p \to K^+\Lambda$ calculated using C and S parameter sets. Experimental data are taken from [18](CLAS) and [17](SAPHIR).

resonance is found to be small in both calculations due to destructive interference with the background process. There are no significant contributions from the spin-$\frac{5}{2}$ resonances to the $\gamma N \to K\Lambda$ reaction.

The calculated photon beam asymmetry Σ_x and recoil polarization P_Λ are shown in Fig. 5 and Fig. 6. Since the beam asymmetry data from the SPring-8 collaboration [21] are available only for energies above 1.94 GeV, these measurements give an insignificant constraint on the model parameters. Therefore, the results for the asymmetry might be regarded as a prediction rather than an outcome of the fit. More information comes from the Λ-polarization data. A good description of the Σ_x and P_Λ data is possible in both the C and S calculations.

5. Summary and Outlook

The importance of a controlled treatment of channel coupling for a quantitative understanding of meson production on the nucleon was pointed out.

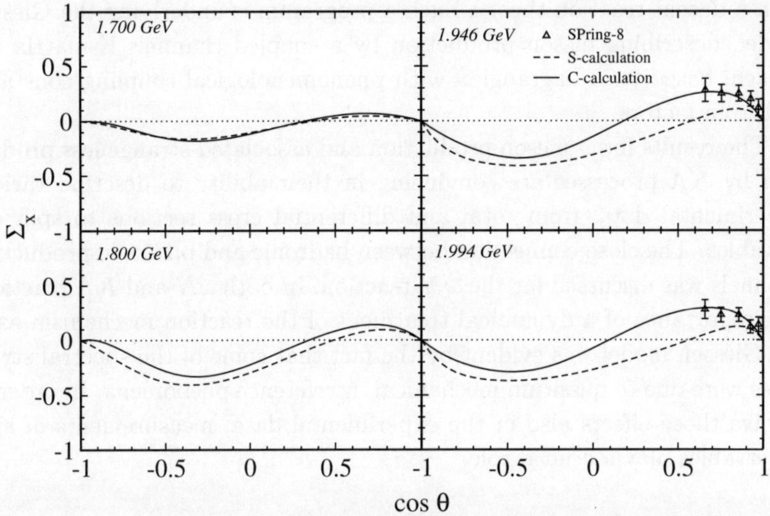

Fig. 5. The calculated photon beam asymmetry. Data are taken from [21]

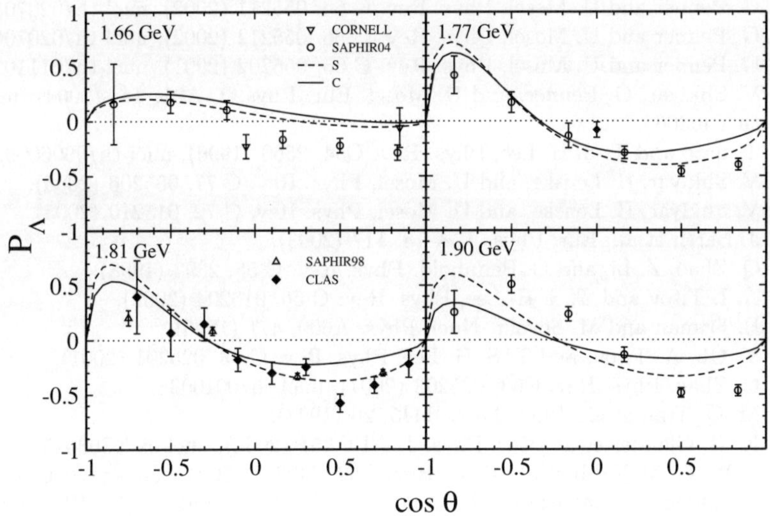

Fig. 6. Λ-polarization in the $\gamma p \to K^+\Lambda$ reaction. Data are from SAPHIR98 [16], SAPHIR04 [17], CLAS [18], CORNELL [19].

An approach, fulfilling the – sometimes delicate – balance between flexibility and generality, is given by using a Lagrangian model in conjunction with a decent reaction theory. Such a programm is underlying the Giessen model, describing meson production by a coupled channels K-matrix approach, based on a Lagrangian with phenomenological coupling constants and from factors.

The results for ω meson production and associated strangeness production by $K\Lambda$ processes are convincing in their ability to describe various experimental data, from total and differential cross sections to spin observables. The close connection between hadronic and photonic production channels was discussed for the ωN reaction. In both ωN and $K\Lambda$ reactions the importance of a dynamical treatment of the reaction mechanism as in the Giessen model was evident by the fact that come of the spectral structures were due to quantum mechanical interference phenomena. In order to resolve those effects also in the experimental data, measurements of spin observables play a crucial role.

References

1. T. Feuster and U. Mosel, Phys. Rev. C58, 457 (1998), nucl-th/9708051;
2. T. Feuster and U. Mosel, Phys. Rev. C 59, 460 (1999), nucl-th/9803057.
3. G. Penner and U. Mosel, Phys. Rev. C 66, 055211 (2002), nucl-th/0207066.
4. G. Penner and U. Mosel, Phys. Rev. C 66, 055212 (2002), nucl-th/0207069.
5. G. Penner and U. Mosel, Phys. Rev. C 65, 055202 (2001), nucl-th/0111023.
6. V. Shklyar, G. Penner, and U. Mosel, Eur. Phys. J. A21, 445 (2004), nucl-th/0403064.
7. T. Sato and T. S. H. Lee, Phys. Rev. C54, 2660 (1996), nucl-th/9606009.
8. V. Shklyar, H. Lenske, and U. Mosel, Phys. Rev. C 71, 055206 (2005).
9. V. Shklyar, H. Lenske, and U. Mosel, Phys. Rev. C 72, 015210 (2005)
10. J. Barth et al., Eur. Phys. J. A 18, 117 (2003).
11. Q. Zhao, Z. Li, and C. Bennhold, Phys. Rev. C 58, 2393 (1998).
12. A. I. Titov and T. S. H. Lee, Phys. Rev. C 66, 015204 (2002).
13. B. Friman and M. Soyeur, Nucl. Phys. A600, 477 (1996).
14. Y. Oh, A. Titov, and T. S. H. Lee, Phys. Rev. C 63, 025201 (2001).
15. Q. Zhao, Phys. Rev. C63, 025203 (2001), nucl-th/0010038.
16. M. Q. Tran et al., Phys. Lett. B445, 20 (1998).
17. K. H. Glander et al., Eur. Phys. J. A19, 251 (2004), nucl-ex/0308025.
18. J. W. C. McNabb et al., Phys. Rev. C69, 042201 (2004), nucl-ex/0305028.
19. H. Thom, E. Gabathuler, E. Jones, B. McDaniel, and W. M. Woodward, Phys. Rev. Lett. 11, 433 (1963).
20. B. Julia-Diaz et al., (2005), nucl-th/0501005.
21. R. G. T. Zegers et al., Phys. Rev. Lett. 91, 092001 (2003), nucl-ex/0302005.

THE IMPORTANCE OF INELASTIC CHANNELS IN ELIMINATING CONTINUUM AMBIGUITIES IN PION-NUCLEON PARTIAL WAVE ANALYSES*

A. ŠVARC, S. CECI, AND B. ZAUNER
Rudjer Bošković Institute
Bijenička Cesta 54, 10000 Zagreb, Croatia
E-mail: svarc@irb.hr

1. Continuum ambiguities in partial wave analyses

In spite of widespread interest in the methods of partial wave analyses (PWA), the concept of "continuum ambiguities", which is closely connected to the analysis of processes in the inelastic region, is nowadays rarely discussed. However, the oldest and very frequently referenced PWAs[1,2] do mention it, but they approach the problem in a particularly cautious way. In Cutkosky et.al.[1] the method of stabilizing the PWA solutions using hyperbolic dispersion relations is utilized, while in Höhler[2], at the very beginning of his monography, a whole sub-chapter is devoted to introducing the problem. In these paragraphs it is carefully argued that the elaborated method of imposing fixed-t analyticity together with isospin invariance is sufficient to produce the unique solution. Both analyses are unfortunately quite vague about the origin of the problem, and the very details of the technical aspect of utilized stabilizing procedures are given only in principle. In order to get the feeling how popular the problem of continuum ambiguities is, we have used the Google WEB search looking for places and times when it has been mentioned. Surprisingly, nothing much after mid 70-es has been found, and at that time the problem has been discussed mostly in mathematical literature[3,4]. So, we believe that we are bound to refresh the basic knowledge on the issue.

*The full size conference presentation is located at
http://hadron.physics.fsu.edu/NSTAR2005/TALKS/Wednesday/Plenary/Svarc.ppt

2. What does it mean "continuum ambiguity"?

The differential cross section itself is not sufficient to determine the scattering amplitude, because if $d\sigma/d\Omega = |F|^2$, then the new function $\tilde{F} = e^{i\Phi}F$ gives exactly the same cross section. It should be remarked that this phase uncertainty has nothing to do with the non-observable phase of wave functions in quantum mechanics; the asymptotic wave functions at large distances from the scattering centre may be written as $\Psi(x) \approx e^{i\cdot k\cdot x} + F(\theta)\frac{e^{i\cdot r\cdot}}{r}$, $r \to \infty$, so the phase of scattering amplitude is the *relative* phase of the incident and scattered wave. This phase has observable consequences in situations where multiple scattering occurs, and causes the continuum ambiguity. In the elastic region the unitarity relates real and imaginary parts of each partial wave, the consequence of the existence of equality relation is constraint which effectively removes the "continuum" ambiguity, and leaves potentially only a discreet one. The partial wave must lie *on* the unitary circle. However, as soon as the inelastic threshold opens, unitarity provides *only* an inequality: $|1 + 2\ i\ F_l|^2 \leq 1 \Longrightarrow \mathrm{ImF}_l = |F_l|^2 + I_l$, where $I_l = \frac{1}{4}(1 - e^{-4\mathrm{Im}\delta_l})$. So each partial wave must lie *upon or inside* its unitary circle, and not on it. A whole family of functions Φ, of limited magnitude but of infinite variety of functional form satisfying the required conditions does exist, but in spite that they contain a continuum number of infinite points they are limited in extent. The ISLANDS OF AMBIGUITY are created. See Fig.1.

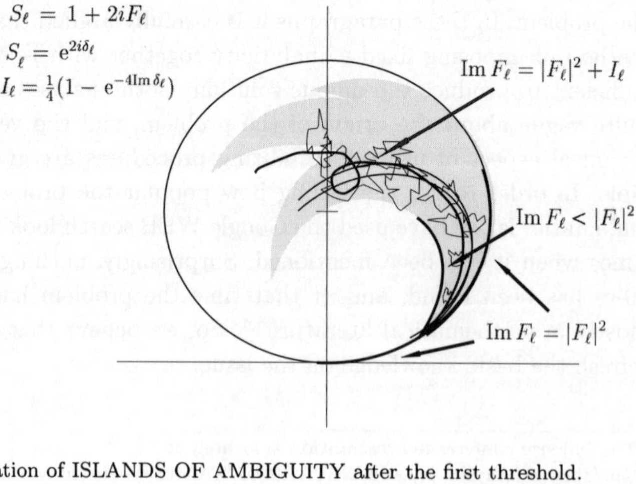

Fig. 1. Creation of ISLANDS OF AMBIGUITY after the first threshold.

Historically, the continuum ambiguity problem has been addressed from two aspects: a. as a mathematical problem (constraining the functional form Φ) and b. as a physics problem (implementation of the partial wave T - matrix continuity; i.e. energy smoothing/search for uniqueness). In Bowcock/Burkhard[4] several smoothing schemes have been suggested and elaborated: i) the shortest path method; ii) explicit analytic parameterization in energy; iii) discrete ambiguities and energy dependence; iv) energy parameterization using dispersion relations; v) partial-wave dispersion relations and vi) fixed-momentum-transfer dispersion relations. For further study we refer the reader to this publication.

3. Continuum ambiguity and coupled channel formalism

Let us formulate the way how we see the continuum ambiguity problem in the language of coupled channel partial wave analyses formalism (CC_PWA). A commonly accepted postulate is that the T matrix is an analytic function of Mandelstam variables s and t. It is well known that each analytic function is fully defined with its poles and cuts. If an analytic function contains a continuum ambiguity it *is not* uniquely defined in the whole complex energy plane, and the direct consequence is that we *do not* possess a complete knowledge about its poles and cuts.
Conclusion:
To eliminate continuum ambiguities in a coupled channel formalism approach it is essential to fully constrain T-matrix poles and cuts.
Basic idea:
We want to demonstrate the role of inelastic channels in fully constraining the poles of the partial wave T-matrix, or alternatively said, we want to show their importance for eliminating continuum ambiguity which arises if only elastic channels are considered.
Implementation:
Supplying scarce information for EACH channel is MUCH MORE CONSTRAINING then supplying perfect information for ONE channel only. We shall use information from as many channels as possible in order to maximally constrain poles and cuts of T matrices in CC_PWA.
In order to familiarize the reader with our concept of looking for poles in the complex energy plane we give:
An attempt of a simple visualization
We are looking for a full set of poles of an analytic function in the complex energy plane while having at our disposal information originating from only restricted number of points on physical axes which we obtain by analyz-

ing experimental data from elastic and inelastic processes. To illustrate our reasoning we present an analogue with a normal, everyday situation. Let us imagine that we are trying to get maximum information about a number of flour bouquets located on the table, not directly looking at them but having at disposal only their images in three mirrors located on three edges - Fig.2. Of course, the mirror closest to the particular bouquet will give the best information about it, it will take us much more effort (a magnifying glass for instance) to get the good information about bouquets further away. Sometimes, if two bouquets are located one behind another, we shall not be able to see the further one at all in some of the mirrors (the bottom right flower bouquets can be seen in the right mirror only). The bouquets in front will completely block our view. Of course, the only proper way is to look at all three mirrors at the same time.

Fig. 2. The illustration of the search for resonances in coupled channel formalism with real world situation.

The same goes for "looking for the resonances in the complex energy plane". Let us identify flour bouquets with resonances, and mirrors with measurements in elastic and inelastic channels. The partial wave analyses which analyzes only elastic channel amplitudes will have to do a very good job to get good information about resonances lying far away from the real axis in the complex energy plane (the magnifying glass have to be very strong), and the information might be incomplete because one resonance might be masked. On the other hand, the formalisms which look at "all three mirrors" at the same time will have a better chance. We claim that the coupled channel formalism (looking at elastic and inelastic channels at the same time) is a method which reveals much more information about resonances then any procedure restricted to only one channel at a time, or differently said it is similar like looking at all three mirrors at the same time in the case of our flour bouquets at the table.

4. Coupled channel formalism

For the collection of formulae we refer the reader either to original paper by Cutkosky et.al[1] or to one of the more recent CC_PWAs; Zagreb[5] or Pittsburgh/ANL[6], but to ease the understanding of the way how the conclusions are reached we give a flow diagram of a Carnagie-Melon-Berkeley (CMU-LBL) type formalism in Fig.3.

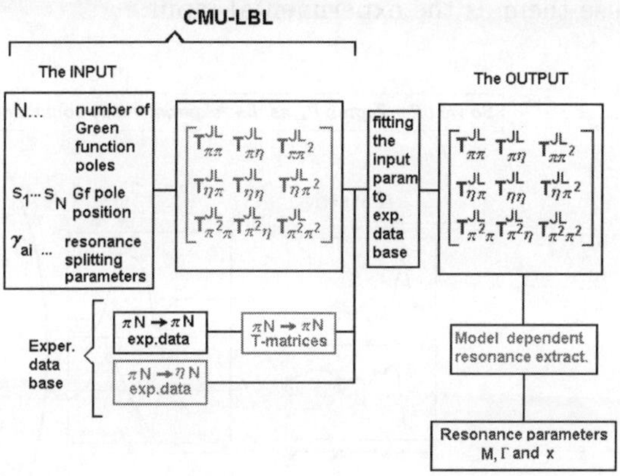

Fig. 3. The flow diagram for the Carnegie-Melon-Berkeley type formalism.

5. The N(1710) P_{11} resonance is a direct consequence of inelastic channels

We shall use the afore introduced CC_PWA formalism to illustrate how the presence of inelastic channel, $\pi N \to \eta N$ data in particular, is imposing the existence of the N(1710) P_{11} state.

Number of channels:

We have simplified the problem in order to demonstrate the genesis of the new P_{11} state by choosing the model with two channels only: πN and the effective two body channel $\pi^2 N$ which is representing all other two/three body processes in a form of a two body process with π^2 being a quasiparticle with a different mass chosen for each partial wave.

The data base:

In principle we should fit experimental data. However, in that case all partial waves have to be simultaneously fitted, and the number of parameters becomes intolerably big. Instead, as the formalism separates individual partial waves, we choose to fit partial waves obtained directly from experiment so the fit can be performed with the reduced number of parameters. In other words, instead of using row experimental data, we choose to represent them as partial waves using *any* form of partial wave analyses (single/multichannel). The only criterion is that they indeed reproduce the experiment correctly. From that moment we treat the obtained results as the optimal amalgamation of different experimental data sets and regardless of their genesis use them as the experimental input.

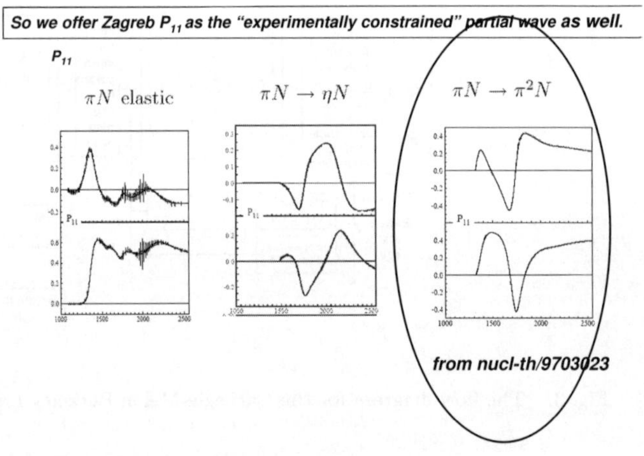

Fig. 4. The inelastic input used to illustrate the existence of the N(1710) P_{11} state.

For P_{11} pion-nucleon elastic partial waves we use the single energy VPI/GWU solution[7,8].

For representing the $\pi N \to \pi^2 N$ data in a form of the P_{11} partial wave, we have followed the way it has been done in ref.[9]. The bottom line of the idea is that the number of experimental data in $\pi N \to \eta N$ channel is insufficient to perform a reasonable single channel partial wave analysis, so a kind of model should be introduced in addition. That is a coupled channel formalism. In that paper the Pittsburgh CC_PWA results[6] are used as representing the *experimentally* constrained S_{11} T-matrix. Following that recipe, we have used the coupled channel curves from the analysis of Batinić et al[10] where the world collection of $\pi N \to \eta N$ data is used to obtain partial wave T matrices, but instead of using the $T_{\pi N,\ \eta N}$ matrices which couple very poorly to the P_{11} partial wave in the 1700 MeV range we have used the $T_{\pi N,\ \pi^2 N}$ part which is much stronger; see - Fig. 4.

The procedure:

In order to show that fitting isolated channels results in different collection of T-matrix poles, we shall start by fitting channel by channel. We start with minimal number of intermediate particles, raise their number as long as the good fit is achieved, and then compare poles. If/when the collection of poles disagree (different collection of poles is needed to fit different channels), we fit all channels simultaneously until the quality of fit can not be improved by increasing the number of exchanged particles.

Elastic channel alone:

Using two physical and two background poles we have fitted only elastic channel, and the obtained results are presented in Fig.5. Only one physical pole is sufficient to achieve the overall agreement of the model with the experimental input of ref.[7,8], and the pole is in the vicinity of 1400 MeV (Roper). Adding new poles is just visually improving the quality of the high energy end of the fit, and we can say that the existence of the second pole near 2100 MeV is only consistent with the data, and not required by them. Inelastic channel is reproduced extremely poorly.

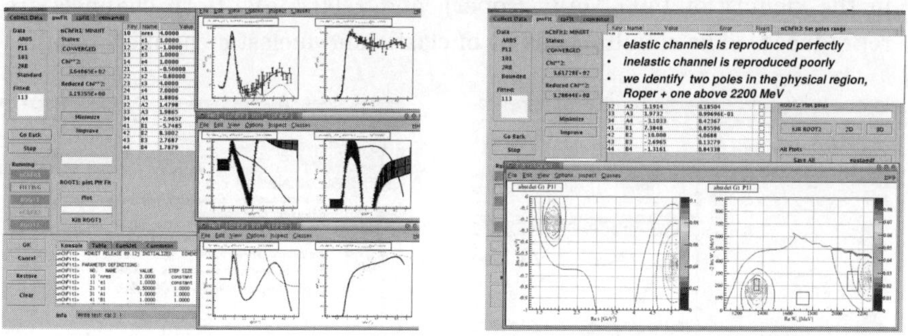

Fig. 5. The $\pi N \to \pi N$, $\pi N \to \pi^2 N$ and $\pi^2 N \to \pi^2 N$ P_{11} T-matrices and pole positions for the two resonant/two background fit of elastic channel alone.

Inelastic channel alone:
Using two physical and two background poles we have fitted inelastic channel alone and the obtained results are presented in Fig.6. At least two physical poles are needed to achieve the overall agreement of the model with the experimental input of ref.[10], and poles are in the vicinity of 1400 MeV (Roper), and 1700 MeV. Elastic channel is reproduced extremely poorly.

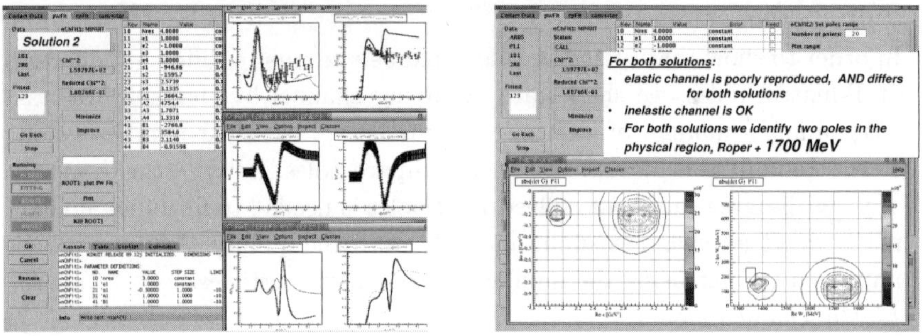

Fig. 6. The $\pi N \to \pi N$, $\pi N \to \pi^2 N$ and $\pi^2 N \to \pi^2 N$ P_{11} T-matrices and pole positions for the two resonant/two background fit of inelastic channel only.

Elastic + inelastic channel:
Using two physical and two background poles we have simultaneously fitted elastic and inelastic channels and the obtained results are presented in Fig.7. At least two physical poles are needed to achieve the overall agreement of the model with the experimental input of refs.[7,10], and poles are in the vicinity of 1400 MeV (Roper), and 1700 MeV. Both channels are reproduced. The simultaneous fit of elastic and inelastic channels requires

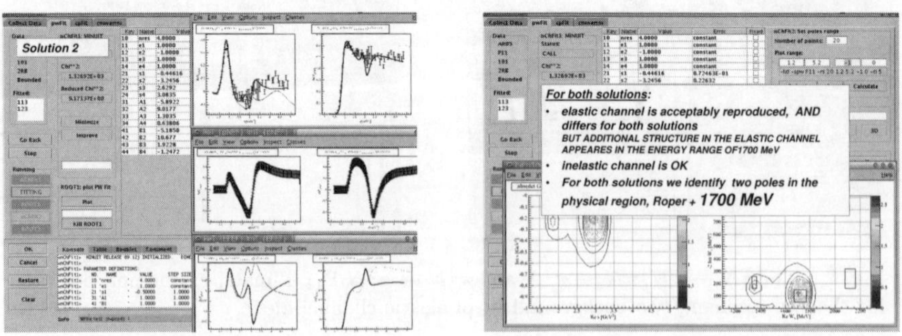

Fig. 7. The $\pi N \to \pi N$, $\pi N \to \pi^2 N$ and $\pi^2 N \to \pi^2 N$ P_{11} T-matrices and pole positions for the two resonant/two background fit of both, elastic and inelastic channels.

that the energy behavior of πN elastic T matrix is not as smooth as in FA02 solution of ref.[8], but has an additional structure, very similar to solutions given in all old PWAs[1,2,11] and new ones[5,6,12]. However, single energy solution offered by the VPI/GWU PWA is consistent with the existence N(1710) P_{11}, because the structure not existing in FA02, and required by everyone else, can be understood as being hidden underneath disproportinably strong error bars which are reported in that solution in the vicinity of 1700 MeV - see Fig.8.

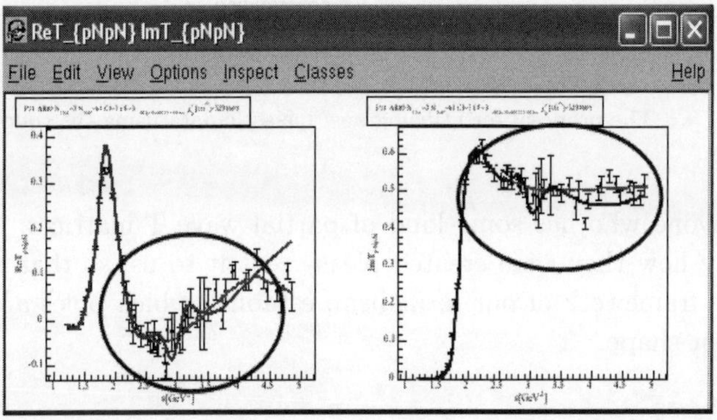

Fig. 8. The $\pi N \rightarrow \pi N$ T matrix of FA02 (grey line) and our solution (dark line) for the two resonant/two background fit compared with SES of ref.[7]

6. Conclusions

T matrix poles, invisible when only elastic channel is analyzed, spontaneously appear in the coupled channel formalism when inelastic channels are added.
It is demonstrated that the N(1710) P_{11} state exists, that the pole is hidden in the continuum ambiguity of VPI/GWU FA02, and that it spontaneously appears when inelastic channels are introduced in addition to the elastic ones.

7. Future prospects for utilizing inelastic channels data

Instead of using raw data we propose to represent them in a form of partial wave T-matrices (single channel PWA, some form of energy smoothing can be as well introduced), and regardless of their genesis, use them in a CC_PWA. The scheme if shown in Fig.9.
A call for help:

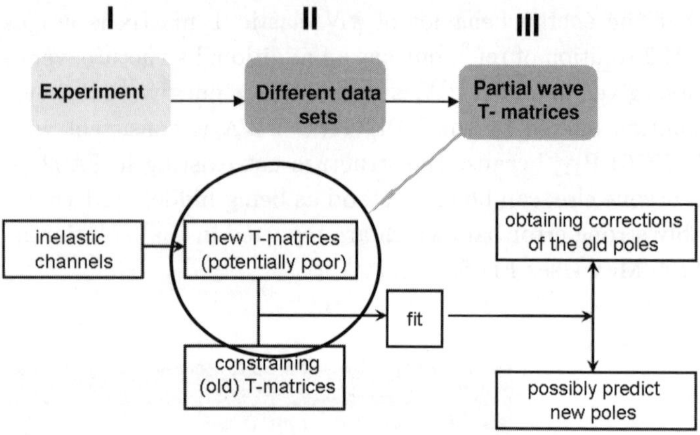

Fig. 9. The proposal for utilizing new inelastic data using the coupled channel formalism.

Anyone who has some kind of partial wave T-matrices, regardless of the way how they were created please sent it to us, so that we could, within the framework of our formalism, establish which poles are responsible for their shape.

References

1. R.E. Cutkosky, C.P. Forsyth, R.E. Hendrick and R.L. Kelly, Phys. Rev. **D20**, 2839 (1979).
2. G. Höhler, in *Pion-Nucleon Scattering*, eddited by H. Schopper, Landolt-Börnstein, Vol **I/9b2** (Springer-Verlag, Berlin, 1983).
3. D. Atkinson, P.W. Johnson and R.L. Warnock, Commun. mat. Phys. **33** (1973) 221.
4. J.E. Bowcock and H. Burkhard, Rep. Prog. Phys. **38** (1975) 1099.
5. M. Batinić, I. Šlaus, A. Švarc and B.M.K. Nefkens, Phys. Rev. **C51**, 2310 (1995); M. Batinić, I. Dadić, I. Šlaus, A. Švarc, B.M.K. Nefkens and T.S.-H. Lee, Physica Scripta **58**, 15 (1998).
6. T.P. Vrana, S.A. Dytman and T.S.-H- Lee, Phys. Rep. **328**, 181 (2000).
7. see: http://gwdac.phys.gwu.edu/analysis/pin_analysis.html.
8. R. A. Arndt et. al., Phys. Rev. **C69**, 035213 (2004).
9. A. Kiswandhi, S. Capstick and S. Dytman, Phys. Rev. **C 69** 025205 (2004).
10. M. Batinić *et al.* arXiv:nucl-th/9703023
11. D.M. Manley and E.M. Saleski, Phys. Rev. **D45**, 4002 (1992).
12. G. Penner and U. Mosel, Phys. Rev. **C 66**, 055211 (2002).

PHENOMENOLOGICAL ANALYSIS OF THE CLAS DATA ON DOUBLE-CHARGED PION PHOTO- AND ELECTROPRODUCTION OFF PROTONS

V. I. MOKEEV, V. D. BURKERT, AND L. ELOUADRHIRI

Jefferson Lab,
12000 Jefferson Ave., Newport News, VA 23606, USA

A. A. BOLUCHEVSKY, G. V. FEDOTOV, E. L. ISUPOV, B. S. ISHKHANOV, AND N. V. SHVEDUNOV

Skobeltsyn Nuclear Physics Institute at Moscow State University,
119899 Vorobevy gory, Moscow, Russia

First comprehensive data on the evolution of nucleon resonance photocouplings with photon virtuality Q^2 are presented for excited proton states in the mass range from 1.4 to 2.0 GeV. N^* photocouplings were determined in phenomenological analysis of CLAS data on 2π photo and electroproduction within the framework of the JLAB-MSU phenomenological model.

1. Introduction

Comprehensive studies of nucleon resonances were carried out in phenomenological analysis of CLAS data on double charged pion production by real and virtual photons [3,4]. The analysis was performed within the framework of the 2005 version of JLAB-MSU phenomenological model in the following referred to as JM05.

2. 2π model. From 2003 to 2005 version.

Electromagnetic production of two pions from proton is sensitive to contributions from both low-lying and high-lying N^* states. This exclusive channel offers an opportunity to determine electromagnetic transition form factors from the nucleon ground state to many excited states. Moreover, it is also a promising channel in the search for so-called "missing" N^* states. These states are predicted from symmetry principles of the symmetric constituent quark models. Many of these states are expected to couple strongly to N$\pi\pi$ final state. However, N^* decays into the 2π channel contribute only a fraction of the total 2π production; a large part is due to non-resonant production mechanisms. In such conditions, the development of reaction models, which provide reasonable background treatment and N^*/background separation, become necessary for the evaluation of N^* parameters.

We have developed a phenomenological model, that incorporates particular meson-baryon mechanisms based on their manifestations in observables: as enhancements in invariant mass distributions, sharp forward/backward slopes in angular distributions. The p$\pi\pi$ final state offers a variety of single differential cross sections for analysis. Any particular meson-baryon mechanism has distinctive reflections in different cross sections. Therefore, a combined analysis of an entire set of single differential cross sections makes it possible to establish the relevant meson-baryon mechanisms from the data fit. Our model is currently limited to the N$\pi^-\pi^+$ final state and incorporates particular meson-baryon mechanisms needed to describe π^+p, $\pi^+\pi^-$, π^-p invariant masses and π^- angular distribution. These cross sections were analyzed in the hadronic mass range from 1.41 to 1.89 GeV and in four Q^2 bins centered at 0., 0.65, 0.95, 1.30 GeV^2. The overall Q^2-coverage ranges from 0. to 1.5 GeV^2.

In the 2003 version (JM03) [1,2] double charged pion production was described by the superposition of quasi-two-body channels with the formation and subsequent decay of unstable particles in the intermediate states:

$$\gamma p \rightarrow \pi^- \Delta^{++} \rightarrow \pi^- \pi^+ p, \tag{1}$$

$$\gamma p \rightarrow \pi^+ \Delta^0 \rightarrow \pi^+ \pi^- p, \tag{2}$$

$$\gamma p \rightarrow \rho^0 p \rightarrow \pi^+ \pi^- p. \tag{3}$$

$$\gamma p \rightarrow \pi^+ D^0_{13}(1520) \rightarrow \pi^+ \pi^- p. \tag{4}$$

Remaining residual mechanisms were parametrized as 3-body phase space with the amplitude fitted to the data. This amplitude was a function of photon virtuality Q^2 and invariant mass of the final hadronic system W only. In this approach we were able to reproduce the main features of integrated cross sections as well as $\pi^+\pi^-$, π^+p invariant masses and π^- angular distributions in CLAS electroproduction data [3,5].

The production amplitudes for the first three quasi-two-body mechanisms (1-3) were treated as sums of N^* excitations in the s-channel and non-resonant mechanisms described in Refs [1,2]. The quasi-two-body mechanism (4) was entirely non-resonant [6]. In reactions (1-3) all well established 4 star resonances with observed decays to the two pion final states were included as well as the 3-star states $D_{13}(1700)$, $P_{11}(1710)$, $P_{33}(1600)$, and

$P_{33}(1920)$. For the $P_{33}(1600)$ a 1.64 GeV mass was obtained in our fit. This value is in agreement with the results of recent analyses of πN scattering experiments. N^* electromagnetic transition form factors were fitted to the data. Hadronic couplings for $N^* \rightarrow \pi\Delta$ and ρp decays were taken from the analyses of experiments with hadronic probes, except for $P_{33}(1600)$, $P_{13}(1720)$, the candidate $3/2^+(1720)$, $D_{13}(1700)$, $P_{33}(1920)$, $F_{35}(1905)$, and $F_{37}(1950)$ states. Poorly known hadronic decay parameters for these states were fitted to the data.

Analysis of CLAS 2π electroproduction data within the framework of this approach revealed the structure around 1.7 GeV, which can not be explained by the contributions from conventional N^* only [3]. We found two possible ways to describe CLAS data around W=1.7 GeV: a) assuming drastically different $\pi\Delta$ and ρp hadronic couplings for the $P_{13}(1720)$ state with respect to the established couplings, or b) keeping hadronic decay parameters for all $N^{*\prime}$s inside established uncertainties, a new baryon state with quantum numbers $3/2^+(1720)$ is needed to describe CLAS data around W=1.7 GeV.

The analysis of preliminary CLAS data on 2π photoproduction [4] within the framework of JM03 revealed shortcomings in the description of π^- angular distributions at backward hemisphere (Fig. 1). Similar incompatibilities were also seen for electroproduction data [3,5]. They are related to the parametrization of remaining mechanisms as 3-body phase, which is incompatible with the steep increase of the measured π^- angular distributions at backward angles. In JM05, the 3-body phase space description was replaced by the set of exchange terms shown in Fig. 2a. This allowed much improved description of the π^- angular distributions (solid lines in Fig. 1) in the entire Q^2 range covered by the CLAS data. Parametrization of these exchange amplitudes is described in [7].

Such modifications for remaining mechanisms allow us to reproduce CLAS data on 2π photo- and electroproduction [3–5] at W$\leq$1.7 GeV reasonably well. Above 1.7 GeV the measured cross sections around Δ^0 mass in the π^-p mass distributions exceed the model cross sections (Fig 3). To reproduce the strength of the Δ^0 peaks (solid lines in Fig. 3) as well as to improve description of the π^+p mass distributions we implemented an additional contact term for isobar channels (1)-(2), shown on Fig. 2c. Parametrization of these mechanisms is presented in [8].

After these improvements we still have shortcomings in the description of π^+p, π^-p mass and π^--angular distributions at W above 1.8 GeV (dashed lines in Fig. 4). Lack of strength in the calculated π^+p, π^-p mass

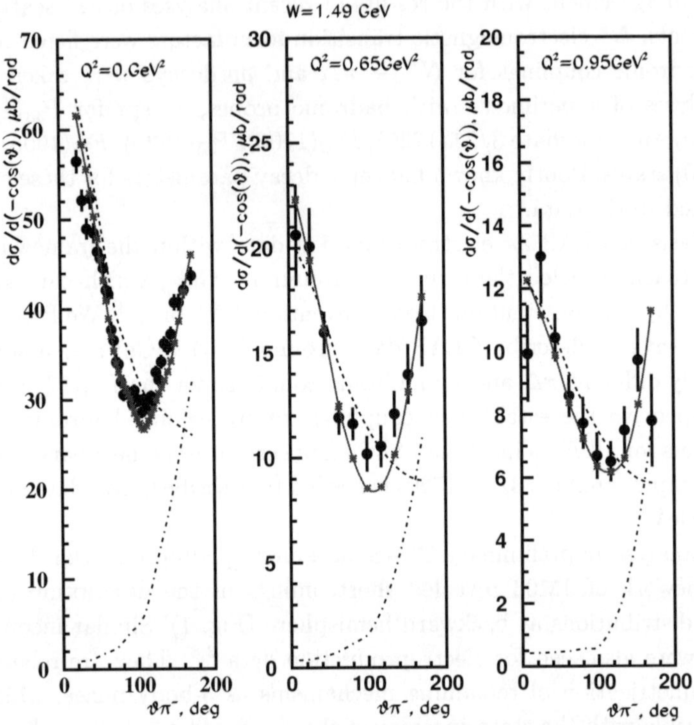

Fig. 1. π^- CM angular distributions at W=1.49 GeV and at three photon virtualities. CLAS data [3-5] are shown in comparison with the JM05 results: solid lines represent full calculations; the contributions from 2π direct production mechanisms are shown by dashed-dotted lines. The full calculations within the framework of the JM03 are shown by dashed lines.

distributions are centered at the masses of $P_{33}(1660)$ (1.64 GeV fitted mass) and $F_{15}(1685)$ resonances, respectively. The gap in angular distributions may be filled by t-channel exchange mechanisms for $\pi^+ F^0_{15}(1685)$ intermediate state. The observed discrepancies thus indicate contributions from the isobar channels:

$$\gamma p \to \pi^+ F^0_{15}(1685) \to \pi^+\pi^- p, \tag{5}$$

$$\gamma p \to \pi^- P^{++}_{33}(1600) \to \pi^-\pi^+ p. \tag{6}$$

Implementation of isobar channels (5)-(6) with amplitudes as outlined in [8] allow us to reproduce the data at W≥ 1.8 GeV reasonably well (solid lines

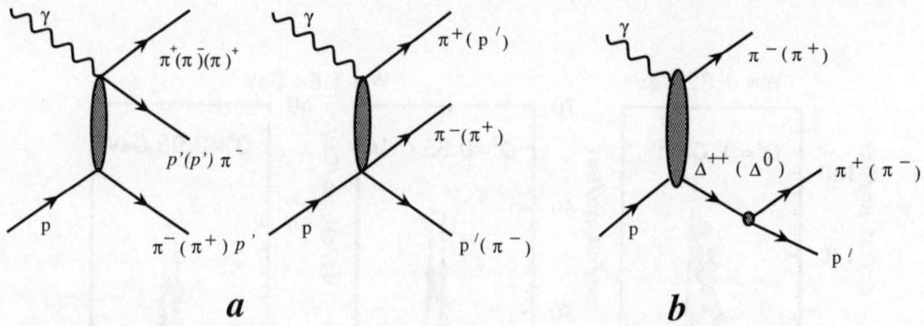

Fig. 2. Direct 2π production mechanisms (a). Complementary contact term in $\pi\Delta$ production (b).

in Fig. 4).

After these modifications, we succeeded in describing all available CLAS data on unpolarized observables in 2π photo and electroproduction. These results are presented in [8]. We found no need for remaining mechanisms of unknown dynamics. Therefore, the quality of the CLAS data allow us to establish all significant mechanisms in 2π production, implementing particular meson-baryon diagrams as determined from the data fit.

The credibility of our description of non-resonant mechanisms and the separation of resonant and non-resonant contributions was tested in a combined analysis of CLAS data on 1π and 2π electroproduction [7]. We found a common set of N^* photocouplings, which allowed us to reproduce all observables measured in these two exclusive channels combined. Since 1π and 2π channels represent two major contributors in N^* excitation region with considerably different background, their successful fit offers compelling evidence for credible background description and N*/background separation achieved in JM05.

3. N^* analysis in 2π photo and electroproduction.

We fit all available CLAS data on 2π photo and electroproduction at W$\leq$.1.9 GeV and Q^2 from 0 to 1.5 GeV^2 within the framework of JM05. N^* photocouplings were sampled according to the normal distribution around the values, obtained in the JM03 [9]. The photocouplings were varied within 0.3 σ from theirs starting values. Poorly known masses and hadronic couplings were also fluctuated inside the uncertainties established in experiments with hadronic probes. Adjustable parameters of non-resonant mechanisms were varied within 0.2 σ. For each trial set of model parameters we calculated all kind of single differential cross sections in all available

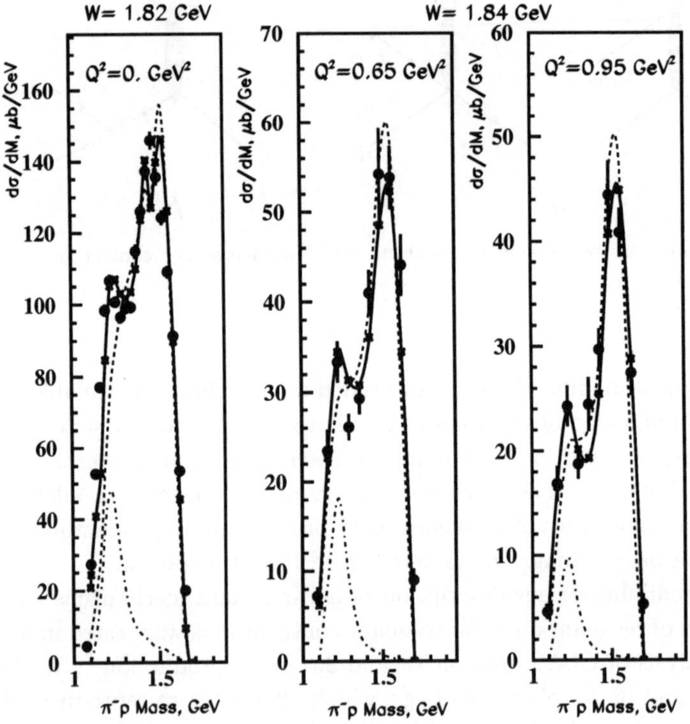

Fig. 3. Evidence for complementary contact term contribution to the $\pi\Delta$ isobar channels. Full calculations in JM05 are shown by solid lines, while evaluations within the framework of JM03 are shown by dashed lines. $\pi^+\Delta^0$ channel contributions estimated in JM05 model are shown by dotted-dashed lines. CLAS 2π photo and electroproduction data are from [3-5]

W and Q^2 bins. From comparison between calculated and measured single differential cross sections χ^2/d.p. were estimated. We isolated a bunch of calculated cross sections inside the data uncertainties, applying restriction $\chi^2 \leq \chi^2_{th}$, where χ^2_{th} is a predetermined maximal allowed value. Integrated 2π cross sections in comparison with selected calculated cross sections are shown in Fig. 5. A reasonable description of all cross sections was achieved.

N^* photocouplings for selected cross sections were averaged and mean values were treated as extracted from the data fit, while dispersions were assigned to photocoupling uncertainties. In this way we obtained the photocouplings for the states: $P_{11}(1440)$, $D_{13}(1520)$,

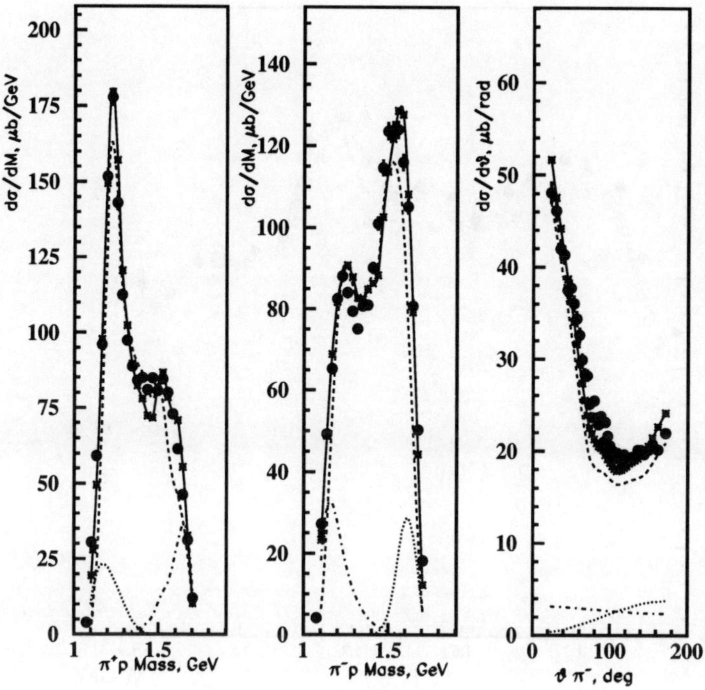

Fig. 4. Manifestation of $\pi^+ F^0_{15}(1685)$ and $\pi^- P^{++}_{33}(1600)$ isobar channels. The photoproduction CLAS data at W=1.86 GeV are compared to the JM05: full calculations (solid lines); isobar channels (5),(6) are absorbed in 3-body phase space (dashed lines). The contributions from channels (5),(6) are shown by dotted and dotted-dashed lines respectively.

$S_{31}(1620)$, $S_{11}(1650)$, $P_{33}(1600)$, $F_{15}(1680)$, $D_{13}(1700)$, $D_{33}(1700)$, candidate $3/2^+(1720)$, $P_{13}(1720)$, $F_{35}(1905)$, $P_{33}(1920)$, and $F_{37}(1950)$ [8].

In Fig. 6 we present the photocouplings for the well studied $D_{13}(1520)$ state in comparison with previously available data. Reasonable overlap between our results and previous ones support the reliability of our procedure for the extraction of N^* photocoupling from 2π data fit. For the first time, we determine the electrocouplings for high lying N^*, which preferably decay with 2π emission: $D_{13}(1700)$, $D_{33}(1700)$, candidate $3/2^+(1720)$, $P_{13}(1720)$, $F_{35}(1905)$, $P_{33}(1920)$, and $F_{37}(1950)$. In Fig. 7 we present photo- and electrocouplings for the $D_{33}(1700)$ and $P_{13}(1720)$ states extracted from the

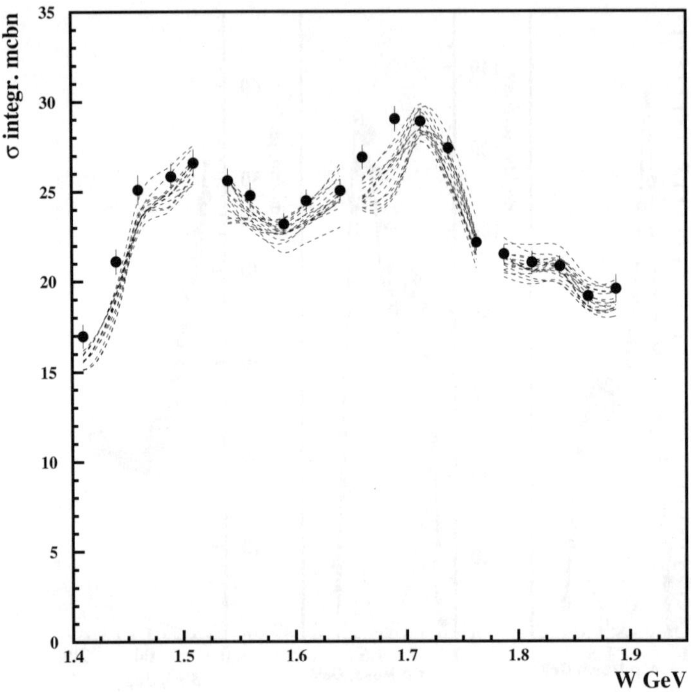

Fig. 5. Total 2π electroproduction cross sections calculated within the framework of JM05 with N^* parameters fitted to the CLAS data [3-5] at photon virtuality 0.65 GeV^2.

CLAS 2π data, as well as couplings obtained in previous studies of 1π production. The analysis of the CLAS 2π data for the first time provides accurate information on electrocouplings for high lying nucleon excitations.

4. Conclusions.

A phenomenological model for the description of 2π production from protons in the nucleon resonance region was developed with most complete accounting for all relevant mechanisms. The reliability of the background treatment and N^*/background separation was confirmed by the reasonable description obtained for of all unpolarized observables in this exclusive channel, as well as in the combined analysis of 1π and 2π production. Electromagnetic transition form factors were extracted at photon virtualities

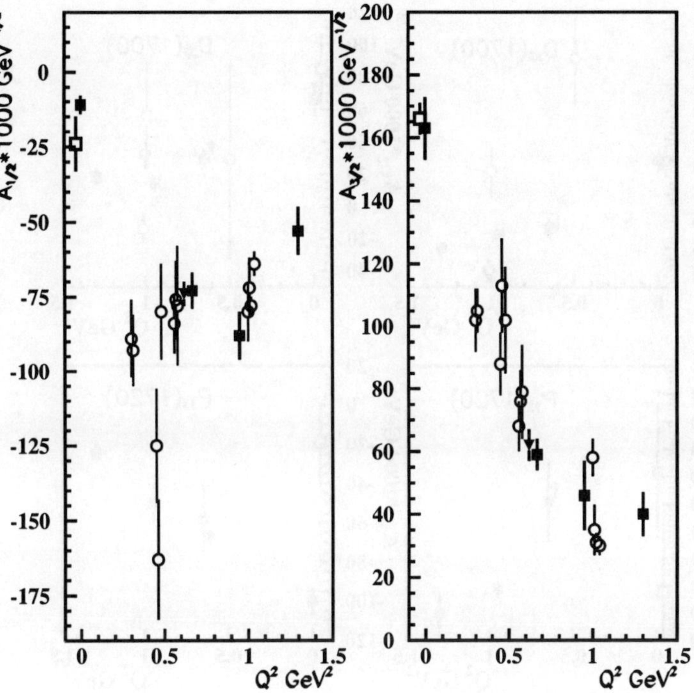

Fig. 6. $D_{13}(1520)$ photocouplings extracted from the analysis of CLAS 2π data [3-5] (filled squares) in comparison with world data [10] (open circles) and results of analysis of the CLAS 1π and 2π data combined [7] (filled triangles).

$Q^2 \leq 1.5\ GeV^2$ for N^* states in the mass range from 1.4 to 2.0 GeV. For the first time transition form factors were obtained for many high lying proton states with major 2π decay.

References

1. V. Mokeev et. al., Phys, of Atom. Nucl. **64**, 1292 (2001).
2. V.D Burkert, et. al., Phys, of Atom. Nucl. **66**, 2199 (2003).
3. M. Ripani et. al., Phys. Rev. Lett. **91**, 022002 (2003).
4. M.Bellis, et.al., Proceedings of NSTAR2004 workshop , March 24-27, 2004, Grenoble, France, World Scientific, ed. by J.-P. Bocquet, V. Kuznetsov, D. Rebreyend, 139.
5. CLAS Physics Data Base, *http : //clasdb3.jlab.org*.
6. V. Mokeev et. al., Proceedings of NSTAR2004 workshop , March 24-27, 2004,

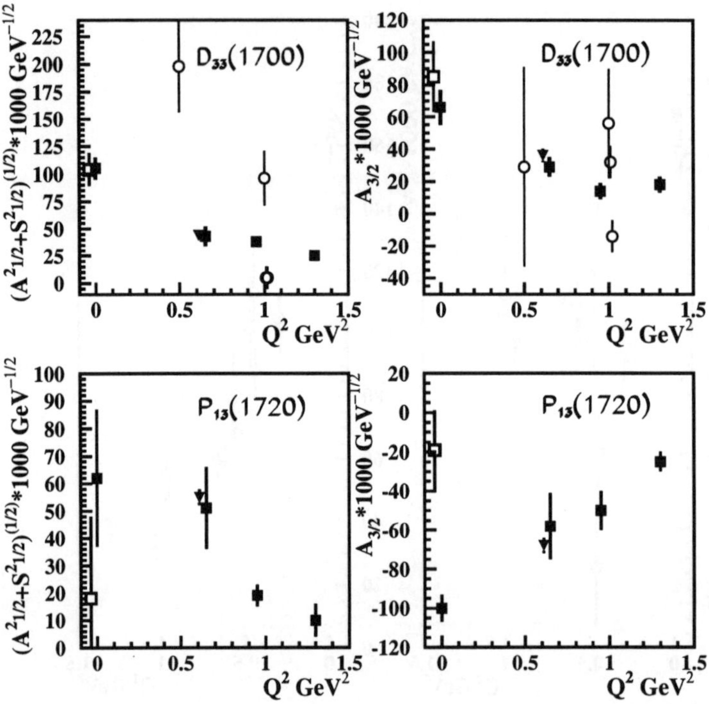

Fig. 7. $D_{33}(1700)$, $P_{13}(1720)$ photocouplings extracted from analysis of CLAS 2π data in comparison with previous results. Symbols are the same as in the Fig. 6.

Grenoble, France, World Scientific, ed. by J.-P. Bocquet, V. Kuznetsov, D. Rebreyend, 317.

7. I. G. Aznauryan et. al., Phys. Rev. **C72**, 045201 (2005).
8. V.I.Mokeev *http://hadron.physics.fsu.edu/nstar/scientificProg.htm*
9. V.D. Burkert, et. al., Proc. of the 17 International UPAP Conference on Few-Body Problems in Physics, Durham. NC, USA, 5-10 June 2003, Elsevier, 2004, ed.by W. G. Glöckle, W. Tornow, S231.
10. V.D. Burkert, et. al., Phys. Rev. **C67**, 035204 (2003).

GAUGE-INVARIANT APPROACH TO MESON PHOTOPRODUCTION INCLUDING THE FINAL-STATE INTERACTION

H. HABERZETTL

Center for Nuclear Studies, Department of Physics, The George Washington University, Washington, DC 20052, USA
E-mail: helmut@gwu.edu

K. NAKAYAMA

Department of Physics and Astronomy, University of Georgia, Athens, GA 30602, USA
E-mail: nakayama@uga.edu

S. KREWALD

Institut für Kernphysik (Theorie) Forschungszentrum Jülich, 52425 Jülich, Germany
E-mail: s.krewald@fz-juelich.de

A gauge-invariant formalism is presented for the practical treatment of photo- and electroproduction of pseudoscalar mesons off nucleons that allows an explicit incorporation of hadronic final-state interactions. The semi-phenomenological approach is based on a field theory developed by one of the authors. It generalizes an earlier approach by allowing for systematic improvement of approximations in a controlled manner. The practical feasibility is illustrated by applying the lowest-order result to the photoproduction of both neutral and charged pions.

1. Introduction

The photo- and electroproduction of mesons off nucleons is one of the primary tools to learn about the excited states of the nucleon. From a theoretical point of view, therefore, what is needed are reaction theories that allow one to disentangle the resonance information from the background in a reliable manner. There exist a number of different approaches that describe the dynamics of the photon-induced production of mesons, but the limited space of the present contribution does not permit a detailed comparison of

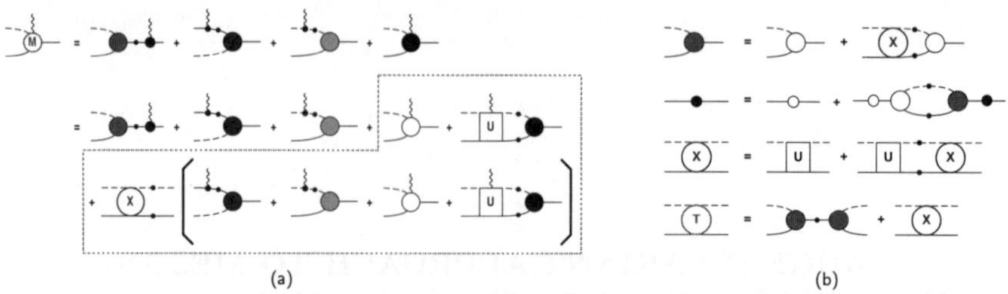

Fig. 1. Diagrammatic summary of the field-theory formalism of Ref. 1. Time proceeds from right to left. (a) Meson production current M^μ. The first line corresponds to Eq. (1) summing up, in that order, the s-, u-, and t-channel diagrams and the interaction current $M^\mu_{\rm int}$, whose dynamical content is exhibited by the diagrams enclosed in the dashed box of the last two lines. This also includes, in the bottom line, the final-state interaction mediated by the nonpolar πN amplitude X that satisfies the integral equation shown in (b). The diagram element labeled U subsumes all exchange currents U^μ contributing to the process (see Fig. 2). The diagram with open circle depicts the bare current $m^\mu_{\rm bare}$ (*i.e.*, the Kroll–Ruderman term). (b) Pion-nucleon scattering with dressed hadrons. The full πN-amplitude is denoted by T, with X subsuming all of its nonpolar (*i.e.*, non-s-channel) contributions. The latter satisfies the integral equation $X = U + UG_0X$ depicted in the third line here, where the driving term U sums up all nonpolar irreducible contributions to πN-scattering, *i.e.*, all irreducible contributions which do not contain an s-channel pole (see Ref. 1 for full details). Diagram elements with open, unlabeled circles describe bare quantities.

their respective merits.

The basic feature of all approaches, however, is that topologically there are four distinct contributions to the amplitude M^μ, as shown in Fig. 1. There are three contributions, referred to as the s-, u-, and t-channel contributions — M^μ_s, M^μ_u, and M^μ_t, respectively — according to the respective Mandelstam variables of the intermediate hadron, where the photon attaches itself to an external leg of the basic underlying πNN vertex, and there is a fourth contribution, the interaction current $M^\mu_{\rm int}$, where the photon interacts *within* the vertex. This breakdown,

$$M^\mu = M^\mu_s + M^\mu_u + M^\mu_t + M^\mu_{\rm int} , \tag{1}$$

thus is generic and reflected in all approaches. There is considerable difference, however, in the way the four basic contributions are implemented dynamically.

The present work is based on the field-theoretical approach given by Haberzettl.[1] The complete formalism takes into account all possible hadronic mechanisms, including the final-state interaction (FSI), and it is gauge-invariant as a matter of course. However, in view of its complexity and high nonlinearity, its practical implementations require that some reaction mechanisms need to be truncated and/or replaced by phenomenological approximations. In doing so, and depending on the level of approximation, the resulting reaction dynamics then become very similar to that of dynamical meson-exchange models of hadronic interactions.

Gauge invariance is one of the fundamental symmetries that must be maintained in any approach involving the electromagnetic interaction with hadronic matter. Our objective here is to preserve full gauge invariance through all levels of approximations. As a matter of course, this must also include the case where the underlying πNN interaction is modeled by *ad hoc* phenomenological form factors and, in particular, gauge invariance must remain true also in the presence of explicit final-state interactions. The latter problem has been treated already in Ref. 2 as a two-step procedure where the gauge-invariant treatment of explicit FSIs was added on to an already gauge-invariant tree-level amplitude that had been constructed according to the prescriptions given in Refs. 3, 4. The present approach instead starts from the full amplitude and derives a single condition for the mechanisms to be approximated that follows directly from the generalized Ward–Takahashi identities given in the next section.[1,5] At the lowest order, we reproduce the earlier results.[2] In addition, we shall provide a general scheme that allows one, in principle, to systematically improve on the initial approximation in a controlled manner.

In the following, for definiteness, we will explicitly consider the production of pions off the nucleon, *i.e.*, $\gamma + N \to \pi + N$, but the formalism will of course apply equally well to the photo- or electroproduction of any pseudoscalar meson. Moreover, at intermediate stages of the reaction, we will ignore other mesons or baryonic states since they are irrelevant for the problem at hand, *i.e.*, how to preserve gauge invariance in the presence of hadronic final-state interaction.

2. Gauge Invariance

The production current M^μ is gauge invariant if its four-divergence satisfies the generalized Ward–Takahashi identity[1,5]

$$\begin{aligned} k_\mu M^\mu = & -[F_s\tau]S_{p+k}Q_iS_p^{-1} + S_{p'}^{-1}Q_fS_{p'-k}[F_u\tau] \\ & + \Delta^{-1}_{p-p'+k}Q_\pi\Delta_{p-p'}[F_t\tau] \ , \end{aligned} \tag{2}$$

where p and k are the four-momenta of the incoming nucleon and photon, respectively, and p' and q are the four-momenta of the outgoing nucleon and pion, respectively, related by momentum conservation $p' + q = p + k$.

S and Δ are the propagators of the nucleons and pions, respectively, with their subscripts denoting the available four-momentum for the corresponding hadron; Q_i, Q_f, and Q_π are the initial and final nucleon and the pion charge operators, respectively. The index x at $[F_x\tau]$ labels the appropriate kinematic situation for πNN vertex in the s-, u-, or t-channel diagrams of Fig. 1 (and τ explicitly denotes the isospin-operator dependence in a schematic manner for later convenience). This is an *off-shell* condition. In view of the inverse propagators appearing in each term here, if all external hadronic legs are on-shell, this reduces to

$$k_\mu M^\mu = 0 \qquad \text{(hadrons on-shell)} , \tag{3}$$

which describes current conservation.

Physically relevant, of course, is only current conservation. However, in a *microscopic* dynamical theory it is not sufficient to simply conserve the overall current. In addition, one must have *consistency* at all level of the hierarchy of mechanisms that provide the details of the reaction dynamics. It is found[1,5] that it is *not* possible to achieve current conservation consistently unless the current satisfies the off-shell condition (2).

The electromagnetic currents for the nucleons and the pions, Γ_N^μ and Γ_π^μ, respectively, satisfy the Ward–Takahashi identities

$$k_\mu \Gamma_N^\mu(p',p) = S_{p'}^{-1} Q_N - Q_N S_p^{-1} , \tag{4a}$$

$$k_\mu \Gamma_\pi^\mu(q',q) = \Delta_{q'}^{-1} Q_\pi - Q_\pi \Delta_q^{-1} , \tag{4b}$$

where the four-momentum relations $p' = p + k$ and $q' = q + k$ hold. It is therefore possible to replace the generalized WT identity (2) by the *equivalent* gauge-invariance condition

$$k_\mu M_{\text{int}}^\mu = -[F_s\tau]Q_i + Q_f[F_u\tau] + Q_\pi[F_t\tau] \equiv -F_s e_i + F_u e_f + F_t e_\pi , \tag{5}$$

where the operators

$$e_i = \tau Q_i , \quad e_f = Q_f\tau , \quad \text{and} \quad e_\pi = Q_\pi\tau \tag{6}$$

describe the respective hadronic charges in an appropriate isospin basis (component indices and summations are suppressed here). Charge conservation for the production process simply reads $e_i = e_f + e_\pi$. In the following, we use the condition (5), instead of (2), together with (4).

3. Formalism

For the present purpose, it is sufficient to summarize the field-theory formalism of Ref. 1 in terms of the diagrams of Figs. 1 and 2. It is seen here that the interaction current $M^\mu_{\rm int}$, where the photon couples inside the vertex, explicitly contains the hadronic FSI. The structure of the interaction current thus is rather complex and we read off the diagrams enclosed by the dashed box of Fig. 1(a) that

$$M^\mu_{\rm int} = m^\mu_{\rm bare} + U^\mu G_0[F\tau] \\ + XG_0\Big(M^\mu_u + M^\mu_t + m^\mu_{\rm bare} + U^\mu G_0[F\tau]\Big)\,, \tag{7}$$

where $m^\mu_{\rm bare}$ is the bare Kroll–Ruderman contact current, U^μ subsumes all possible exchange currents (see Fig. 2), G_0 describes the intermediate πN two-particle propagation, and the FSI is mediated by the nonpolar part X of the πN T matrix. The notation $[F\tau]$ is used for the dressed initial $N \to \pi N$ vertex. The equation for this dressed vertex is given in Fig. 1.

The preceding expression is still exact. In practical applications, however, one will not be able to calculate all mechanisms that contribute to the full reaction dynamics and one must make some approximations. This is particularly true for the complex mechanisms that enter U^μ, as depicted in Fig. 2. Approximations should preserve the gauge invariance of the amplitude. To see how this can be done, let us define

$$M^\mu_a = m^\mu_{\rm bare} + U^\mu G_0[F\tau]\,, \tag{8}$$

thus write

$$M^\mu_{\rm int} = M^\mu_a + XG_0\Big(M^\mu_u + M^\mu_t + M^\mu_a\Big)\,, \tag{9}$$

and recast the gauge-invariance condition (5) as a condition for M^μ_a. Denoting the *non-transverse* parts of the u- and t-channel currents M^μ_u and M^μ_t by m^μ_u and m^μ_t, respectively, *i.e.*,

$$k_\mu(M^\mu_u - m^\mu_u) = 0 \quad \text{and} \quad k_\mu(M^\mu_t - m^\mu_t) = 0 \tag{10}$$

holds true as an *off-shell* property, this immediately leads to

$$k_\mu M^\mu_a = (1 - UG_0)\left[-F_s e_i + F_u e_f + F_t e_\pi\right] - k_\mu U G_0(m^\mu_u + m^\mu_t) \tag{11}$$

as the necessary condition that M^μ_a must satisfy so that $M^\mu_{\rm int}$ yields the gauge-invariance condition (5). Note that no approximation has been made up to here.

3.1. *Approximating* M^μ_a

The structure of the preceding condition suggests the following approximation strategy. The condition evidently is satisfied if we now approximate

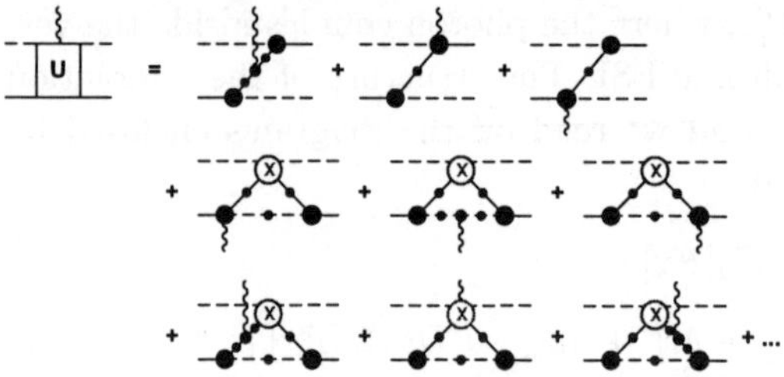

Fig. 2. Exchange-current contributions subsumed in U^μ. The three contributions in the top row are referred to as E^μ in the Eq. (21).

M_a^μ by

$$M_a^\mu = (1 - UG_0)M_c^\mu - UG_0(m_u^\mu + m_t^\mu) + T^\mu \ , \tag{12}$$

where M_c^μ can be *any* contact current satisfying

$$k_\mu M_c^\mu = -F_s e_i + F_u e_f + F_t e_\pi \tag{13}$$

and T^μ is an undetermined transverse *contact* current that is unconstrained by the four-divergence (11). With the choice (12), the corresponding approximate M_{int}^μ is then easily found from (9) as

$$M_{\text{int}}^\mu = M_c^\mu + T^\mu + XG_0\big[(M_u^\mu - m_u^\mu) + (M_t^\mu - m_t^\mu) + T^\mu\big] \ . \tag{14}$$

In this scheme, therefore, the choice one makes for M_c^μ (and T^μ) corresponds to an implicit approximation of the full dynamics contained in the right-hand side of Eq. (8). Moreover, beyond this actual choice, the only explicit effect of the FSI X is from explicitly *transverse* loop contributions, which is precisely the same result that was found in Ref. 2. Thus it follows that

$$k_\mu M_{\text{int}}^\mu = k_\mu M_c^\mu \tag{15}$$

and this approximate interaction current then obviously satisfies the gauge-invariance condition (5).

3.2. *Phenomenological choice for M_c^μ*

The phenomenological choice that we make here for M_c^μ is a variant of the procedure proposed in Refs. 1, 3 that is more general than what was suggested in 2. We parameterize the πNN vertices by

$$F_x = g_\pi \gamma_5 \left[\lambda + (1-\lambda)\frac{\not{q}_\pi}{m+m'}\right] f_x \ , \tag{16}$$

where q_π is the outgoing pion four-momentum, $x = s$, u, or t indicate the kinematic context, g_π is the physical coupling constant, m and m' are the nucleon masses before and after the pion is emitted/absorbed and the parameter λ allows for the mixing of pseudoscalar (PS: $\lambda = 1$) and pseudovector (PV: $\lambda = 0$) contributions. For simplicity, the functional dependence f_x of the vertex (which depends on the squared four-momenta of all three legs) is chosen as common to both PS and PV and it is normalized to unity if all vertex legs are on-shell. We define then an auxiliary current

$$C^\mu = -e_\pi \frac{(2q-k)^\mu}{t-q^2}(f_t - \hat{F}) - e_f \frac{(2p'-k)^\mu}{u-p'^2}(f_u - \hat{F}) - e_i \frac{(2p+k)^\mu}{s-p^2}(f_s - \hat{F}) \,, \tag{17}$$

where

$$\hat{F} = 1 - \hat{h}\,(1-\delta_s f_s)(1-\delta_u f_u)(1-\delta_t f_t) \,. \tag{18}$$

The factors δ_x are unity if the corresponding channel contributes to the reaction in question, and zero otherwise. In principle, the parameter $\hat{h}$ may be an arbitrary (complex) function, $\hat{h} = \hat{h}(s,u,t)$, subject to crossing-symmetry constraints.[4] However, in the application discussed in the next section, we simply take $\hat{h}$ as a fit constant. With this choice for $\hat{F}$, the auxiliary current C^μ is manifestly nonsingular, *i.e.*, it is a *contact* current, and in view of charge conservation, $e_\pi + e_f - e_i = 0$, its four-divergence evaluates to

$$k_\mu C^\mu = e_\pi f_t + e_f f_u - e_i f_s \,. \tag{19}$$

Using the vertex parametrization (16) and writing out the gauge-invariance condition (13) explicitly, we may then extract the ansatz

$$M_c^\mu = -g_\pi \gamma_5 \frac{(1-\lambda)\gamma^\mu}{m'+m} e_\pi f_t + g_\pi \gamma_5 \left[\lambda + \frac{(1-\lambda)\not{k}}{m'+m}\right] C^\mu \,. \tag{20}$$

In this lowest order, therefore, the bare $\gamma_5\gamma^\mu e_\pi$ Kroll–Ruderman coupling is dressed by the t-channel form factor $e_\pi \to e_\pi f_t$. In addition, there is an auxiliary current given by (17), with the same coupling structure as the underlying πNN vertex (16).

Obviously, the above choice of M_c^μ is not unique, for we can always add another transverse current to it. We emphasize that such a transverse contact current in M_c^μ should not be confused with the transverse contact current T^μ appearing in Eq. (14). In particular, M_c^μ does not contribute to the FSI loop integral, but T^μ does.

3.2.1. *Next-order approximation*

The approximation made so far was to replace M_a^μ of (8) by (12), with an undetermined transverse contribution T^μ. This approximation may be systematically improved by explicitly accounting for more of the features of U^μ. We mention without derivation that the next order in this scheme reads

$$T^\mu = \left(E^\mu - \tilde{E}^\mu\right) G_0[F\tau] + T'^\mu \ , \tag{21}$$

which replaces T^μ in Eq. (12) by an explicit contribution due to the single-particle exchange current E^μ (see Fig. 2), still leaving an undetermined transverse current T'^μ. $\tilde{E}^\mu$ is the *non-transverse* part of E^μ, *i.e.*, one has

$$k_\mu \left(E^\mu - \tilde{E}^\mu\right) = 0 \tag{22}$$

as an off-shell property, similar to (10). This scheme can be systematically extended to all orders of mechanisms contained in U^μ.

3.3. *First application: pion photoproduction*

In this section, as a first feasibility study, we apply the approach developed in the preceding section at its lowest order to the photoproduction reaction $\gamma + N \rightarrow \pi + N$. We restrict ourselves to photon energies up to about 400 MeV. Therefore, in addition to the basic nucleons and pions discussed in the preceding section, our model also incorporates intermediate Δs in the s- and u-channels, and we include the ρ, ω, and a_1 meson-exchanges in the t-channel. Note here that transition currents between different hadronic states are transverse individually and therefore play no role for the issue of gauge invariance. For the πN FSI, we employ the πN T-matrix developed by the Jülich group.[6] Full details of this first application will be given elsewhere. We only mention here that the present model has only three free parameters that are adjusted to reproduce the pion photoproduction cross section data, as shown in Fig. 3.

Generally, for the total cross section, the agreement with the data is very good except for energies above $T_\gamma \sim 360$ MeV, where the prediction tends to underestimate the data. In particular, around $T_\gamma \approx 390$ MeV, the discrepancy is about 10%. We also see that the FSI loop contribution is relatively small compared to the Born contribution. However, it plays a crucial role in reproducing the observed energy dependence through its interference with the dominant Born term. For differential cross sections for neutral and charged pion productions at various energies, we also see that, overall, the data are reproduced quite well. The dashed curves in the top row correspond to the results represented by the solid curves multiplied by an arbitrary factor of 1.1 to facilitate visualizing that the shape of the angular distribution is well reproduced in spite of the absolute normalization being underestimated at this energy.

4. Summary

By exploiting the generalized Ward-Takahashi identity for the production amplitude and total charge conservation, we have constructed a fully gauge

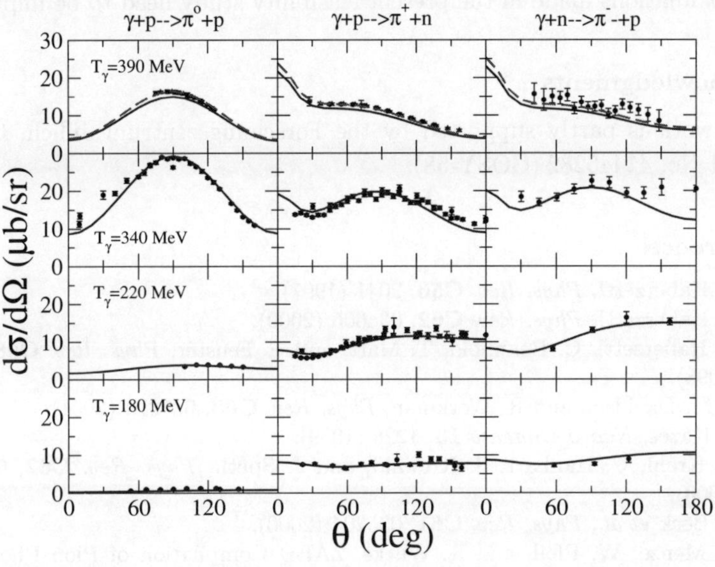

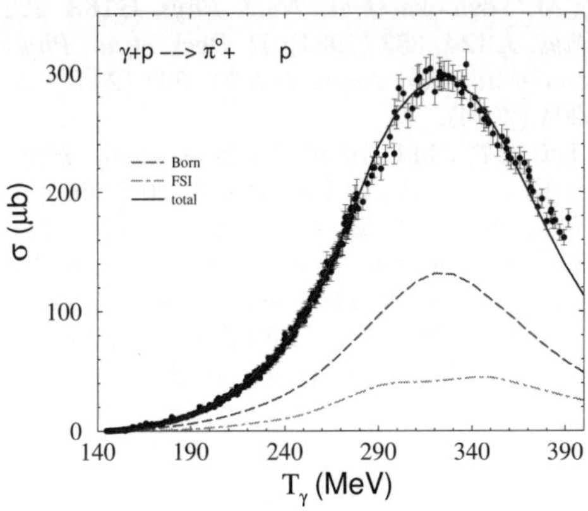

Fig. 3. Differential and total cross sections. In the top row of the differential cross sections, the dashed curves correspond to the results represented by solid curves multiplied by an arbitrary factor of 1.1 (Data: Refs. 7, 8.) The total cross section for the reaction $\gamma + p \rightarrow \pi^0 + p$ is given as a function of photon incident energy T_γ. The dashed curve corresponds to the Born contribution and the dash-dotted curve to the FSI loop contribution. The solid curve is the total contribution. (Data: Ref. 9.)

invariant (pseudoscalar) meson photoproduction amplitude which includes the hadronic final-state interaction explicitly. The method based on an earlier field-theoretical approach[1] is quite general and can be readily extended to any other meson photo- and electroproduction reactions. This method should be particularly relevant for the latter reaction. As an example of application of the present approach in its lowest order, we have calculated both the neutral and charged pion photoproduction processes off nucleons up to about 400 MeV photon incident energy which illustrates the feasibility of the present method. Obviously, for a more quantitative calculation, including not only cross sections but also other observables, some of the approximations made in the present feasibility study need to be improved.

Acknowledgments

This work is partly supported by the Forschungszentrum Jülich, COSY Grant No. 41445282 (COSY-58).

References

1. H. Haberzettl, *Phys. Rev.* **C56**, 2041 (1997).
2. H. Haberzettl, *Phys. Rev.* **C62**, 034605 (2000).
3. H. Haberzettl, C. Bennhold, T. Mart, and T. Feuster, *Phys. Rev.* **C58**, R40 (1998).
4. R. M. Davidson and R. Workman, *Phys. Rev.* **C63**, 025210 (2001).
5. E. Kazes, *Nuovo Cimento* **13**, 1226 (1959).
6. O. Krehl, C. Hanhart, S. Krewald, and J. Speth, *Phys. Rev.* **C62**, 025207 (2000).
7. R. Beck *et al.*, *Phys. Rev.* **C61**, 035204 (2000).
8. D. Menze, W. Pfeil, and R. Wilcke, ZAED Compilation of Pion Photoproduction Data, Univ. Bonn (1977); M. Yoshioka *et al.*, *Nucl. Phys.* **B168**, 222 (1980); J.-L. Faure *et al.*, *Nucl. Phys.* **A424**, 383 (1984); R. Beck *et al.*, *Phys. Rev. Lett.* **78**, 606 (1997); J. Ahrens *et al.*, *Eur. Phys. J.* **A21**, 323 (2004); A. Shafi *et al.*, *Phys. Rev.* **C70**, 035204 (2004).
9. E. Mazzucato *et al.*, *Phys. Rev. Lett.* **57**, 3144 (1986); R. Beck *et al.*, *Phys. Rev. Lett.* **65**, 1841 (1990); M. Fuchs *et al.*, *Phys. Lett.* **B368**, 20 (1996); J. C. Bergstrom *et al.*, *Phys. Rev.* **C53**, 1052 (1996); A. Schmidt *et al.*, *Phys. Rev. Lett.* **87**, 232501 (2001); B. B. Govorkov *et al.*, *Sov. J. Nucl. Phys.* **6**, 370 (1968); W. Hitzeroth *et al.*, *Nuovo Cimento* **A60**, 467 (1969); J. Ahrens *et al.*, *Phys. Rev. Lett.* **84**, 5950 (2000); R. G. Vasilkov *et al.*, *JETP* **10**, 7 (1960); F. Härter, PhD Thesis, Univ. Mainz (1996); M. MacCormick *et al.*, *Phys. Rev.* **C53**, 41 (1996).

THE STATUS OF PENTAQUARK BARYONS

VOLKER D. BURKERT
Jefferson Laboratory
12000 Jefferson Avenue, Newport News, VA23606, USA
E-mail: burkert@jlab.org

The status of the search for pentaquark baryon states is reviewed in light of new results from the first two dedicated experiments from CLAS at Jefferson Lab and of new analyses from several labs on the $\Theta^+(1540)$. Evidence for and against the heavier pentaquark states, the $\Xi(1862)$ and the $\Theta_c^0(3100)$ observed at CERN and at HERA, respectively, are also discussed. I conclude that the evidence against the latter two heavier pentaquark baryons is rapidly increasing making their existence highly questionable. I also conclude that the evidence for the Θ^+ state has significantly eroded with the recent CLAS results, and just leaves room for a possible state with an intrinsic width of $\Gamma < 0.5$ MeV. Preliminary new evidence from various experiments will be discussed as well.

1. Introduction

The announcement in 2003 of the discovery of the $\Theta^+(1540)$, a state with flavor exotic quantum numbers and a minimum valence quark content of $(uudd\bar{s})$[1], generated a tremendous amount of excitement in both the medium-energy nuclear physics and the high energy physics communities. Within less than one year the initial findings were confirmed by similar observations in nine other experiments [2–10], both in high energy and in lower energy measurements. These results seemed to beautifully confirm the theoretical prediction, within the chiral soliton model by D. Diakonov, M. Petrov, and M. Polyakov [11], of the existence of a state with strangeness $S = +1$, a narrow width of < 15 MeV, and a mass of about 1.53 GeV. This state was predicted as the isosinglet member of an anti-decuplet of ten states, three of which (Θ^+, Ξ^{--}, Ξ^+) with exotic flavor quantum numbers that experimentally can be easily distinuished from ordinary 3-quark baryons. Two observations of heavier pentaquark candidates at CERN [12] and at HERA [13] added to the expectation that a new avenue of research in hadron structure and strong QCD had been opened up. Yet, more than two

Table 1. Initial positive observations of the Θ^+, Ξ_5, and Θ_c^0 pentaquark candidates.

Experiment	Reaction	Energy (GeV)	Mass (MeV/c^2)	significance
LEPS	$\gamma^{12}C \to K^- X$	$E_\gamma \approx 2$	1540 $\pm$ 10	4.6σ
DIANA	$K^+ Xe \to pK_s^0 X$	$E_{K^+} < 0.5$	1539 $\pm$ 2	4σ
CLAS(d)	$\gamma d \to pK^- K^+ n$	$E_\gamma < 3.8$	1542 $\pm$ 5	5.2
SAPHIR	$\gamma p \to K_s^0 K^+ n$	$E_\gamma < 2.65$	1540 $\pm$ 4 $\pm$ 2	4.4σ
CLAS(p)	$\gamma p \to \pi^+ K^- K^+ n$	$E_\gamma = 4.8 - 5.5$	1555 $\pm$ 10	7.8σ
νBC	$\nu A \to pK_s^0 X$	range	1533 $\pm$ 5	6.7σ
ZEUS	$ep \to epK_s^0 X$	$\sqrt{s} = 320$	1522 $\pm$ 1.5	4.6σ
HERMES	$ed \to pK_s^0 X$	$E_e = 27.6$	1528 $\pm$ 2.6 $\pm$ 2.1	5.2σ
COSY	$pp \to \Sigma^+ pK_s^0$	$P_p = 3$	1530 $\pm$5	3.7σ
SVD	$pA \to pK_s^0 X$	$E_p = 70$	1526 $\pm$3 $\pm$ 3	5.6σ
NA49	$pp \to \Xi^- \pi^- X$	$E_p = 158$	1862 $\pm$ 2	4σ
H1	$ep \to D^{*-} pD^{*+} \bar{p} X$	$\sqrt{s} = 320$	3099 $\pm$3 $\pm$ 5	5.4σ

years after the initial announcement was made, the status of pentaquark baryons is less clear than it seemed two years back. I am here to address the question if pentaquark states have really been observed.

2. The positive sightings of the Θ^+

A summary of the published experimental evidence for the Θ^+ and the heavier pentaquark candidates is given in table 1. In most cases the published width is limited by the experimental resolution. The observed masses differ by up to more than 20 MeV. The quoted significance S in some cases is based on a naive, optimistic evaluation $S = signal/\sqrt{background}$, while for unknown background a more conservative estimate is $S = signal/\sqrt{signal + background}$, which would result in lowering the significance by one or two units. Despite of this, these observations presented formidable evidence for a state at a mass of 1525-1555 GeV. A closer look at some of the positive observations begins to reveal possible discrepancies.

2.1. *A problem with the width and production ratios Θ^+/Λ^*?*

The analysis of K^+A scattering data showed that the observed Θ^+ state must have an intrinsically very narrow width. Two analyses of different data sets found a finite width of $\Gamma = 0.9 \pm 0.3$ MeV [16,17], while others came up with upper limits of 1 MeV [18] to several MeV [19,20]. When compared with the production ratio of the $\Lambda^*(1520)$ hyperon with intrinsic width of $\Gamma_{\Lambda^*} =$ 15.9 MeV the rate of the total cross section for the formation of the two

Table 2. The ratio R_{Θ^+,Λ^*} measured in various experiments. The first 5 experiments claimed a Θ^+ signal, while the others give upper limits. The last two results are from the most recent CLAS measurements that give very small upper limits.

Experiment	Energy(GeV)	Θ^+/Λ^* (%)
LEPS	$E_\gamma \approx 2$	~ 40
CLAS(d)	$E_\gamma = 1.4 - 3.8$	~ 20
SAPHIR	$E_\gamma = 1.4 - 3$	10
ZEUS	$\sqrt{s} = 320$	5
HERMES	$E_\gamma \approx 7$	~ 200
CDF	$\sqrt{s} = 1960$	< 3
HERA-B	$\sqrt{s} = 42$	< 2
SPHINX	$\sqrt{s} = 12$	< 2
Belle	$\sqrt{s} \approx 2$	< 2.5
BaBar	$\sqrt{s} = 12$	< 3
CLAS-2(p)	$E_\gamma = 1.4 - 3.8$	< 0.2
CLAS-2(d)	$E_\gamma = 1.4 - 3.6$	
K^+A analysis	$\sqrt{s} = 1.54$	~ 1.5

states is expected to be $R_{\Theta^+,\Lambda^*} \equiv \sigma_{tot}(\Theta^*)/\sigma_{tot}(\Lambda^*) = 0.014$ for a 1 MeV width of the Θ^+ [21]. Although this relationship holds strictly for resonance formation at low energies only, dynamical models for the photoproduction of the Θ^+ show that the production cross section at modest energies of a few GeV strongly depends on the width of the state [25–28]. One therefore might expect R_{Θ^+,Λ^*} to remain small in the few GeV energy range. The published data however, suggested otherwise: Much larger ratios were observed than expected from the estimate based on the Θ^+ width. These results, together with the upper limits obtained from experiments with null results, including the most recent results from CLAS are summarized in table 2.

3. Non-observations of the Θ^+.

Something else that happened in 2004 and 2005 was a wave of high energy experiments presenting high statistics data that did not confirm the existence of the Θ^+ state. There are two types of experimental results, one type of experiments studies the decays of intermediate states produced in e^+e^- collisions, and gives limits in terms of branching ratios. The other type of experiments searched for the Θ^+ in fragmentation processes. Several experiments give upper limits for the R_{Θ^+,Λ^*} ratio. More detailed discussions of experiments that claimed sightings of a pentaquark candidate state, as well as those that generated null results are presented in recent reviews [14,15].

4. Are these results consistent ?

It is difficult to compare the low energy experiments with the high energy experiments. Low energy experiments study exclusive processes where completely defined final states are measured, and hadrons act as effective degrees of freedom. At high energies we think in terms of quark degrees of freedom, and fragmentation processes are more relevant. How can these different processses be compared quantitatively? The only invariant quantities for a resonance are quantum numbers, mass, and intrinsic width. In the absence of a resonance signal we can only place an upper limit on its width as a function of the invariant mass in which the resonance signal is expected to occur. It is therefore the total resonance width, or an upper limit on it, that allows us to compare processes for different reactions. It is not unreasonable to assume that a narrow width of the Θ^+ will result in much reduced production cross section compared to broader states such as the Λ^* both at low and at high energies. A similar conclusion may be drawn if one considers quark fragmentation as the main source of hadron production at high energies, e.g. in $e^+e^- \to q\bar{q}$, where hadrons are generated through the creation of $q\bar{q}$ pairs from the vacuum via glue string breaking. In a scenario of independent creation of a number of $q\bar{q}$ pairs starting with a single $q\bar{q}$ pair produced in e^+e^- annihilation, or a single quark knocked out of a target nucleon in deep inelastic scattering, four additional $q\bar{q}$ pairs from the vaccuum are needed to form a $(uudd\bar{s})$ 5-quark object. This should be much less likely to occur than the creation of a 3-quark baryon such as the Λ^*. The latter requires creation of only two additional $q\bar{q}$ pairs.

If we take the estimate $R_{\Theta^+,\Lambda^*} \sim 0.015$ for a 1 MeV width of the Θ^+ from low energy resonance formation as a guide, we have a way of relating high energy and low energy processes. Comparing the limits for that ratio from table 2 one can make several observations: 1) The first set of experiments claiming sighting of the Θ^* show very large ratios. 2) The second set of experiments quoting upper limits are not below the ratio extracted from low energy K^+A analysis. 3) The recent CLAS results are an order of magnitude below that value. This is to be contrasted with the very large ratios measured in the first set of experiments in table 2. The focus of new experimental investigations should be to verify that these initial results are indeed correct.

5. New results - mostly against the existence of pentaquark states.

During the past six months much new evidence against and some in favor of the existence of pentaquark baryons have emerged. There are new high statistics results from the CLAS detector at Jefferson Lab. New analyses of previously published data have become available from ZEUS and the SVD-2 collaborations, and LEPS studied a new channel with claimed Θ^+ sensitivity. The Belle and BaBar collaborations have generated high statistics data that test the lower energy photoproduction results, and high energy experiments at Fermilab, HERA and CERN confront the claims for the Ξ_5 and Θ_c^0 pentaquark candidates. New evidence for a doubly charged Θ^{++} comes from the STAR detector at RHIC, while a high statistics CLAS experiment finds a stringent upper limit for $\Theta^{+}+$ production on protons. These new results will be discussed in the following sections.

5.1. *New results from CLAS.*

The CLAS collaboration has recently completed the first two dedicated high statistics experiments aimed at verifying previously reported observations of the Θ^+.

The first experiment measured the reaction $\gamma p \rightarrow K_s^0 K^+(n)$, where the neutron is reconstructed using 4-momentum conservation. No Θ^+ signal is seen, and an upper limit for the cross section is derived by fitting the data with a polynomial background distribution and a sliding Gaussian that represents the experimental resolution. In the mass range from 1.525 to 1.555 GeV, a limit of (0.85 - 1.3) nb (95% c.l.) is derived. What does this result tell us? There are several conclusions that can be drawn from the CLAS result on the proton.

1) It directly contradicts, by two orders of magnitude in cross section, the SAPHIR experiment[4] that claimed a significant signal in the same channel and in the same energy range, and published a cross section of 300nb for Θ^+ production.

2) Together with the extracted Λ^* cross section it puts an upper limit on the Θ^+/Λ^* ratio in table 2 that is an order of magnitude lower than the value from the K^+A analyses.

3) It puts a very stringent limit on a possible production mechanism. For example, it implies a very small coupling Θ^+NK^* which in many hadronic models was identified as a major source for Θ^+ production.

4) If there is no large isospin asymmetry in the elementary process, the γD

and γA experiments at lower statistics should not be able to see a signal. Possible mechanisms to obtain a large isospin asymmetry have been discussed in the literature[31,32] following the first announcement of the new CLAS data.

The second new CLAS experiment measured the reaction $\gamma D \rightarrow pK^-K^+(n)$, where the neutron again is reconstructed from the overdetermined kinematics. This experiment represents a dedicated measurements to verify a previous CLAS result that claimed more than 4.6σ significance for the Θ^+ in the same channel and same energy range. The aim is to measure the elementary production on neutrons through $\gamma n \rightarrow K^-K^+n$. To avoid the complication of precise neutron detection the recoil proton is measured instead, requiring momenta of greater than 0.35 GeV/c for the proton to be detected in CLAS. This reduces the acceptance for the exclusive reaction by a large factor. No significant signal is seen in a data sample with about seven times the statistics of the previous result. From this result an upper limit of 5nb (95% c.l.) is derived for the elementary cross section on the neutron. The limit is somewhat model-dependent as rescattering effects in the deuteron must be taken into account. The result clearly contradicts the previous lower statistics data. In order to understand the discrepancy the older data have been reanalyzed with a background distribution extracted from the new high statistics data set. The results show an underestimation of the background normalization in the original analysis. A new fit with the improved background yields a signal with a significance of 3σ compared to the published value of $(5.2 \pm 0.6)\sigma$.

What is the impact of the combined CLAS data on proton and neutron? For this we compare the cross section limits with various dynamical model calculations [25–28]. In hadro-dynamical models, the cross section is computed based on an effective Lagrangian approach. The comparison is shown in table 3 for the $J^P = 1/2^+$ assignment and $\Gamma_{\Theta^+} = 1$ MeV. The upper limit for the combined proton and neutron targets would be less than 0.5 MeV for at least one of the targets in each model.

5.2. *BaBar study of quasi-real photoproduction*

The BaBar collaboration also studied the quasi-real photoproduction of $e + Be \rightarrow pK^0_s + X$ [34]. In this case electrons with energies of $\sim$ 9 GeV resulting from small angle scattering off the positron beam interact with the beryllium beam pipe. The scattered electron is not detected, and the invariant mass of final state inclusive pK^0_s is studied for possible contributions from $\Theta^+ \rightarrow pK^0_s$. There is no evidence for a signal. The data can be

Table 3. Limit on Θ^+ width from recent CLAS results. Upper limits for the total cross section on protons of 1.25nb, and on neutrons of 4nb are used to determine the limit on the width. The first line in each row is for γp, the second line is for γn. The cross section is computed for a $J^P = 1/2^+$ assignment of the Θ^+ and a width of 1 MeV.

Publication	$\sigma(\gamma N)$ (nbarn)	Γ_{Θ^+} (MeV)
S. Nam et al. [27]	2.7 2.7	< 0.5 < 1.7
Y. Oh et al., [28]	~ 1.6 ~ 8.7	< 0.8 < 0.5
C.M Ko et al., [26]	15 15	< 0.08 < 0.25
W. Roberts [25]	5.2 11.2	< 0.24 < 0.4

directly compared to the HERMES results which were taken in quasi-real photoproduction kinematics from deuterium at higher electron beam energies. The absence of any signal in the very high statistics BaBar data calls the signal observed by HERMES at much lower statistics into question. BaBar also compare their null results with the ZEUS signal observed at $Q^2 > 20$ GeV^2. However, since ZEUS sees no signal at low Q^2 and BaBar only probes the quasi-real photoproduction kinematics, this comparison is indeed misleading.

5.3. *LEPS at SPring-8*

The LEPS experiment originally claimed the discovery of the Θ^+ in photoproduction from a carbon target in the inclusive reaction $\gamma C \to K^- X$ plotting the Fermi-momentum corrected missing mass M_X. The experiment has been repeated with a liquid deuterium target and higher statistics. A peak at 1530 MeV is observed. The data also show a large ratio of Θ^+/Λ^* (see table 2). Since these data are obtained at energies similar to the new CLAS data on deuterium, they need to be confronted with the recent exclusive CLAS data taken on deuterium in the reaction $\gamma d \to K^- p K^+ n$, and the resulting cross section limit for the elementary cross section on neutrons. This will require extraction of a normalized cross section from the LEPS data. There are also new results from LEPS on the channel $\gamma d \to \Lambda^* X$ where a narrow peak near 1530 MeV/c^2 with 5σ significance is claimed. The signal emerges only when events are selected with $M_{K^-p} \sim M_{\Lambda^*}$, indicating that the process $\gamma D \to \Lambda^* \Theta^+$ may be observed. The mass distribution also shows an excess of events near 1600 MeV/c^2.

5.4. *Results from Belle*

New results by the Belle collaboration have been presented recently[23]. Belle uses hadrons created in high energy e^+e^- collisions and reconstructs the hadron interaction with the vertex detector materials. The momentum spectrum is sufficiently low so that resonance formation processes such as $K^+n \to \Theta^+ \to pK^0_s$ can be studied. No Θ^+ signal is seen. The upper limit on the formation cross section can be used to extract an upper limit for the Θ^+ width. At a specific mass of 1539 MeV, an upper limit of $\Gamma_{\Theta^+} < 0.64$ MeV (90% c.l.) is derived. The mass corresponds to the Θ^+ mass claimed by the DIANA experiment [2]. If one allows the entire mass range for the Θ^+ from 1525 to 1555 MeV claimed by experiments, the upper limit is $\Gamma_{\Theta^+} < 1$ MeV (90% c.l.). The latter value confirms the limit derived in previous analyses.

5.5. *BaBar results in quark fragmentation*

The BaBar collaboration at SLAC searches for the Θ^+ as well as the Ξ^{--} pentaquark states directly in e^+e^- collisions [24], mostly in the quark fragmentation region. With high statistics no signal is found for either $\Theta^+(1540)$ or $\Xi^{--}(1862)$, and upper limits are placed on their respective yields. The limit on the production rates are 8 or 4 times lower than the rates of ordinary baryons at the respective masses. It is, however, not obvious what this result implies. The slope measured for the production of pseudoscalar mesons is $d(event\ rate)/d(mass) = 10^{-2}$/GeV. For 3-quark baryons it is 10^{-4}/GeV, i.e. the rate drops by a factor of 10,000 per one unit of GeV in mass. In the quark fragmentation region, if we extrapolate from mesons where only one $q\bar{q}$ pair must be created to form a meson starting with one of the initial quarks in the e^+e^- annihilation, and baryons where two $q\bar{q}$ pairs are needed, to pentaquarks where four $q\bar{q}$ pairs are needed, the slope for pentaquark production in fragmentation would be 10^{-8}/GeV. Since there is no rate measured for a pentaquark state there is no normalization point available. If we normalize the pentaquark line at 1 GeV to the baryon and meson rates, the line falls one order of magnitude below the upper limit for the Θ^+ and several orders of magnitude below the upper limit for the Ξ^{--} assuming a mass of 1862 MeV for the latter. The sensitivity of quark fragmentation to 5-quark baryon states is thus questionable. Moreover, the limit for $R_{\Theta^+,\Lambda^*} < 0.02$ at 95% c.l. is not in contradiction with the ratio estimated at low energy assuming a width of $\Gamma_{\Theta^+} = 1$ MeV [17].

The ZEUS collaboration has extended the analysis of their Θ^+ signal and studied possible production mechanisms[33]. The signal emerges at $Q^2 >$ 20 GeV^2 and remains visible at $Q^2 > 50$ GeV^2. The Θ^+ and $\bar{\Theta}^-$ signals are nearly equally strong, however, the signal is present only at forward rapidity $\eta^{lab} > 0$ and not visible at backward rapidity $\eta^{lab} < 0$. There is currently no production mechanism proposed that would generate such a pattern. The ZEUS collaboration extracted the Q^2-dependence of the ratio R_{Θ^+,Λ^*} which shows a weak dependence on Q^2.

5.6. *New results from high energy hadronic interaction experiments*

The SVD-2 collaboration has reanalyzed their published data[43] with much improved event reconstruction efficiency. The experiment measured the reaction $pA \to pK_s^0X$ using a 70 GeV incident proton beam. The main component in the detector system is the silicon vertex detector (SVD). Events are divided into two samples: events with the K_s^0 decaying inside and events with the decay outside the SVD. The two distributions both show a significant peak at the mass of 1523 MeV/c^2. A combined significance for two independent data sets of $\sim 7.5\ \sigma$ is obtained. The strangeness assignment in the pK_s^0 channel is not unique, and could also indicate excitation of a Σ^* resonance. In this case one would expect a decay $\Sigma \to \Lambda\pi$, which is not observed. Therefore, an exotic $S = +1$ assignment of that peak is likely should it be a resonant state. The SVD-2 results have been challenged by the WA89 collaboration that measured the process $\Sigma^- A \to pK_s^0X$ in comparable kinematics [44]. Their mass distribution does not exhibit any signal in the mass range of the Θ^+ candidate. The WA89 collaboration claims that their results are incompatible with the SVD results.

6. An isovector Θ^{++} candidate?

Inspired by the prediction of Diakonov et al., of an anti-decuplet of 5-quark states, with the Θ^+ being an isoscalar, the focus of the search for the lowest mass pentaquark was on an isoscalar baryon with $S = +1$. However, searches have also been conducted for a possible isovector baryon state with charge $Q = +2$. The final state to study is pK^+. No signal was seen in any of these searches. However, recently the STAR collaboration at RHIC presented data indicating a small but significant Θ^{++} candidate [45]. A peak with a significance of 5σ is seen at a mass of about 1530 MeV/c^2 in the d-Au collision sample. The Λ^* signal is also clearly visible.

If the Θ^{++} signal is real, then there must be also a signal in the singly charged channel, i.e. a Θ^+ . A small peak with relatively low significance appears in the $K_s^0 p$ invariant mass spectrum, however shifted by about 10 MeV/c^2 to higher mass values. High statistics CLAS results on $\gamma p \rightarrow pK^+K^-$ show no evidence for for a Θ^{++} candidate [46]. An upper limit of 150 picobarn (95% c.l.) is derived. Using a model description for a possible production mechanism an upper limit for the width of $\Gamma_{\Theta^{++}} < 75$ KeV (95% c.l.) is obtained. Clearly, such a narrow width cannot be consistent with a hadronically decaying state.

So far I have focussed in my talk on the Θ^+, as without the evidence for the Θ^+ there would not have been any search for other pentaquark states within the anti-decuplet. However, much effort has been put recently into the search for the two heavier pentaquark candidates claimed in two high energy experiments.

7. Status of Ξ_5 and Θ_c^0

A candidate for a 5-quark Ξ_5 has been observed in the $\Xi^-\pi^-$ final state by the CERN NA49 experiment, and a candidate Θ_c^0 for the charmed equivalent of the Θ^+ has been claimed by the H1 experiment at HERA in the channel D^*p. In contrast to the Θ^+, which has been claimed in at least ten experiments, the heavier candidates have not been seen in any other experiment. The Ξ_5 state of NA49 has been searched for by several experiments[35–42]. The ratios $\Xi_5/\Xi(1530)$ determined by several experiments are shown in table 4. However, the highest energy experiments probe production through quark fragmentation, and may not be directly comparable to the NA49 results. The FOCUS photoproduction experiment and the COMPASS muon scattering experiment are closer to the kinematics of NA49. Neither experiment observed a signal. FOCUS derived an upper limit nearly two orders lower than the signal seen by NA49. A summary of results on the Ξ_5 search is given in table 4.

The $\Theta_c^0(3100)$ pentaquark candidate so far has only been seen by the H1 experiment at HERA. Several other experiments came up empty-handed [47], and ZEUS and FOCUS both claim incompatibility of their results with the H1 findings.

Table 4. Results of searches for the Ξ_5 pentaquark state.

Experiment	Initial state	Energy (GeV)	$\Xi_5^{--}/\Xi(1530)$
NA49	pp	$E_p = 158$	0.24
COMPASS	$\mu^+ A$	$E_\mu = 160$	< 0.046
ALEPH	e^+e^-	$\sqrt{s} = M_Z$	< 0.075
BaBar	e^+e^-	$m_{\Upsilon(4s)}$	< 0.0055
CDF	$p\bar{p}$	$\sqrt{s} = 1960$	< 0.03
E690	pp	$E_p = 800$	< 0.003
FOCUS	γp	$E_\gamma < 300$	< 0.003
HERA-B	pA	$E_p = 920$	< 0.04
HERMES	eD	$E_e = 27.6$	< 0.15
WA89	$\Sigma^- A$	$E_\Sigma = 340$	< 0.013
ZEUS	ep	$\sqrt{s} = 310$	not seen

8. Summary and conclusions

Over the past year the evidence for the existence of pentaquark baryons has clearly lost much of its original significance.

The evidence for the two heavy pentaquark candidate states, the $\Xi_5(1862)$ and the $\Theta_c^0(3100)$, observed at unexpected masses, and each seen in one experiment only, has been drastically diminished. Several experiments with high sensitivity to the relevant processes have found no indication of these states. In the face of overwhelming evidence against these states, experiments claiming positive sightings should either explain why other experiments are not sensitive, or should re-evaluate their own results.

The situation with the Θ^+ state observed at masses near 1540 MeV is less clear, although evidence for the state has also diminished significantly. So far more than ten experiments claimed to have observed a narrow state with exotic flavor quantum number $S = +1$. Two of the initial results (SAPHIR and CLAS(d)) have been superseded by higher statistics measurements from CLAS [29,30] conducted at same energies and with same or overlapping acceptances. No signal was found in either case. In addition, the HERMES results are being challenged by new high statistics data from BaBar [34]. It is remarkable that experiments claiming a Θ^+ signal, measured Θ^+/Λ^* ratios much above values naively expected from K^+A scattering analysis (see table 2). HERMES even measures a Θ^+ cross section that is significantly higher than the cross section for Λ^* production.

The Belle experiment [23] studying K^+A scattering, is beginning to challenge the DIANA results, the second experiment claiming observation of the Θ^+. Belle extracted an upper limit of 0.64 MeV (90% c.l.) for the width of any Θ^+ signal at a mass of 1539 MeV. In the larger mass range of 1525-1555 an upper limit of 1 MeV has been extracted. This is to be contrasted with the width extracted from the DIANA experiment as well as from K^+D scattering of $\Gamma_{\Theta^+} = 0.9 \pm 0.3$ MeV. However, the Belle limit is not (yet) in strict contradiction to the DIANA results at this point. It will be interesting to see if the Belle limit can be further reduced with higher statistics.

The new CLAS results on protons and neutrons also challenge the value of the Θ^+ width. Using hadronic models, upper limits of 0.1 to 0.6 MeV are obtained for proton targets, and the $J^P = 1/2^+$ assignment. For neutron targets limits from 0.26 and 1.7 MeV are obtained. Much smaller limits of < 0.1 MeV are obtained for $J^P = 3/2^-$, while for $J^P = 1/2^-$, limits of 1 to 2.5 MeV are extracted. Although these limits are model-dependent, taken together they still present formidable constraints on the Θ^+ width. Hadronically decaying resonances with total decay widths of less than a few MeV would seem unusual, but limits on the width of less than a few hundred KeV would make the existence of the state highly unlikely. The CLAS results on a possible Θ^{++} state present very stringent upper limits of $<$ 75 KeV on the width. In order to have quantitative tests of the LEPS results, which is the only remaining low energy photoproduction experiment with a positive signal, the old and new results from LEPS should be turned into normalized cross sections and compared to the CLAS data on deuterium.

When the dust will have settled on the issue of narrow pentaquark baryons, we will have learned a lot about the physics of hadrons, no matter what the final outcome will be.

References

1. T. Nakano et al. (LEPS),*Phys. Rev. Lett.* 91:012002 (2003)
2. V.V. Barmin et al. (DIANA), *Phys. Atom. Nuclei* 66, 1715 (2003)
3. S. Stepanyan et al. (CLAS),*Phys. Rev. Lett.* 91:25001 (2003)
4. J. Barth et al. (SAPHIR), *Phys. Lett. B* 572, 127 (2003)
5. A. E. Asratyan, A.G. Dolgolenko, M.A. Kubantsev, *Phys. Atom. Nuclei* 67, 682 (2004)
6. V. Kubarovsky et al. (CLAS), *Phys. Rev. Lett.* 92:032001 (2004)
7. A. Airapetian, et al. (HERMES), *Phys. Lett. B* 585, 213 (2004)
8. S. Chekanov et al., (ZEUS), *Phys. Lett. B* 591, 7 (2004)
9. M. Abdel-Barv et al. (COSY-TOF), *Phys. Lett. B* 595, 127 (2004)
10. A. Aleev et al. (SVD), *hep-ex/0401024*
11. D. Diakonov, P. Petrov, M. Polyakov,*Z. f. Phys, A* 359, 305 (1997)
12. C. Alt et al. (NA49), *Phys. Rev. Lett.* 92:042003 (2004)
13. A. Aktas et al. (H1), *Phys. Lett. B* 588, 17 (2004)
14. K. Hicks, *hep-ph/0504027*
15. M. Danilov, *hep-ex/0509012*

16. R.N. Cahn and G.H. Trilling, *Phys. Rev. D* 69:011501 (2004)
17. W. Gibbs, *Phys. Rev. C* 70:045208 (2004)
18. A. Sibirtsev et al., *Eur.Phys.J. A* 23:491 (2005)
19. R. Arndt, I. Strakovsky, R. Workman, *Phys. Rev. C* 68:042201, 2003, Erratum-ibid.C69:019901 (2004)
20. R. Gothe and S. Nussinov, *hep-ph/0308230*
21. W. Gibbs, private communications
22. T. Nakano, private communications.
23. K. Abe et al. (BELLE), *hep-ex/0507014*
24. Tetiana Berger-Hryn'ova (BaBar), *hep-ex/0510043*
25. W. Roberts; *Phys. Rev. C* 70, 065201 (2004)
26. C.M. Ko and W. Liu, *nucl-th/0410068*
27. S. Nam et al.; *hep-ph/0505134*
28. Y. Oh, K. Nakayama, and T.-S.H. Lee, *hep-ph/0412363*
29. M. Battaglieri et al. (CLAS), hep-ex/0510061, submitted to *Phys. Rev. Lett.*
30. S. Stepanyan et al. (CLAS), paper contributed to this conference
31. S. Nam, A. Hosaka, and H. Kim, *hep-ph/0508220*
32. Marek Karliner and Harry J. Lipkin, *hep-ph/0506084*
33. ZEUS collaboration, submitted to XXII International Symposium on Lepton Photon Interactions at High Energies, July 2005, Uppsala, Sweden.
34. K. Goetzen (BaBaR), *hep-ex/0510041*; J. M. Izen, Talk presented at the EINN workshop on "New hadrons: Facts and Fancy", Milos, Greece, September 19-20, 2005
35. B. Aubert et al. (BaBar), *hep-ex/0502004*
36. D.O. Litvintsev (CDF), *hep-ex/0410024*
37. K. Stenson (FOCUS), *hep-ex/0412021*
38. K.T. Knöpfle et al. (HERA-B), *J.Phys.G* 30:S1363 (2004); J. Spengler (HERA-B), *Acta Phys. Polon. B36:2223 (2005)*
39. A. Airapetian et al. (HERMES), *Acta Phys. Polon. B* 36:2213 (2005)
40. M.I. Adamavich et al. (WA89), *Phys. Rev. C* 70:022201 (2004)
41. S. Chekanov et al.(ZEUS), paper submitted to LP2005, DESY-05-18 (2005)
42. E.S. Ageev et al.(COMPASS), *Eur. Phys. J. C* 41, 469 (2005);
43. A. Aleev et al. (SVD-2), *hep-ex/0509033*
44. M.I. Adamovich et al. (WA89), *hep-ex/05100013*
45. H. Z. Huang, *nucl-ex/o509037*
46. V. Kubarovsky et al. (CLAS), Talk presented at Pentaquark 2005, November 3-5, 2005, Jefferson Lab.
47. J.M. Link et al. (FOCUS); *hep-ex/0506013*; S. Chekanov et al. (ZEUS), paper submitted to LP2005, DESY-04-164(2004); A. Airapetian et al. (HERMES), *Phys.Rev. D* 71:032004 (2005)

RECENT BES RESULTS FROM J/ψ DECAYS

ZIJIN GUO
(For the BES Collaboration)
University of Hawaii,
2505 Correa Road,
Honolulu, HI 96822, USA
E-mail: guozj@phys.hawaii.edu

The studies on the multi-quark candidates, excited baryon states at BES are presented, based on 58 million J/ψ data colleted with BES II detector. We also report the measurements of $J/\psi \to p\bar{p}$, $\Lambda\bar{\Lambda}$, and $\Sigma^0\bar{\Sigma}^0$.

1. Introduction

The Beijing Spectrometer (BES) is a general purpose solenoidal detector at the Beijing Electron Positron Collider (BEPC). BEPC operates in the center of mass energy range from 2 to 5 GeV with a luminosity at the J/ψ energy of approximately 5×10^{30} cm^{-2}s^{-1}. BES (BESI) is described in detail in Ref. 1, and the upgraded BES detector (BESII) is described in Ref. 2. The 58 million J/ψ events have been accumulated with BES II, which provides a good laboratory for the study of the non-$q\bar{q}$ states and hadron spectroscopy.

2. Multi-quark search and study

2.1. *Near $p\bar{p}$ threshold enhancement in $J/\psi \to \gamma p\bar{p}$*

There is an accumulation of evidence for anomalous behavior in the $p\bar{p}$ system near $2m_p$ mass threshold. We analyze $J/\psi \to \gamma p\bar{p}$ with BES II J/ψ data. [3] Figure 1 shows the $p\bar{p}$ invariant mass distribution for selected events. Except for a peak near $M_{p\bar{p}} = 2.98$ GeV/c^2 that is consistent in mass, width, and yield with expectations for $J/\psi \to \gamma\eta_c$, $\eta_c \to p\bar{p}$ and a broad enhancement around $M_{p\bar{p}} \sim 2.2$ GeV/c^2, there is a narrow, low-mass peak near the $p\bar{p}$ mass threshold.

The low mass region of the $p\bar{p}$ distribution is fitted by an acceptance-

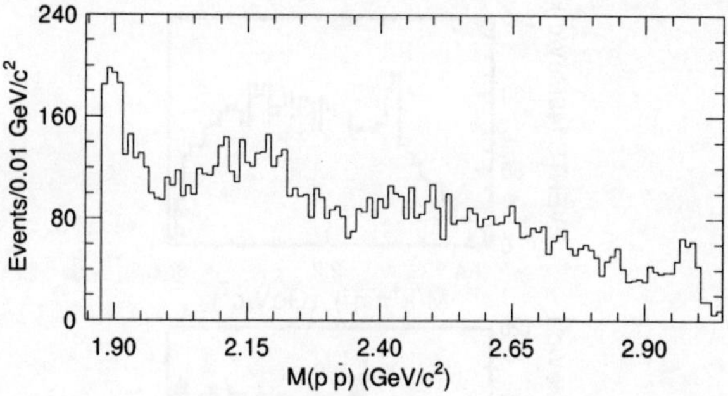

Fig. 1. The $p\bar{p}$ invariant mass distribution for the $J/\psi \to \gamma p\bar{p}$-enriched event sample.

weighted S-wave Breit-Wigner (BW) function and $f_{bkg}(\delta)$ which represnet the low-mass enhancement and the background, repectively. The fit yields 928 ± 57 events in the BW function with a peak mass of $M = 1859\ ^{+3}_{-10}\ ^{+5}_{-25}$ MeV/c^2 and a full width of $\Gamma < 30$ MeV/c^2 at the 90% confidence level.

2.2. *New observation of X(1835)*

In the analysis of $J/\psi \to \gamma\pi^+\pi^-\eta'$, [4] η' is tagged in two decay modes, $\eta' \to \pi^+\pi^-\eta$ ($\eta \to \gamma\gamma$) and $\eta' \to \gamma\rho$. Figure 2 shows the $\pi^+\pi^-\eta'$ invariant mass spectrum for the combined $J/\psi \to \gamma\pi^+\pi^-\eta'$ ($\eta' \to \pi^+\pi^-\eta$) and $J/\psi \to \gamma\pi^+\pi^-\eta'$ ($\eta' \to \gamma\rho$) samples. This spectrum is fitted with a BW function convolved with a Gaussian mass resolution function to represent the $X(1835)$ signal plus a smooth polynomial background function. The mass and width obtained from the fit (shown in the bottom panel of Figure 2) are $M = 1833.7 \pm 6.1 \pm 2.7$ MeV/c^2 and $\Gamma = 67.7 \pm 20.3 \pm 7.7$ MeV/c^2. The statistical significance is 7.7σ. The mass and width of the $X(1835)$ are not compatible with any known meson resonance. [5] Using MC-determined selection efficiencies of 3.72% and 4.85% for the $\eta' \to \pi^+\pi^-\eta$ and $\eta' \to \gamma\rho$ modes, respectively, we determine a product branching fraction of

$$B(J/\psi \to \gamma X(1835)) \cdot B(X(1835) \to \pi^+\pi^-\eta') = (2.2 \pm 0.4 \pm 0.4) \times 10^{-4}.$$

We examine the possibility that the $X(1835)$ is responsible for the $p\bar{p}$ mass threshold enhancement observed in radiative $J/\psi \to \gamma p\bar{p}$ decays. [3] The $X(1835)$ resonance might be a prime candidate for the source of the $p\bar{p}$ mass threshold enhancement in $J/\psi \to \gamma p\bar{p}$ process. In this case, the J^{PC} and I^G of the $X(1835)$ could only be 0^{-+} and 0^+, which can be tested in future experiments. Since decays to $p\bar{p}$ are kinematically allowed only for a small portion of the high-mass tail of the resonance and have very

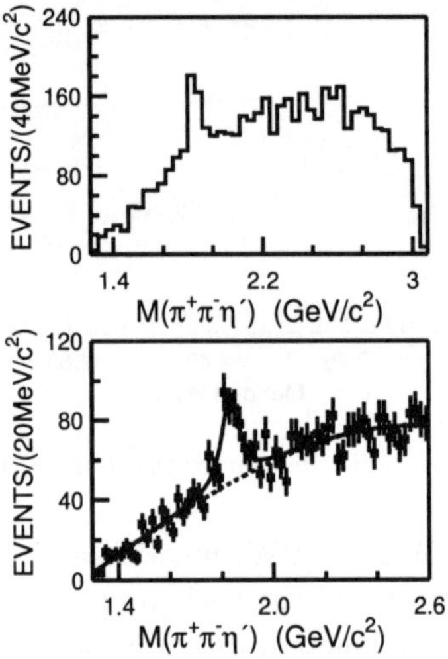

Fig. 2. The $\pi^+\pi^-\eta'$ invariant mass distribution for selected events from both the $J/\psi \to \gamma\pi^+\pi^-\eta'$ ($\eta' \to \pi^+\pi^-\eta$, $\eta \to \gamma\gamma$) and $J/\psi \to \gamma\pi^+\pi^-\eta'$ ($\eta' \to \gamma\rho$) analyses. The bottom panel shows the fit (solid curve) to the data (points with error bars); the dashed curve indicates the background function.

limited phase space, the large $p\bar{p}$ branching fraction implies an unusually strong coupling to $p\bar{p}$, as expected for a $p\bar{p}$ bound state. [6] However, other possible interpretations of the $X(1835)$ that have no relation to the $p\bar{p}$ mass threshold enhancement are not excluded.

2.3. *Enhancement in the* $p\bar{\Lambda}$ *invariant mass spectrum in* $J/\psi \to pK^-\bar{\Lambda}$ *and in* $\psi(2S) \to pK^-\bar{\Lambda}$ *decays*

The $J/\psi \to pK^-\bar{\Lambda}$ candidate events are required to have four good charged tracks with total charge zero. Events are subjected to a four-constraint (4C) kinematic fit with the corresponding mass assignments for each track. For events with ambiguous particle identification, all possible 4C combinations are formed, and the combination with the smallest χ^2 is chosen. A sample of 5421 $J/\psi \to pK^-\bar{\Lambda}$ candidates survive the final selection. Monte Carlo (MC) studies indicate that the background in the selected event sample is at the $1 \sim 2\%$ level.

The $p\bar{\Lambda}$ invariant mass spectrum for the selected events is shown in Figure 3(a), where an enhancement is evident near the mass threshold. No corresponding structure is seen in a sample of $J/\psi \to pK^-\bar{\Lambda}$ MC events generated with a uniform phase space distribution. The $pK^-\bar{\Lambda}$ Dalitz plot is

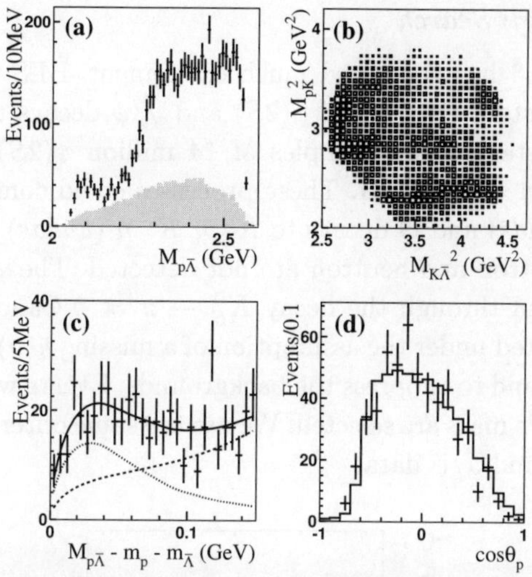

Fig. 3. (a) The points with error bars indicate the measured $p\bar{\Lambda}$ mass spectrum; the shaded histogram indicates phase space MC events (arbitrary normalization). (b) The Dalitz plot for the selected event sample. (c) A fit (solid line) to the data. The dotted curve indicates the BW signal and the dashed curve the phase space 'background'. (d) The $\cos\theta_p$ distribution under the enhancement, the points are data and the histogram is the MC (normalized to data)

shown in Figure 3(b). In addition to bands for the well established $\Lambda^*(1520)$ and $\Lambda^*(1690)$, there is a significant N^* band near the $K^-\bar{\Lambda}$ mass threshold, and a $p\bar{\Lambda}$ mass enhancement, isolated from the Λ^* and N^* bands, in the right-upper part of the Dalitz plot.

This enhancement can be fit with an acceptance weighted S-wave BW function, together with a function describing the phase space contribution, as shown in Figure 3(c). The fit gives a peak mass of $m = 2075 \pm 12 \pm 5$ MeV/c^2 and a width $\Gamma = 90 \pm 35 \pm 9$ MeV/c^2. The significance is about 7σ. A P-wave BW resonance functions can also fit the enhancement. The $\cos\theta_p$ distribution, shown in Figure 3(d), where θ_p is the decay angle of p in the $p\bar{\Lambda}$ CM frame, agrees well with that of a MC sample of $J/\psi \to KX \to Kp\bar{\Lambda}$. Since the MC $\cos\theta_p$ distribution is generated as a uniform S-wave distribution and the detected MC distribution agrees with data in Figure 3(d), the observed distribution for the enhancement is consistent with S-wave decays to $p\bar{\Lambda}$.

Evidence of a similar enhancement is observed in $\psi(2S) \to pK^-\bar{\Lambda}$ when the same analysis is performed on the $\psi(2S)$ data sample. More detail can be found in Ref. 7.

2.4. *Pentaquark Search*

The pentaquark [8] has generated much excitement. BES has searched for the pentaquark state $\Theta(1540)$ in $\psi(2S)$ and J/ψ decays to $K^0_S p K^- \bar{n}$ and $K^0_S \bar{p} K^+ n$ final states using samples of 14 million $\psi(2S)$ and 58 million J/ψ events taken with BES II. These processes could contain Θ decays to $K^0_S p$, $K^+ n$ $(uudd\bar{s})$ and $\bar{\Theta}$ decays to $K^0_S \bar{p}$, $K^- n$ $(\bar{u}\bar{u}\bar{d}\bar{d}s)$.

The anti-neutron and neutron are not detected. The K^0_S meson in the event is identified through the decay $K^0_S \to \pi^+\pi^-$. Candidate events are kinematically fitted under the assumption of a missing $\bar{n}(n)$ to obtain better mass resolution and to suppress the backgrounds. Events with missing mass close to the $\bar{n}(n)$'s mass are selected. We use the same criteria and treatment for both $\psi(2S)$ and J/ψ data.

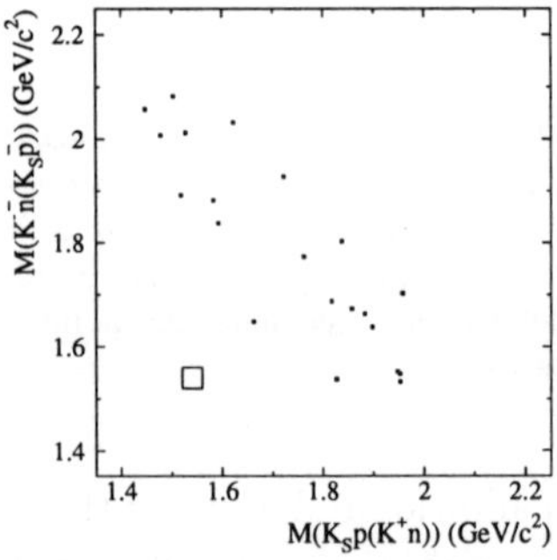

Fig. 4. Scatter plot of $K^- n$ $(K^0_S \bar{p})$ versus $K^0_S p$ $(K^+ n)$ for $\psi(2S) \to K^0_S p K^- \bar{n} + K^0_S \bar{p} K^+ n$ modes.

The scatter plot of $K^- n$ $(K^0_S \bar{p})$ versus $K^0_S p$ $(K^+ n)$ for $\psi(2S) \to K^0_S p K^- \bar{n} + K^0_S \bar{p} K^+ n$ modes is shown in Figure 4. Zero events fall within the signal region, shown as a square centered at (1.540, 1.540) GeV/c^2, and we set an upper limit at the 90% confidence level (C.L.) on the branching fraction:

$$B(\psi(2S) \to \Theta\bar{\Theta} \to K^0_S p K^- \bar{n} + K^0_S \bar{p} K^+ n) < 0.84 \times 10^{-5}.$$

Another possibility is that the $\psi(2S)$ decays to only one Θ or $\bar{\Theta}$ state. To determine the number of $\Theta(1540)$ events from single Θ or $\bar{\Theta}$ production, we count the number of events within regions of 1.52 - 1.56 GeV/c^2 in the projections of Figure 4 and set upper limits, shown in Table 1.

For the decays of $J/\psi \to K^0_S p K^- \bar{n}$ and $K^0_S \bar{p} K^+ n$, we use the same criteria and analysis method as those used for the $\psi(2S)$ data to study

Table 1. Summary of upper limits.

Decay mode	$\psi(2S)$	J/ψ
$\to \Theta\bar{\Theta} \to K_S^0 pK^-\bar{n}x + K_S^0\bar{p}K^+n$	0.88×10^{-5}	1.1×10^{-5}
$\to \Theta K^-\bar{n} \to K_S^0 pK^-\bar{n}$	1.0×10^{-5}	2.1×10^{-5}
$\to \bar{\Theta} K^+n \to K_S^0\bar{p}K^+n$	2.6×10^{-5}	5.6×10^{-5}
$\to K_S^0 p\bar{\Theta} \to K_S^0 pK^-\bar{n}$	0.60×10^{-5}	1.1×10^{-5}
$\to K_S^0\bar{p}\Theta \to K_S^0\bar{p}K^+n$	0.70×10^{-5}	1.6×10^{-5}

possible $\Theta(1540)$ production. There is no significant $\Theta(1540)$ signal, and we determine upper limits on the branching fractions at the 90% C.L., shown in Table 1. Full details may be found in Ref. 9.

3. Study of the excited baryon states

3.1. $J/\psi \to p\bar{n}\pi^- + c.c.$

The nucleon is the simplest system in which the three colors of QCD can combine to form a colorless object, and the essential nonabelian character of QCD is manifest. It is necessary to understand the internal quark-gluon structure of the nucleon and its excited N^* states before we can claim to really understand the strong interaction. The πN system in decays of $J/\psi \to \bar{N}N\pi$ is limited to be isospin 1/2 by isospin conversion. This provides a big advantage in studying $N^* \to \pi N$ compared with πN and γN experiments which mix isospin 1/2 and 3/2 for the πN system.

For the decay $J/\psi \to p\pi^-\bar{n}$, the anti-neutron cannot be detected directly. We select the p and π^- from two prong events with oppositely charged tracks and require the missing mass to be consistent with the $\bar{n}$ mass. The missing mass distribution (Figure 5) shows a clear $\bar{n}$ peak.

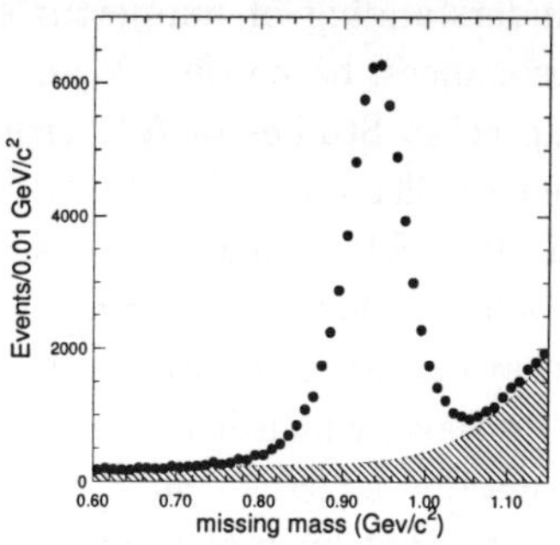

Fig. 5. The missing mass distribution for $J/\psi \to p\pi^-\bar{n}$ with the fitted background shown by the shaded area.

Figure 6 shows the πN invariant mass spectrum from $J/\psi \to p\bar{n}\pi^-$. Besides two well known N^* peaks at 1500 MeV/c^2 and 1670 MeV/c^2, there are two new, clear N^* peaks in the $p\pi$ invarinat mass spectrum around 1360 MeV/c^2 and 2030 MeV/c^2. They are the first direct observation of the $N^*(1440)$ peak and a long-sought "missing" N^* peak above 2 GeV/c^2 in the πN invariant mass spectrum. A simple BW fit gives the mass and width for the $N^*(1440)$ peak as $1358 \pm 6 \pm 16$ MeV/c^2 and $179 \pm 26 \pm 50$ MeV/c^2, and for the new N^* peak above 2 GeV/c^2 as $2068 \pm 3^{+15}_{-40}$ MeV/c^2 and $165 \pm 14 \pm 40$ MeV/c^2, respectively. [10]

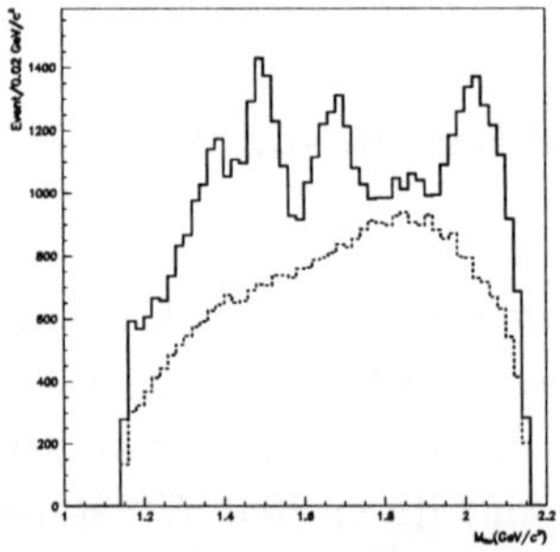

Fig. 6. The invariant mass spectrum of πN. The dashed histogram is the phase space from MC simulation.

3.2. $\psi(2S) \to p\bar{p}\pi^0$ *and* $p\bar{p}\eta$

The J/ψ and $\psi(2S)$ decays into $p\bar{p}\pi^0$ and $p\bar{p}\eta$ are expected to be dominated by two-body decays involving excited nucleon states. These states play an important role in the understanding of nonperturbative QCD. However, our knowledge on N^* resonances, based on πN and γN experiments, [5] is still very limited and imprecise. Studies of N^* resonances have also been performed using J/ψ events collected at the BEPC [11,12]. Based on 58 million J/ψ events collected by BES II, a new N^* peak with a mass at around 2065 MeV/c^2 was observed [10]. Due to its large mass, the production of this $N^*(2065)$ in J/ψ decays is rather limited in phase space, and a similar search for it in $\psi(2S)$ decays, which has a larger phase space available may be helpful. Experimentally, $\psi(2S) \to p\bar{p}\pi^0$ was studied by Mark-II in 1983, [13] and $\psi(2S) \to p\bar{p}\eta$ has not been observed before.

The data used in this analysis were taken with the BES II detector. The data sample corresponds to a total of $(14.0 \pm 0.6) \times 10^6$ $\psi(2S)$ decays, as determined from inclusive hadronic events. For these two decay modes $\psi(2S) \to p\bar{p}\pi^0(\eta)$, the final states are the same $p\bar{p}\gamma\gamma$, and the numbers of signal events are obtained by fitting the $\gamma\gamma$ invariant mass distribution in the selected events with $p\bar{p}\gamma\gamma$ final state. [14] The branching fractions for $\psi(2S) \to p\bar{p}\pi^0$ and $\psi(2S) \to p\bar{p}\eta$ are:

$$B(\psi(2S) \to p\bar{p}\pi^0) = (13.2 \pm 1.0 \pm 1.5) \times 10^{-5},$$

$$B(\psi(2S) \to p\bar{p}\eta) = (5.8 \pm 1.1 \pm 0.7) \times 10^{-5}.$$

For $\psi(2S) \to p\bar{p}\pi^0$, the error is much smaller than the previous measurement by Mark-II. [13]

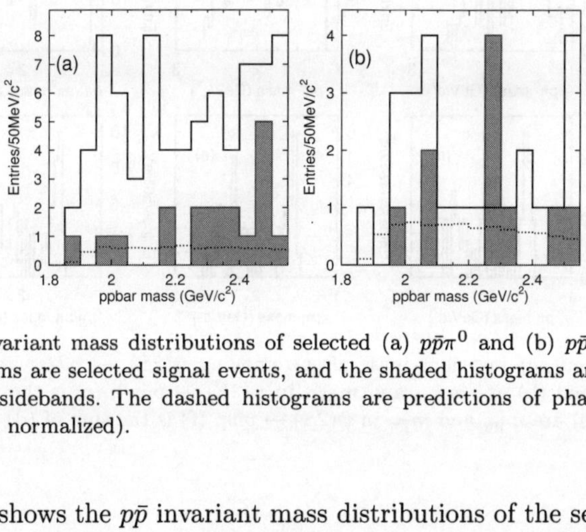

Fig. 7. $p\bar{p}$ invariant mass distributions of selected (a) $p\bar{p}\pi^0$ and (b) $p\bar{p}\eta$ events. The blank histograms are selected signal events, and the shaded histograms are events from π^0 or η mass sidebands. The dashed histograms are predictions of phase space with S-wave $p\bar{p}$ (not normalized).

Figure 7 shows the $p\bar{p}$ invariant mass distributions of the selected $p\bar{p}\pi^0$ and $p\bar{p}\eta$ events, together with the expected background estimated from π^0 or η mass sidebands. There are indications of some enhancement around 2 GeV/c^2 in both channels; the probabilities that the enhancements are produced by background fluctuations, estimated using the sideband events, are 1.2×10^{-11} in the $p\bar{p}\pi^0$ mode and 2.9×10^{-4} in the $p\bar{p}\eta$ mode. Fitting the enhancement with an S-wave BW and a linear background, with phase space and a mass dependent efficiency correction, yields a mass around 2.0 GeV/c^2 in the $p\bar{p}\pi^0$ mode and 2.06 GeV/c^2 in $p\bar{p}\eta$, with the width in both channels around 30-80 MeV/c^2, and significances around 2.7σ. The nature of the enhancements is not clear, and the statistics are too low to allow a detailed study. The enhancements in the two channels cannot be the same since they have different isospin.

Figure 8 shows the $p\pi$ (or $p\eta$) invariant mass after removing $\psi(2S) \to J/\psi + X$ backgrounds and the possible $p\bar{p}$ mass threshold enhancements. There is a faint accumulation of events in the $p\pi$ invariant mass spectrum at around 2065 MeV/c^2, but it is not statistically significant. The enhancement between 1.4 and 1.7 GeV/c^2 may come from $N^*(1440)$, $N^*(1520)$, $N^*(1535)$, etc. We do not attempt a partial wave analysis due to the limited statistics. There is a clear enhancement with $p\eta$ mass at (1549 ± 13) MeV/c^2, which is possibly the $N^*(1535)$.

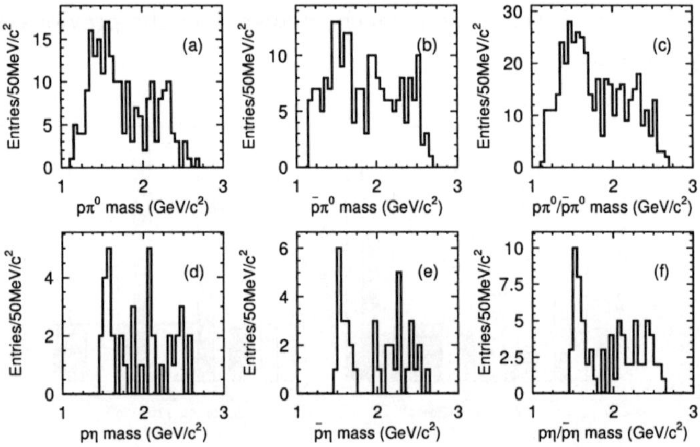

Fig. 8. The $p\pi(p\eta)$ invariant mass after removing $\psi(2S) \to \eta J/\psi$ and the possible $R(2000)$. (a) and (b) are $m_{p\pi^0}$ and $m_{\bar{p}\pi^0}$ in $\psi(2S) \to p\bar{p}\pi^0$; (c) is the sum of (a) and (b); (d) and (e) are $m_{p\eta}$ and $m_{\bar{p}\eta}$ in $\psi(2S) \to p\bar{p}\eta$; (f) is the sum of (d) and (e).

4. $J/\psi \to p\bar{p}$, $\Lambda\bar{\Lambda}$, and $\Sigma^0\bar{\Sigma}^0$

With 58 million produced J/ψ events, we also analyze the decays $J/\psi \to p\bar{p}$, $\Lambda\bar{\Lambda}$, and $\Sigma^0\bar{\Sigma}^0$. [15,16] The branching fractions are measured to be $B(J/\psi \to p\bar{p}) = (2.26 \pm 0.01 \pm 0.14) \times 10^{-3}$, $B(J/\psi \to \Lambda\bar{\Lambda}) = (2.05 \pm 0.03 \pm 0.11) \times 10^{-3}$, and $B(J/\psi \to \Sigma^0\bar{\Sigma}^0) = (1.40 \pm 0.03 \pm 0.07) \times 10^{-3}$. The angular distribution is of the form $\frac{dN}{d\cos\theta} = N_0(1 + \alpha\cos^2\theta)$, with $\alpha = 0.676 \pm 0.036 \pm 0.042$ for $J/\psi \to p\bar{p}$, $\alpha = 0.65 \pm 0.12 \pm 0.08$ for $J/\psi \to \Lambda\bar{\Lambda}$, and $\alpha = -0.22 \pm 0.17 \pm 0.09$ for $J/\psi \to \Sigma^0\bar{\Sigma}^0$. The α values for $J/\psi \to p\bar{p}$ and $J/\psi \to \Lambda\bar{\Lambda}$ obtained by BES II are consistent with previous measurements. The value of α for $J/\psi \to \Sigma^0\bar{\Sigma}^0$ is negative, which disagrees with the theoretical expectation by about 3σ [17 18]. Although the values of

α for $J/\psi \to \Sigma^0\overline{\Sigma}^0$ are positive in Mark-II [19] and DM2 [20] experiments, they have large errors and differ from BES-II only at about one sigma.

Acknowledgments

I wish to acknowledge the efforts of my BES colleagues on all the results presented here. I also want to thank the organizers for the opportunity to present these results at NSTAR2005.

References

1. BES Collaboration, J.Z. Bai *et al.*, Nucl. Instrum. Methods Phys. Res., Sect. **A344**, 319 (1994).
2. BES Collaboration, J.Z. Bai *et al.*, Nucl. Instrum. Methods Phys. Res., Sect. **458**, 627 (2001).
3. BES Collaboration, J.Z. Bai *et al.*, Phys. Rev. Lett., **91** (2003) 022001.
4. BES Collaboration, M. Ablikim *et al.*, hep-ex/0508025.
5. S. Eidelman *et al.* (Particle Data Group), Phys. Lett. **B592**, 1 (2004).
6. S.L. Zhu and C.S. Gao, hep-ph/0507050.
7. BES Collaboration, M. Ablikim *et al.*, Phys. Rev. Lett. **93**, 112002 (2004).
8. LEPS Collaboration, T. Nakano *et al.*, Phys. Rev. Lett. **91**, 012002 (2003).
9. BES Collaboration, M. Ablikim *et al.*, Phys. Rev. **D70**, 012004 (2004).
10. BES Collaboration, M. Ablikim *et al.*, hep-ex/0405030.
11. B.S. Zou, Nucl. Phys. **A675**, 167C (2000).
12. BES Collaboration, J.Z. Bai *et al.*, Phys. Lett. **B510**, 75 (2001).
13. Mark-II Collaboration, M.E.B. Franklin. *et al.*, Phys. Rev. Lett. **51**, 963 (1983).
14. BES Collaboration, M. Ablikim *et al.*, Phys. Rev. **D71**, 072006 (2005).
15. BES Collaboration, J.Z. Bai *et al.*, Phys. Lett. **B591**, 42 (2004).
16. BES Collaboration, M. Ablikim *et al.*, hep-ex/0506020.
17. M.Claudson, S. L. Glashow and M. B. Wise, Phys. Rev. **D25** 1345 (1982).
18. C. Carimalo, Int. J. Mod. Phys. **A2** 249 (1987).
19. Mark-II Collaboration, M.W. Eaton *et al.*, Phys. Rev. **D29** 804 (1984).
20. DM2 Collaboration, D. Pallin *et al.*, Nucl. Phys. **B292** 653 (1987).

RESULTS FROM THE GDH EXPERIMENT AT MAINZ AND BONN

A. BRAGHIERI*

INFN, Sezione di Pavia
via Bassi 6, 27100 Pavia, Italy
E-mail: Alessandro.Braghieri@pv.infn.it

An extensive experimental program was carried out jointly at MAMI (Mainz) and ELSA (Bonn) to measure for the first time pion photo-production reactions on the nucleon in double polarization and to investigate the Gerasimov-Drell-Hearn Sum Rule. We obtained results for the proton in the energy range from about 200 MeV to 3000 MeV and for the neutron, using a deuteron target, from about 200 MeV to 1900 MeV. Preliminary results on the total absorption cross sections are presented and discussed. In addition most of the partial channels have been also measured.

1. Introduction

The GDH sum rule[1,2] relates static properties of the nucleon (the anomalous magnetic moment κ , the charge e, and the mass M) to the difference of the spin dependent photo-absorption cross sections for circularly polarized photons on longitudinally polarized nucleons. It is written as:

$$\int_0^\infty \frac{\sigma_{3/2}(\nu) - \sigma_{1/2}(\nu)}{\nu} d\nu = \frac{\pi e^2}{2M^2}\kappa^2 \tag{1}$$

where $\sigma_{3/2}$ and $\sigma_{1/2}$ are the photo-absorption cross sections for the total helicity states 3/2 and 1/2, respectively (parallel or anti-parallel photon-nucleon spin configurations) and ν is the photon energy.

The GDH sum rule is based on fundamental physics principles: Lorentz and gauge invariance, unitarity and causality applied to the forward Compton scattering amplitude. It was derived under the no-subtraction assump-

*On behalf of the GDH and A2 Collaborations

tion for the dispersion relation connected to the spin-dependent part of the Compton scattering amplitude.

Its experimental check is not only a validation of these principles but also provides important information on the spin structure of the Nucleon. The GDH sum rule is the counter-piece of the Bjorken sum rule[3] and of the Ellis-Jaffe sum rule[4]. While the first deals with the absorption of real photons, at Q^2=0, the other two give predictions at Q^2 infinite.

Therefore an additional interest in the GDH sum rule was stimulated in late eighties when deep inelastic scattering experiments at SLAC and at CERN reported the apparent failure of the Ellis-Jaffe sum rule, giving rise to the so called 'spin crisis'. Anselmino, Ioffe and Leader[5] evoked the GDH sum rule as a possible interpretation of these data. They attempted for the first time to extend the GDH sum rule in the domain of the virtual photons, pointing out the importance of the region at low Q^2.

The experimental check of the GDH sum rule requires the measurement of the difference $\Delta\sigma = \sigma_{3/2} - \sigma_{1/2}$ over a wide photon energy range (left side in Eq. 1) and had never been realized so far because highly polarized photon beams and target were not available.

These technical requirements have been recently fulfilled and the GDH Collaboration has developed an extensive experimental program to study doubly polarized photo-absorption.

Several other experiments have been planned at SLAC, JLAB, SRING-8, LEGS and GRAAL, some of them to study the Q^2 dependence of the sum rule (virtual photon absorption).

Our experimental program was carried out in two steps: at MAMI up to 800 MeV and extended at ELSA up to 3 GeV. Circularly polarized photons were produced by bremsstrahlung of linearly polarized electrons. A frozen spin target [6] was developed providing longitudinally polarized protons or deuteron (as a neutron target). Two different detector set-up were used for the two energy ranges. Both the devices have an acceptance close to 4π sr and are complemented by a Čerenkov counter detector in the forward direction, allowing a very good rejection of the electromagnetic background.

The description of the devices as well as the data analysis are fully reported in references [8–11].

2. The GDH sum rule on the Proton

An inclusive method of data analysis has been developed to determine the total absorption cross section.

A large fraction of the cross section can be directly accessed by measuring the number of events with charged hadrons in the final state detected inside the detector acceptance. Most of the remaining part is deduced from the neutral pions events measured. In such a way corrections and extrapolation to the whole acceptance are kept to a minimum and both statistical and systematical errors are optimized.

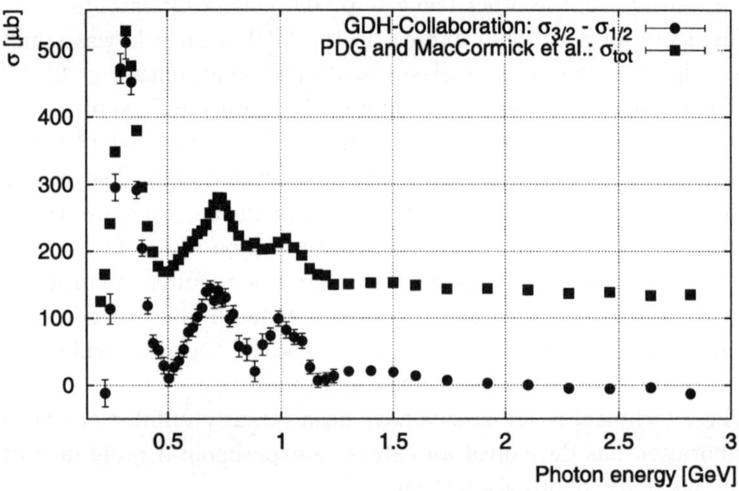

Fig. 1. The Helicity-dependent total photo-absorption cross sections on the proton (circles) compared to the unpolarized cross section (squares).

The helicity difference $\Delta\sigma = \sigma_{3/2} - \sigma_{1/2}$ is shown by circles in Figure 1 and compared to the unpolarized total cross section (squares). Resonances are evident in both the plots, but the non-resonant background mostly disappears in the polarized observable $\Delta\sigma$. In addition $\Delta\sigma$ changes it sign and becomes negative at about $E_\gamma \simeq 2.1$ GeV.

This behavior at high energy is confirmed by preliminary results from JLAB[12]. Negative values are also predicted by models based on a Regge fit to Deep Inelastic Scattering data extrapolated to the real photon point.

The experimental check of the GDH sum rule is based on values of Figure 1 and integrated according to the left side of Eq. 1. In the measured range between 200 MeV and 3 GeV we get the major contribution to the sum rule. The contribution of the unmeasured part below the detection threshold (about 200 MeV) was estimated to be -27.5 μb by the MAID[13] model.

The GDH integral, as a function of the upper photon energy limit, is represented in Figure 2 and clearly indicates how fast the sum rule is saturated. The horizontal dashed line is the value of the sum rule.

The contribution below 200 MeV was already included. The first steep corresponds to the contribution of the $\Delta(1232)$ resonance, than at higher photon energy the slope decreases as a consequence of two effects: (*i*) the helicity difference decreases; (*ii*) the denominator in the integrand increases.

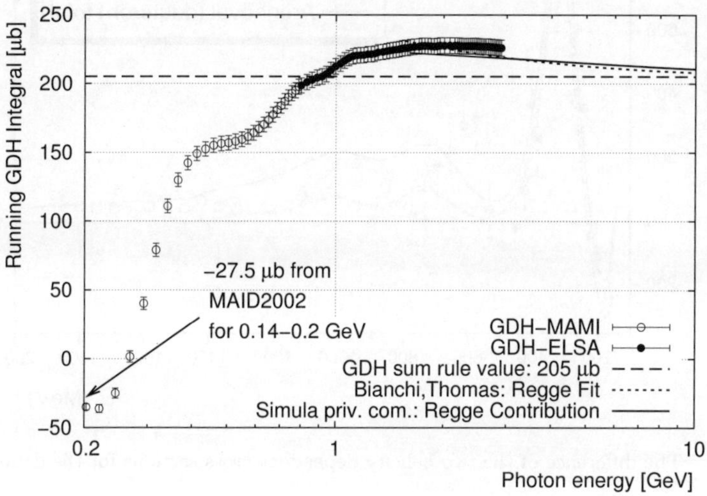

Fig. 2. The running GDH integral.

The experimental data end at about 3 GeV. For higher energy the predictions of the Regge model in ref.[15,16], represented by the dotted and solid line respectively, were used to estimate the contribution to -14 μb.

As a consequence the estimate value of the sum rule was I=212±5±12 μb, to be compared with the theoretical value of I=205 μb as calculated from the right side of Eq. 1. These two values are in good agreement within the experimental and model uncertainties.

3. The GDH sum rule on the Neutron

In order to measure the helicity dependent cross sections on the neutron, a deuteron target was used. Then data should be extracted from the bound neutron, taking into account kinematics and nuclear effects. Preliminary results for the deuteron are shown in Figure 3. The data from MAMI are still preliminary and were obtained during a pilot run in 1998. The high statistics data of 2003 are currently under analysis. On the contrary, analysis of ELSA data has been concluded and the results have been already published[11].

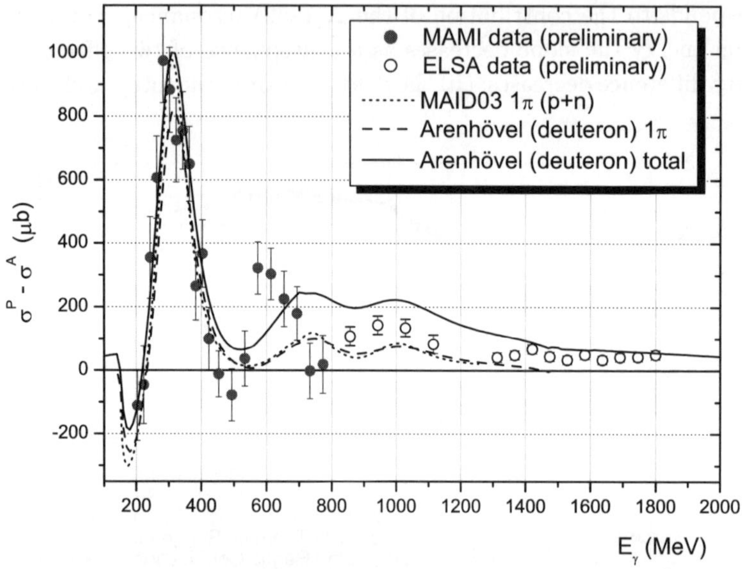

Fig. 3. The difference of the two helicity dependent cross sections for the deuteron.

The dotted curve represents $\Delta\sigma$ calculated in the frame of the MAID[13] model. Only production of single pions on the free proton and on the free neutron were considered. This result is not significantly different from the calculation of single pion production on the deuteron performed by Arenhövel and represented by the dashed curve.

A first calculation was made by Arenhövel, Fix and Schwamb[17] to also take into account the nuclear effects on the deuteron. The predicted helicity difference is shown in by the solid line.

According to this calculation, at high energy nuclear effects play a minor role, so the difference between the deuteron (Figure 3) and the proton (Figure 1) can be taken as an estimate for the spin asymmetry of the neutron. The resulting cross section on the neutron is shown in Figure 4 by black circles and compared to that on the proton (open circles).

The proton and the neutron have, within statistical errors, a similar behavior, but while the single pion contribution correctly describes the proton data in the 3$^{\rm rd}$ resonance region at about 1000 MeV (dashed line), there is a significance discrepancy for the neutron (solid line).

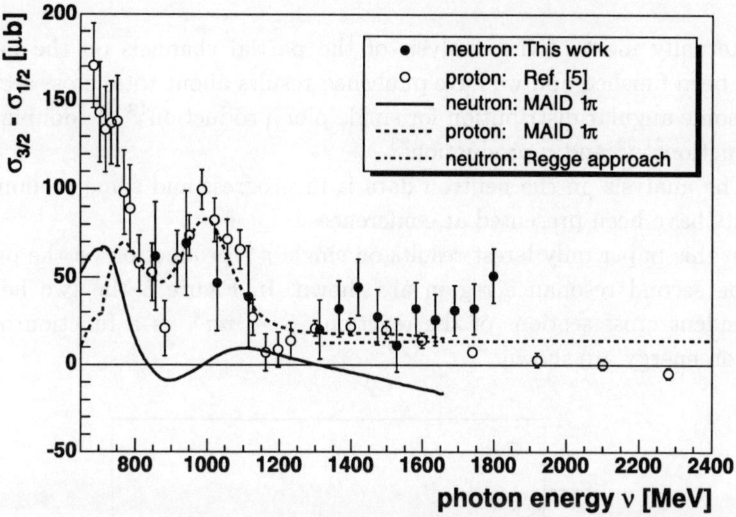

Fig. 4. The neutron (black circles) and proton (open circles) helicity difference in the ELSA energy range.

Therefore multi-pion final states are supposed to play an important role for the neutron and this open question could only be solved by separating exclusively the partial reaction channels.

4. Exclusive Reaction Channels

At MAMI the large solid angle DAPHNE detector[7] was used, complemented by additional devices to increase the forward acceptance. The detector is able to track and discriminate charged particles, to reconstruct their energy and to provide a signature for neutral pions with a good efficiency. Therefore it was possible not only to measure the inclusive cross section, but also to separate all the exclusive reaction channels up to 800 MeV and measure total cross section or angular distributions.

The measurement of the exclusive reactions is a very important item, besides the GDH sum rule, because it allows to analyze some contributions from single resonances. For example precise photo-decay amplitudes can be extracted by partial wave analysis of single-meson production.

Up to now only unpolarized data and only very few measurements with linearly polarized photons and unpolarized nucleon target have been obtained. Therefore new measurements in double polarization represent a unique tool to study resonances and set new constraints in multipole analysis.

Actually most of the analysis of the partial channels on the proton have been finished and we have published results about total cross sections and some angular distribution for single pion production[18–20], double pion production[21,22] and η production[23].

The analysis on the neutron data is in progress and some preliminary results have been presented at conferences.

In this paper only latest results on single π^+ production on the proton in the second resonance region are shown. In Figure 5 the two helicity dependent cross sections of the reaction $\vec{\gamma}\vec{p} \rightarrow n\pi^+$ as a function of the photon energy are shown.

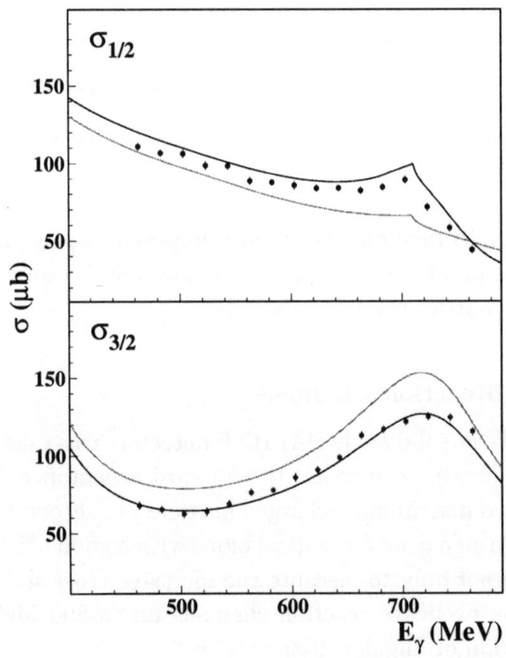

Fig. 5. The two helicity dependent cross sections for $\vec{\gamma}\vec{p} \rightarrow n\pi^+$. See text for details.

The well defined cusp appearing in the helicity 1/2 cross section (upper plot) corresponds to the opening of the η threshold (E_γ=705 MeV) and is due to unitarity, as expected.

The dark and ligth curves are predictions from MAID[13] and SAID[14] models respectively. It is important to notice that both the models well reproduce the unpolarized cross section, while there is a substantial discrepancy for the polarized cross sections, especially for the SAID calculation. This is due to the contribution of the E_{0^+} and E_{2^-} multipoles, which contribute with opposite sign in the helicity difference.

5. Summary and Outlook

The Gerasimov-Drell-Hearn sum rule on the proton was investigated for the first time and verified at about 10% level. Preliminary results on the neutron, using a deuteron target, are available, but a carefull analysis must be performed to take into account nuclear effects at low photon energy.

On the other hand at low energy all partial channels were also measured and first results showed a net improvement in our understanding of multipoles and resonances.

References

1. S. B. Gerasimov, *Sov. J. Nucl. Phys.* **2**, 430 (1966)
2. S. D. Drell and A. C. Hearn, *Phys. Rev. Lett.* **16**, 908 (1966)
3. J.D. Bjorken, *Phys. Rev.* **148**, 1467 (1966)
4. J.R. Ellis and R.L. Jaffe, *Phys. Rev.* **D9**, 1444 (1974)
5. M. Anselmino et al., *Sov. J. Nucl. Phys.* **49**, 136 (1989)
6. C. Bradtke et al., *Nucl. Instr. Meth.* **A 436**, 430 (1999)
7. G. Audit et al. *Nucl. Instr. Meth.*, **A 301**, 473 (1991)
8. J. Ahrens et al., *Phys. Rev. Lett.* **84**, 5950 (2000)
9. J. Ahrens et al., *Phys. Rev. Lett.* **87**, 022003 (2001)
10. H. Dutz et al., *Phys. Rev. Lett.* **91**, 192001 (2003)
11. H. Dutz et al., *Phys. Rev. Lett.* **93**, 032003 (2004)
12. D. Sober et al, Proceedings GDH2002, Genova (2002)
13. D. Drecshel et al., *Nucl. Phys.* **A 570**, 580 (1999)
14. R.A. Arndt et al., *Phys. Rev.* **66**, 55213 (2002)
15. N. Bianchi and E. Thomas, *Phys. Lett.* **B 450**, 439 (1999)
16. S. Simula et al., *Phys. Rev.* **D 65**, 034017 (2002)
17. H Arenhovel, A. Fix and M. Schwamb, *Phys Rev. Lett.* **93**, 202301 (2004)
18. J. Ahrens et al, *Phys. Rev. Lett.* **84**, 5950 (2000)
19. J. Ahrens et al, *Phys. Rev. Lett.* **88**, 232002 (2002)
20. J. Ahrens et al.*Eur. Phys. J.* **A 21**, 323 (2004)
21. J. Ahrens et al.*Phys. Lett.* **B 551** 49, (2003)
22. J. Ahrens et al.*Phys. Lett.* **B 624** 173, (2005)
23. J. Ahrens et al.*Eur. Phys. J.* **A 17**, 241 (2003)

THE STRANGENESS PHYSICS PROGRAM AT CLAS

D. S. CARMAN

Department of Physics, Ohio University
Athens, OH 45701, USA
E-mail: carman@ohio.edu
(for the CLAS Collaboration)

An extensive program of strange particle production off the proton is currently underway with the CEBAF Large Acceptance Spectrometer (CLAS) in Hall B at Jefferson Laboratory. This talk will emphasize strangeness photo- and electroproduction in the baryon resonance region between W =1.6 and 2.5 GeV, where indications of s-channel structure are suggestive of high-mass baryon resonances coupling to kaons and hyperons in the final state. Precision measurements of cross sections and polarization observables are being carried out with both electron and real photon beams, both of which are available with high polarization at energies up to 6 GeV.

1. Introduction

A key to understanding the structure of the nucleon is to understand its spectrum of excited states. However understanding nucleon resonance excitation provides a serious challenge to hadronic physics due to the non-perturbative nature of QCD at these energies. Recent symmetric quark model calculations predict more states than have been seen experimentally[1]. Mapping out the spectrum of these excited states will provide for insight into the underlying degrees of freedom of the nucleon.

Most of our present knowledge of baryon resonances comes from reactions involving pions in the initial and/or final states. A possible explanation for the so-called missing resonance problem could be that pionic coupling to the intermediate N^* or Δ^* states might be weak. This suggests a search for these hadronic states in strangeness production reactions. Beyond different coupling constants (e.g. g_{KNY} vs. $g_{\pi NN}$), the study of the exclusive production of $K^+\Lambda$ and $K^+\Sigma^0$ final states has other advantages in the search for missing resonances. The higher masses of the kaon and hyperons, compared to their non-strange counterparts, kinematically favor a two-body decay mode for resonances with masses near 2 GeV, a situation that is experimentally advantageous. In addition, baryon resonances have large widths and are often overlapping. Studies of different final states can provide for important cross checks in quantitatively understanding the contributing amplitudes. Note that although the two ground-state hyperons have the same valence quark structure (uds), they differ in isospin, such that intermediate N^* resonances can decay strongly to $K^+\Lambda$ final states, while both N^* and Δ^* decays can couple to $K^+\Sigma^0$ final states.

The search for missing resonances requires more than identifying features in the mass spectrum. QCD cannot be directly tested with N^* spectra without a model for the production dynamics[2]. The s-channel contributions are known to be important in the resonance region in order to reproduce the invariant mass (W) spectra, while t-channel meson exchange is also necessary to describe the diffractive part of the production and u-channel diagrams are necessary to describe the backward-going processes. Thus measurements that can constrain the phenomenology for these reactions are just as important as finding one or more of the missing resonances.

Theoretically, there has been considerable effort during the past decade to develop models for the KY photo- and electroproduction processes. However, the present state of understanding is still limited by a sparsity of data. Model fits to the existing cross section data are generally obtained at the expense of many free parameters, which leads to difficulties in constraining existing theories. Moreover, cross section data alone are not sufficiently sensitive to fully understand the reaction mechanism, as they probe only a small portion of the full response. In this regard, measurements of spin observables are essential for continued theoretical development in this field, as they allow for improved understanding of the dynamics of this process and provide for strong tests of QCD-inspired models.

In this talk I focus on the strangeness physics program in Hall B at Jefferson Laboratory using the CLAS detector[3]. Presently there is very limited knowledge of $N^*, \Delta^* \to KY$ couplings. With the existing CLAS program, the present lack of data will be remedied with a wealth of high quality measurements spanning a broad kinematic range.

2. KY Photoproduction

Photoproduction measurements for $K^+\Lambda$ and $K^+\Sigma^0$ made with CLAS have provided both differential cross sections and hyperon polarizations. The data shown here were collected at electron beam energies of 2.4 and 3.1 GeV. This gives rise to measurements spanning photon energies from threshold

E_γ=0.911 GeV (W=1.61 GeV) up to E_γ=2.95 GeV (W=2.53 GeV). The final state hyperons were reconstructed from the (γ, K^+) missing mass. Detection of the decay proton from the hyperon was also required. The average hyperon mass resolution was σ=8.5 MeV. The Λ and Σ^0 events were separated from the pion mis-identification background using lineshape fits to the missing mass spectra in each bin of photon energy and kaon angle.

CLAS has already published photoproduction data from another analysis where only the final state K^+ was detected[4]. The results shown here represent a new analysis[5] with very different systematics. The two separate analyses are now in good agreement within the associated uncertainties, giving us full confidence in the CLAS results.

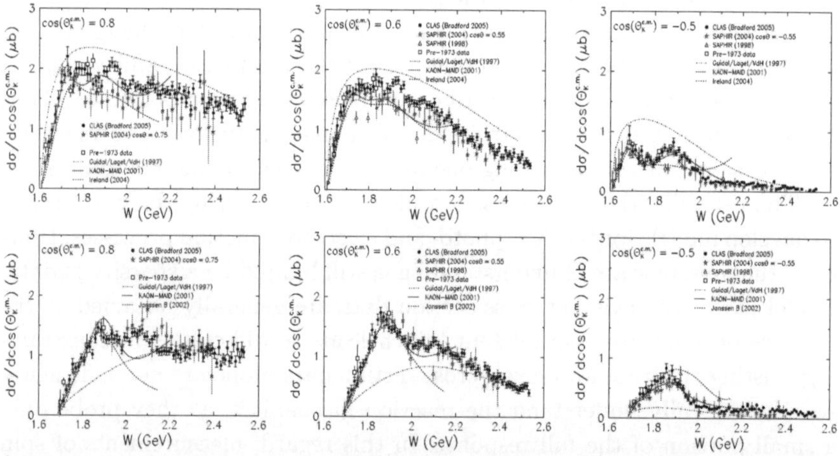

Fig. 1. Differential cross sections for $K^+\Lambda$ (top) and $K^+\Sigma^0$ (bottom) vs. W for three kaon angle bins[5] (solid circles). Data from Bonn/SAPHIR[6] (open triangles) are also shown. The curves are calculations from KAON-MAID[7] (solid), Ireland[8] (dashed), Janssen[9] (dashed), and Guidal[10] (dot-dashed).

Figure 1 shows a sample of the CLAS differential cross sections for $K^+\Lambda$ and $K^+\Sigma^0$ photoproduction as a function of the invariant energy W for angle bins at $\cos\theta_K^*$=0.8, 0.6, and -0.5. The different angle bins allow us to vary the relative contributions to the s, t, and u reaction channels. Existing data from SAPHIR at Bonn[6] are also included. The data are compared with effective Lagrangian calculations from Mart/Bennhold[7], Ireland[8], and Janssen[9], which are based on adding the non-resonant Born terms with a number of resonances and leaving their coupling constants as free parameters bounded loosely by SU(3) predictions. These models have been developed from fits to the Bonn data, however they only reproduce the threshold region of the data. Much beyond about 200 MeV above threshold the calculations do not reflect the CLAS data. The data are also compared with a Reggeon exchange model[10] that uses only K and K^* exchanges, with no resonance contributions. The prediction was made using a model that fit higher energy kaon electroproduction data well.

For the $K^+\Lambda$ data the broad structure just above the threshold region is typically accounted for by the known $S_{11}(1650)$, $P_{11}(1710)$, and $P_{13}(1720)$ resonances. Centered at roughly 1.9 GeV is another broad structure, first seen in the Bonn data, that remains unexplained, whose position and width vary with kaon angle. This has been interpreted by Mart and Bennhold[11] as evidence for a missing $D_{13}(1900)$ resonance, where the assignment was consistent with the measured angular distributions, as well as a predicted quark model state[1]. However, other groups have shown that the same data can also be explained by accounting for u-channel hyperon exchanges[12] or with an additional P-wave resonance[9]. Interestingly, the Regge model fully saturates the strength in the reaction, leaving no room for significant s and u channel contributions.

For the $K^+\Sigma^0$ data, there is a single peak in the differential cross sections at about 1.9 GeV. This has been associated with a cluster of Δ resonances in this mass range. However both isospin 1/2 (N^*) and isospin 3/2 (Δ^*) resonances can contribute to this final state. These data, as well as the Bonn data, show evidence for resonant decays to $K^+\Sigma^0$. The Regge model here now provides only a fraction of the reaction strength with significant contributions possible from the s and u reaction channels.

Another part of the photoproduction analysis program is to measure the induced hyperon polarization with an unpolarized beam and target. An attractive feature of the hyperon decay is its well known self-analyzing nature. The hyperon polarization is revealed by the asymmetry in the angular distribution of the protons from the mesonic decay of the hyperon. From parity conservation, the only allowed polarization component is along the axis perpendicular to the K^+Y reaction plane. Measurement of this observable is important since it is related to interferences of the imaginary part of resonant amplitudes with other amplitudes, including Born terms. These data are shown in Fig. 2 as a function of W for two kaon angle bins at $\cos\theta_K^*$=0.3 and -0.3[4]. The CLAS data provide the first precision data for this observable. The data show a sizeable negative Λ polarization for forward-going kaons, and an equally sizeable positive polarization when the kaons go backward. The basic trend is reversed in the Σ^0 data. The CLAS results are consistent with some older data points from Bonn[6]. CLAS has also completed measurements of the beam recoil polarization observables C_x and C_z[13].

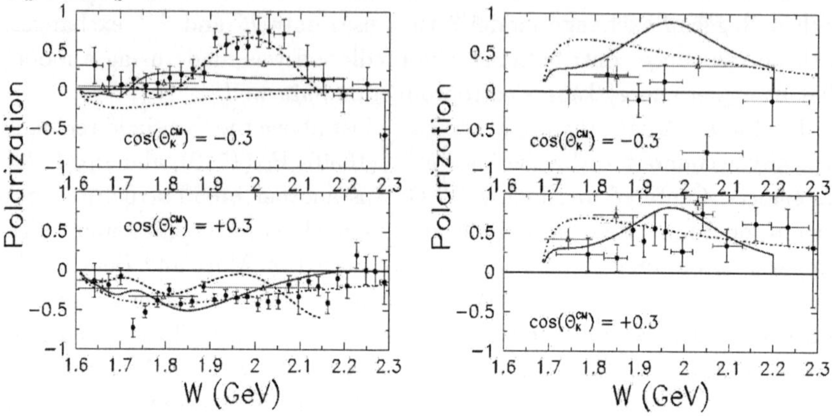

Fig. 2. Induced polarization of the Λ and Σ^0 hyperon as a function of W for two kaon angle bins[4]. The curves are calculations from KAON-MAID[7] (solid), Janssen[9] (dashed), and Guidal[10] (dot-dashed).

Neither the hadrodynamic nor Regge calculations reproduce the magnitudes or the trends seen in the hyperon polarization data across the broad kinematic region covered. The significant discrepancies between the calculations and the data imply that these data can serve to provide for significant new constraints on the model parameters.

A recent coupled-channels analysis[14] of photoproduction data from SAPHIR and CLAS, as well as beam asymmetry data from SPring-8/LEPS for $K^+\Lambda$[15] and data from π and η photoproduction, reveals evidence for new baryon resonances in the high W mass region. The full suite of data can only be satisfactorily fitted by including a new P_{11} state at 1840 MeV and two D_{13} states at 1870 and 2130 MeV. The only Δ state that contributed significantly to the $K^+\Sigma^0$ final state is the $D_{33}(1940)$. This analysis has certain ambiguities that can be resolved or better constrained by incorporating the expansive set of electroproduction data from CLAS.

3. *KY* Electroproduction

CLAS has measured exclusive $K^+\Lambda$ and $K^+\Sigma^0$ electroproduction on the proton for a range of momentum transfer Q^2 from 0.5 to 4.5 (GeV/c)^2 with electron beam energies from 2.6 to 5.7 GeV. For this talk I will focus attention on our 2.6 GeV data set. The final state hyperons were reconstructed from the $(e, e'K^+)$ missing mass. The average hyperon resolution was about

8 MeV, similar to what was found for photoproduction. The hyperon yields were extracted using Monte Carlo templates with a background determined from the data associated with pions misidentified as kaons.

The most general form for electroproduction cross section of the kaon from an unpolarized-proton target is given by:

$$\frac{d^4\sigma}{dQ^2 dW d\Omega_K^*} \equiv \sigma_0 = \Gamma_v[\sigma_T + \epsilon\sigma_L + \epsilon\sigma_{TT}\cos 2\Phi + \sqrt{2\epsilon(\epsilon+1)}\sigma_{LT}\cos\Phi].$$

In this expression, the cross section is decomposed into four structure functions, σ_T, σ_L, σ_{TT}, and σ_{LT}, which are in general functions of Q^2, W, and θ_K^* only. Γ_v represents the virtual photon flux factor, ϵ is the virtual photon polarization, and Φ is the angle between the electron scattering and hadronic reaction planes. One of the goals of the electroproduction program is to provide a detailed tomography of the structure functions vs. Q^2, W, and $\cos\theta_K^*$. In a first phase of the analysis at CLAS, we have measured the unseparated cross section ($\sigma_U = \sigma_T + \epsilon\sigma_L$) and, for the first time in the resonance region away from parallel kinematics, the interference cross sections σ_{TT} and σ_{LT}. At the amplitude level, these interference responses are related to real photon measurements of the polarized beam asymmetry, and so they are sensitive to some of the same structure information. Exploiting the Φ dependence of the reaction allows us to extract these responses from the CLAS data. The Q^2 dependence of the data provides sensitivity to the associated form factors.

A small sample of the available results from this analysis is shown in Fig. 3 vs. W for each of our six angle bins for the kaon. The kinematic dependence of the unpolarized structure functions shows that Λ and Σ^0 hyperons are produced very differently. σ_U at forward angles for $K^+\Lambda$ is dominated by a structure at W=1.7 GeV. For larger kaon angles, a second structure emerges at about 1.9 GeV, consistent with the signature in photoproduction. σ_{TT} and σ_{LT} are clearly non-zero and reflect the structures in σ_U. The fact that σ_{LT} is non-zero is indicative of longitudinal strength. For the $K^+\Sigma^0$ final state, σ_U is centrally peaked, with a single broad structure at 1.9 GeV. This is consistent with the photoproduction data. σ_{TT} reflects the features of σ_U, with σ_{LT} consistent with zero everywhere, indicative of σ_L being consistent with zero.

To date we have completed analysis of data sets at 2.6 and 4.2 GeV and have performed a Rosenbluth separation for several W bins over the full kaon angular range for a single bin at Q^2=1.0 (GeV/c)^2 where the data sets overlap. A crucial part in this analysis of extracting absolute cross sections

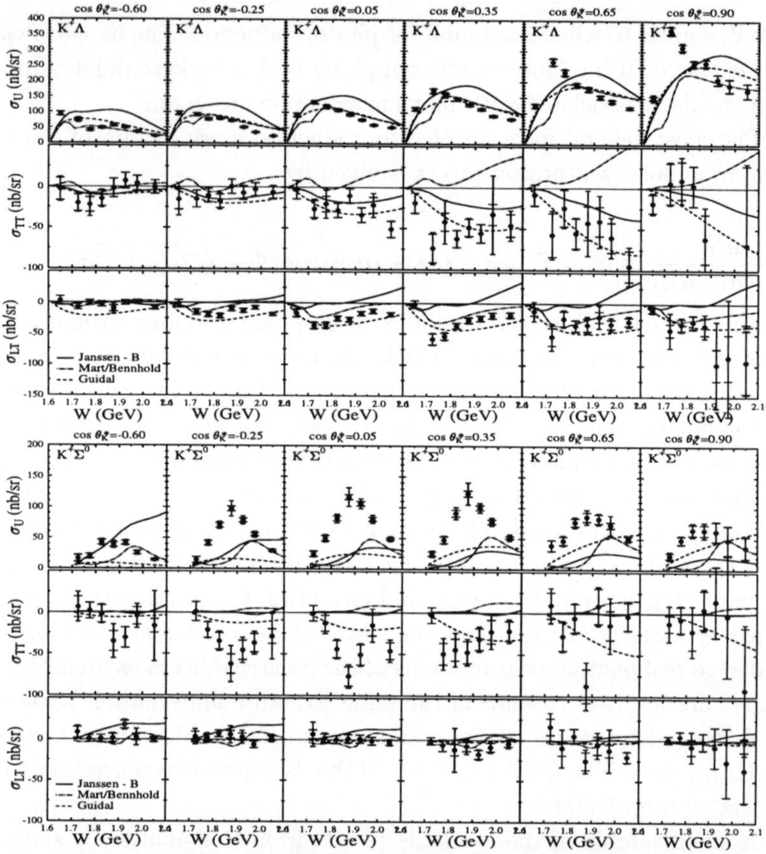

Fig. 3. Preliminary separated structure functions σ_U, σ_{LT}, and σ_{TT} for the $K^+\Lambda$ (top) and $K^+\Sigma^0$ final state (bottom) at 2.6 GeV and Q^2=0.65 (GeV/c)^2. The curves correspond to the calculations of Janssen[9], Guidal[10], and Mart/Bennhold[11].

is to minimize the physics model dependence of the detector acceptance function, radiative corrections, and bin-centering factors. We estimate an average absolute systematic uncertainty on these data points of about 15%.

The polarized-beam asymmetry provides access to the fifth structure function $\sigma_{LT'}$. This observable probes imaginary parts of the interfering L and T amplitudes (as opposed to the real parts of the interference from σ_{LT}). These imaginary parts vanish identically if the resonant state is determined by a single complex phase, which is the case for an isolated resonance. A representative sample of our data at 2.6 GeV and Q^2=0.65 (GeV/c)^2 is shown in Fig. 4 for the $K^+\Lambda$ final state[16]. The calculations shown are not able to reproduce the features seen in the data.

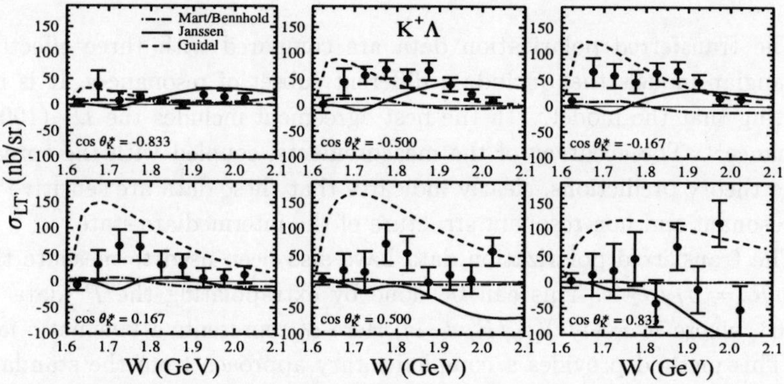

Fig. 4. Preliminary CLAS results[16] for the $K^+\Lambda$ structure function $\sigma_{LT'}$ vs. W for six kaon angle bins for Q^2=0.65 (GeV/c)^2. The curves are from the model calculations from Janssen[9], Guidal[10], and Mart/Bennhold[11].

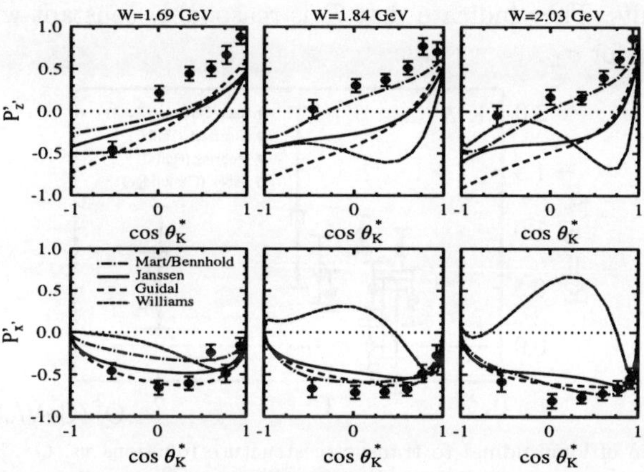

Fig. 5. CLAS transferred polarization[19] from $\vec{e}p \to e'K^+\vec{\Lambda}$ vs. $\cos\theta_K^*$ at 2.6 GeV for three different W bins summed over Φ and Q^2. The curves are for effective Lagrangian calculations of Janssen[9], Guidal[10], Bennhold[17], and WJC[18].

The first measurements of spin transfer from a longitudinally polarized electron beam to the Λ hyperon produced in the exclusive $p(\vec{e}, e'K^+)\vec{\Lambda}$ reaction have recently been completed at CLAS[19]. A sample of the results highlighting the angular dependence of P' summed over all Q^2 for three different W bins is shown in Fig. 5 at 2.6 GeV. The polarization along the virtual photon direction $P'_{z'}$ decreases with increasing θ_K^*, while the orthogonal component in the hadronic reaction plane $P'_{x'}$ is constrained to be zero at $\cos\theta_K^* = \pm 1$ due to angular momentum conservation, and reaches a minimum at $\theta_K^* \sim 90°$. The component normal to the hadronic reaction plane $P'_{y'}$ is statistically consistent with zero as expected.

The transferred polarization data are compared with three effective Lagrangian models that include a different subset of resonances. It is interesting that the model with the best agreement includes the $D_{13}(1900)$ resonance[17]. The accuracy of the measurements, coupled with the spread in the theory predictions, clearly indicates that these data are sensitive to the resonant and non-resonant structure of the intermediate state.

The transferred polarization data have also been used to measure the ratio $\mathcal{R} = \sigma_L/\sigma_T$[20]. This can be done by extrapolating the P' data to $\theta_K^*=0°$, where $\mathcal{R} = (1/\epsilon)(c_0/P'_{z'} - 1)$. Here c_0 represents a kinematic factor. This method provides a complementary approach from the standard Rosenbluth measurement in a situation with different systematics. Existing data from Hall C have remained controversial. The new results, shown in Fig. 6, are consistent with, although systematically smaller than, the latest Hall C results. They indicate that $\mathcal{R}$ is reasonably constant with Q^2 with small values for σ_L.

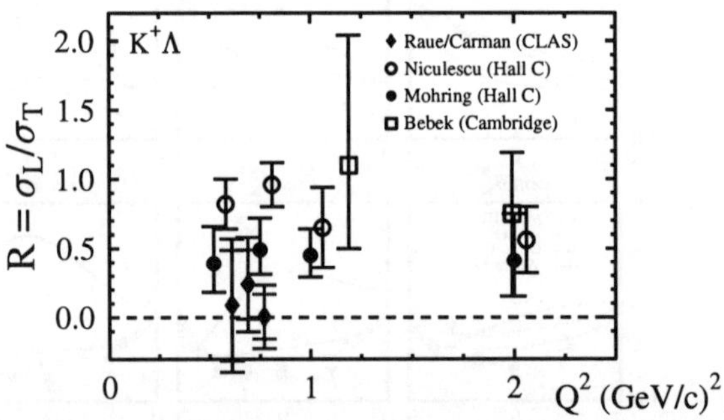

Fig. 6. Ratio of longitudinal to transverse structure functions vs. Q^2. The Niculescu results[21], which were superseded by the Mohring results[22], are offset in Q^2 for clarity.

4. Summary and Conclusions

In this talk I have reviewed some of the key reasons why the photo- and electroproduction processes of open-strangeness production are important for the investigation of baryonic structure and missing quark model states. I have discussed several aspects of the CLAS strangeness physics program highlighting the breadth and quality of our data sets. Results were presented for cross sections and single and double-polarization observables for $K^+\Lambda$ and $K^+\Sigma^0$ photo- and electroproduction. Our analyses indicate that the data are highly sensitive to the ingredients of the models, including the specific baryonic resonances included, along with their associated form factors and coupling constants. The production dynamics for $K^+\Lambda$ and $K^+\Sigma^0$ are also seen to be very different.

Work on publication of the full set of hyperon cross section and polarization data sets reported here is in progress. The main qualitative conclusion seems clear: these data show significant unexplained baryon resonance structure at higher masses. While the comparison of the effective Lagrangian calculations to the data is illustrative to highlight the present deficiencies in the current models and their parameter values, the next step in the study of the reaction mechanism is to include our data in the available data base and to refit the set of coupling strengths. Additionally new amplitude-level analyses are called for to more fully unravel the contributions to the intermediate state.

This work has been supported by the U.S. Department of Energy and the National Science Foundation.

References

1. S. Capstick and W. Roberts, Phys. Rev. D **58**, 74011 (1998).
2. T.-S.H. Lee and T. Sato, Proceedings of the N*2000 Conference, eds. Burkert *et al.*, (World Scientific, Singapore, 2001), p. 215.
3. B.A. Mecking *et al. (CLAS Collaboration)*, Nucl. Inst. and Meth. A **503**, 513 (2003).
4. J. McNabb *et al. (CLAS Collaboration)*, Phys. Rev. C **69**, 042201 (R) (2004).
5. R. Bradford *et al. (CLAS Collaboration)*, nucl-ex/0509033, submitted to Phys. Rev. C, (2005).
6. M.Q. Tran *et al.*, Phys. Lett B **445**, 20 (1998).
7. T. Mart *et al*, "KaonMAID 2000" at www.kph.uni-mainz.de/MAID.
8. D.G. Ireland, S. Janssen, and J. Ryckebusch, Nucl. Phys. A **740**, 147 (2004).
9. S. Janssen *et al.*, Phys. Rev. C 65, 015201 (2002).
10. M. Guidal, J.M. Laget, and M. Vanderhaegen, Nucl. Phys. **A627**, 645 (1997).
11. T. Mart and C. Bennhold, Phys. Rev. C **61**, 012201 (2000).
12. B. Saghai, nucl-th/0105001, (2001).
13. R. Bradford, see talk in these proceedings.
14. A.V. Sarantsev *et al.*, hep-ex/0506011, submitted to Eur. Phys. J, (2005).
15. R.G.T. Zegers *et al. (LEPS Collaboration)*, Phys. Rev. Lett. **91**, 092001 (2003).
16. R. Nasseripour *et al. (CLAS Collaboration)*, to be submitted for publication.
17. H. Haberzettl *et al.*, Phys. Rev. C **58**, R40 (1998).
18. R.A. Williams, C. Ji, and S.R. Cotanch, Phys. Rev. C **46**, 1617 (1992).
19. D.S. Carman *et al. (CLAS Collaboration)*, Phys. Rev. Lett. **90**, 131804 (2003).
20. B.A. Raue and D.S. Carman, Phys. Rev. C **71**, 065209 (2005).
21. G. Niculescu *et al.*, Phys. Rev. Lett. **81**, 1805 (1998).
22. R.M. Mohring *et al.*, Phys. Rev. C **67**, 055205 (2003).

RECENT RESULTS FROM THE CRYSTAL BARREL EXPERIMENT AT ELSA

U. THOMA

HISKP, Bonn University
Nussallee 14 - 16
53115 Bonn, Germany
E-mail: thoma@hiskp.uni-bonn.de

Experiments with electromagnetic probes are promising to search for new baryon states and to determine the properties of baryon resonances. The Crystal Barrel experiment at ELSA is very well suited to investigate final states with neutral mesons decaying into photons. Interesting results on baryon resonances have been obtained. In particular a new $D_{15}(2070)$ decaying into $p\eta$ was recently observed. In multi-meson final states such as $p\pi^0\pi^0$ and $p\pi^0\eta$ baryon cascades occur. Baryon resonances decaying not only into $\Delta\pi$ but also into $D_{13}(1520)\pi$, $S_{11}(1535)\pi$, and via a higher mass state around 1660 MeV are observed.

1. Introduction

At medium energies, our present understanding of QCD is limited. Here, in the energy regime of meson and baryon resonances the strong coupling constant is large and perturbative methods can no longer be applied. One of the key issues in this energy regime is to identify the relevant degrees-of-freedom and the effective forces between them. A necessary step towards this aim is undoubtedly a precise knowledge of the experimental spectrum of baryon resonances and of their properties.

Quark models are in general amazingly successful in describing the spectrum of existing states. However, constituent quark models usually predict many more resonances than have been observed so far, especially at higher energies[1,2]. In addition the mass of some states are difficult to understand within a constituent quark model, such as e.g. the low mass of the $P_{11}(1440)$ or the $P_{33}(1600)$. Another example are the negative parity Δ^*-states around 1900 MeV, which at present still need experimental confirmation. An alternative description of the baryon spectrum in which baryon resonances are

generated dynamically from their decays is developed in [3], evoking different degrees of freedom as the relevant ones.
Experiments need to provide a detailed knowledge on the spectrum but also of the properties of baryon resonances to test predictions from the different models and from lattice QCD-calculations. Experiments using electromagnetic probes, such as the Crystal Barrel experiment at ELSA, are not only able to search for unknown states but also to determine the properties of resonances like photo-couplings and partial widths. The properties of a resonance are also of big importance for an interpretation of its nature. One immediate debate is e.g. whether the $P_{11}(1440)$ is really a 3-quark state. A good understanding of production and decay properties may help to elucidate the nature of a state. In the following, different final states investigated with the Crystal Barrel detector at ELSA will be discussed.

2. The $\gamma p \to p\eta$-channel

New data on η-photoproduction has been taken by the CB-ELSA experiment in Bonn [4]. Due to its electromagnetic calorimeter consisting of 1380 CsI(Tl) crystals covering 98% of the 4π solid angle, the CB-ELSA detector is very well suited to measure photons. The η is observed either in its $\gamma\gamma$- or $3\pi^0$- decay. The two or six photons are detected in the calorimeter and the proton is identified in a 3-layer scintillating fiber detector. The invariant masses show a clear η signal over an almost negligible background (Fig. 1, top right). The differential and the total cross sections are shown in Fig. 1 in comparison to the TAPS[5], GRAAL[6] and CLAS[7] data. The new CB-ELSA data extends the covered angular and energy range significantly compared to previous measurements. The total cross section is obtained by integrating the differential cross sections. The extrapolation to forward and backward angles uses the result of the partial wave analysis (PWA) discussed below. The result of the PWA is shown as solid line Fig. 1. The formalism used in the PWA to extract the contributing resonances from the data is summarized in [15]. In the fit [10,17] the following following data sets were included in addition to the CB-ELSA data on $\gamma p \to p\eta$: The CB-ELSA data on $\gamma p \to p\pi^0$ [8], the TAPS data on $\gamma p \to p\eta$ [5], the beam asymmetries $\Sigma(\gamma p \to p\pi^0)$ and $\Sigma(\gamma p \to p\eta)$ from GRAAL[9], and $\Sigma(\gamma p \to p\pi^0)$, $\gamma p \to n\pi^+$ from SAID and data on $K\Lambda$ and $K\Sigma$ from SAPHIR[11,12], CLAS[13], and LEPS[14]. In the PWA a new state coupling strongly to the η-channel was found, an $D_{15}(2070)$ with a mass of (2060 ± 30) MeV and a width of (340 ± 50) MeV. The χ^2-minimum of the

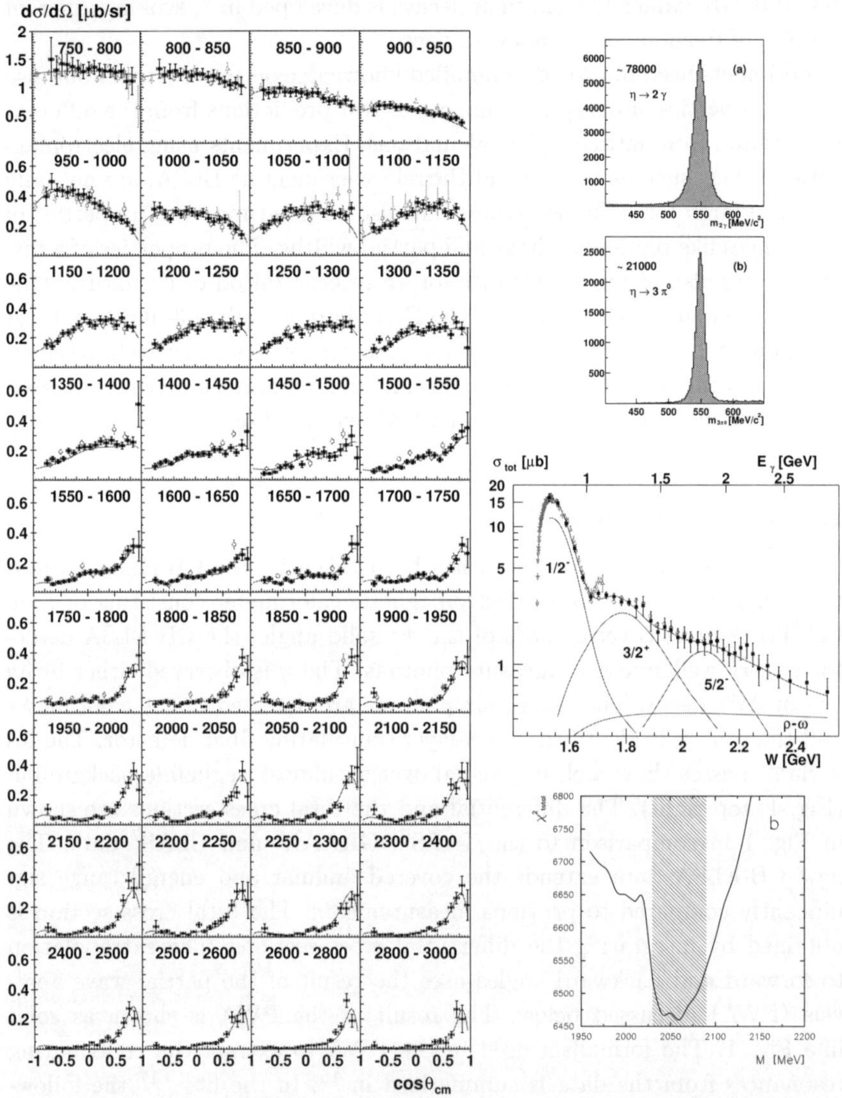

Fig. 1. Left: Differential cross sections for $\gamma\,\mathrm{p} \to \mathrm{p}\,\eta$, for $E_\gamma = 750\,\mathrm{MeV}$ to $3000\,\mathrm{MeV}$: CB-ELSA(black squares)[4], TAPS[5], GRAAL[6] and CLAS[7] data (in light gray). The solid line represents the result of our fit. Right column: Invariant $\gamma\gamma$ and $3\pi^0$ invariant mass, total cross section (logarithmic scale) for the reaction $\gamma\,\mathrm{p} \to \mathrm{p}\,\eta$, χ^2 distribution for different masses of the D_{15} imposed in the fit. For further details see [10,4].

overall fit is shown in Fig. 1. A clear minimum is observed. The $D_{15}(2070)$ contributes rather strongly to the data set as shown in Fig.1. The PDB[16] includes a 2-star $D_{15}(2200)$. Its mass is rather badly determined and scatters from 1900 MeV to 2230 MeV the same is true for measurement of its width. Presently it is unclear whether all signals seen are due to the same state. Additional evidence for further new states found in the combined PWA is discussed in [10,17]. No evidence was found for a third S_{11} for which claims have been reported at masses of 1780 MeV [18] and 1846 MeV [19].

3. The $\gamma p \to p\pi^0\pi^0$-channel

At higher energies multi-meson final states play a role of increasing importance. Above 1900 MeV the spectrum and the properties of resonances are rather badly known, while this is, according to quark model predictions, the energy regime where many new states should occur. Predictions indicate that many of the so far unobserved states should have a significant $\Delta\pi$-coupling[1]. Within the different $\gamma p \to N2\pi$-channels the $\gamma p \to p\pi^0\pi^0$-channel is the one best suited to investigate the $\Delta\pi$ decay of baryon resonances. Compared to other isospin-channels, many non-resonant-"background" amplitudes do not contribute like the diffractive ρ-production or the Δ-Kroll-Rudermann term. In addition, the number of possible Born terms and t-channel processes is strongly reduced; π-exchange is e.g. not possible. This leads to a high sensitivity of the $\gamma p \to p\pi^0\pi^0$-channel on baryon resonances decaying into $\Delta\pi$. The $\gamma p \to p\pi^0\pi^0$ cross section as measured by TAPS[20] in the low energy range and by GRAAL[21] up to an incoming photon energy of about 1500 MeV is shown in Fig. 2; two peak-like structures are observed. The data has been interpreted within the Laget-[22] and Valencia model[23], resulting in very different interpretations. In the Valencia-model, which is limited to the low energy region, the $D_{13}(1520)$ decaying into $\Delta(1232)\pi$ dominates the lower energy peak, while in the Laget-model the $P_{11}(1440)$ decaying into σp is clearly the dominant contribution. This shows that even-though both models lead to a reasonable description of the total cross section their interpretation of the data is in disagreement. Recently data on $\gamma p \to p\pi^0\pi^0$ has also been taken by the CB-ELSA experiment in Bonn extending the covered photon energy range up the E_γ=3.0 GeV ($\sqrt{s}$=2.6 GeV). To investigate the reaction $\gamma p \to p\pi^0\pi^0$, events with 4 photons are selected. In Fig 3 the invariant $p\pi^0$-mass is shown for two $\sqrt{s}$-bins. At low $\sqrt{s}$ only a peak due to the $\Delta(1232)$ is observed, at higher energies additional structures become visible. In ad-

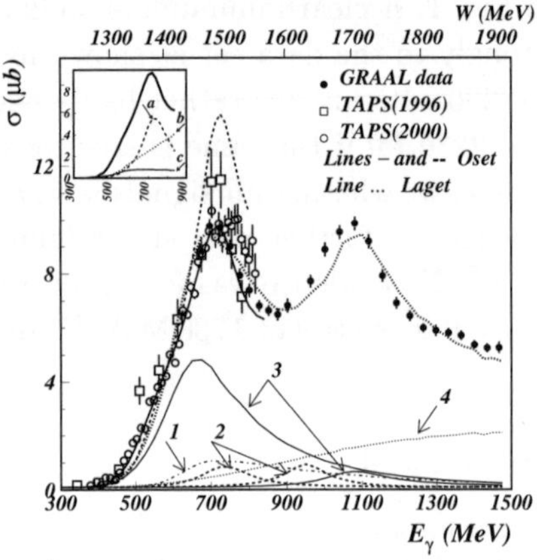

Fig. 2. Total $\gamma p \to p 2\pi^0$-photoproduction cross sections measured by GRAAL[21] and TAPS[20], shown together with the interpretation as given by Laget[22] and within the Valencia-model[23]. In the lower part if the figure the partial contributions as found in the Laget model are given: (1) $\gamma p \to P_{11}(1440) \to \Delta\pi$, (2) $\gamma p \to D_{13}(1520), D_{13}(1710) \to \Delta\pi$, (3) $\gamma p \to P_{11}(1440), P_{11}(1710) \to p\sigma$, (4) $\gamma p \to p\sigma$. In the inset the Valencia-model calculation is shown; lines (a), (b), (c) for $D_{13}(1520)$, Δ, and $P_{11}(1440)$ intermediate states, respectively. The figures are taken from[21].

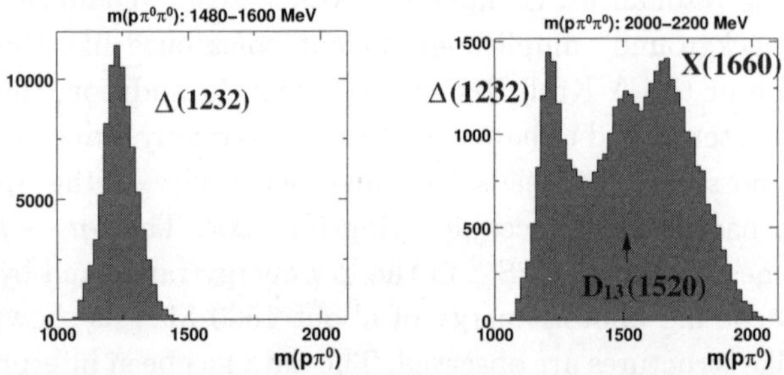

Fig. 3. Left: $p\pi^0$ invariant mass for events with $\sqrt{s}$=1.48-1.6 GeV (left), and $\sqrt{s}$= 2.-2.2 GeV (right).

dition to the $\Delta(1232)$ contributions from the $D_{13}(1520)$, and a further state around 1660 MeV is observed. For further details on the data as well as for differential and total cross sections see [24]. To extract the contributing resonances, their quantum numbers and their properties from the data, a partial wave analysis (PWA) has been performed. The formalism used is summarized in [15]. For the parametrisation of s-channel resonances either Breit-Wigners or K-matrices are used. t-channel processes are described by the exchange of Regge-trajectories. In addition u-channel amplitudes and Born-terms are included. An event-based, unbinned maximum-likelihood fit was performed; it takes all the correlations between the five independent variables correctly into account. The fits include the preliminary TAPS data[25] in the low energy region in addition to the CB-ELSA data. The latter was taken using two beam energy settings (E_{e^-}=1.4 GeV, 3.2 GeV). Resonances with different quantum numbers were introduced in various combinations allowing, so far, for the following decay modes: $\Delta(1232)\pi$, $N(\pi\pi)_s$, $P_{11}(1440)\pi$, $D_{13}(1520)\pi$ and $X(1660)\pi$. For a good description of the data resonances like e.g. the $P_{11}(1440)$, the $D_{13}(1520)$, the $D_{13}/D_{33}(1700)$, the $P_{13}(1720)$, the $F_{15}(1680)$ as well as several additional states and background amplitudes are needed. An example for the quality of the data description reached is shown in Fig. 4. One preliminary result of the PWA is a dominant contribution of the $D_{13}(1520) \to \Delta\pi$ amplitude in the energy range where the first peak in the cross section occurs. Fig. 5 shows the $p\pi^0$ invariant mass and angular distributions in the $\sqrt{s}$-range from 1450 MeV to 1550 MeV. The $\Delta(1232)$ clearly dominates the $p\pi^0$ invariant mass. The PWA attributes most of these events to the $\gamma p \to D_{13}(1520) \to \Delta\pi$ amplitude. This interpretation gets further substantiated by looking at the angular distributions. The angular distribution for the $D_{13}(1520) \to \Delta\pi$-amplitude has a very similar shape as the data. This is not the case for the $P_{11}(1440)$. The preliminary result of our analysis is therefore in contradiction to the interpretation given within the Laget model, where the $P_{11}(1440) \to p\sigma$-amplitude dominates in this energy range. Another result of the PWA is the observation of baryon cascades. Baryon resonances not only decaying into $\Delta\pi$ but also via $D_{13}(1520)\pi$ and $X(1660)\pi$ are observed for the first time. The according enhancements are already visible in the

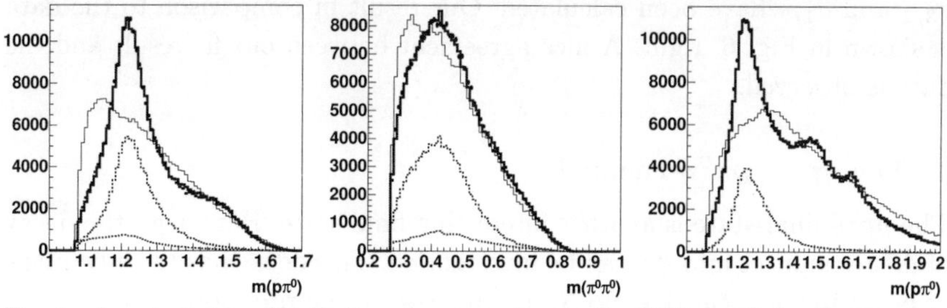

Fig. 4. Invariant mass distributions in comparison to the PWA-result for the lower energy CB-ELSA data (without acceptance correction). Each plot shows the experimental data (points with error bars), and the result of the PWA (solid gray curve). The contribution of the $D_{13}(1520)$ and the $P_{11}(1440)$ are shown as dashed black (upper) and dashed gray (lower) curves, respectively. The thin black line represents the phase space distribution. The description of the angular distributions are of similar quality. Right: $p\pi^0$ invariant mass for the higher energy data set (E_γ=0.8-3.0 GeV) in comparison to the result of the PWA. (preliminary)

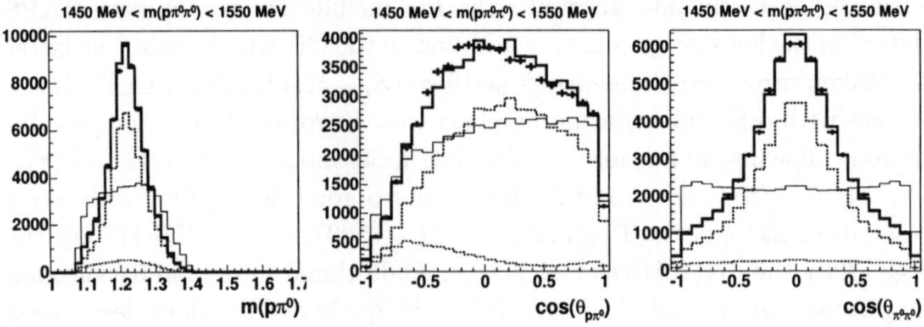

Fig. 5. $p\pi^0$ invariant mass and angular distributions for $\sqrt{s}$=1450-1550MeV. See Figure 4 for the meaning of the different curves.

corresponding $p\pi^0$ invariant mass (Fig. 3). The observation of baryon cascades is also interesting in respect to the search for states which might not couple to πN and γp; they still could be produced in such baryon cascades. Presently, a combined analysis of this data together with the single meson photoproduction data discussed in Section 2 and the Crystal Ball data on $\pi^- p \to n\pi^0\pi^0$ [26] is in progress.

Recently the result of the $\gamma p \to p\pi^0\pi^0$ double polarization experiment performed by the GDH-collaboration at MAMI has been published [27]. It was found that the $\sigma_{3/2}$ component which cannot be due to the $P_{11}(1440)$ dominates in the energy region up to about 800 MeV. According to the Valencia-model[23,28] the $D_{13}(1520)$-resonance is largely responsible for the observed dominance of the $\sigma_{3/2}$ cross section. The non-negligible $\sigma_{1/2}$ component observed is presently underestminated by the Valencia-model (Fig. 6, left). Using the result of our PWA obtained by fitting the unpolarized data only, $\sigma_{3/2}$ and $\sigma_{1/2}$ have been calculated. Our result in comparison to the data is shown in Fig. 6, right. A nice agreement between our fit result and the data is observed.

4. The $\gamma p \to p\pi^0\eta$-channel

The $p\pi^0\eta$-final state is another interesting final state. Here, e.g., the decay of Δ^* resonances into $\Delta\eta$ can be investigated. This decay has the advantage of being isospin selective; no N* resonances can be produced.

This reaction is e.g. well suited to investigate the existence of the negative parity Δ^*-states around 1900 MeV. These states would, if they exist, pose a problem for quark model calculations because of their low mass. But so far the evidence for their existence is weak. Only one of the three states, $S_{31}(1900)$, $D_{33}(1940)$, and $D_{35}(1930)$ is a 3-star resonance, the $D_{35}(1930)$. The 1-star $D_{33}(1940)$ resonance can decay with orbital angular momentum

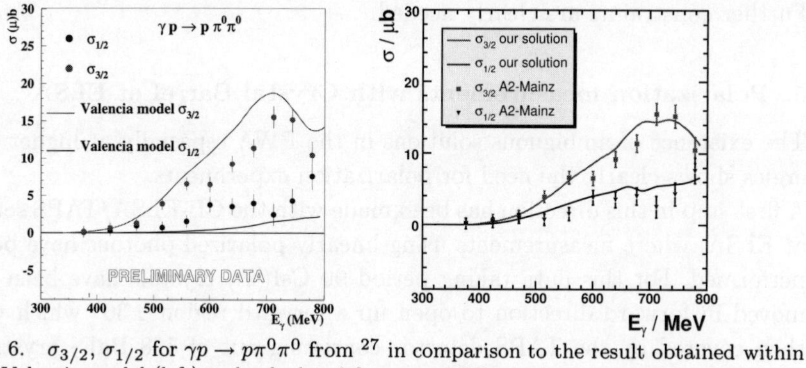

Fig. 6. $\sigma_{3/2}$, $\sigma_{1/2}$ for $\gamma p \rightarrow p\pi^0\pi^0$ from [27] in comparison to the result obtained within the Valencia model (left) and calculated from our PWA-result determined by fitting the unpolarized data only (right).

zero into $\Delta\eta$, which makes the $p\pi^0\eta$-final state a good place to investigate the existence of this resonance.

With the Crystal Barrel detector at ELSA this final state has been investigated from threshold up to photon energies of 3.0 GeV. The $p\pi^0$ invariant mass (Fig. 7) shows that the data set is indeed dominated by $\Delta\eta$ events. In addition to the $\Delta(1232)$ there are other interesting structures observed, such as the $a_0(980)$ in the $\pi^0\eta$ invariant mass or the $S_{11}(1535)$ in the Dalitz plot. To extract the resonance contributions from the data a PWA of the $p\pi^0\eta$-final state has been performed. Three ambiguous solutions were found. All three result in a similar likelihood being based at the same time on quite different sets of contributing amplitudes. All three solutions do need a $D_{33}(\sim 1900)$ and show indications baryon cascades, e.g. decays of higher mass resonances not only via $\Delta\eta$ but also via $S_{11}(1535)\pi$. In addition there seem to be hints for a new higher mass resonance. But based on ambiguous solutions it is of course impossible to make a more definite statement.

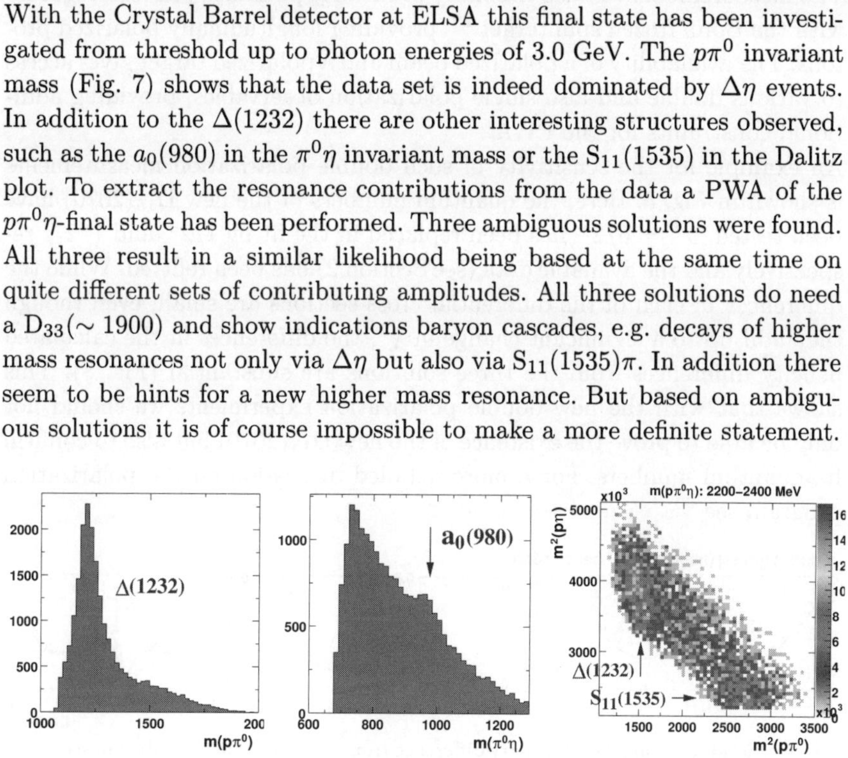

Fig. 7. $\gamma p \rightarrow p\pi^0\eta$-events: $p\pi^0$ (left) and $\pi^0\eta$ (middle) invariant mass for $E_\gamma = E_{\text{threshold}} - 3.0$ GeV. Right: Dalitz plot for $\sqrt{s}$= 2.2-2.4 GeV.

Further constraints are clearly needed.

5. Polarization measurements with Crystal Barrel at ELSA

The existence of ambiguous solutions in the PWA especially at higher energies shows clearly the need for polarization experiments.
A first step in this direction has been made with the CB-ELSA/TAPS setup at ELSA, where measurements using linearly polarized photons have been performed. For this data taking period 90 CsI(Tl) crystals have been removed in forward direction to open up a forward region ±30° which was then covered by the TAPS detector consisting out of 528 BaF_2 crystals. The analysis of this data to determine the beam asymmetries of various final states is presently in progress.
Presently the experimental setup at ELSA is modified for the new double polarization experiments. New forward detector components are installed. The measurements will use circularly or linearly polarized photons together with the Bonn frozen spin target [29], providing longitudinally polarized protons. The availability of a polarized beam and a polarized target gives access to various double and also single polarization observables, providing additional constraints for the PWA.
An example for the sensitivity of such double polarization measurements is shown in Fig. 8. Here, the quantum numbers of the new $D_{15}(2070)$ have been tested. $J^P = 5/2^-$ has been replaced in the fit by $1/2^+$ and $1/2^-$, respectively and the available data (see Section 2) has been refitted. While the differences in each of the differential cross sections are small, even though they add up to a significant change in χ^2, the differences in the calculated helicity differences from the three solutions are substantial (Fig. 8). This shows that with the new double polarization experiments we should not only be able to prove the existence of the new $D_{15}(2070)$ but also to confirm its quantum numbers. For a more detailed discussion on the polarization program see [30].

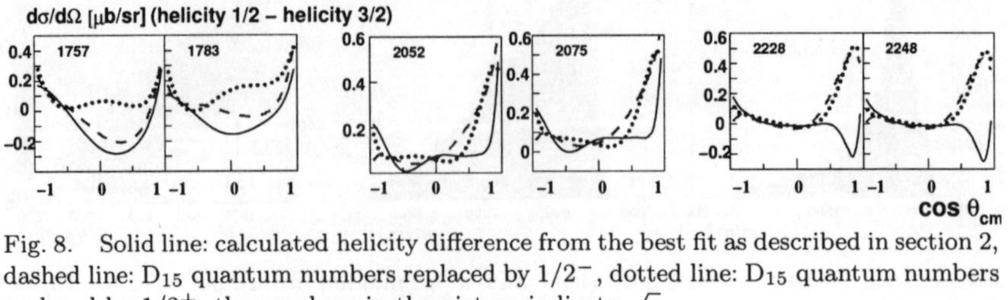

Fig. 8. Solid line: calculated helicity difference from the best fit as described in section 2, dashed line: D_{15} quantum numbers replaced by $1/2^-$, dotted line: D_{15} quantum numbers replaced by $1/2^+$, the numbers in the picture indicate $\sqrt{s}$.

Acknowledgments

This work is supported by the Deutsche Forschungsgemeinschaft (DFG) within the SFB/TR16. The author acknowledges an Emmy Noether grant from the DFG.

References

1. S. Capstick and W. Roberts, Phys. Rev. **D47**, 1994 (1993), Phys. Rev. **D49**, 4570 (1994), S. Capstick, Phys. Rev. **D46**, 2864 (1992)
2. U. Loring, B. C. Metsch and H. R. Petry, Eur. Phys. J. A **10**, 395 (2001)
3. M.Lutz, this conference
4. V. Crede *et al.* [CB-ELSA Collab.], Phys. Rev. Lett. **94**, 012004 (2005)
5. B. Krusche *et al.*, Phys. Rev. Lett. **74**, 3736 (1995).
6. F. Renard *et al.* [GRAAL Collab.], Phys. Lett. B **528**, 215 (2002).
7. M. Dugger *et al.* [CLAS Collab.], Phys. Rev. Lett. **89**, 222002 (2002).
8. O. Bartholomy *et al.* [CB-ELSA Collab.], Phys. Rev. Lett. **94**, 012003 (2005)
9. J. Ajaka *et al.* [GRAAL collaboration], Phys. Rev. Lett. **81**, 1797 (1998), O. Bartalini *et al.* submitted to Eur. Phys. J. A.
10. A. V. Anisovich, A. Sarantsev, O. Bartholomy, E. Klempt, V. A. Nikonov and U. Thoma, Eur. Phys. J. A **25**, 427 (2005)
11. K. H. Glander *et al.*, Eur. Phys. J. A **19**, 251 (2004)
12. R. Lawall *et al.*, Eur. Phys. J. A **24**, 275 (2005)
13. J. W. C. McNabb *et al.* [The CLAS Collab.], Phys. Rev. C **69**, 042201 (2004)
14. R. G. T. Zegers *et al.* [LEPS Collab.], Phys. Rev. Lett. **91**, 092001 (2003)
15. A. Anisovich, E. Klempt, A. Sarantsev and U. Thoma, Eur. Phys. J. A **24**, 111 (2005)
16. S. Eidelman *et al.* [Particle Data Group], Phys. Lett. B **592** (2004) 1.
17. A. V. Sarantsev, V. A. Nikonov, A. V. Anisovich, E. Klempt and U. Thoma, Eur. Phys. J. A **25**, 441 (2005)
18. B. Saghai and Z. Li, Proceedings of NSTAR 2002, nucl-th/0305004.
19. G. Y. Chen, S. Kamalov, S. N. Yang, D. Drechsel and L. Tiator,
20. M. Wolf *et al.*, Eur. Phys. J. **A9**, 5 (2000). F. Härter *et al.*, Phys. Lett. **B401**, 229 (1997).
21. Y. Assafiri *et al.*, Phys. Rev. Lett. **90**, 222001 (2003).
22. J.-M. Laget, L. Y. Murphy, shown in [21].
23. J. A. Gomez Tejedor *et al.*, Nucl. Phys. **A600**, 413 (1996).
24. M. Fuchs, this conference
25. M. Kotulla, private communication.
26. S. Prakhov *et al.* [Crystal Ball Collab.], Phys. Rev. C **69**, 045202 (2004).
27. J. Ahrens *et al.* [GDH and A2 Collab.], Phys. Lett. B **624**, 173 (2005), H.-J. Arends, talk at NSTAR 2004.
28. J. Nacher *et al.*, Nucl. Phys. **A695**, 295 (2001), **A697**, 372 (2002).
29. Ch. Bradtke *et al.*, Nucl. Instr. and Meth. **A436**, 430 (1999).
30. H. Schmieden, this conference

$K\Lambda$ AND $K\Sigma$ PHOTOPRODUCTION IN A COUPLED CHANNELS FRAMEWORK

O. SCHOLTEN AND A. USOV

Kernfysisch Versneller Instituut, University of Groningen, 9747 AA, Groningen, The Netherlands

A coupled channels analysis, based on the K-matrix approach, is presented for photo-induced kaon production. It is shown that channel coupling effects are large and should not be ignored and that contact terms in the analysis, associated with short range correlations, are important. A brief view is given on topics that require major theoretical model building.

1. Introduction

In this talk a calculation is presented of kaon photoproduction in a coupled-channels framework based on the K-martix formalism. We will show that channel coupling effects are large for most observables. This implies that loop contributions beyond the tree-level diagrams are important. Our model obeys gauge invariance which is guaranteed through contact terms added to the model Lagrangian. We observe that the choice for these contact terms, which have a large amount of model dependence, does affect some of the calculated cross sections in a significant way. From a pessimistic point of view this sensitivity to gauge restoration scheme implies that the data are sensitive to other terms in the Lagrangian besides the usual Born terms and resonance couplings. One should thus be very reluctant using simplified models to extract resonance properties. From an optimistic point of view this implies that one is sensitive to QCD dynamics.

A major theoretical challenge is to build a model that, in addition to all symmetries obeyed by the K-matrix model, also obeys causality. This topic is addressed in the last chapter.

2. Coupled Channels K-matrix

A basic exposé of K-matrix formalism is given in ref.[1]. The essence is that the scattering matrix is written as $T = \frac{K}{1-iK} = K + iK \times K + \cdots$, such that

$S = 1 + 2iT$ is unitary. From the expansion in powers of the kernel K it is seen that the formalism includes a non-perturbative summation over an infinity of loop corrections. The strength of the K-matrix approach is that, besides unitarity, also co-variance, gauge-invariance, and crossing symmetry are obeyed. Even though it may seem complicated it is in a certain aspect simpler that a tree-level calculation. Since all reaction channels are coupled, the parameters are subject to many more constraints and the number of free parameters for an additional reaction channel is generally rather small.

The kernel is taken as the sum of tree level diagrams, including resonance contributions, supplemented with contact terms to guarantee gauge invariance.

The aforementioned expansion in powers of the kernel generates, however, only the pole-part (the imaginary part) of the complete loop corrections. The real part of the loop corrections is taken into account in an approximate way through the introduction of form factors or vertex functions[2,1]. Since at this level the self-consistency between the real and imaginary contributions is missing a symmetry, causality, is violated. In a later chapter this point will be addressed in some more detail.

2.1. *Kaon Production, Model Ingredients*

Full details of the present model are given in ref.[3]. Here we mention only the most important aspects for obtaining an appreciation of the calculations. The channel space is formed by the following baryon-meson states: (N+γ), (N+π), (N+η), (N+ρ), (N+Φ), (Λ+K), and (Σ+K). All possible spin and iso-spin states are taken into account for each channel. In the kernel a complete set of s- & u-type diagrams is included with N, Λ, Σ, $S_{11} \times 2$, S_{31}, $P_{11} \times 2$, P_{31}, P_{13}, $P_{33} \times 2$, $D_{13} \times 2$, and D_{33} as intermediate states. The t-channel contributions due to π, η, ρ, ω, σ, K, and K^* exchanges have been incorporated where allowed by the quantum numbers.

At all three-point vertices form factors have been introduced with cut-off parameters of the order of 1 GeV. For this reason also contact terms (four point vertices) are introduced as part of a gauge restoration scheme. As this scheme is far from unique, as discussed in the following, we have allowed for a systematic treatment of the gauge-restoration ambiguities through the introduction of a rather general set of contact terms, following the philosophy used in chiral-perturbation theory.

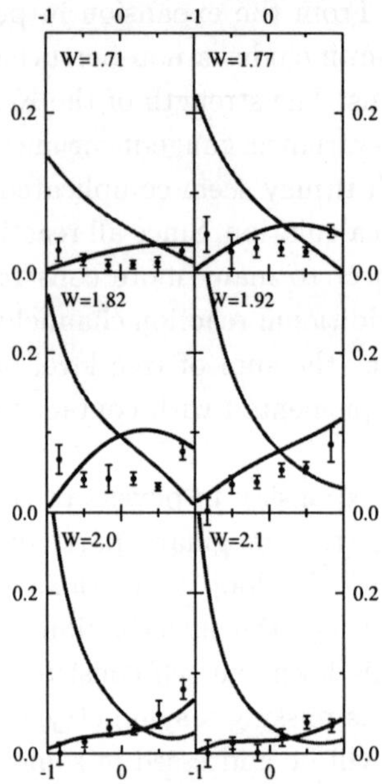

Fig. 1. Effect of suppression of the u-type diagrams by means of a modified form factor (red curve) on the calculated cross section for $p + \gamma \rightarrow \Sigma^+ + K^0$.

2.2. *u-channel Vertex Function*

In the model the three-point vertices are dressed with form factors. Apart from the aforementioned theoretical reasons for this dressing also phenomenology requires the introduction of form factors. These depend on the invariant mass of the off-shell leg of the three-point diagram, i.e. for s-type diagrams $F(s) = \frac{\Lambda^2}{\Lambda^2+(s-m^2)^2}$, with ($\Lambda > 1$ GeV^2). In the calculation we observe that the same functional form does not give sufficient suppression of u-type diagrams. For phenomenological reasons a modified form factor, $H(u) = \frac{u}{m^2}F(u)$ is preferred. In Fig. 1 the effects of the different choices for the form factor are compared.

2.3. *Gauge Restoration not Unique*

The use of form factors necessitates the introduction of contact terms to restore gauge-invariance. For the construction of these additional terms different procedures can be used commonly named after their inventors.

Ohta: Simple minimal substitution[6].

DW: Davidson-Workman[7], Ohta plus additional contact terms.
JR: Janssen-Ryckebusch[8], DW plus other contact terms.

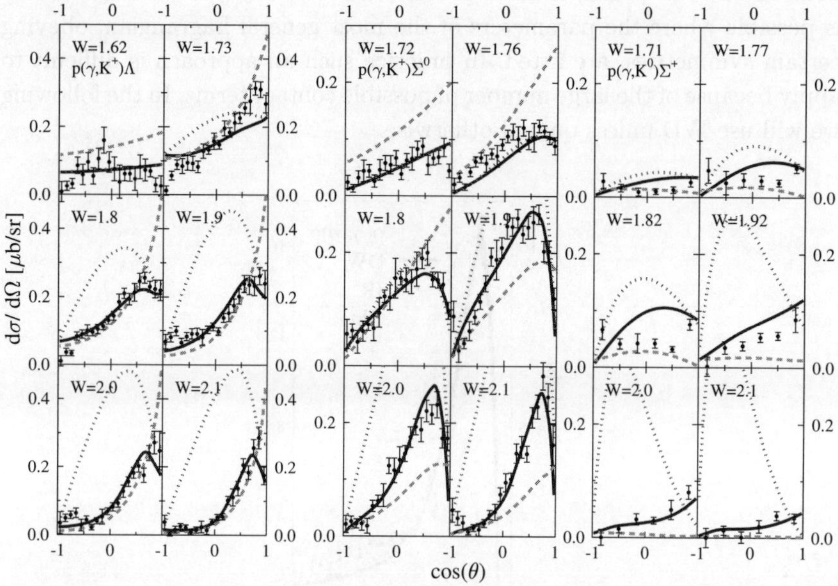

Fig. 2. Effects of different gauge-restoration schemes (drawn: DW, dotted: Ohta, dashed: JR ×1/3) on differential cross sections, the data are from ref.[5].

In Fig. 2 the calculations following the different gauge-restoration procedures are compared. The parameters in the DW calculation have been fitted to the data, which explains why this calculation shows the best agreement. Independent of this it can be shown that in the Ohta procedure the convection current is not suppressed which gives rise to a strongly increasing cross section with energy. This is in flagrant disagreement with the data. In the DW procedure the convection current is suppressed which allows for a good fit to the data. In the JR procedure there is a different form factor for the suppression of magnetic contributions and has been shown to allow for a good agreement with the data in tree-level calculation[8]. Important to notice is that the dependence on the gauge restoration scheme is large and it will strongly affect the values of the model parameters extracted from the data.

A criticism of the DW procedure is that it is obtained following a rather ad-hoc procedure. In ref.[3] it has been shown that also this form can be obtained following a minimal substitution procedure provided the terms are arranged appropriately.

By treating gauge invariance restoration via counter (contact) terms as done in ref.[3], in principle an approach like that of χ-perturbation theory is possible where the parameters of the most general Lagrangian, obeying certain symmetries, are fitted. In practice such an approach is difficult to apply because of the large number of possible contact terms. In the following we will use WD unless quoted otherwise.

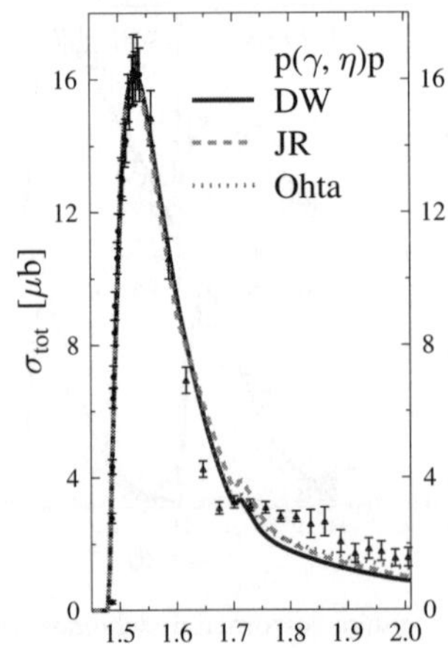

Fig. 3. Effects of contact terms on the calculated cross section for η production[9].

The strong dependence of the cross section on the gauge restoration scheme is not the same for all reactions. The cross section for $(\gamma+p \to \eta+p)$, see Fig. 3, is almost independent of the procedure followed. The reason is that this reaction is resonance dominated and contact terms, which are of the same order of magnitude as the Born terms, are of little importance.

2.4. *Challenge of Photon Coupling at Higher Energies*

For kaon production the particularities of the gauge-restoration scheme are clearly very important. In general the choice of contact terms, or equivalently, the gauge restoration scheme, is an issue if $E_\gamma \geq M_p$. On the one hand this implies a strong model dependence in the calculation of the continuum part of the spectrum which makes it more difficult to extract resonance parameters. On the other hand this dependence can be translated into sensitivity to physics that is apparently not well under control. This physics is clearly related to the same physics that necessitates the introduction of form factors, short range effects and loop corrections[2]. In principle the loop corrections can be calculated since they are dependent on the same couplings as enter in the tree-level model. This offers the possibility to address short-range physics, related to QCD quark-loop contributions. At energies $E_\gamma \geq M_p$ this should not be surprising.

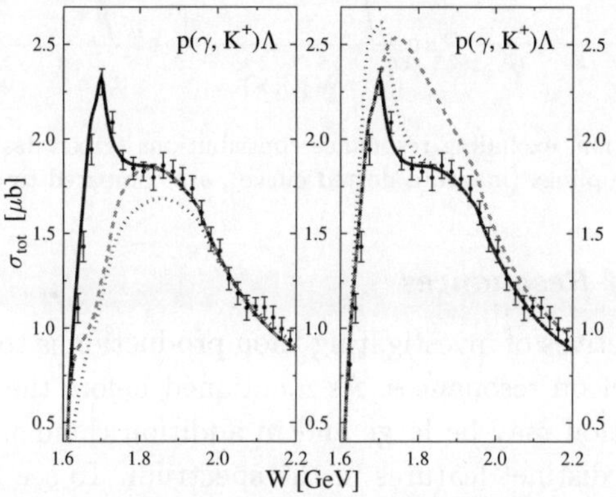

Fig. 4. Examples showing the importance of channel coupling effects as discussed in the text. The data are taken from ref.[5].

2.5. *Sensitivity to Coupled Channels*

The importance of channel-coupling effects is illustrated in Fig. 4 using the $(\gamma + p \to \Lambda + K)$ reaction as an example. The drawn curve gives the results of the default calculation. The green dashed curve on the left panel is a calculation in which the $g_{N\gamma S_{11}}$ coupling is decreased by a factor 10, at the same time increasing $g_{K\Lambda S_{11}}$ by a factor 10 to keep the same matrix element for the tree level contribution to $(\gamma + p \to K + \Lambda)$. For the magenta dotted curve the coupling $g_{NK^*\Lambda}$ is set to zero. Since the K^* hardly contributes at the direct tree level, this also shows the importance of channel coupling. For the green dashed curve on the right panel the $\rho - N$ final state has been excluded from the calculations. The magenta dotted curve shows a calculation in which the sign of the η-coupling to the S_{11} is reversed. All these results show clearly that channel coupling effects are large and, more importantly, that they may induce non-trivial structures in the spectrum which can easily be mistaken for the result of resonance contributions.

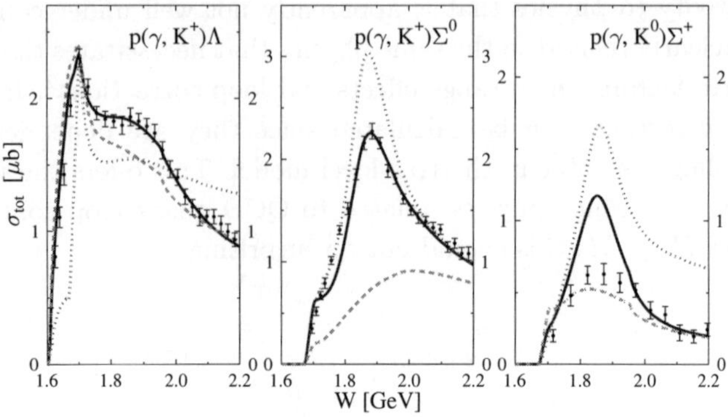

Fig. 5. Cross section, excluding resonance contributions (green dashed curve) and excluding channel couplings (magenta dotted curve), are compared to data[5].

2.6. *Effects of Resonances*

One of the objectives of investigating kaon production is to learn about the structure of nucleon resonances. As mentioned before the continuum part of the cross section may be large and in addition channel coupling effects can give rise to distinct features in the spectrum. To see the contribution of resonances to the cross section we have performed a calculation in which the resonance contributions are excluded (green dashed curve in Fig. 5). The importance of resonances clearly depends on the reaction channel. Excluding channel coupling effects (magenta dotted curve in Fig. 5) shows an effect of similar order of magnitude.

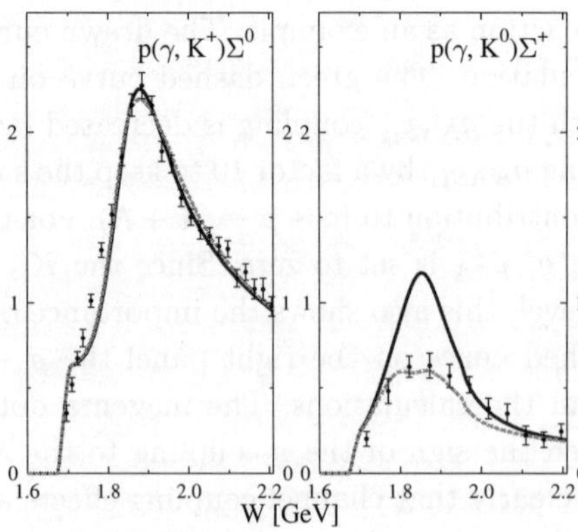

Fig. 6. An additional P_{13} resonance brings good agreement with the data[5].

The large discrepancy seen for Σ^+ production can be easily corrected by adding an extra P_{13} resonance to the calculation (green dashed curve in Fig. 6). The fact that the results of this calculation almost coincide (for Σ^+ production) with the calculation excluding all resonances (green dashed curve in Fig. 5) implies that one should be very careful claiming evidence for this resonance.

3. Beyond K-matrix

As mentioned in the introduction, the strength of the K-matrix approach is that it incorporates many symmetries of nature in a computationally convenient approach. There is,however, one symmetry which is violated, namely that of causality which is equivalent to analyticity of the scattering amplitude. This results in a violation of dispersion relations and is a topic of much concern as is discussed in detail in the contribution of Mathias Lutz[10] to this conference. In the framework of the K-matrix approach we are investigating two approaches to restore analyticity of the scattering amplitude without spoiling the existing symmetries.

Analyticity of the scattering amplitude can be important in the vicinity of a threshold for the opening of a reaction channel. Since the imaginary part of the amplitude changes rapidly at this energy, the real part has a cusp which may show as a rapidly varying cross section. As was shown in ref.[1] analyticity is also crucial for a correct calculation of polarizabilities. In addition dynamic resonances can be predicted[10].

3.1. *The "Dressed K-matrix" approach*

One of the earliest procedures proposed to restore analyticity of the amplitude is the 'Dressed K-matrix' approach[11]. In this approach the real vertex and self-energy functions are constructed from the cut-loop (or imaginary or pole) contributions using Hilbert Transforms (dispersion relations). Since a modified vertex implies a modification of the total cross section which is reflected in the cut-loop contribution, an iterative procedure is necessary.

The 'Dressed K-matrix' approach has been applied to the case of Compton scattering incorporating the effect of the pion-nucleon channel. It is the first approach where nucleon polarizabilities, high energy Compton scattering, pion photoproduction and pion nucleon scattering are calculated in a single consistent approach with results that are in agreement with the data. The draw-back of this method is that it is computationally involved and difficult to extent to the case of more open channels.

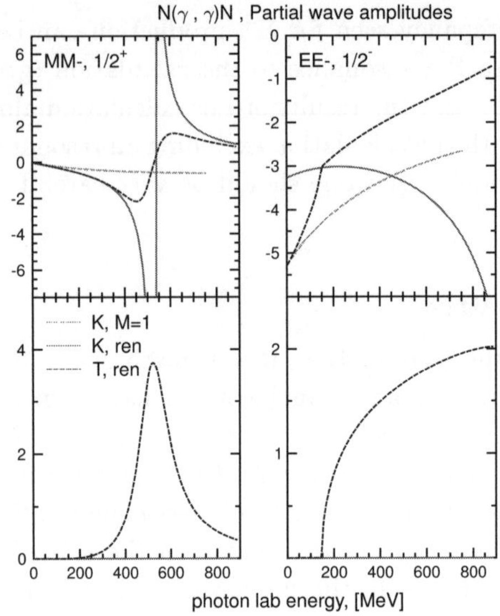

Fig. 7. $j^\pi = 1/2^\pm$ partial wave amplitudes for Compton scattering calculated in the Imaginary K formalism.

3.2. *The "Imaginary K-matrix" approach*

In the "Imaginary K-matrix" approach the basic idea is to make an analytic extension of the K-matrix elements for the channels that are not kinematically allowed. The momentum is taken as imaginary for those channels that have been extended in the un-physical regime. By construction the calculated amplitude is analytic. The price one has to pay is that the amplitude of open channels is affected by coupling to channels for which one is below threshold. This makes it necessary to renormalize the three-point vertices and to introduce bare masses. In addition one has to face the problem that this procedure violates crossing symmetry.

To see the possibilities of this approach we have applied it to a simplified model for Compton and pion scattering in which the kernel is build from s-type diagrams only[12]. The green dashed curve in Fig. 7 shows the result of a pure tree-level calculation for the different amplitudes. The blue dashed curve shows the results of the full calculation where the parameters have been renormalized such as to reproduce the 'green' calculation at low energies. The calculation shows the well-known cusp structure in the EE, $1/2^-$ amplitude at the pion productions threshold. In addition a resonance structure is generated for the magnetic $1/2^+$ amplitude.

4. Conclusions

It has been shown that coupled-channels are important for the interpretation of kaon production. In addition one should be very careful in the interpretation of the continuum part of the cross section since the there is a high sensitivity to the gauge-restoration procedure which is followed.

The K-matrix formalism is very efficient to account for the channel coupling effects. It has been shown that the K-matrix calculation can even be incorporated in a multi-parameter & multi-channel Monte-Carlo and genetic-algorithm fitting calculation[13]. A good fit to the data for different channels can be obtained in a consistent model.

The theoretical challenges are, among others, the incorporation of analyticity (while keeping the other symmetries) in the calculation and to model short range correlations based on QCD.

References

1. S. Kondratyuk and O. Scholten PRC62(2000)025203; Acta Phys. Pol. B33(2002)847.
2. A.Yu. Korchin and O. Scholten, PRC68(2003)045206.
3. A. Usov and O. Scholten, Phys. Rev. C **72**, 025205 (2005).
4. CLAS collaboration, nucl-ex/0509033.
5. SAPHIR collaboration, EPJ **A19**, 251 (2004); **A24**, 275 (2005).
6. K. Ohta, Phys. Rev. C **40**, 1335 (1989).
7. R.M. Davidson and R. Workman, Phys. Rev. C **63**, 025210 (2001).
8. S. Janssen, J. Ryckebusch, D. Debruyne, and T. Van Cauteren, Phys. Rev. C **66**, 035202 (2002), and S. Janssen, private communication.
9. V. Crede et al. (CB-ELSA Collaboration), Phys. Rev. Lett. **94**, 012004 (2005).
10. M.F.M. Lutz and E.E. Kolomeitsev, Nucl. Phys. A **700**, 193 (2002), and the contribution to this conference.
11. S. Kondratyuk and O. Scholten, Phys.Rev.C64(2001)024005; Nucl.Phys.A677(2000)396.
12. O. Scholten, Setsuo Tamenaga and Hiroshi Toki, work in progress.
13. A. Usov, D. Ireland, and O. Scholten, work in progress.

CASCADE PHYSICS: A NEW WINDOW ON BARYON SPECTROSCOPY

JOHN W. PRICE

California State University, Dominguez Hills
E-mail: jprice@csudh.edu

The Ξ, or Cascade, hyperons have the unique features of double strangeness and narrow widths. Typically, $\Gamma_{\Xi^*} \sim 10 - 20\,\text{MeV}$, which is 5–30 times narrower than N^*, Δ, Λ, or Σ states. These features, combined with its isospin of 1/2, make possible a wide-ranging program centered on the physics of the cascade hyperon and its excited states using the photoproduction reaction $\gamma p \to K^+K^+\Xi^-$. The photoproduction cross section is large enough to consider a coarse survey of cascade-proton scattering. We present the physics motivations for a systematic study of the Cascade hyperons, showing recent results from the CLAS Collaboration, and describe recent developments within the physics community to identify the possibilities for future work in this field.

1. Introduction

The structure and dynamics of the nucleon and its excited states has been the subject of much inquiry since the discovery of the Δ in the early 1950's. Despite over fifty years of research, there are still many questions left unanswered. More to the point, it is becoming increasingly difficult to make further progress due to the nature of the work involved. Typically, research into nucleon structure entails acquiring a large data set using data from several experiments which must then be subjected to a detailed computer analysis; not only can we not find new states by looking at simple spectra, we cannot even improve our knowledge of the existing states. Because the states in the excited N^* spectrum are broad and overlapping, the details of the individual states can only be determined via a partial-wave analysis.

To alleviate this situation, we would ideally search for a particle that does not have the problematic large width of the excited nucleon, and yet has properties similar to those of the nucleon. Such a particle would open a "new window" on baryon spectroscopy, in that it could provide a complementary approach to the study of the excited baryons.

2. SU(3) flavor symmetry

An indication of where we might find such a particle is provided by the flavor symmetry of $SU(3)$, $SU(3)_F$. This is due to our ability to separate the QCD lagrangian into two parts: $\mathcal{L}_{QCD} = \mathcal{L}_0 + \mathcal{L}_m$. The first part, $\mathcal{L}_0$, depends only on the quark and gluon fields, and has no mass dependence. The entire mass dependence lies in the second part, $\mathcal{L}_m$. To the extent that we can treat the mass term as a perturbation (*i.e.*, if the quark masses are small in comparison to the baryon masses), we find that $\mathcal{L}_0$ sets the mass scale for a given $SU(3)_F$ multiplet. In this limit, the masses of the baryons are degenerate; $\mathcal{L}_m$ breaks this symmetry, setting the mass splitting within each multiplet, leading to the multiplet structure we are familiar with.

$SU(3)_F$ symmetry is observed experimentally in the properties of the baryons. Static properties, such as the masses of the baryons, are found to obey this symmetry in that the mass difference $\Delta m[\Lambda(1600)\frac{1}{2}^+ - \Lambda(1115)\frac{1}{2}^+]$ is very similar to the mass difference $\Delta m[N(1440)\frac{1}{2}^+ - N(939)\frac{1}{2}^+]$. Additionally, the dynamic properties of the baryons exhibit $SU(3)_F$ symmetry, as shown both by the similarity of the near-threshold cross sections for the processes $\pi^- p \to \eta n$ and $K^- p \to \eta\Lambda$[1] and by the similarity in the Dalitz plots for the processes $\pi^- p \to \pi^0\pi^0 n$ and $K^- p \to \pi^0\pi^0\Lambda$.[2,3]

Given this evidence for the usefulness of $SU(3)_F$ symmetry, we turn to the baryon multiplets. Each multiplet is identified by its spin and parity J^P. $SU(3)_F$ symmetry implies that the members of the multiplets differ only in their quark makeup, and that the basic properties of the baryons should be similar. The octets consist of a N^*, Λ^*, Σ^*, and a Ξ^* state. We thus expect that for every N^* state, there should be a corresponding Ξ^* state *with similar properties.* Additionally, since the decuplets consist of Δ^*, Σ^*, Ξ^*, and Ω^* states, we expect to find a decuplet Ξ^* with properties similar to every Δ^* state. This observation forms the basis of our program in Cascade physics.

3. The Cascade: Gross features

While the higher mass and double strangeness of the Ξ^*'s make them harder to produce in the laboratory than their N^* and Δ^* cousins, this same double strangeness contributes to their narrow width, which makes isolating one cascade from another a realistic possibility. The cascades thus provide a complementary approach to the study of N^* physics.

Figure 1 shows the widths of all the N^* and Ξ^* states in the RPP.[4]

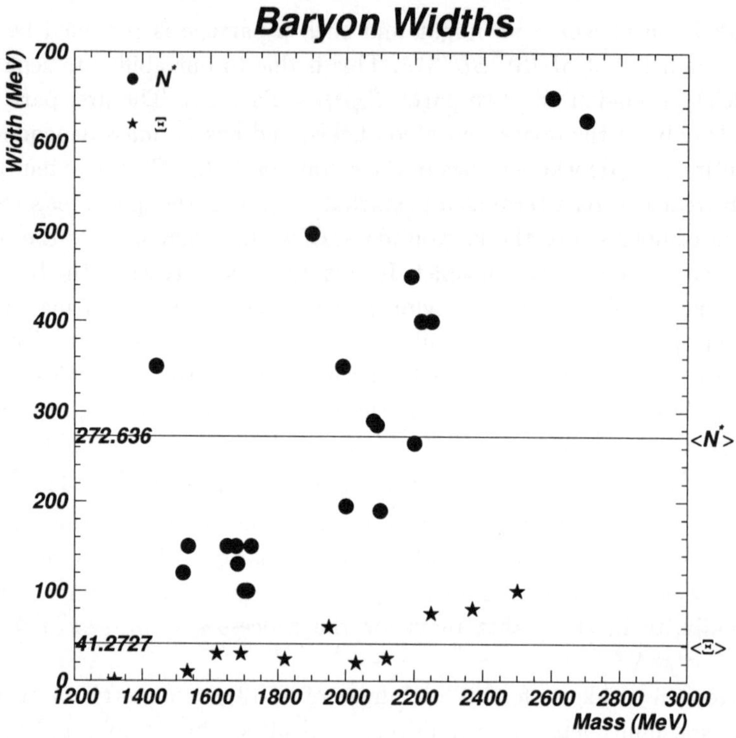

Fig. 1. The widths of all the N^* (circles) and Ξ^* (stars) states from the RPP[4] plotted against their mass.

As seen in the figure, the Cascade hyperons are much narrower than the nucleons. An explanation for this narrowness was first given by Riska,[5] and is related to the number of light quarks in the baryon. Specifically, the width is proportional to the square of the number of light quarks, hence we expect to find the ratio $\Gamma_{N^*,\Delta^*} : \Gamma_{\Lambda^*,\Sigma^*} : \Gamma_{\Xi^*} \approx 9 : 4 : 1$. This prediction agrees very well with experiment. An alternative explanation for narrow Cascades, based on their structure and decay modes, is given by Capstick.[6]

Because the Cascades are narrow, we can get at least a first glimpse at their properties by inspection of a simple spectrum, either of an invariant mass plot in an experiment that detects the decay products of the Cascade (*e.g.* $\Xi^- \to \pi^-\Lambda$, followed by $\Lambda \to \pi^- p$), or in a missing mass spectrum, such as that from a process such as $K^- p \to K^+ X(\Xi^-)$.

4. Cascade Production

The bulk of the data that has been collected on the Cascades has used the process $K^-p \to K^+\Xi^-$. This process would be an excellent choice, as the cross section is relatively large. There are, however, currently no suitable kaon beams available, and other means must be found.

The process we have chosen is the photoproduction process $\gamma p \to K^+K^+\Xi^-$. The simplest analysis involves detecting the two positive kaons, and inferring the Cascade in the missing mass. This reaction has the advantage that below 1.6 GeV, there is no physics background. This process may be used without modification to study the entire Ξ^- spectrum. To study the neutral Ξ^0 states, we may use the process $\gamma p \to K^+K^+\pi^-\Xi^0$.

Hall B at Jefferson Laboratory is suitable for this work. The CLAS detector can be used to detect the multiple charged particles in the final state with good efficiency, and the tagged photon beam has both excellent resolution and intensity. Recent results from the CLAS collaboration indicate that the cross section for Ξ^- photoproduction is both sufficiently large and clean enough for a program in Cascade physics.[7] Figure 2 shows two missing mass spectra from the original CLAS publication. In Fig. 2a, the ground-state $\Xi^-(1321)$ is clearly seen; using a phase-space model to calculate the acceptance, a cross section of $3.5 \pm 0.5(stat.) \pm 1.0(syst.)$ is obtained. High-statistics work with CLAS data has already begun, and is being reported on at this workshop.[8] These results imply the production of several thousand ground-state Ξ^- per week, which is sufficient to consider a program built around the physics of the Cascade.

5. Cascade Physics Program

There are many different topics that may be addressed with a dedicated program in Cascade physics. A few are listed below.

5.1. *Search for missing/Study of known Ξ states*

The RPP lists 22 N^* states, of which 14 have either three or four stars (a somewhat arbitrary criterion used here to indicate that they are "well-established").[4] Similarly, it lists 22 Δ^* states, of which 10 are well-established. By contrast, there are only eleven Ξ^* states listed, of which only six are well-established. This implies, from our earlier discussion, that there are somewhere between 13–33 Ξ states that have not yet been found.

Finding the missing Ξ states is of great importance to our understanding of baryon structure. As previously mentioned, $SU(3)_F$ symmetry *requires*

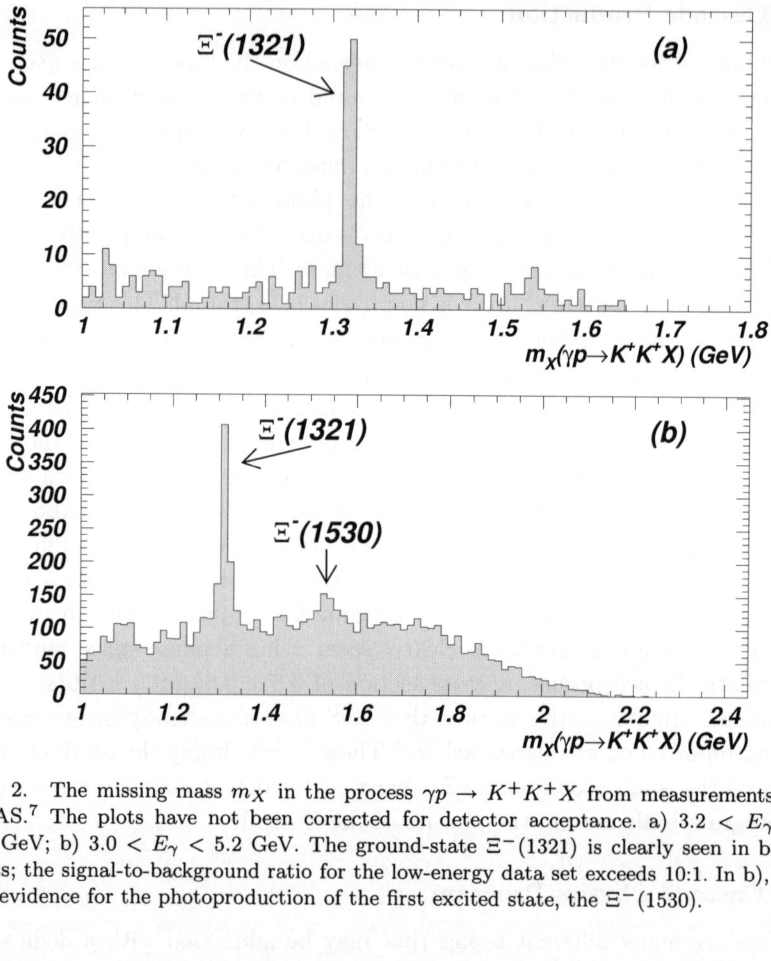

Fig. 2. The missing mass m_X in the process $\gamma p \to K^+K^+X$ from measurements at CLAS.[7] The plots have not been corrected for detector acceptance. a) $3.2 < E_\gamma < 3.9$ GeV; b) $3.0 < E_\gamma < 5.2$ GeV. The ground-state $\Xi^-(1321)$ is clearly seen in both plots; the signal-to-background ratio for the low-energy data set exceeds 10:1. In b), we see evidence for the photoproduction of the first excited state, the $\Xi^-(1530)$.

the existence of these states; if they do not exist, then our model of baryon structure is fundamentally flawed.

We may use our knowledge of the other baryons to guide our search for new Ξ states. For instance, the RPP lists three baryons that decay primarily via the emission of an η: the $N(1535)\frac{1}{2}^-$, the $\Lambda(1670)\frac{1}{2}^-$, and the $\Sigma(1750)\frac{1}{2}^-$. Based on this, Tuan has predicted the existence of a $\Xi(1870)\frac{1}{2}^-$ that decays in large part to $\Xi(1321)\eta$.[9] Finding such a state would suggest that these baryons belong in the same octet.

It is not sufficient to search for new Ξ states. Our knowledge of the Ξ^* spectrum is limited; the states we have found thus far are poorly-measured compared to the other baryon families. To improve this situation, it is

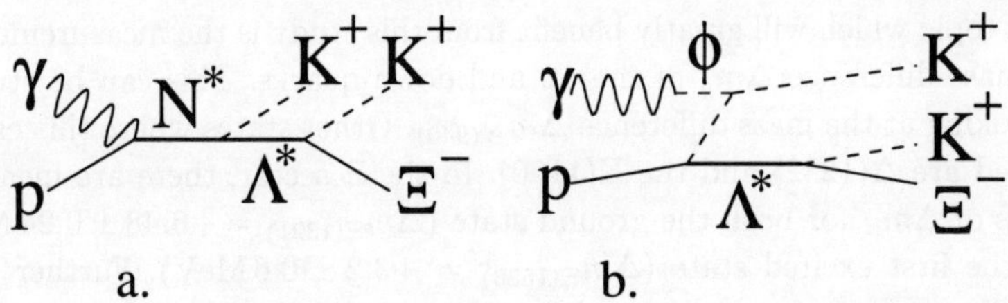

Fig. 3. Two possible mechanisms for the photoproduction of the Ξ hyperons. a) N^* excitation; b) K-exchange.

important to do a systematic survey of the known Ξ states.

An example of why this is necessary is the Ξ(1950). This state has three stars, which makes it a well-established state. However, its J^P value has never been measured, and its width is anomalously large in comparison with the other Ξ states in its vicinity. Additionally, there is a large disagreement among the fourteen separate measurements as to what the mass is, ranging from about 1.90 to 1.96 GeV. The Particle Data Group summarizes by saying "...there may be more than one Ξ near this mass."[4] A high-statistics study will help resolve the status of this state.

5.2. Ξ *production mechanism*

Unlike the N^*, Δ^*, Λ^*, and Σ^* states, the Ξ^* states cannot be produced directly in a reaction, but must instead be produced via some intermediate state. Two examples of possible production mechanisms are shown in Fig. 3. Fig. 3a shows an *s*-channel mechanism, in which the photon excites an N^* or Δ^* which decays to K^+Y^*, followed by $Y^* \to K^+\Xi^-$. Figure 3b shows another possibility, a *t*-channel mechanism in which the photon exchanges a kaon with the proton, which then becomes a Y^* and subsequently decays to $K^+\Xi^-$. These mechanisms will have different angular distributions and energy dependences; understanding this will require a very large data sample, along with substantial theoretical input.

5.3. *Search for the* Ξ^0 *states*

Detection of the Ξ^0 is accomplished using a reaction similar to that for the Ξ^-: $\gamma p \to K^+K^+\pi^-\Xi^0$. The π^- can come from any of the vertices in Fig. 3, so it will be difficult to isolate the production mechanism. Since the mass of the Ξ^0 must be near the mass of the associated Ξ^-, this will serve as a strong test of the existence of any Ξ.

A topic which will greatly benefit from this study is the measurement of the mass difference Δm_q of the up and down quarks. This can be studied by looking at the mass difference $\Delta m_{N(940)}$. Other states where this can be studied are $\Delta(1232)$ and the $\Sigma(1190)$. In the Ξ sector, there are measurements of Δm_Ξ for both the ground state ($\Delta m_{\Xi(1321)} = +6.48 \pm 0.24\,\text{MeV}$) and the first excited state ($\Delta m_{\Xi(1530)} = +3.2 \pm 0.6\,\text{MeV}$). Further work will allow a test of medium modifications, as we measure this quantity as a function of J^P.

5.4. Ξ *decays*

The decay modes of both the charged and neutral ground-state Ξ are fairly well-known. Both decay to $\Lambda\pi$ with a branching ratio of nearly 100%. There are several weak decays that may be used to test the $\Delta S = \Delta Q$ and the $\Delta S \neq 2$ rules, and the radiative decay may be used to test chiral perturbation theory.

All of the excited Ξ^*'s decay strongly. The $\Xi(1530)$ decays primarily to $\Xi\pi$; an upper limit of 4% has been measured for the branching ratio $BR[\Xi(1530) \rightarrow \Xi\gamma]$. There is not much that can be learned by improving on this limit.

In looking at the higher-mass Ξ^*'s, the $\Xi\pi\pi$ channel opens up at 1585 MeV, the $\overline{K}\Lambda$ channel at 1608 MeV, and the $\overline{K}\Sigma$ channel at 1682 MeV. By looking at all of these decay modes simultaneously, we can improve our ability to distinguish states close to each other in mass.

5.5. Ξ J^P *measurements*

Although the parity of the ground-state $\Xi(1321)$ has never been measured, it is assumed to be positive. Aside from the ground state, only the $\Xi(1530)\frac{3}{2}^+$ and the $\Xi(1820)\frac{3}{2}^-$ have J^P measurements. The other eight Ξ states lack such identification (although measurements indicate that the spin of the $\Xi(2030)$ is $\geq 5/2$. This determination is an important part of the study of the Ξ spectrum; only after the J^P is measured can a state be placed within an $SU(3)_F$ multiplet.

Partial-wave techniques have not been applied to this problem, although this may change in the next few years. Instead, there are various spin tests that have been applied. These make use of the angular distributions of the decay products, and were developed in the 1960s.

The method of Byers and Fenster[10] developed a technique to study the decay $B^* \to B\pi$, where B^* is an excited baryon, and B has spin-1/2. This method uses the angular distribution of the decay of the B^* followed by a measurement of the polarization of the ground-state B to determine J^P of the original B^*. This method was used to measure the spin and parity of the $\Xi(1530)$.[11]

Button-Shafer[12] extended this method to look at excited baryons that decay to a spin-3/2 baryon. This method was used to make the first estimate of the J^P of the $\Xi(1820)$, via the decay $\Xi(1820) \to \Xi(1530)\pi$.

Minnaert[13] took this a step further, by looking at multiple decay modes for a given baryon. This method exploits the fact that the spin of the excited baryon is independent of the decay mode. This method was used to improve the confidence in the J^P determination of the $\Xi(1820)$.

The amount of data anticipated should be sufficient to provide new information on the J^P values of several of the Ξ^* states. More theoretical input is needed to define the optimal procedure for the extraction of J^P for the Cascades.

5.6. *s-d quark mass difference*

A major issue in the quark model is the values of the input parameters, for instance the quark masses. A key parameter is the mass difference between the strange and the down quarks. This mass difference is built into the various mass relations for the baryons: the Gell-Mann-Okubo relation for the octet states $[2(m_N + m_\Xi) = 3m_\Lambda + m_\Sigma]$, and the Gell-Mann equal-spacing relation for the decuplet states $(m_\Sigma - m_\Delta = m_\Xi - m_\Sigma = m_\Omega - m_\Xi)$. Because there are only three Cascades with reliable J^P determinations, we can only test these relations with three $SU(3)_F$ multiplets. The main question of interest here is whether the mass difference $m_s - m_d$ is the same for the excited baryons as it is for the ground-state baryons.

5.7. *Cascade-proton scattering*

There has been very little work on Cascade-proton scattering. The main reason for this is the difficulty of extracting a signal. With a dedicated program in Cascade physics, however, one can make use of the relatively long lifetime of the Ξ ($c\tau_{\Xi^-} = 4.91\,\text{cm}$; $c\tau_{\Xi^0} = 8.71\,\text{cm}$) in conjunction with a long target to make detailed measurements of this process. There are two main reasons why this should be pursued.

The process $\Xi^- p \to \Lambda\Lambda$ has strong implications for hypernuclear physics. One of the currently interesting lines of research is in double hypernuclei, for which this process will provide much-needed input.

By comparing the process $\Xi^- p \to \Lambda\Lambda$ with the processes $\Xi^- p \to \Xi^0 n$ and $\Xi^- p \to \Lambda\Lambda\gamma$, we may be able to determine the parity of the ground-state Ξ relative to that of the nucleon.[14] This would be the first such measurement of this important quantity.

There is a reasonable measurement of these cross sections at high energy from CERN[15], but the data at lower energies are scarce.[16–19] A recent result from KEK[20] detected two candidate $\Xi^- p \to \Xi^- p$; from this, they calculate a 90% C.L. upper limit for the elastic scattering process to be 24 mb. Using three candidate $\Xi^- p \to \Lambda\Lambda$ events, they arrive at a cross section of $4.3^{+6.3}_{-2.7}$ mb. Both measurements used an initial sample of 6000 ground-state Ξ^-'s. With a larger data sample, a correspondingly better measurement could be achieved.

6. Summary

A systematic study of the Ξ^* states represents a "new window" on baryon spectroscopy. With the new results from CLAS, we have opened the window a crack, and what we have seen so far looks very encouraging. Whether one's interests lie in the nature of the Ξ^* spectrum, in Ξ^* production mechanisms or decay modes, or in other areas such as quark-mass differences or Ξp scattering, there are many topics in this field to keep the experimentalists busy for quite some time.

On the theoretical side, there is also much to do. While the experiments are sufficiently interesting to stand on their own, the results would have much more impact if there was a body of theoretical work available to be tested. Thus far, there has been little progress made in understanding the nature of the Ξ^* states. In particular, theoretical input is urgently needed on the production models of the Ξ^* states; without this input, it is very difficult to understand detector acceptances, which in turn makes the calculation of cross sections problematic. Another possible avenue of theoretical research lies in the lattice community, as the presence of two strange quarks in the Ξ^* may ease the calculation of reliable results.

To explore these and other topics related to Cascade physics, a workshop was organized at Jefferson Laboratory in early December 2005. As of this writing, this workshop has taken place, and was very successful. Over fifty people registered for the workshop, and a white paper on the issues related to Cascade physics is in preparation. It is hoped that we can form a core group of physicists, both experimental and theoretical, to carry out and extend the program described here.

References

1. S. Prakhov *et al.* (Crystal Ball Collaboration), Phys. Rev. C **72**, 015203 (2005).
2. S. Prakhov *et al.* (Crystal Ball Collaboration), Phys. Rev. C **69**, 042202 (2004).
3. B.M.K. Nefkens, these Proceedings.
4. S. Eidelman *et al.*, Phys. Lett. **B592**, 1 (2004), and 2005 partial update for the 2006 edition available on the PDG WWW pages (URL: http://pdg.lbl.gov/).
5. D.-O. Riska, in Proceedings of the 7th Workshop on Electron Nucleus Scattering, Isola d'Elba, Italy (24–28 June 2002), Eur. Phys. J. **A17**, 297 (2003).
6. S. Capstick, these Proceedings.
7. J.W. Price *et al.* (CLAS Collaboration), Phys. Rev. C **71**, 058201 (2005).
8. L. Guo, these Proceedings.
9. S.F. Tuan, Phys. Rev. D **46**, 4095 (1992).
10. N. Byers and S. Fenster, Phys. Rev. Lett. **11**, 52 (1963).
11. P.E. Schlein *et al.*, Phys. Rev. Lett. **11**, 167 (1963).
12. J. Button-Shafer, Phys. Rev. **139**, B607 (1965).
13. P. Minnaert, Phys. Lett. **B66**, 157 (1977).
14. S.B. Treiman, Phys. Rev. **113**, 355 (1959).
15. S.F. Biagi *et al.*, Z. Phys. **C17**, 113 (1983).
16. G.R. Charlton, Phys. Lett. **B32**, 720 (1970).
17. R.D.A. Dalmeijer *et al.*, Nuovo Cimento Letters **4**, 373 (1970).
18. R.A. Muller, Phys. Lett. **B38**, 123 (1972).
19. J.M. Hauptman *et al.*, Nucl. Phys. **B125**, 29 (1977).
20. J.K. Ahn *et al.*, arXiv:nucl-ex/0502010.

POLARIZATION OBSERVABLES IN THE PHOTOPRODUCTION OF TWO PSEUDOSCALAR MESONS

W. ROBERTS[1,2]

[1]*Department of Physics, Old Dominion University, Norfolk, VA 23529, USA*
and
Continuous Electron Beam Accelerator Facility
12000 Jefferson Avenue, Newport News, VA 23606, USA
[2] *On leave at the Office of Nuclear Physics, Department of Energy*
19901 Germantown Road, Germantown, MD 20874.

The many polarization observables that can be measured in process like $\gamma N \rightarrow M_1M_2B$, where M_1 and M_2 are pseudoscalar mesons and B is a spin-1/2 baryon, are discussed. The relationships among these observables, their symmetries, as well as inequalities that they satisfy are briefly discussed. Within the context of a particular model for $\gamma N \rightarrow NKK$, some of the observables are calculated, and their sensitivity to the ingredients of the model, and hence to the underlying dynamics of the process, are discussed.

1. Introduction and Motivation

Polarization asymmetries are an essential ingredient in the interpretation of various meson production reactions in terms of the various resonances that contribute to the processes as real or virtual intermediate states. For instance, much of the information that we have on the light baryon resonances has been garnered from pion-nucleon and kaon-nucleon scattering experiments. In addition, photoproduction experiments have augmented the database of measurements that provide information on these resonances. The differential and total cross sections for these reactions, together with various polarization observables, are used to extract the amplitudes for the process, and these are then interpreted as arising from a number of resonant and non-resonant contributions [1,2].

In this note, we briefly discuss a few aspects of polarization observables for processes in which two pseudoscalar mesons are photoproduced from a nucleon target. In the next section we briefly discuss the observables, the relationships among them, and the measurements needed for a 'complete' experiment. In section 3, we examine observables for $\gamma N \to K\bar{K}N$ in a model constructed using the phenomenological Lagrangian approach. In section 4, we present our conclusion and outlook.

2. Observables in $\gamma N \to M_1 M_2 B$

2.1. *Kinematics*

The center of mass kinematics for the process $\gamma N \to M_1 M_2 B$ are shown in fig. 1. k is the four-momentum of the incident photon, p that of the target nucleon and p' that of the recoil baryon. $\vec{k}$ is the three-momentum of the photon, which defines the z-axis, and the three-momenta $\vec{k}$ and $\vec{p}'$ define the $x-z$ plane. q_i is the four-momentum of meson M_i. The z''-axis is defined along the sum of the momenta $\vec{q}_1 + \vec{q}_2$. The angles Θ and Φ define the momentum $\vec{q}_1$ in the $x''y''z''$ coordinate system (the y- and y''-axes coincide).

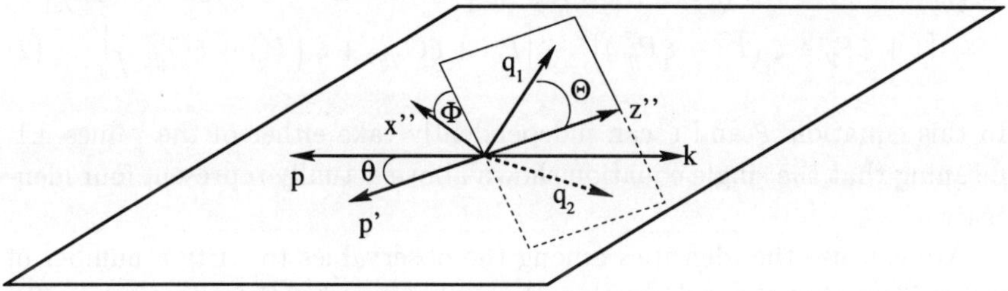

Fig. 1. Kinematics for the proces $\gamma N \to M_1 M_2 B$. The variables are explained in the text.

2.2. *Reaction Rate and Relationships Among Observables*

The reaction rate I for a process $\gamma N \to M_1 M_2 B$ can be written

$$\begin{aligned}\rho_f I = I_0 \Big\{ &\Big(1 + \vec{\Lambda}_i \cdot \vec{P} + \vec{\sigma} \cdot \vec{P}' + \Lambda_i^\alpha \sigma^{\beta'} \mathcal{O}_{\alpha\beta'}\Big) \\ &+ \delta_\odot \Big(I^\odot + \vec{\Lambda}_i \cdot \vec{P}^\odot + \vec{\sigma} \cdot \vec{P}^{\odot\prime} + \Lambda_i^\alpha \sigma^{\beta'} \mathcal{O}^\odot_{\alpha\beta'}\Big) \\ &+ \delta_\ell \Big[\sin 2\beta \Big(I^s + \vec{\Lambda}_i \cdot \vec{P}^s + \vec{\sigma} \cdot \vec{P}^{s\prime} + \Lambda_i^\alpha \sigma^{\beta'} \mathcal{O}^s_{\alpha\beta'}\Big) \\ &+ \cos 2\beta \Big(I^c + \vec{\Lambda}_i \cdot \vec{P}^c + \vec{\sigma} \cdot \vec{P}^{c\prime} + \Lambda_i^\alpha \sigma^{\beta'} \mathcal{O}^c_{\alpha\beta'}\Big)\Big]\Big\}, \end{aligned} \tag{1}$$

where $\vec{P}$ represents the polarization asymmetry that arises if the target nucleon is polarized, $\rho_f = \frac{1}{2}\left(1 + \vec{\sigma} \cdot \vec{P}'\right)$ is the density matrix of the recoiling nucleon, and $\mathcal{O}_{\alpha\beta'}$ is the observable if both the target and recoil polarization are measured. The primes indicate that the recoil observables are measured with respect to a set of axes x', y', z', in which z' is along the direction of motion of the recoiling nucleon, and $y' = y$ (the z'-axis is opposite the z''-axis, and the x'-axis is opposite the x''-axis). $\delta_\odot$ is the degree of circular polarization in the photon beam, while δ_ℓ is the degree of linear polarization, with the direction of polarization being at an angle β to the x-axis. The expressions for the observables in terms of the helicity or transversity amplitudes for the process are given in Ref. 3.

There are 28 relations that arise among the 64 polarization observables of Eqn. (1) from consideration of the absolute magnitudes of the helicity or transversity amplitudes, and another 21 that arise from consideration of their phases, leaving 15 independent quantities. One example of such a relationship is

$$\left[P_{x'} + \xi\mathcal{O}_{yx'} + \zeta\left(P_{x'}^{\odot} + \xi\mathcal{O}_{yx'}^{\odot}\right)\right]^2 + \left[P_{z'} + \xi\mathcal{O}_{yz'} + \zeta\left(P_{z'}^{\odot} + \xi\mathcal{O}_{yz'}^{\odot}\right)\right]^2$$
$$= \left[1 + \xi P_y + \zeta\left(I^{\odot} + \xi P_y^{\odot}\right)\right]^2 - \left[P_{y'} + \xi\mathcal{O}_{yy'} + \zeta\left(P_{y'}^{\odot} + \xi\mathcal{O}_{yy'}^{\odot}\right)\right]^2. \quad (2)$$

In this equation, ξ and ζ can independently take either of the values ± 1, meaning that the single equation shown above actually represent four identities.

We can use the identities among the observables to write a number of inequalities that the polarization observables must satisfy. From eqn. (2), for example, we obtain

$$\left|1 + \xi P_y + \zeta\left(I^{\odot} + \xi P_y^{\odot}\right)\right| \geq \left\{\left|P_{y'} + \xi\mathcal{O}_{yy'} + \zeta\left(P_{y'}^{\odot} + \xi\mathcal{O}_{yy'}^{\odot}\right)\right|,\right.$$
$$\left.\left|P_{x'} + \xi\mathcal{O}_{yx'} + \zeta\left(P_{x'}^{\odot} + \xi\mathcal{O}_{yx'}^{\odot}\right)\right|, \quad \left|P_{z'} + \xi\mathcal{O}_{yz'} + \zeta\left(P_{z'}^{\odot} + \xi\mathcal{O}_{yz'}^{\odot}\right)\right|\right\} \quad (3)$$

where ξ and ζ can again independently take either of the values ± 1. This inequality can be manipulated further to lead to

$$1 + P_y^2 + \left(I^{\odot}\right)^2 + \left(P_y^{\odot}\right)^2 \geq \left\{P_{y'}^2 + \mathcal{O}_{yy'}^2 + \left(P_{y'}^{\odot}\right)^2 + \left(\mathcal{O}_{yy'}^{\odot}\right)^2,\right.$$
$$\left.P_{x'}^2 + \mathcal{O}_{yx'}^2 + \left(P_{x'}^{\odot}\right)^2 + \left(\mathcal{O}_{yx'}^{\odot}\right)^2, \quad P_{z'}^2 + \mathcal{O}_{yz'}^2 + \left(P_{z'}^{\odot}\right)^2 + \left(\mathcal{O}_{yz'}^{\odot}\right)^2\right\}. \quad (4)$$

The full set of identities satisfied by the observables, as well as the various inequalities among them, are given in Ref. 3.

2.3. *Required Measurements*

We can examine which observables need to be measured in order to extract the helicity or transversity amplitudes. As there are eight such amplitudes, a minimum of eight measurements must be made at each kinematic point (each observable depends on 5 kinematic variables) to obtain the absolute magnitudes of the helicity or transversity amplitudes. In terms of the transversity basis used in Ref. 3, the measurements needed are the differential cross section, along with P_y, $P_{y'}$, $\mathcal{O}_{yy'}$, $I^{\odot}$, $P^{\odot}_y$, $P^{\odot}_{y'}$ and $\mathcal{O}^{\odot}_{yy'}$.

The eight phases of the transversity amplitude mean that there are seven independent phase differences that can be extracted, and a minimum of seven measurements are needed for this. For instance, four of the relative phases require measurement of any four of the eight observables $P_{x'}$, $P_{z'}$, $\mathcal{O}_{yx'}$, $\mathcal{O}_{yz'}$, $P^{\odot}_{x'}$, $P^{\odot}_{z'}$, $\mathcal{O}^{\odot}_{yx'}$ and $\mathcal{O}^{\odot}_{yz'}$. Two more phases may then be extracted from measurement of any two observables from among P_x, P_z, $\mathcal{O}_{xy'}$, $\mathcal{O}_{zy'}$, $P^{\odot}_x$, $P^{\odot}_z$, $\mathcal{O}^{\odot}_{xy'}$ and $\mathcal{O}^{\odot}_{zy'}$. The remaining independent phase can then be extracted from one of the observables that arise from linearly polarized photons. A 'complete' set of experiments will therefore require measurement of single, double and triple polarization observables, using both linearly and circularly polarized photon beams, in addition to the differential cross section.

2.4. *Parity Invariance*

Unlike the case in which a single pseudoscalar meson is produced, the invocation of parity invariance does not reduce the number of independent helicity (or transversity) amplitudes needed to describe the process. The relationships that are satisfied by the helicity amplitudes for $\gamma N \to M_1 M_2 B$ are

$$\mathcal{M}^{-\lambda_\gamma}_{-\lambda_N -\lambda'_N}(\theta,\Theta,\Phi) = (-1)^{\lambda_\gamma-\lambda_N+\lambda'_N}\mathcal{M}^{\lambda_\gamma}_{\lambda_N\lambda'_N}(\theta,\Theta,2\pi-\Phi). \qquad (5)$$

These relations can not be used to decrease the number of independent helicity amplitudes, since they relate different helicity amplitudes at different kinematic points. However, these relationships can be used to determine which observables are even or odd under the transformation $\Phi \leftrightarrow 2\pi - \Phi$.

Examination of the full set of observables yields,

$$\begin{array}{llll} I_0 = -\mathcal{O}^c_{yy'} & P_y = -P^c_{y'} & P_{y'} = -P^c_y & \mathcal{O}_{xx'} = -\mathcal{O}^c_{zz'} \\ \mathcal{O}_{xz'} = \mathcal{O}^c_{zx'} & \mathcal{O}_{yy'} = -I^c & \mathcal{O}_{zx'} = \mathcal{O}^c_{xz'} & \mathcal{O}_{zz'} = -\mathcal{O}^c_{xx'} \\ P^{\odot}_x = \mathcal{O}^s_{zy'} & P^{\odot}_z = -\mathcal{O}^s_{xy'} & P^{\odot}_{x'} = -\mathcal{O}^s_{yz'} & P^{\odot}_{z'} = \mathcal{O}^s_{yx'} \\ P^s_x = -\mathcal{O}^{\odot}_{zy'} & P^s_z = \mathcal{O}^{\odot}_{xy'} & P^s_{x'} = \mathcal{O}^{\odot}_{yz'} & P^s_{z'} = -\mathcal{O}^{\odot}_{yx'} \end{array} \qquad (6)$$

all at $\Phi = 0$, $\Phi = \pi$, $\Phi = 2\pi$. All of the observables in Eqn. (6) are even under the transformation in Φ. All other observables are odd under the transformation in Φ, and vanish at $\Phi = 0$, $\Phi = \pi$ and $\Phi = 2\pi$.

3. Model Calculation

We now examine two aspects of the observables for a specific process, in a specific model, in an attempt to demonstrate their rich structure and their sensitivity to the underlying dynamics of the process. The process we treat is $\gamma N \to K\bar{K}N$, in a phenomenological Lagrangian approach discussed in detail in Ref. 4.

The polarization observables for this process are five-fold differential, which means that there are a number of different ways in which they can be displayed. Since it is not obvious how to display five-fold differential quantities, it is usual to integrate over some of the independent variables. In the following subsections, we integrate over some of the kinematic variables, showing the resulting observables as curves for different values of Φ.

We note that since the observables are either even or odd under the transformation $\Phi \leftrightarrow 2\pi - \Phi$, we do not integrate over this variable. In all the plots that follow, the observables are shown for four values of Φ: $\pi/6$, $\pi/4$, $\pi/3$ and $\pi/2$. We note that observables that are even in the Φ transformation could be displayed as Dalitz plots.

3.1. *Sub-threshold Resonance*

The $\Lambda(1405)$ resonance is one of the relatively well-established hyperons. It lies just below the NK threshold, so its coupling (to NK) is not very well known. We examine the sensitivity of a number of observables to the presence of this state in the model calculation, and show the results for one observable, $\mathcal{O}^c_{yz'}$, in fig. 2. The process process is $\gamma p \to nK^+\overline{K}^0$. In this figure, the results when the $\Lambda(1405)$ is excluded are shown in the left panel. The right panel shows the results we obtain when this state is included.

In this figure, the curves with the $\Lambda(1405)$ included are very different from those without it, especially near the lowest values of $m_{N\overline{K}}$. This feature serves to illustrate that in calculations such as this, 'small' contributions may not affect the cross section much, but can have significant effects in polarization observables.

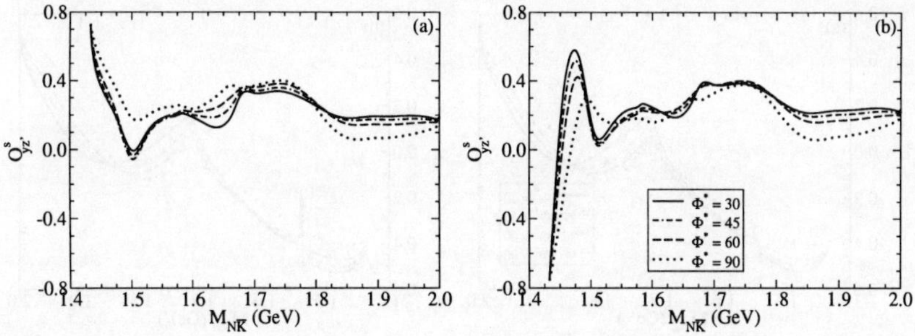

Fig. 2. The observable $\mathcal{O}^c_{yz'}$, showing its sensitivity to the sub-threshold resonance $\Lambda(1405)$. (a) results with the $\Lambda(1405)$ excluded in the calculation. (b) results when this state is included. All curves are shown as functions of $m_{N\overline{K}}$.

3.2. *Pentaquark Search*

One very interesting question regarding the polarization observables is their possible sensitivity to exotic resonances, such as the Θ^+ [5]. If observables are found that show sensitivity to this state, they can be used to confirm its existence (or otherwise), assuming production mechanisms like those presented in Ref. 4. One of the disadvantages of using the differential cross section to search for states like this is that one state (or a few states) may provide a very large background against which a small signal must be sought. In the case of the pentaquark searches, the large backgrounds are provided by the $\Lambda(1520)$, along with other non-exotic hyperons, as well as mesons like the ϕ. With polarization observables, large 'backgrounds' are not necessarily a problem, and the curves that we show illustrate what might be possible in pentaquark (or similar) searches.

In the model that we use for these calculations, the production cross section for the Θ^+ is of the order of a few nanobarns, consistent with the upper limit recently announced by researchers at JLab [5]. In the calculation, this size of cross section assumes that the pentaquark has $J^P = 1/2^+$, and that mechanisms involving the K^* are not important. In the same framework, the cross section obtained is significantly smaller if the state has $J^P = 1/2^-$. In either case, the cross sections for producing the non-exotic hyperons, particularly the $\Lambda(1520)$ are several hundred times larger, and would contribute to the difficulty of extracting a Θ^+ signal, if the state were to exist.

Figs. 3 to 5 ($P_x^\odot$, $P_z^\odot$ and $I^\odot$, respectively) show the curves that result when there is no Θ^+ in the calculation (the curves in (a)), and when a Θ^+ with $J^P = 1/2^+$ is included (the curves in (b)). In the case of Fig.

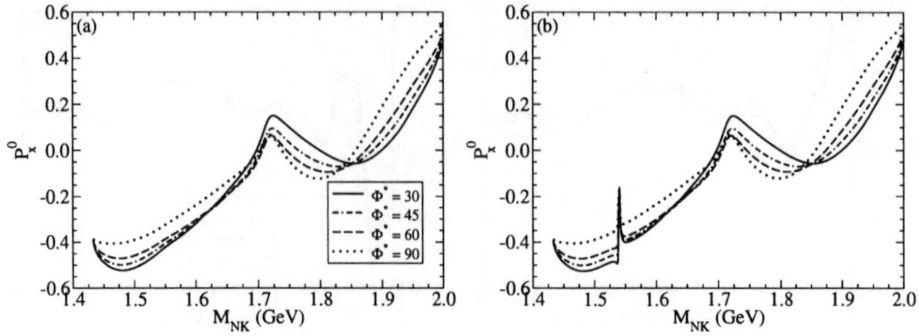

Fig. 3. The observable $P_x^{\odot}$, showing its sensitivity to the exotic resonance Θ^+. (a) results when the Θ^+ is excluded from the calculation. (b) results when a Θ^+ of positive parity is included in the calculation. All curves are shown as functions of m_{NK}.

5, the curves in (c) result when a Θ^+ with $J^P = 1/2^-$ is included in the calculation. In each case, the process is $\gamma p \to nK^+\overline{K}^0$. In fig. 5, the helicity asymmetry without the pentaquark is small, but recent work has demonstrated that even such a small observable is measurable with high precision at JLab [6]. When the pentaquark is included in the calculation, this observable remains small, except for a structure in the region of the invariant mass of the pentaquark.

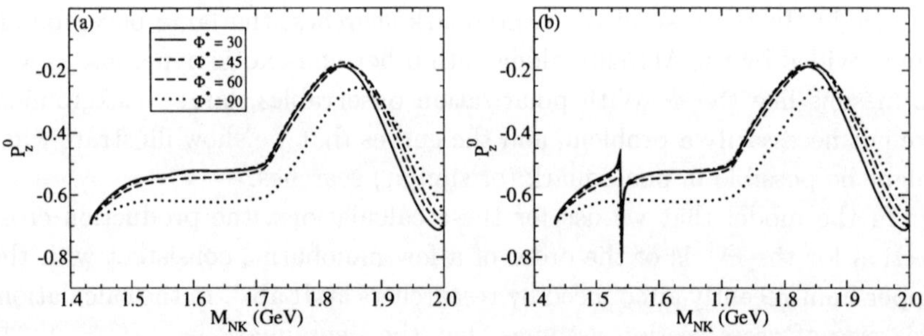

Fig. 4. The observable $P_z^{\odot}$, showing its sensitivity to the exotic resonance Θ^+. (a) results when the Θ^+ is excluded from the calculation. (b) results when a Θ^+ of positive parity is included in the calculation. All curves are shown as functions of m_{NK}.

For a pentaquark of positive parity, the signal in Fig. 5 is significant, but extraction could still be a challenge, as the width of the 'structure' is similar to the width of pentaquark (here, we are using a width of 1 MeV). For a pentaquark with negative parity, the theoretical curves also show a significant structure, but it is somwehat less so than for the case of a positive parity pentaquark. One striking feature here is the difference in the 'sign' of the signal between the positive and negative parity cases, suggesting that this observable could act as an excellent parity filter for the pentaquark. Note that this asymmetry has already been measured at JLab for $\gamma p \to p\pi^+\pi^-$ [6]. Thus, it may be possible to measure it for $\gamma N \to NK\overline{K}$ relatively quickly.

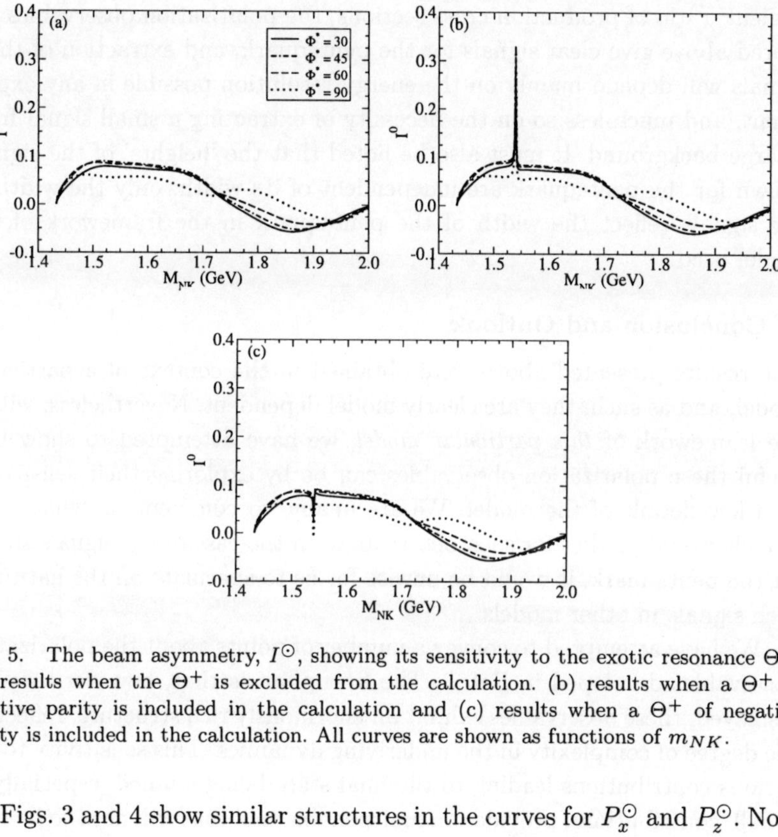

Fig. 5. The beam asymmetry, $I^{\odot}$, showing its sensitivity to the exotic resonance Θ^+. (a) results when the Θ^+ is excluded from the calculation. (b) results when a Θ^+ of positive parity is included in the calculation and (c) results when a Θ^+ of negative parity is included in the calculation. All curves are shown as functions of m_{NK}.

Figs. 3 and 4 show similar structures in the curves for $P_x^{\odot}$ and $P_z^{\odot}$. Note that in all cases, the structures stand out clearly for two reasons. The first is that the pentaquark is a narrow state, and the 'width' of any structure that might be observed will be similar to that of the state giving rise to the structure. The second reason is that the Θ^+ is the *only* resonance in the nK^+ channel. All other resonances are in the $n\overline{K}^0$ channel. The kinematic reflections from these resonances will show up, as can be seen in figs 3 and 4, but the presence of the Θ^+ in this channel has a marked effect. Note that in fig. 4, this observable is predicted to be large (in the framework of the model used) and negative. A number of other observables utilizing linearly polarized photons (not shown here) show similar strctures for the pentaquark.

It must be emphasized here that the potential signals shown above arise in the full model, including the contribution of the $\Lambda(1520)$. In the model, the cross section for production of this state is several hundred times larger than the cross section for producing the pentaquark. Despite this 'inconvenient' ratio of production cross sections, the polarization observables explored above give clear signals for the pentaquark, and extraction of these signals will depend mainly on the energy resolution possible in any experiment, and much less so on the necessity of extracting a small signal from a large background. It must also be noted that the 'heights' of the signals shown for the pentaquark are independent of its width: only the width of the signals reflect the width of the pentaquark in the framework of the model used.

4. Conclusion and Outlook

The results presented above were obtained in the context of a particular model, and as such, they are clearly model dependent. Nevertheless, within the framework of *this particular model*, we have attempted to show how useful these polarization observables can be by exploring their sensitivity to a few details of the model. We are unable to comment on what other models would predict for such observables. In the case of the signals shown for the pentaquark, it would be unwise for us to speculate on the nature of such signals in other models.

We have attempted to convey a number of points about the polarization observables developed in Ref. 3. The first point is that, however they are displayed, these observables exhibit an enormously rich structure, reflecting the degree of complexity in the underlying dynamics. This sensitivity to the various contributions leading to the final state being studied, especially to 'small' contributions, provides an indispenable tool that will need to be fully exploited in our attempts to understand processes like the ones discussed herein. Such processes are expected to be among the primary sources of information in the on-going attempts to understand the dynamics of soft QCD.

A number of these observables should be accessible in the near future at existing facilities, in a number of different processes. The obvious applications are to the process discussed herein, $\gamma N \to NK\overline{K}$, and to $\gamma N \to N\pi\pi$. However, final states like $N\eta\pi$, $N\eta\eta$, $YK\pi$ (where Y is a Λ or Σ), $YK\eta$, and even $KK\Xi$, will require the same kinds of measurements in order to disentangle the various contributions leading to them. In the processes that produce hyperons in the final states, their various self-analysing decays provide access to recoil polarization measurements, thus opening up more possibilities. Many of these opportunities will have to be seized for continued progress to be made in our understanding of baryon spectroscopy.

Acknowledgements

This work was supported by the Department of Energy through contract DE-AC05-84ER40150, under which the Southeastern Universities Research Association (SURA) operates the Thomas Jefferson National Accelerator Facility (TJNAF).

References

1. See, for example, B. H. Bransden and R. G. Moorhouse, *The Pion Nucleon System*, Princeton University Press, New Jersey, 1973.
2. See, for example, J. S. Hyslop, R. A. Arndt, D. Roper and R. L. Workman, *Phys. Rev.* **D46**, 961 (1992);
   B. R. Martin, *Nucl. Phys.* **B94**, 413 (1975).
3. W. Roberts and T. Oed, *Phys. Rev.* **C71**, 055201 (2005).
4. W. Roberts, *Phys. Rev.* **C70**, 065201 (2004).
5. See for example, V. Burkert, these Proceedings, and references therein;
   M. Battaglieri, *et al.* [CLAS Collaboration], arXiv:hep-ex/0510061.
6. S. Strauch *et al.* [CLAS Collaboration], *Phys. Rev. Lett.* **95** 162003 (2005).

THE POLARISATION PROGRAMME AT ELSA *

H. SCHMIEDEN
University of Bonn, Germany
E-mail: schmieden@physik.uni-bonn.de

One of the central issues of contemporary hadron physics is the understanding of the strongly interacting many–body structure of proton and neutron which is closely related to the nucleons excitation spectrum. In order to disentangle the broad, overlapping states, polarisation observables are indispensable in photoproduction experiments. Their importance is discussed along with selected projects at ELSA using the Crystal Barrel detector in conjunction with polarised beams and targets.

1. Introduction

Similar to the role of atomic spectroscopy at the threshold to the era of quantum mechanics almost a century ago, the ultimate goal of baryon spectroscopy today is to reveal the inner dynamics and to identify the relevant degrees of freedom of the strongly bound quark systems, especially protons and neutrons. This is an unsolved matter despite the common belief that QCD represents the basic underlying theory of quark and gluon interactions — the non-linear and strong couplings provide an unsurmountable obstacle yet, despite the first successes of Lattice calculations.

Constituent quark models using different residual interactions have basic problems in common. Individual states are not well described — examples are (i) the level inversion of the lowest-lying $N = 1$ and 2 harmonic oscillator states, the famous $N_{1/2^+}$ (1440) P_{11} and $N_{1/2^-}$ (1535) S_{11} nucleon excitations, and (ii) the large splitting between the angular momentum partners $\Lambda(1520)$ and $\Lambda(1405)$. The second major puzzle concerns the number of states. Currently, much more are predicted than experimentally manifest. It is unclear whether this points to a serious deficiency of the models, since correlations or channel couplings are missing, or to incomplete experiments,

*This work is supported by SFB/TR–16 of the *Deutsche Forschungsgemeinschaft*

since nucleon spectroscopy is very much based on pionic reactions. States with weak π–N coupling might thus have remained undetected yet.

In contrast to atomic spectroscopy, the experimental challenge in nucleon spectroscopy is to disentangle a large number of broad and overlapping states. This can only be accomplished using partial wave analyses. However, their unambiguousness is very much limited by the number of available independent observables. A major step ahead is expected from the measurement of double polarisation observables in several meson-photoproduction channels even though the "complete set of experiments" will be very difficult to achieve even for the simplest reactions.

2. The Role of Polarisation Observables

Several new resonances have already been discussed in the photoproduction of non-pionic final states, e.g. $D_{13}(1900)$ in $K^+\,\Lambda(1116)$ [1] or $S_{11}(1780)$ in ηp [2]. The latter reaction seems particularly interesting, since in the vicinity of the threshold it is dominated by two S_{11} s–channel resonances at 1535 and 1650 MeV mass, and t–channel contributions are small. However, as in case of $K^+\,\Lambda$, the partial wave analyses were not sufficiently constrained for definitive conclusions about the existence of the third S_{11} at 1780 MeV.

Table 1. Observables in the photoproduction of pseudoscalar mesons off the nucleon. σ_0 is the unpolarised cross section. The photon–beam asymmetry Σ exhibits with linearly polarised photon beam. The other polarisation observables require a polarised target in the initial state (target) or the observation of the recoil baryon polarisation in the final state (recoil) or both (target–recoil). x, y, z denote the cartesian directions with regard to the photon beam as z–axis. In the primed frame the z'–axis is aligned along the direction of the recoiling nucleon.

photon		target			recoil			target-recoil			
		x	y	z	x'	y'	z'	x' x	x' z	z' x	z' z
unpol.	σ_0	0	T	0	0	P	0	$T_{x'}$	$-L_{x'}$	$T_{z'}$	$L_{z'}$
lin.pol.	$-\Sigma$	H	-P	-G	$O_{x'}$	-T	$O_{z'}$	$-L_{z'}$	$T_{z'}$	$-L_{x'}$	$-T_{x'}$
circ.pol.	0	F	0	-E	$-C_{x'}$	0	$-C_{z'}$	0	0	0	0

This also holds for the η photoproduction off the proton measured at ELSA. Based on differential and total cross sections up to $E_\gamma = 3$ GeV, no evidence was found for the third S_{11} but, instead, for a yet unobserved $N(2070)D_{15}$ state at 2070 ± 22 MeV with a width of $\Gamma = 295 \pm 40$ MeV

[3,4]. As will be shown below, double polarisation observables with linear polarised photon beam and longitudinally polarised proton target are expected to greatly reduce the ambiguities of the partial wave analysis.

In general, photoproduction of pseudoscalar mesons, as for example $\gamma + p \rightarrow \eta + p$, is characterised by $2 \times 2 = 4$ possible helicity combinations in the initial state and $1 \times 2 = 2$ in the final state, hence 8 complex amplitudes which reduce to 4 due to parity conservation. To fully determine the reaction up to an arbitrary phase 7 independent quantities need thus to be determined. To circumvent sign ambiguities from quadratic contributions to the cross sections, at minimum 8 carefully selected observables must be measured which include 4 beam–target and target–recoil quantities [5].

Polarisation observables for photoproduction of a single pseudoscalar meson are summarised in Table 1. In addition to the unpolarised cross section, σ_0, the photon–beam asymmetry Σ is accessible with linearly polarised photon beam. A polarised target in combination with linear and circular polarised beams provides sensitivity to six further observables, usually labelled [6] T, H, P, G, F, and E, and the addition of recoil polarimetry enables, in principle, measurement of 14 further single and double polarisation observables. According to the helicity considerations above, the observables in Table 1 can not all be independent [7].

The "complete experiment" with regard to a PWA seems out of reach for photo–reactions beyond production of single pseudoscalar mesons. Double pseudoscalar meson photoproduction already requires more than 15 independent observables to be measured, photoproduction of vector mesons more than 23. But even a complete set of observables does not guarantee an unambiguous PWA solution in an individual reaction, see [8]. To resolve continuum ambiguities the cross relation between different coupled channels must be considered. Thus, the unambiguous determination of resonance contributions remains a formidable task in all photoreactions, and in most — if not all — cases will be impossible to achieve. Nevertheless, polarisation observables are the only tool to reduce the remaining ambiguities to the minimum, maybe completely in the simplest cases.

3. Realisation of Experiments at ELSA

In order to realise close-to-complete experiments in the sense of the preceding section, it is necessary to provide a photon beam of sufficient energy to cover the full nucleon resonance–region, linear and circular polarisation including beam polarimetry, a polarised target in conjunction with a 4π detector setup, and recoil polarimetry. While the latter is only available for weak decays in the final state, e.g. in the $\gamma + p \rightarrow K_s^0 + \Sigma^+$ reaction [9], all other experimental tools are routinely available, or will be shortly at disposal at the Electron Stretcher Accelerator ELSA of the University of Bonn.

3.1. *The Electron Accelerator Facility*

The layout of ELSA is shown in Figure 1. After 20 MeV pre-acceleration in dedicated LINACs, unpolarised or polarised electrons are injected into the Bonn booster synchrotron. The pulsed beam continuously fills the stretcher ring, where the circulating beam can be ramped up in energy to maximum 3.5 GeV. It is then slowly extracted with typically several nA current and duty factors of about 80 % to one of two experimental areas.

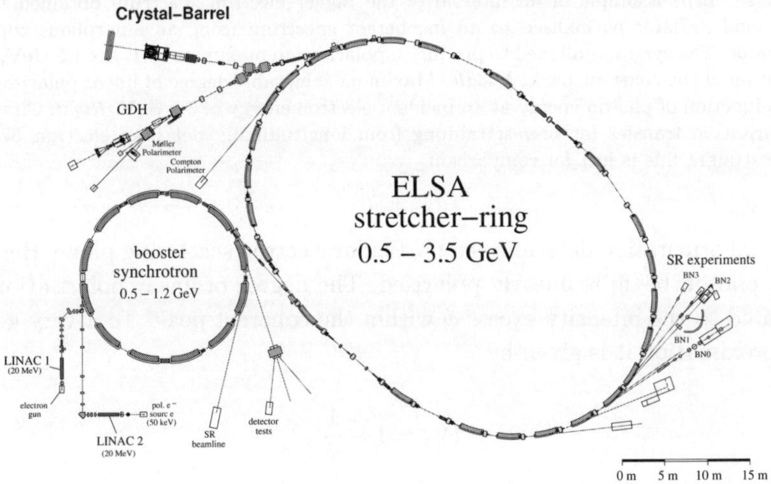

Fig. 1. Layout of the ELectron Stretcher Accelerator ELSA at Bonn.

3.2. *Polarised Photon Beams*

Impinging on a radiator target, the extracted electron beam produces a bremsstrahlung photon beam. The photon energy is tagged by the momentum specific detection of the decelerated electrons by means of a dipole spectrometer. This tagging system covers the energy range of 18 – 95 % of the incoming electron beam energy [10] with a resolution of $\delta E \simeq 15$ MeV. Linear polarisation is obtained using a diamond crystal as radiator [11]. Similar to the Mössbauer effect, under certain kinematic conditions depending on the crystal alignment relative to the electron beam, the crystal lattice instead of individual nuclei takes the bremsstrahlung recoil. This generates a coherence effect in the otherwise continuous $1/E_\gamma$ bremsstrahlung spectrum (see Figure 2 left). Within the energy region of the coherent peak the

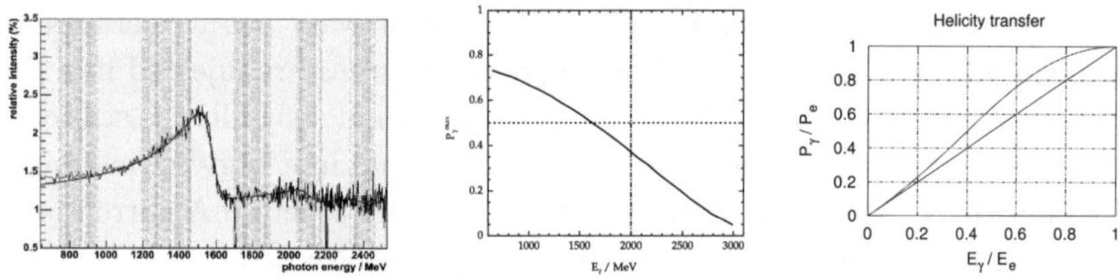

Fig. 2. *Left:* Example of an interval of the tagger electron spectrum obtained with diamond radiator normalised to an incoherent spectrum using an amorphous copper radiator. The cyrstal is aligned to produce a polarisation maximum at $E_\gamma \simeq 1.5$ GeV, the position of the coherent peak. *Middle:* Maximum achievable degree of linear polarisation as a function of photon energy at an incident electron energy of 3.5 GeV. *Right:* Circular polarisation transfer for bremsstrahlung from longitudinally polarised electron beam. The straight line is just for comparison.

crystal orientation determines a particular electron scattering plane. Hence, the photon beam is linearly polarised. The degree of linear polarisation is related to the intensity excess ϕ within the coherent peak. To a very good approximation it is given by

$$P_{\text{lin}} = 1 - \frac{1}{\phi} , \tag{1}$$

provided that only one reciprocal lattice vector contributes and the photon beam is not collimated. The maximum achievable degree of linear polarisation depends on the energies of incident electron and outgoing photon. It is plotted in the middle part of Figure 2 for 3.5 GeV electron beam. Experimentally, the degree of linear polarisation is essentially determined from the tagger electron–spectrum via Equation 1. To measure the polarisation independently, the setup of a prototype photon–polarimeter is planned [12].

Circularly polarised photon beam is obtained from bremsstrahlung of longitudinally polarised electrons. The helicity transfer depends on the ratio

E_γ/E_e of outgoing photon to incoming electron energy [13]:

$$\frac{P_{\gamma,\mathrm{circ}}}{P_e} = \frac{E_\gamma}{E_e} \frac{1+\frac{1}{3}(1-E_\gamma/E_e)}{1-\frac{2}{3}(1-E_\gamma/E_e)+(1-E_\gamma/E_e)^2}. \tag{2}$$

It is depicted in Figure 2 (right).

Using Equation 2, the degree of circular polarisation can be calculated once the polarisation P_e of the electron beam is known. The latter will be measured by a Møller polarimeter integrated into the tagging spectrometer [14]. Alternatively, in double polarisation experiments it is possible to directly determine the product of beam and target polarisation, $P_{\gamma,\mathrm{circ}} \cdot P_T$, through the measurement of the beam–helicity asymmetry in $\gamma + p \to \eta + p$. The analysing power of this reaction is close to 1 in a sufficiently wide energy interval above threshold, due to the dominance of S_{11} intermediate resonance states [15].

3.3. *The Bonn Frozen Spin Target*

The Bonn frozen spin target [16] provides longitudinally polarised protons or neutrons. The target material is 2 cm thick solid butanol [$H(CH_2)_4OH$] with an effective density of 0.57 g/cm^3, deuterated to obtain polarised neutrons. It is contained in a horizontal $^3He/^4He$ dilution cryostat. Using the process of dynamic nuclear polarisation, the free protons/deuterons are polarised up to 80 % in a strong magnetic field of about 6 T. At a temperature reduced to 55 mK relaxation times of approximately 200 hours and average polarisations around 70 % have already been achieved. In this frozen spin mode the external polarisation magnet is replaced by an internal superconducting 0.42 T "holding coil" which provides virtually no obstruction for outgoing particles within the full 4π detector acceptance.

3.4. *The planned Double Polarisation Setup at ELSA*

The complete setup as it is planned for double polarisation experiments at ELSA is shown in Figure 3. Polarised or unpolarised electrons from the right hit the radiator target, diamond crystal or amorphous copper, in the vacuum chamber of the goniometer. The electrons subsequently are horizontally deflected in the Tagger dipole magnet. Coincident detection in a 480 channel scintillating–fibre array and 96 overlapping scintillator strips of variable width ensures a resolution in photon energy of $\simeq$ 15 MeV. A further dipole magnet behind the tagger deflects the exiting electron beam.

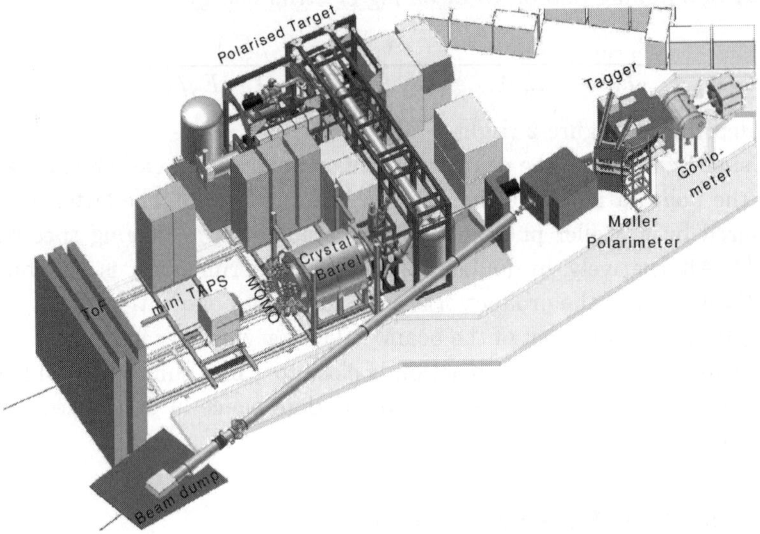

Fig. 3. Planned setup for double polarisation experiments at ELSA.

After collimation and cleaning from charged particles in a sweeping magnet, the photon beam hits the polarised target at the centre of the Crystal Barrel detector. In this setup the central barrel covers the polar angular range from 30 to 156 degrees over the full azimuth. It consists of 1230 individual CsI(Tl) crystals of 16 radiation lengths with photodiode readout. Energy and angular resolutions of $\delta E/E = 2.5\,\%/\sqrt[4]{E/\mathrm{GeV}}$ and $\delta\Theta \simeq 2\,\mathrm{deg}$ are achieved.

The open 30 degree forward cone allows experiment-specific detector extensions. This region will be covered by two further photon detectors during the first round of double polarisation experiments: (1) A CsI(Tl) array of 90 crystals of the central barrel type, but with photomultiplier readout and individual plastic vetos for charged particle recognition [17], and (2) a 216 element BaF_2 array, the *Mini–TAPS* detector, also equipped with photomultipliers and individual plastic vetos. In contrast to the central barrel itself, this forward configuration allows fast first level triggering on neutral and charged particles. Charged particles are also detected in a three-layer cylinder of scintillating fibres which surrounds the target in the polar angular range of the central barrel [18], and in the plane scintillating-fibre detector MOMO in forward direction [19]. At a distance of 5 m from the target, the time-of-flight of neutrons and punch-through charged particles is determined in a 3×3 m^2 array of plastic scintillators.

4. Selected Experiments

4.1. *η–Photoproduction off Proton and Neutron*

The recently evidenced D_{15} state at 2070 MeV (c.f. section 2) is expected to exhibit its existence much clearer in double polarisation observables than in differential and total cross section measurements [20]. As an example, Figure 4 (left) shows the G–asymmetry, related to longitudinally polarised target and linearly polarised photon beam (c.f. Table 1). Further sensitivity is provided by the helicity–asymmetry discussed in reference [4].

Off the neutron, η photoproduction exhibits a significant enhancement over the proton target at centre-of-mass energies around 1675 MeV, c.f. Figure 4 (middle). In the MAID parametrisation such an effect is attributed to the specific coupling to the $D_{15}(1675)$ nucleon resonance. Its pronounced effect on the angular distributions, however, seems only partly reproduced in experiment [21]. As illustrated in Figure 4 (right), the measurement of G provides a very high sensitivity to investigate the role of this resonance [22].

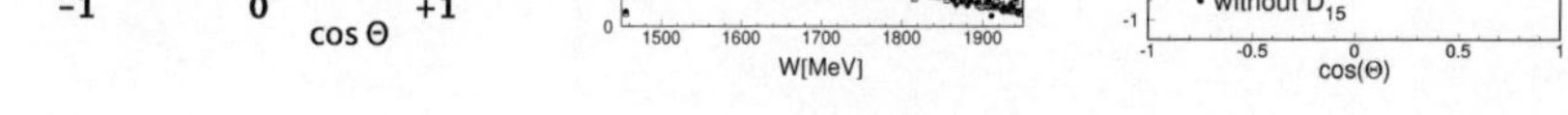

Fig. 4. *Left:* G–asymmetry in η–photoproduction off the proton at $E_{cm} = 2075$ MeV. Full curve is with $D_{15}(2070)$, dotted (dashed) with $1/2^{-(+)}$ partial waves instead. *Middle:* Total cross section of the reaction $\gamma + n \to \eta + n$ using a deuteron target (full points) in comparison to proton target (open points). *Right:* MAID based simulation of the G–asymmetry with and without the $D_{15}(1675)$ state.

4.2. *Helicity–Difference in $\pi^0\pi^0$ Photoproduction*

Unpolarised Crystal Barrel data of the $\gamma + p \to \eta + p$ reaction [23], analysed in conjunction with $2\,\pi^0$ photoproduction [24], showed some evidence for a

$\Delta(1940)D_{33}$ resonance contributing to the cross sections. This is a particular interesting state, since it belongs to a series of negative parity Δ's in the vicinity of 1900 MeV which seem to form (at least partially) mass–degenerate doublets with positive parity states. It is under debate whether this is due to a restoration of chiral symmetry in this high mass range or to features of the intra–quark dynamics.

Experimentally, the existence of these states is by no means firm. It turns out that the helicity difference in $2\,\pi^0$ photoproduction is very sensitive to this resonance, as is illustrated in Figure 5 (left) [25].

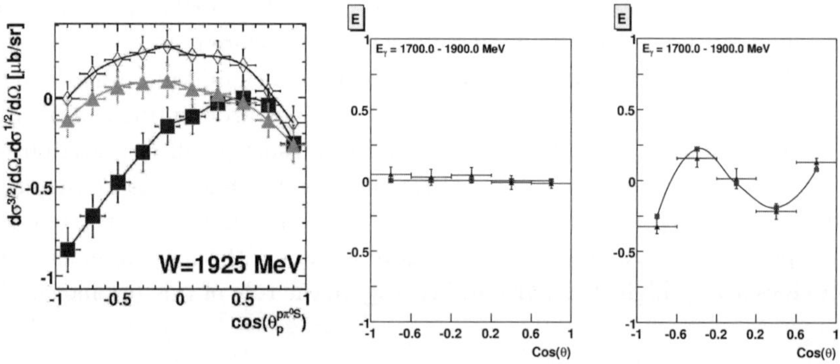

Fig. 5. *Left:* Helicity $3/2 - 1/2$ difference in the $2\,\pi^0$ photoproduction off the proton. Errors are projected to 800 hours of beamtime. Diamonds correspond to full PWA, triangles without $P_{33}(1920)$ and squares without $D_{33}(1940)$. *Middle and Right:* Beam–target asymmetry in ω photoproduction off the proton in the energy interval $E_\gamma = 1700 - 1900$ MeV for a pure π^0–exchange reaction mechanism (middle) and involving resonance contributions (right). Errors are expected for $\simeq 2000$ h beamtime.

4.3. *Photoproduction of ω Mesons*

The photoproduction of vector mesons provides some discovery potential for so far unobserved "missing" resonances. ω production appears particularly suited, since the reaction mechanism is best understood of all vector meson channels. The complicated helicity structure of vector–meson photoproduction prohibits a "complete" experiment in the sense of section 2, in this respect however only little different from multi–pseudoscalar final states like $p\;2\,\pi^0$ or $p\;\pi^0\,\eta$ as discussed above. It is possible to identify polarisation observables with very high sensitivity to both the basic reaction mechanism and, in particular, the contribution of s–channel resonances [15].

For one example this is illustrated in Figure 5. On basis of the partial wave analysis [20] of existing ω cross section data [26], the beam–target asymmetry, E, is shown in the photon energy interval $E_\gamma = 1700 - 1900$ MeV for two situations: (a) pure π^0–exchange mechanism without resonance contributions [27] (middle) and (b) involving resonant $5/2^+$ and $1/2^+$ partial waves (right). In both scenarios the cross sections are equally well reproduced, but the differences in the beam–target asymmetry E are significant.

5. Summary

To experimentally disentangle resonances in the final states of photoreactions the measurement of single and double polarisation observables is absolutely required. The Crystal Barrel setup at ELSA is presently extended to perform such experiments with polarised photon beams and longitudinally polarised proton and neutron targets. Measurements of several polarisation observables are intended in the single and double photoproduction of neutral pseudoscalar mesons as well as of vector mesons.

References

1. T. Mart and C. Bennhold, *Phys. Rev.* **C61**, 012201 (2000)
2. B. Saghai and Zhen-ping Li, NSTAR2002, nucl-th/0305004
3. V. Crede et al., *Phys. Rev. Lett.* **94**, 012004 (2005)
4. U. Thoma, these proceedings
5. W.–T. Chiang and F. Tabakin, *Phys. Rev.* **C55**, 2054 (1997)
6. D. Drechsel and L. Tiator, *J. Phys. G: Nucl. Part. Phys.* **18**, 449 (1992)
7. G. Knöchlein, D. Drechsel and L. Tiator, *Z. Phys.* **A 352**, 327 (1995)
8. A. Svarc, these proceedings
9. R. Castelijns, PhD. thesis, KVI Groningen (2005)
10. K. Fornet–Ponse, doctoral thesis, Bonn, in preparation
11. D. Elsner, doctoral thesis, Bonn, in preparation
12. H. Schmieden, *Eurotag* objective O6 of EU framework programme FP6–I3HP
13. H. Olsen and L.C. Maximon, *Phys. Rev.* **114**, 887 (1959)
14. S. Kammer, doctoral thesis, Bonn, in preparation
15. H. Schmieden, proposal ELSA-4/2005 (2005)
16. Ch. Bradtke et al., *Nucl. Instr. Meth.* **A 436**, 430 (1999)
17. C. Wendel, V. Sokhoyan, doctoral theses, Bonn (in preparation)
18. G. Suft et al., *Nucl. Instr. Meth.* **A 538**, 416 (2005)
19. R. Joosten, doctoral thesis, Bonn (1996)
20. A. Sarantsev, priv. comm.
21. I. Jaegle, these proceedings
22. B. Krusche et al., proposal ELSA/03-2005 (2005)
23. I. Horn, doctoral thesis, Bonn (2004),
24. M. Fuchs, these proceedings

25. U. Thoma et al., proposal ELSA/06-2005 (2005)
26. J. Barth et al., *Eur. Phys. J.* **A 18**, 117 (2003)
27. B. Friman and M. Soyeur, *Nucl. Phys.* **A 600**, 477 (1996)

EXPERIMENTS WITH FROZEN-SPIN TARGET AND POLARIZED PHOTON BEAMS AT CLAS

F. J. KLEIN

Catholic University of America, Washington, DC 20064

Investigations on N^* production generally have to face large t- and u-channel background and broad, overlapping resonances. In order to disentangle the large variety of contributions it is necessary to employ not only unpolarized cross section data but also polarization data. Comparisons of PWA solutions and model calculations show that the presence of particular resonances in the s-channel have large effects on polarization observables. At Jefferson Lab, a major program has been proposed to perform a large set of experiments using linearly and circularly polarized photon beams and longitudinally and transversely polarized targets together with the large acceptance spectrometer (CLAS) in Hall B.

1. Motivation

The nature of QCD confinement continues to provide an ongoing challenge to our understanding of QCD in the non-perturbative regime. The investigation of the proton excitation spectrum provides a major tool to understand the dynamics and relevant degrees-of-freedom within hadrons in the non-perturbative regime of QCD. A large number of baryon resonances has been found through analyses of experimental data, only a few of them, however, are well established as unambiguous states. Resonances like $\Delta(1232), N(1520), N(1680)$, are observed as strong signals with Breit-Wigner shape in single-pion production. However, this is the exception: almost all resonant states occur as broad, overlapping structures that are difficult to disentangle – even by means of extensive partial wave analyses. Moreover, quark models based on SU(6)⊗O(3) symmetry predict many more states, which have either not been detected or produced only weak signals in the existing data. In order to address this question, the properties of established resonances have to be determined more precisely. The existing data sets for single-pseudoscalar photoproduction consist to more than 70 % of unpolarized cross section data, the remainers are almost exclusively single-polarization observables. These data sets are insufficient to precisely determine resonance parameters, since unpolarized data do not provide access to relative phases. Resonance parameters for higher-lying states as well as the parametrization of nonresonant background are largely model dependent. A comparison of MAID [1] and SAID [2] solutions for single-pion photoproduction shows an excellent agreement in the description of cross section data, however large discrepancies in polarization observables, especially those which have hardly been measured.

It is worthwhile noticing that removing contributions from specific resonances, even well-established ones, hardly alters the differential cross section, but changes significantly the description of polarization observables, as shown in Fig. 1 for the $p(\gamma,\pi^+)n$ reaction at W=1500 MeV.

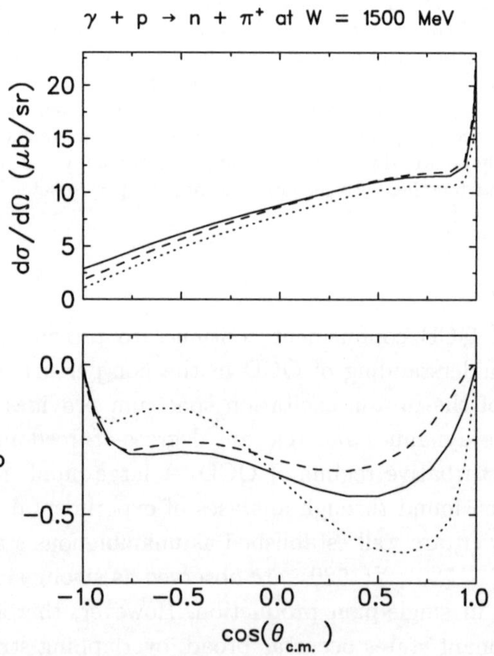

Fig. 1. MAID [1] calculations of the differential cross section and double-polarization observable G in $\gamma p \to \pi^+ n$ at W=1500 MeV. Solid line: full calculation; dashed line: $P_{11}(1440)$ removed; dotted line: $S_{11}(1535)$ removed.

A similar behavior can be found in η and kaon photoproduction showing that unique extraction of resonance parameters cannot be achieved by partial wave decomposition of cross section data, but requires precise measurements of various polarization observables, see [3] for details. The reactions $\gamma p \to \eta p$ and $\gamma p \to K^+\Lambda$ are additionally of special interest in the context of resonance extraction due to their isospin selectivity.

Photoproduction of multi-pion and vector meson channels require significantly more measurements of polarization observables to uniquely determine the resonant couplings. A first step in this direction can be the measurement of all polarization observables obtainable from any combination of beam and target polarization in order to provide crucial constraints on the analysis of $\pi\pi N$ and the extraction of N^*-couplings to $\Delta\pi$, $p\rho$, and $p\omega$.

2. The FROzen-Spin Target (FROST) program

The proposed program to measure single and double polarization observables at Jefferson Lab will make use of the large-acceptance spectrometer (CLAS) [4] and the photon beam facility [5] in Hall B, providing circularly polarized photons from highly polarized electrons and linearly polarized photons produced via coherent bremsstrahlung [6].

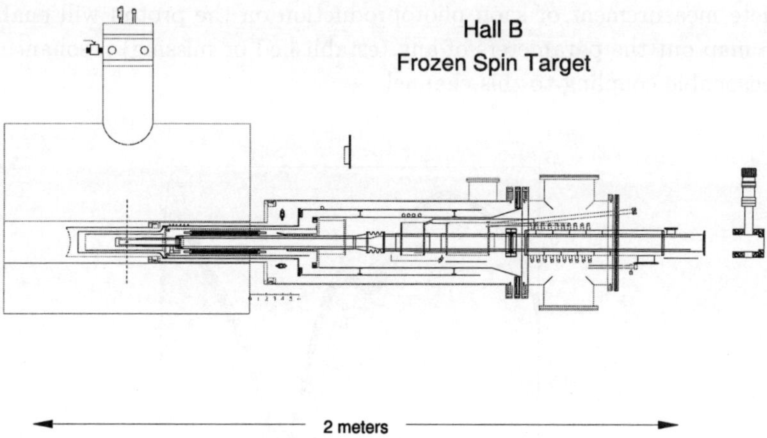

Fig. 2. Schematics of the Hall-B frozen spin target, here inserted into the 5 T polarizing magnet (left).

A new butanol target, which will be operated in frozen-spin mode, is under construction at Jefferson Lab. The target project, sketched in Fig. 2, is especially challenging as not only the material surrounding the target cell has to be minimized, but also a horizontal cryostat that fits into the

CLAS detector and a thin saddle coil that provides the transverse holding field have to be built. Information on the status of the hardware projects can be found at the FROST webpage [7].

In the framework of helicity amplitudes, it can easily be shown that cross section and single-spin observables (P, T, Σ) only fix the moduli of the helicity amplitudes and that at least four double-polarization observables are necessary to determine the phases. Moreover, these double-polarization observables have to be chosen from different classes of double-polarization experiments (beam–target, beam–recoil, target–recoil) to form a complete set of measurements. CLAS does not provide the possibility to measure the proton recoil asymmetry, thus only allowing for measurements of all observables obtained from combinations of polarized beams and targets (observables $d\sigma/d\Omega, \Sigma, T, P, E, F, G, H$), which will provide *almost complete* measurements for pion and η photoproduction. However, the weak decay of Λ and Σ^+ allows for extracting the recoil asymmetry from the distribution of their decay particles, thus allows for performing a *complete set of measurements.* Moreover, measurements of observables from all three above mentioned classes in kaon production will provide a powerful tool to check and minimize the systematics of the experiment. We expect that the complete measurement of kaon photoproduction on the proton will enable us to map out the parameters of any (established or missing) resonance with reasonable coupling to this channel.

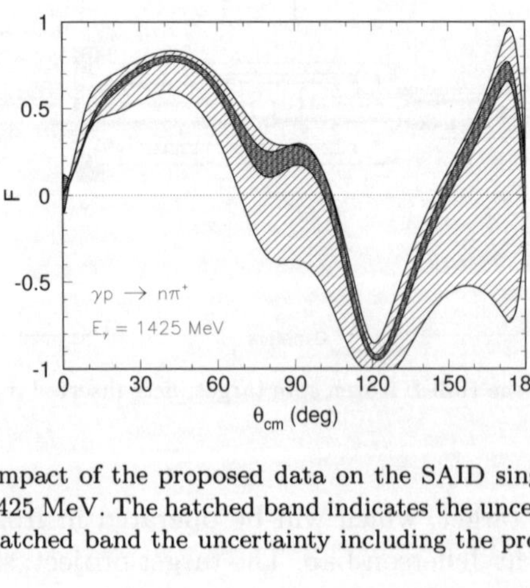

Fig. 3. Expected impact of the proposed data on the SAID single-energy solution for $\gamma p \to \pi^+ n$ at E_γ=1425 MeV. The hatched band indicates the uncertainties in the current solution, the dark hatched band the uncertainty including the proposed data.

Beam time for all four different settings of beam and target polarizations

– circularly/linearly polarized photon beams on longitudinally/transversely polarized targets – have been requested such that all observables can be determined to ±0.05 to ±0.07 over a large range in production angle and energy. Data on multi-pion channels will be taken simultaneously, uncertainties in the corresponding double-polarization observables are expected to be of the same order as for single (pseudoscalar) meson production.

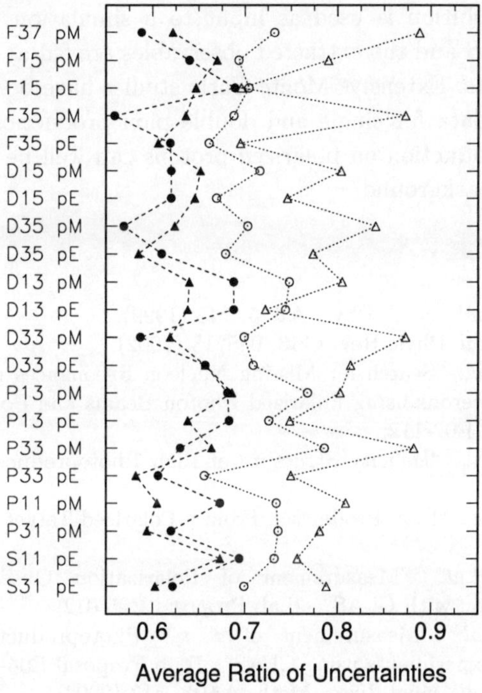

Fig. 4. Average ratio of the uncertainties of amplitudes with and without the proposed data for the four settings of beam and target polarization: circ.beam–long.target (*open triangle*), lin.beam–long.target (*open circle*) circ.beam–trans.target (*filled triangle*), lin.beam–trans.target (*filled circle*).

An example for the expected impact of the proposed experiments on SAID single-energy solutions in pion photoproduction is depicted in Fig. 3, where the hatched area represents the uncertainty of the current SAID solution for the double-polarization observable F and the dark hatched band the projected uncertainty using the data of the proposed measurement with circularly polarized photons on a transversely polarized target. The improved description of observables translates to a considerable improvement of extracted multipoles. Figure 4 shows the averaged ratio of uncertainties in the SAID solutions for all four combinations of beam and target polarization, normalized by the current solution. The comparison shows that double-polarization data on transversely polarized target (*filled markers*) have a greater impact than data on longitudinally polarized target (*open markers*).

The projected results are based on Monte-Carlo studies, in which the current SAID solution is used as input to a simulation of the detector response of CLAS and the extracted observables are fed as additional data into the SAID fits. Extensive Monte-Carlo studies have been performed to verify that the data for single and double pion production as well as ηp and $K\Lambda, K\Sigma$ production on polarized protons can well be separated from bound nucleon background.

References

1. D. Drechsel *et.al.*, Nucl. Phys. **A645**, 145 (1999).
2. R.A. Arndt *et.al*, Phys. Rev. **C66**, 055213 (2002).
3. F.J. Klein *et.al.*, "Search for Missing Nucleon Resonances in the Photoproduction of Hyperons using Polarized Photon Beams and Polarized Targets", JLab Proposal E02-112;
   D.I. Sober *et.al.*, "Helicity Structure of Pion Photoproduction", JLab Proposal E03-102;
   S. Strauch *et.al.*, "Pion Production From a Polarized Target", JLab Proposal E03-105;
   E. Pasyuk *et.al.*, "Measurement of Polarization Observables in eta-photoproduction with CLAS", JLab Proposal E05-012;
   V. Crede *et.al.*, "Measurement of $\pi^+\pi^-$ Photoproduction in Double-Polarization Experiments using CLAS", JLab Proposal E06-013.
4. B. Mecking *et.al.*, Nucl. Instr. Meth. **A503**, 513 (2003).
5. D.I. Sober *et.al.*, Nucl. Instr. Meth. **A440**, 263 (2000).
6. F.J. Klein *et.al.*, submitted to Nucl. Instr. Meth. A.
7. Web page: *http://clasweb.jlab.org/frost.*

THE CRYSTAL BALL AT MAMI *

D. P. WATTS FOR THE CrystalBall@MAMI COLLABORATION†

School of Physics, University of Edinburgh, UK
E-mail: daniel.watts@ed.ac.uk

The Crystal Ball and TAPS detector systems provide a highly segmented photon and hadron calorimeter for use with the Glasgow Tagger at the upgraded MAMI electron beam facility. The facility will provide very high quality photonucleon and photonuclear data for incident E_γ up to ~1.4 GeV. Some preliminary results will be presented and future plans for double-polarisation measurements in meson photoproduction will be outlined.

1. MAMI

1.1. *The MAMI facility*

The Crystal Ball is now the central detector facility to be used in experiments with the Glasgow Tagger at the MAMI electron microtron in Mainz, Germany. The MAMI-B[1] electron beam facility produces a 0.85 GeV high quality ~100% duty factor electron beam. The facility is presently being upgraded to MAMI-C[2] which will involve the inclusion of a further microtron stage to give an output beam energy of 1.5 GeV. The MAMI electron beam can be passed to 4 experimental halls. The Crystal Ball[3] and

*This research is part of the EU integrated infrastructure initiative hadronphysics project under contract number RII3-CT-2004-506078

†The CrystalBall@MAMI Collaboration: J. Ahrens, S. Altieri, J.R.M. Annand, H.J. Arends, R. Beck, V. Bekrenev, C. Bennhold, B. Boilllat, A. Braghieri, D. Branford, W. Briscoe, J. Brudvik, S. Cherepnya, R. Codling, D. Drechsel, E. Downie, L. Fil'kov, K. Foehl, S.B. Gerasimov, D. Glazier, P. Grabmayr, R. Gregor, D. von Harrach, T. Hehl, F. Hjelm, A. Koulbardis, M. Korolija, N. Kozlenko, D. Hornidge, V. Kashevarow, J. Kellie, M. Kotulla, D. Krambrich, S. Kruglov, B. Krusche, M. Lang, K. Livingston, S. Lugert, V. Lisin, M. Manley, I.J.D. MacGregor, J.C. McGeorge, V. Metag, K. Makonyi, B.M.K. Nefkens, R. Novotny, R. Owens, P. Pedroni, R. Kondratiev, T. Pinelli, S.N. Prakhov, D. Protopopescu, A. Polonski, J.W. Price, G. Rosner, S. Scherer, S. Schumann, D. Sober, A. Starostin, H. Staudenmaier, I. Supek, C.M. Tarbert, A. Thomas, L. Tiator, M. Thiel, D. Trnka, M. Unverzagt, Yu.A. Usov, M. Vanderhaeghen, T. Walcher, D.P Watts, F. Zehr.

TAPS[4]calorimeters are used in the real photon (A2) hall where the electron beam produces an intense ($\sim 10^8\ \gamma\ \text{sec}^{-1}$) beam of real photons through bremsstrahlung in a thin metal foil radiator. By employing a crystalline radiator (diamond) linear polarisation of up to ~90% can be achieved. Circular polarisation of up to ~85% can be achieved using longitudinally polarised electrons from MAMI. Following the bremsstrahlung process the scattered electrons are momentum analysed in the Glasgow Tagger, a magnetic spectrometer[5] which provides a determination of the energy of the associated bremsstrahlung photon with a resolution of ~2 MeV. Improved resolution and flux can be achieved for smaller sub ranges of photon energy by using a more highly segmented focal plane microscope subdetector system[6]. The MAMI photon beam compares favourably with other facilities available in this E_γ range as can be seen in Fig.1

	E_γ^{max} (GeV)	I_γ^{max} ($s^{-1}MeV^{-1}$)	ΔE_γ^{FWHM} (MeV)	Pol_γ^{lin} (%)	Pol_γ^{circ} (%)
ELSA	3.5	10^4	5	70	80
Graal	1.5	10^3	15	100	100
Jefferson Lab	5.4	10^4	5	70	80
Legs	0.45	10^3	5	100	100
MAMI B C	**0.8** **1.5**	**10^5**	**1**	**70**	**80**
SPring-8	3.0	10^3	30	100	100

Fig. 1. Comparison of the major gamma beam facilities

1.2. *The Crystal Ball*

The Crystal Ball (Fig. 2) was conceived and built in the mid 1970s at SLAC and was the central detector system in experiments at SPEAR, DORIS and Brookhaven National Laboratory before arriving at MAMI. The Crystal Ball is a 672 element NaI detector covering 94% of 4π. Each element is shaped like a a truncated pyramid ~41cm long. Photons incident on the ball produce an electromagnetic shower which generally deposits energy in a number of crystals (typically 98% of the deposited energy is contained in a cluster of 13 crystals). From analysing the centre of gravity of the shower angular resolutions for the photon of 2-3° in theta and 2°/sinθ in phi are achieved. The scintillation light deposited in each crystal is read by its own 2 inch Photomultiplier tube (PMT) and the high light output of NaI results in a good determination of the photon energy (σ/E ~1.7%/$E^{0.4}$(GeV)).

There has been a complete overhaul of the electronics for the Crystal Ball with its move to Mainz[7]. The output signal from each PMT is fed to a split-delay module. One output branch feeds to the trigger logic, a second feeds individual multi-hit TDCs and scalers via a dual threshold discriminator and a third is fed to a 40 MHz sampling ADC. The ADC modules also have an 80 MHz sampling capability which may be exploited in the future. The sampling of different regions of the pulse shape enables correction for remnant light present in the crystals before the event of interest, which improves the energy resolution.

A detector subsystem consisting of two cylindrical MWPCs and a 24 element barrel of plastic scintillators gives a good separation of protons and pions and gives track resolutions of σ_θ~1.5° and σ_ϕ~1.3°.

1.3. *TAPS*

The TAPS detector system covers the angular range $\theta = 1° - 20°$ which is not covered by the Crystal Ball. This is an important part of the phase space for fixed target experiments with high beam momenta. TAPS consists of 510 hexagonally shaped BaF2 detectors each 25 cm long corresponding to 12 radiation lengths. TAPS is a versatile detector system and has been employed at MAMI, GSI, Ganil, CERN and most recently with the Crystal Barrel at Bonn. BaF_2 scintillator has two scintillation components, a fast component which lasts for ~20ns and a slow component which lasts for ~2μs. Different particle species produce different relative light yields in the two scintillation components and this additional pulse shape information can be used along with time-of-flight and ΔE information in particle identification. The energy resolution for the detector is of similar order to that obtained in the Crystal Ball (σ/E ~ 0.59 $E^{-1/2}$(GeV) + 1.9%). New VME based readout electronics for the TAPS detectors have been constructed[8]. The new design improves significantly on the previous system by allowing the signal information to be digitised close to the detectors which eliminates problems associated with passing analogue signals the long distance to the control room such as attenuation, shaping and cross talk.

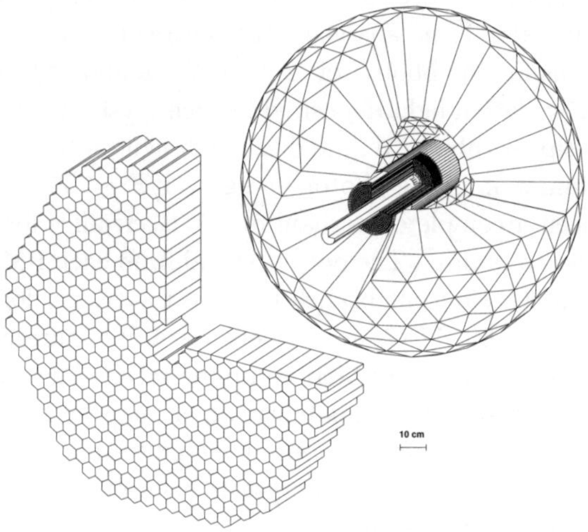

Fig. 2. The Crystal Ball and TAPS detectors in the A2 hall at Mainz (with sections removed to allow visibility of the detector elements)

1.4. *Targets*

The MAMI facility has a good range of targets either presently available or in development for use in the Crystal Ball programme. Cryogenic target systems capable of liquefying both hydrogen and deuterium are presently available. A system for accurately mounting solid targets has been operated successfully. A Mainz frozen spin target is due to be completed in 2007 which can use either butanol or deuterated-butanol targets to give a source of longitudinally polarised protons or neutrons respectively. In 2008 a polarised ^{3}He gas target will additionally become available and will offer an alternative source of polarised neutrons.

2. Experimental programme

The experimental programme at MAMI primarily exploits the 4π detection capability and uniquely intense photon flux available at MAMI. This powerful combination is particularly well suited to making measurements of small cross sections or measurements where very accurate and comprehensive determinations are necessary to extract the important physics.

The main themes of the MAMI programme include:

- Precision determinations of nucleon resonance properties
- Tests of low energy theorems
- Tests of fundamental symmetries
- In medium properties of hadrons
- Nuclear structure

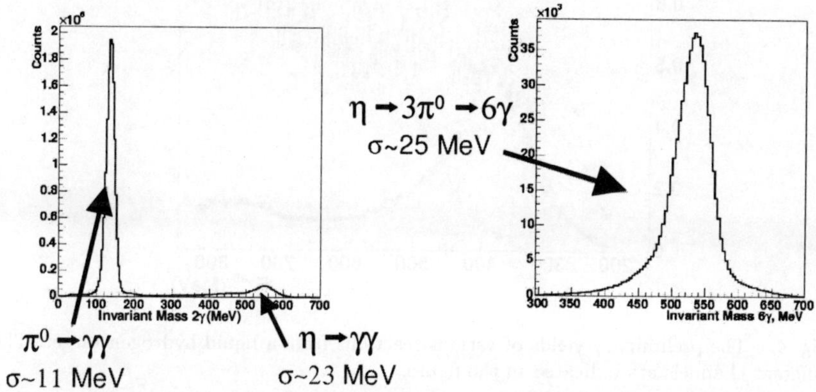

Fig. 3. Reconstructed invariant mass spectra (preliminary).

3. Experiments in the first round

The first round of experiments employed H_2, D_2 and nuclear targets and was successfully completed in the period June 2004 to April 2005. Data taking will resume in 2006 following the completion of the upgrade of the MAMI facility and the Glasgow photon tagging spectrometer. The Crystal Ball experiment took ~3600 hours of beamtime in the first round of experiments, giving data for 10 approved proposals and providing the thesis data for 13 Ph.D students. A brief outline of the experiments already completed and some preliminary analyses are presented in the following sections.

3.1. *Liquid Hydrogen Target*

The high granularity and almost complete kinematic coverage of the Crystal Ball and TAPS results in good quality reconstruction of π^0 and η mesons as can be seen from the preliminary spectra in Fig. 3. The majority of

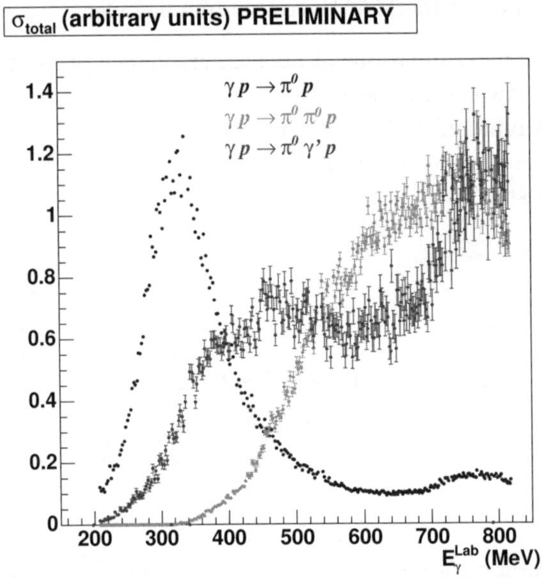

Fig. 4. The preliminary yields of various reactions from a liquid hydrogen target. The different channels are indicated in the figure.

proposals completed in the first round measured meson photoproduction reactions from the liquid hydrogen target. The first experiment comprised a measurement of π^o photoproduction in which the crystalline (diamond) radiator was oriented to produce a high degree of linear polarisation for E_γ around pion production threshold, giving accurate asymmetry and cross section data to extract the S and P wave strength and test predictions from chiral perturbation theory (χPT). The later experiments were carried out with high degrees of linear polarisation in the E_γ=0.4-0.5 GeV region, above the peak of the Δ resonance and in the threshold region of 2π production. The new data for the 2π reaction near threshold will also test χPT, in particular pion loop mechanics, and will allow detailed investigation of the resonances in the second resonance region. The analysis of the large sample of $\eta \to \pi^0\gamma\gamma$ will test the third order terms in the χPT momentum expansion and the $\eta \to 3\pi^0$ data will also give important tests of the theory.

A further important physics goal of the first round programme is to obtain an accurate determination of the magnetic dipole moment of the first excited state of the proton, the $\Delta(1232)$, which will provide an elegant test of various theoretical descriptions of the hadron. Sensitivity to the dipole moment arises from the contribution of the $\Delta^+ \to \Delta^+\gamma$ self-decay process in the reaction amplitude.of the $\gamma + p \rightarrow p\pi^0\gamma$ reaction. As well as increasing the data set for $\gamma + p \rightarrow p\pi^0\gamma$ by over an order of magnitude the new measurement will give a determination of reaction asymmetries for both linear and circular incident photon polarisations, which are predicted to give additional sensitivities to the value of μ_{Δ^+}. The experiment also allowed a simultaneous determination of the $\gamma + p \rightarrow n\pi^+\gamma$ reaction which gives important additional tests of the modelling of the background bremsstrahlung processes as well as giving a further independent measure of μ_{Δ^+}. The analysis of the data taken with the liquid H_2 target is presently in progress but some very preliminary yields[9] are shown in Fig. 4.

3.2. *A≥2 targets*

The first round of data taking also included ^{7}Li, ^{12}C, H_2O, ^{40}Ca and ^{208}Pb targets. The 2π production from these targets will be analysed to investigate modification of the π-π interaction in the I=J=0 channel. The ^{7}Li data will be used to search for evidence of η-mesic nuclear states.

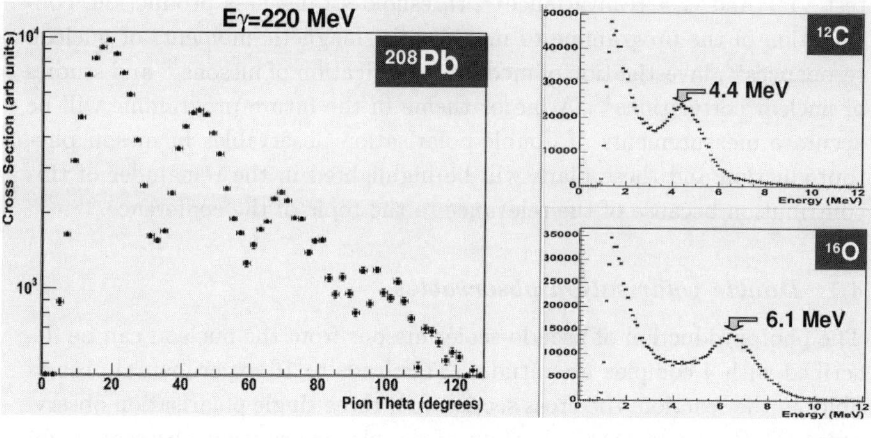

Fig. 5. Left; Preliminary π^0 angular distribution from the ^{208}Pb(γ, π^0) reaction. Right: Coincident low energy events in the Crystal Ball showing γ from the decay of the 2^+ state at 4.4 MeV in ^{12}C and states around 6.1 MeV in ^{16}O

Analysis of single π^0 production is also in progress, with the primary motivation to more firmly establish the matter distribution of nuclei. In the coherent production of a π^0 from the nucleus the amplitudes for production from all nucleons in the nucleus adds coherently and the diffraction pattern in the angular distribution of the outgoing π^0 contains information on the matter form factor. The high detection efficiency and high uniformity of the Crystal Ball will eliminate many of the systematic errors in extracting this coherent strength from the incoherent events which leave the nucleus in an excited state. The diffraction spectra for the ^{208}Pb are visible even with a first crude analysis as can be seen in Fig 5. Also shown in the figure are the energies of the low energy uncharged clusters occurring in coincidence with the π^0. It is clear that valuable information on low energy nuclear decay gamma rays are visible in the Crystal Ball. Their analysis will allow the coherent events to be extracted over a wider E_γ range and allow the incoherent strength to some low lying discrete residual states to be studied in detail for the first time. The capability for detection of decay gammas with the CB will also have applications in other future experiments on nuclei.

4. Future Programme

Present plans for the future programme for the Crystal Ball at MAMI-C include further detailed measurements of η and η' branching ratios to test χPT and C, CP invariance[11], threshold strangeness production, continuation of the programme to measure the magnetic moments of nucleon resonances[12], investigation of medium modification of mesons[13] and studies of nuclear correlations[14]. A major theme in the future programme will be accurate measurements of double-polarisation observables in meson photoproduction and these plans will be highlighted in the remainder of this contribution because of the relevance to the topic of the conference.

4.1. *Double polarisation observables*

The photoproduction of pseudo-scaler mesons from the nucleon can be described with 4 complex amplitudes. This leads to 16 experimental observables in the reaction: the cross section (σ), three single polarisation observables (Σ, P, T) and three quartets of double polarisation observables for beam-target (E,F,G,H) beam-recoil ($C_{x'}$, $C_{z'}$, $O_{x'}$, $O_{z'}$) and target-recoil ($T_{x'}$, $T_{z'}$, $L_{x'}$, $L_{z'}$). The need to measure double-polarisation observables to improve our knowledge of the amplitudes has become increasingly apparent in recent years. For example a first measurement of E for $\vec{\gamma} + \vec{p} \to p\pi^0$ showed the need for a significant revision of the helicity amplitudes even for the "well established" $D_{13}(1520)$ resonance[10].

The availability of an intense polarised photon beam, polarised nucleon targets and a $\sim 4\pi$ detector system at Mainz will allow accurate determi-

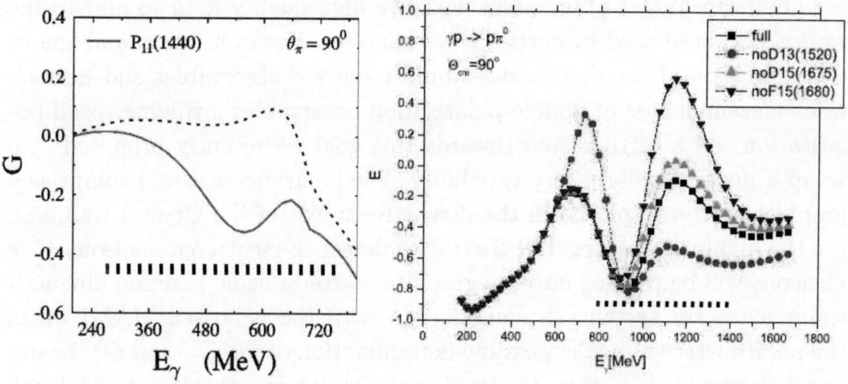

Fig. 6. LEFT: The vertical bands show the expected accuracy in the determination of the G observable for $p(\gamma, p)\pi^0$ in an E_γ bin of ± 10 MeV, a θ_π bin of 10° and a 600 hour beamtime. Curves show results from the MAID PWA with and without the roper resonance RIGHT: Vertical bands give the expected accuracy for the E observable in a 250 hour beamtime with the same E_γ, θ_π binning. The curves show the values of the E variable extracted from MAID

nation of a range of double polarisation observables. The E observable is accessed with a longitudinally polarised nucleon target and circularly polarised photons. New measurements with MAMI-C will greatly improve the statistical accuracy and the E_γ range of the experimental data for π^0, η and 2π production channels, and assess their contributions to the Gerasimov Drell Hearn sum rule. The expected accuracy of the proposed E measurements[15] for the $p(\vec{\gamma}, \pi^0)p$ reaction and indications of the sensitivity of the variable to different resonances based on the MAID partial wave analysis (PWA) are shown in Fig.6. The measurement on the neutron target will allow a complementary determination of E to test the isospin structure of the resonances and allow investigation of resonances which only couple strongly to the neutron. The determination of the $p(\vec{\gamma}, \eta)p$ reactions, will be particularly interesting because of the sensitivity expected to the $D_{15}(1675)$ which is one possible explanation of the structure in the cross sections measured at GRAAL.

The G observable will be obtained[16] for $p(\vec{\gamma}, \pi^0)p$ and $p(\vec{\gamma}, \pi^+)n$ reactions from a linearly polarised photon beam with a longitudinally polarised target. The suitability of the observable for detailed studies of the poorly established $P_{11}(1440)$ (Roper) resonance is indicated from the MAID PWA predictions shown in Fig. 6. The expected accuracy of the experimental data is also indicated in the figure.

The beam-target observables will give high quality data to add to the worlds database used in partial wave analyses. However, an unambiguous determination of the amplitudes would require 8 observables and necessitates measurements of double-polarisation observables involving recoil polarisation. As a further step towards this goal we recently proposed[17] to setup a nucleon polarimetry capability. The polarimeter would comprise a graphite scatterer, placed in the downstream exit of the Crystal Ball, and use the highly segmented TAPS array to detect the scattered nucleons. The nucleons will be tracked into the graphite scatterer using reaction kinematics or a tracker system. In combination with the polarised MAMI beam the polarimeter will make possible determination of the $C_{x'}$ and $O_{x'}$ beam-recoil observables. Further, the single polarisation observables, P and T will also be accessible (T can be extracted as the y component of recoil polarisation with a linearly polarised photon beam). The expected experimental accuracy, estimated using realistic polarimeter parameters for a 7cm thick graphite scatterer, is shown in Fig.7.

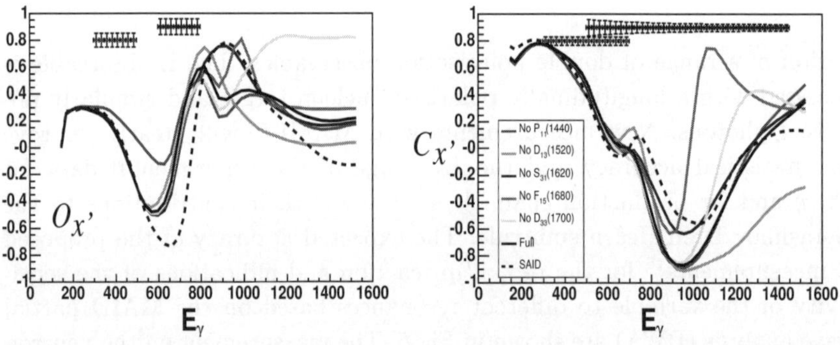

Fig. 7. The data points show the expected statistical accuracy of the $O_{x'}$ (Left) and $C_{x'}$ (Right) measurements in a θ_π bin of $\pm10°$ for a 300 hour run with 0.85 GeV beam energy (red - lower points) and a 600 hour run at 1.5 GeV (blue - upper points). The curves show the values extracted from the SAID and MAID PWA (see key in the figure)

In conclusion, the Crystal Ball at MAMI programme is successfully underway and this powerful combination can be expected to contribute high quality data on meson photoproduction up to E_γ ~1.4 GeV, most with polarisation information.

References

1. http://www.kph.uni-mainz.de

2. A. Jankowiak et. al., Proc. EPAC2002, Paris (2002)
3. E.D. Bloom and C.W. Peck, Ann. Rev. Nucl. Part. Sci, vol. 33 ,p143 (1983)
4. R. Novotny et. al., IEEE Trans Nucl. Sci 38 p379 (1991).
5. I. Anthony et. al., Nucl. Instr. And Meth., A301, p230, (1991)
6. A. Reiter et. al., http://wwwa2.kph.uni-mainz.de/microscope/
7. D. Krambrich, Ph.D thesis (Mainz 2005) in preparation; D.P. Watts, Calorimetry in particle physics, Perugia 2004, p116-123
8. P. Drexler et. al., IEEE Trans. Nucl. Sci 50 p.969 (2003).
9. A. Starostin, *private communication.*
10. J. Ahrens *et. al.*, Phys. Rev. Lett. 88 232002 (2002); *ibid.* 84, 5950 (2000).
11. A. Starostin *et. al.*MAMI/A2/2-05, B. Nefkens *et. al.*MAMI/A2/3-05
12. M. Kotulla *et. al.* MAMI/A2/4-05
13. V. Metag *et. al.* MAMI/A2/1-05
14. I. Macgregor *et. al.* MAMI/A2/6-05
15. P. Pedroni *et. al.* MAMI/A2/9–05; B. Krusche *it. al.* MAMI/A2/10-05
16. R. Beck *et. al.* MAMI/A2/8-05
17. D. Watts *et. al.* MAMI/A2/5-05

THE GRAAL COLLABORATION: RESULTS AND PROSPECTS

CARLO SCHAERF

FOR THE GRAAL COLLABORATION*

Università di Roma "Tor Vergata"
and INFN - Sezione di Roma Tor Vergata,
I-00133 Roma, Italy
E-mail: schaerf@roma2.infn.it

We present preliminary results obtained by the GRAAL collaboration in the photoproduction of π^0 and η on the proton and the neutron in Hydrogen and Deuterium in the energy range 600-1500 MeV. The beam polarization asymmetries are similar for the free proton in Hydrogen and the bound proton in Deuterium. At energies up to 1.2 GeV, for the bound proton and bound neutron in Deuterium, the asymmetries are similar for the photoproduction of η and different for π^0.

1. Introduction

The Graal beamline is devoted to the study of hadronic physics with the polarized and tagged gamma-ray beam produced by the backward scattering of laser light on the high energy electrons circulating in the ESRF storage ring. The gamma-ray-beam energy covers the interval between 600 and 1500 MeV and its energy resolution is 16 MeV (FWHM). The Graal beam belongs to the family of Ladon beams first developed on the storage ring Adone at Laboratori Nazionali di Frascati of INFN[1]. Table I indicates the main characteristics of Ladon beams. The last four are in operation and cover the gamma-ray spectrum from a few MeV to 2.4 GeV[23], as shown in Figure 1. The main advantages of gamma-ray Ladon beams are the high de-

---

*O. Bartalini, V. Bellini, J.P. Bocquet, M. Castoldi, A. D'Angelo, J-P. Didelez, R. Di Salvo, A. Fantini, D. Franco, G. Gervino, F. Ghio, B. Girolami, A. Giusa, M. Guidal, E. Hourany, V. Kouznetsov, R. Kunne, A. Lapik, P. Levi Sandri, A. Lleres, D. Moricciani, A. Mushkarenkov, V. Nedorezov, L. Nicoletti, C. Randieri, D. Rebreyend, N.V. Rudnev, G. Russo, C. Schaerf, M.L. Sperduto, M.C. Sutera, A. Turinge gree of polarization and the almost flat energy spectrum. The gamma-rays with the highest energy have a polarization very close to that of the laser light and therefore their polarization can be easily rotated, or changed to circular, changing the polarization of the laser light with standard optical components. We have used the Green (514 nm) and UV (351 nm) lines of a Argon(Ion) laser to obtain two different gamma-ray spectra with different maximum energies and different polarizations as shown in Figure 2.

TABLE I

Project name	Ladon°	Taladon⁺	ROKK$^{\Delta}$-1M	LEGS*	LEGS-2	Graal**	LEPS†	HIGS$^{\oplus}$
Location	Frascati	(LNF-INFN)	Novosibirsk	Brookhaven	(BNL)	Grenoble	Harima	Durham
Storage ring	Adone	Adone	VEPP-4M	NSLS	NSLS	ESRF	SPring-8	TUNL-FELL
Energy defining method	collimation	internal tagging	tagging	external tagging	external tagging	internal tagging	internal tagging	collimation
Electron energy GeV	1.5	1.5	1.4 - 5.3	2.5	2.8	6.04	8	1.0
Photon energy eV	2.45	2.45	1.17-3.51	3.53	4.71	3.53	3.53	8.2
γ-ray energy MeV	5-80	35-80	100-1200	180-320	285-470	550-1470	1500-2400	5-225
	variable	simultaneous						variable
Energy resolution %	1.4-10	5	--	1.6	1.1	1.1	1.25	1
(FWHM) MeV	0.07-8	4-2	--	5	5	16	30	
Electron current A	0.1	0.1	0.1	0.2	0.2	0.2	0.1	100
Gamma intensity s⁻	10^5	$5\ 10^5$	$2\ 10^6$	$4\ 10^6$	$2\ 10^6$	$2\ 10^6$	$2\ 10^6$	10^6-10^8
Year of operation	1978	1989	1993	1987	1999	1996	1999	≈2000

° Laser ADONe, ⁺TAgged LADON, $^{\Delta}$ROKK is a russian abbreviation for Backscattered Compton Gamma, * Laser Electron Gamma Source, ** GRenoble Anneau Accelerateur Laser, † Laser-Electron Photons at SPring-8, $^{\oplus}$ High Intensity Gamma-ray Source.

Λαδων (Ladon). "riviere de Grece, au Péloponnèse dans l'Arcadie. ...... les Mythologistes firent le Ladon pere de la nymphe Daphné & de la nymphe Syrinx. .... Il étoit couvert de magnifiques roseaux, dont Pan se servit pour sa flûte à sept tuyaux."
(M. Diderot and M. D'Alembert, Encyclopédie, a Paris MDCCLVII)

2. The GRAAL apparatus

The Graal experimental apparatus consists of a Crystal Ball made of 480 BGO crystals plus plastic scintillators and wire chambers to cover almost the entire solid angle. A sketch of the experimental apparatus is represented in Figure 3 . To measure differential cross sections and beam polarization asymmetries in the photoproduction of mesons, Compton photon scattering and deuteron photodisintegration we have used liquid Hydrogen and Deuterium targets. This has allowed a detailed comparison of the results on the free proton (in Hydrogen) with those on the proton and the neutron bound in a deuteron.

Very precise data have been produced for the photoproduction in Hydrogen

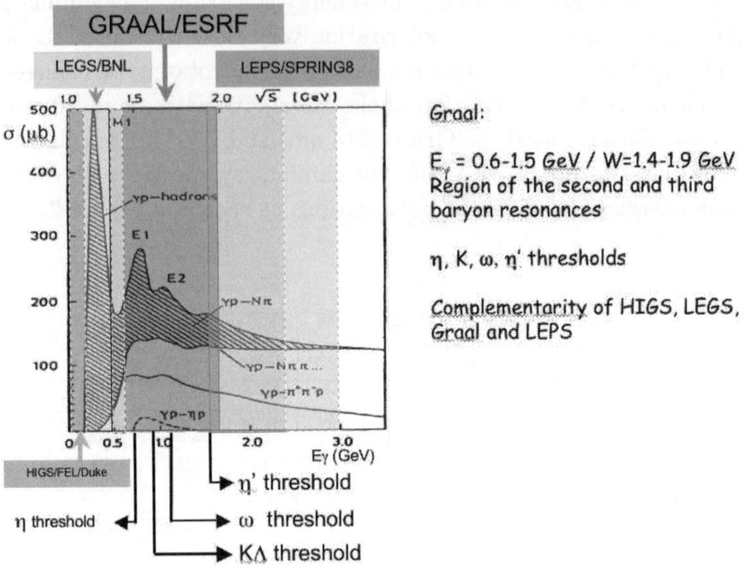

Fig. 1. Ladon beams in the world and the total photoabsorption cross section on proton.

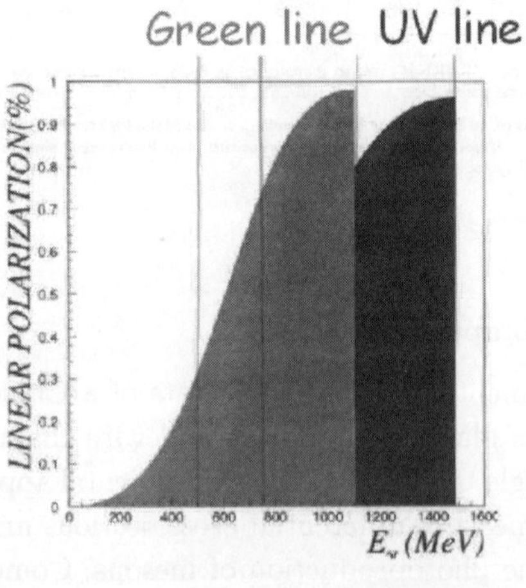

Fig. 2. Beam polarization degree as a function of the incident gamma energy for the Green (514 nm) and UV (351 nm) laser lines. In the overlapping region, we collect data with the same gamma energy but different degrees of polarization.

of π^0 and π^+ and for η as indicated for example in Figure 4.

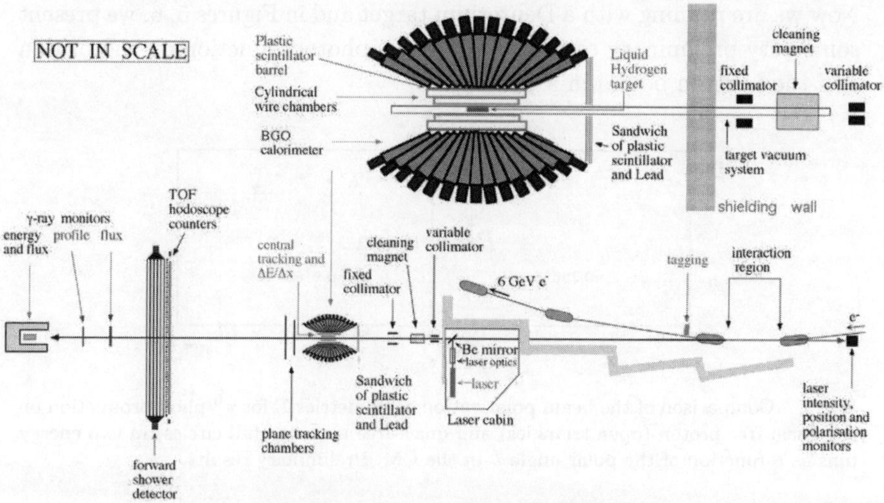

Fig. 3. The GRAAL apparatus.

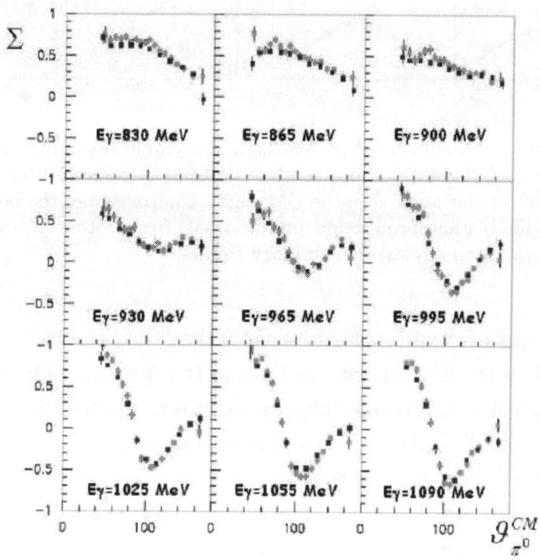

Fig. 4. Beam polarization asymmetries Σ for π^0 photoproduction on the proton in Hydrogen (square points for UV laser line and circles for green laser line). Preliminary results.

Now we are running with a Deuterium target and in Figures 5, 6, we present some very preliminary results on π^0 and η photoproduction on the proton and the neutron bound in a deuteron.

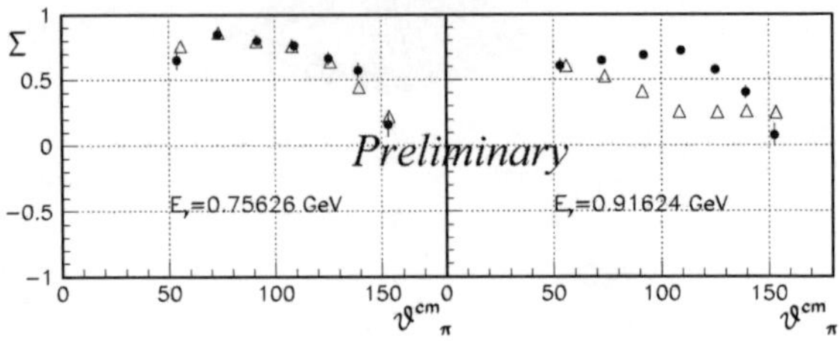

Fig. 5. Comparison of the beam polarization asymmetries Σ for π^0 photoproduction on the quasi-free proton (open triangles) and quasi-free neutron (full circles) in two energy bins as a function of the polar angle ϑ in the CM. Preliminary results.

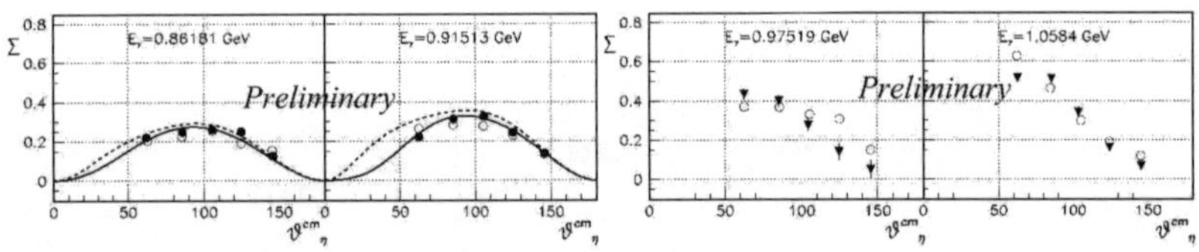

Fig. 6. Left: comparison of the beam polarization asymmetries Σ for η photoproduction on the free proton (full circles) and quasi-free proton (open circles) for two energy bins as a function of the polar angle ϑ in the CM; right: comparison of the beam polarization asymmetries Σ for η photoproduction on the quasi-free proton (black triangles) and quasi-free neutron (open circles). Preliminary results.

In conclusion our preliminary results show that the beam polarization asymmetries are very similar for free and quasi-free protons. On quasi-free protons and quasi-free neutrons they are similar, up to 1200 MeV, for the photoproduction of η, that has isospin 0, but are different for the photoproduction of π^0, that has isospin 1.
Some very preliminary results on the polarization asymmetries for the photoproduction of K mesons with Λ and Σ are shown in Figures 7 and 8.
In η photoproduction in deuterium we have noticed an anomalous peak in the invariant mass of the η-n system, as shown in Figure 9. It appears in the distribution of events with the neutron in the forward detector, large η angles in the CMS, and events with the neutron in the BGO, small η angles in the CMS. The width of the peak is consistent with our present experi-

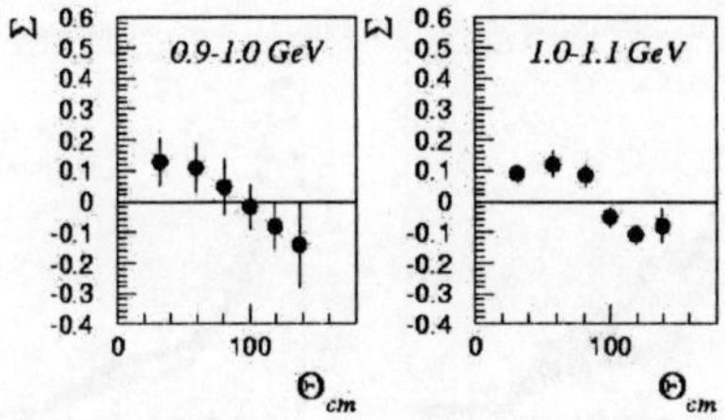

Fig. 7. Beam polarization asymmetries for the photoproduction of KΛ. Preliminary results.

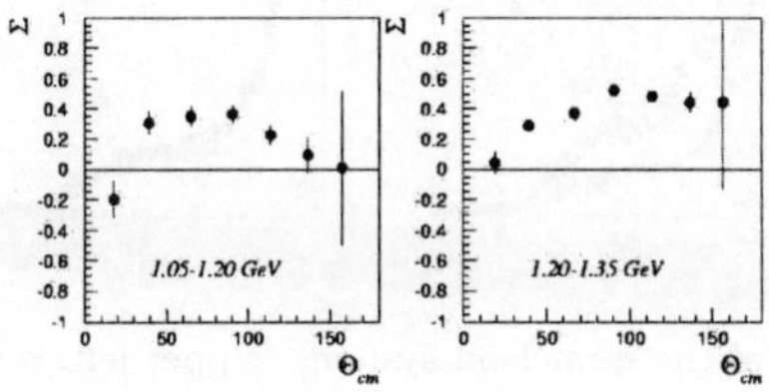

Fig. 8. Beam polarization asymmetries for the photoproduction of KΣ. Preliminary results.

mental resolution but the statistical significance is still modest. A similar peak does not appear in the invariant η-p mass as indicated in Figure 9.

The large amount of gamma-ray spectra collected over the years by the Graal collaboration has been used to probe the light speed anisotropy with respect to the "absolute" reference frame provided by the cosmic microwave background radiation dipole. Figure 10 represents two typical gamma-ray spectra obtained with the green (514 nm) and UV (330, 351 and 363 nm) laser lines. Analysing our collection of similar curves we obtain the maximum gamma-ray energy with a precision: $\Delta E_\gamma/E_\gamma \simeq 10^{-4}$. The maximum gamma-ray energy is approximately proportional to the square of the electron relativistic factor γ and at the ESRF energy of 6030 MeV we have: $\gamma = 11800$ and $\gamma^2 \simeq 1.4\ 10^8$. The revolution of the earth around the sun, at a speed of 30 km/s ($\beta_e \simeq 10^{-4}$), rotates the direction of the Graal vector (the direction of the Graal gamma ray beam) with respect to any "absolute" coordinate system with a period of one year. In Figure 11 we have indicated our measurements of the Compton edge for each day of the year and in Figure 12 we have combined all measurements taken in the same month irrespective of the year. In conclusion taking into account all sources of systematic errors that we could think of, we have estimated an

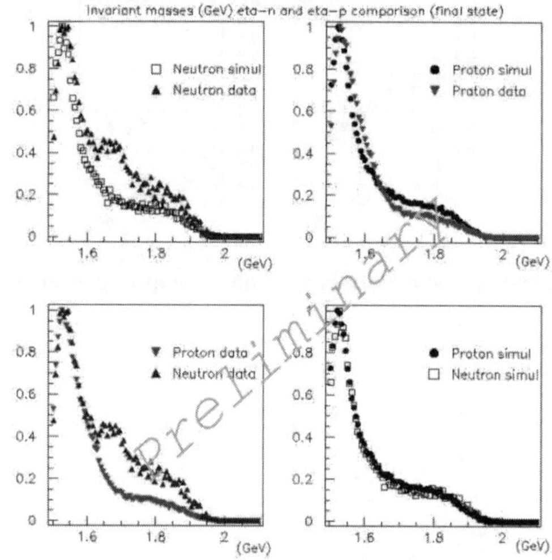

Fig. 9. Invariant mass of the η-nucleon systems. Upper left: η-neutron invariant mass, data and simulation. Upper right: η-proton invariant mass, data and simulation. Lower left: data for η-neutron and η-proton. Lower right: simulation for η-neutron and η-proton. Preliminary results.

upper limit on the anisotropy of light of [4]:

$$\frac{\Delta c}{c} \simeq 3 \cdot 10^{-12}$$

This result is almost three orders of magnitude better of the ones achieved with space probes.

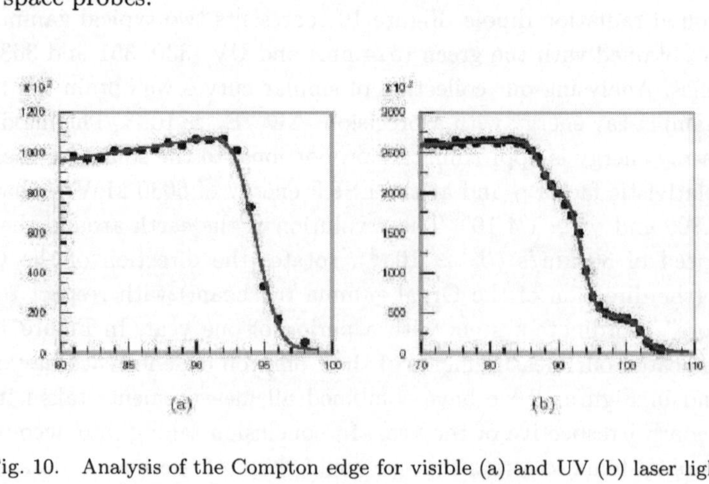

Fig. 10. Analysis of the Compton edge for visible (a) and UV (b) laser light.

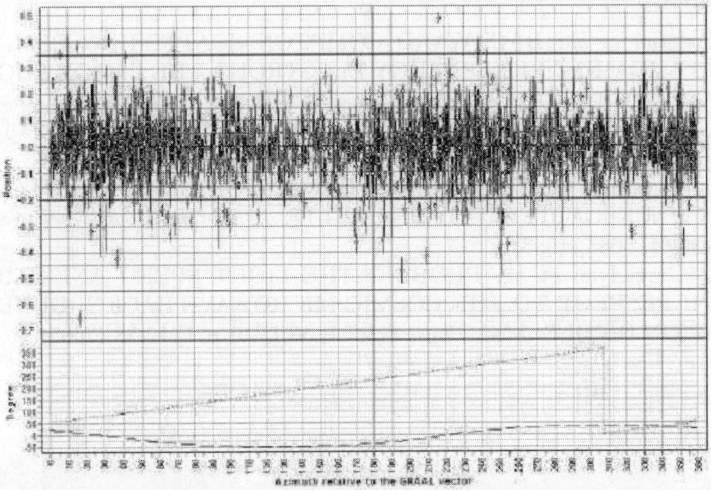

Fig. 11. Experimental data plotted as a function of the azimuth (above); below the variation of the angle between the beam and the CMB dipole decomposed to azimuth (dotted) and declination (dashed) angles is shown.

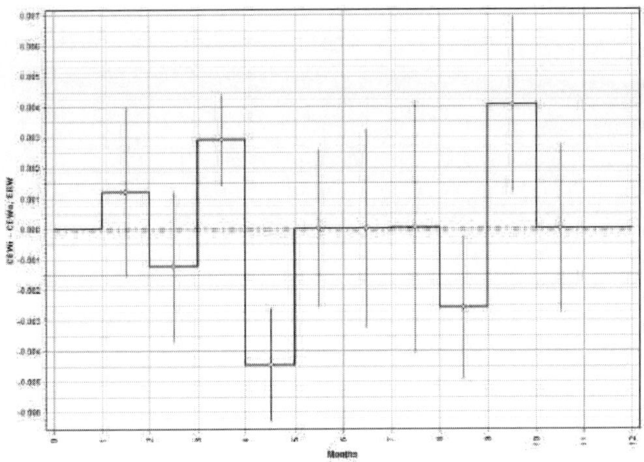

Fig. 12. Experimental data over 12-month period averaged by month. the average is found very close to zero.

References

1. L. Federici et al.: BACKWARD COMPTON SCATTERING OF LASER LIGHT AGAINST HIGH ENERGY ELECTRONS: THE LADON PHOTON BEAM AT FRASCATI *Il Nuovo Cimento* **59B**, 247 (1980).
2. R. Caloi et al.: THE LADON PHOTON BEAM WITH THE ESRF-5 GEV MACHINE. *Lettere al Nuovo Cimento*, **27**, 339 (15 Marzo 1980).
3. A.M. Sandorfi et al.: HIGH ENERGY GAMMA RAY BEAMS FROM COMPTON BACKSCATTERED LASER LIGHT. *IEEE Trans. on Nucl. Science*, **NS-30/4**, 3083 (1983).
4. V.G. Gurzadyan et al.: PROBING THE LIGHT SPEED ANISOTROPY WITH RESPECT TO THE COSMIC MICROWAVE BACKGROUND RADIATION DIPOLE. *Mod. Phys. Lett. A*, **Vol. 20 n.1**, 19-28 (2005).

CLAS: DOUBLE-PION BEAM ASYMMETRY

S. STRAUCH FOR THE CLAS COLLABORATION

University of South Carolina
Department of Physics and Astronomy
712 Main Street
Columbia, SC 29208, USA
E-mail: strauch@sc.edu

Beam-helicity asymmetries for the $\vec{\gamma}p \to p\pi^+\pi^-$ reaction have been measured for center-of-mass energies between 1.35 GeV and 2.30 GeV at Jefferson Lab with the CEBAF Large Acceptance Spectrometer using circularly polarized tagged photons. The beam-helicity asymmetries vary with kinematics and exhibit strong sensitivity to the dynamics of the reaction, as demonstrated in the comparison of the data with results of various phenomenological model calculations. These models currently do not provide an adequate description of the data over the entire kinematic range covered in this experiment. Additional polarization observables are accessible in an upcoming experiment at Jefferson Lab with polarized beam and target.

1. Introduction

The properties of the excited states of baryons reflect the dynamics and relevant degrees of freedom within them. The study of nucleon resonances is thus an important avenue to learn about the strong interaction. To disentangle and study experimentally these many, mostly weak, and largely overlapping resonances is a challenging task, especially at higher energies. Many nucleon resonances in the mass region above 1.6 GeV decay predominantly through either $\pi\Delta$ or ρN intermediate states into $\pi\pi N$ final states (see the Particle-Data Group review[1]). This is the region where resonances are predicted by symmetric quark models, but have not been observed in the πN channel (the so-called "missing" resonances); yet may couple strongly to channels like $\pi\Delta$.[2] This makes electromagnetic double-pion production an important tool in the investigation of the structure of the nucleon and the most promising approach is to employ polarization degrees of freedom in the measurement. There exist a rather large amount of unpolarized cross-section data of double-pion photo- and electroproduction on the proton;[3]

the amount of polarization observables in these reactions, however, remains quite sparse.[4]

The CLAS Collaboration published recently beam-helicity-asymmetry data in the $\vec{\gamma}p \to p\pi^+\pi^-$ reaction for energies W between 1.35 GeV and 2.30 GeV in the center of mass, where the photon beam is circularly polarized and neither target nor recoil polarization is specified.[5] These novel data combine the study of the double-pion final state with the sensitivity of polarization observables. The beam-helicity asymmetry is one of 64 polarization observables in the $\vec{\gamma}p \to p\pi^+\pi^-$ reaction and is defined as[6]

$$I^{\odot} = \frac{1}{P_\gamma} \cdot \frac{\sigma^+ - \sigma^-}{\sigma^+ + \sigma^-}, \tag{1}$$

where P_γ is the degree of circular polarization of the photon and $\sigma^\pm$ are the cross sections for the two photon-helicity states $\lambda_\gamma = \pm 1$.

In this contribution, we give a brief overview of our data, compare with results of phenomenological models, demonstrate the sensitivity of this observable to the dynamics of the reaction, and give an outlook on further studies of other polarization observables.

2. Experiment

The $\vec{\gamma}p \to p\pi^+\pi^-$ reaction was studied with the CEBAF Large Acceptance Spectrometer (CLAS)[7] at Thomas Jefferson National Accelerator Facility (Jefferson Lab). A schematic view of the reaction, together with angle definitions, is shown in Fig. 1. Longitudinally polarized electrons with an energy of 2.445 GeV were incident on the thin radiator of the Hall-B Photon Tagger[8] and produced circularly-polarized tagged photons in the energy range between 0.5 GeV and 2.3 GeV. The collimated photon beam irradiated a liquid-hydrogen target. The circular polarization of the photon beam was determined from the electron-beam polarization and the ratio of photon and incident electron energy.[9] The degree of photon-beam polarization varied from ≈ 0.16 at the lowest photon energy up to ≈ 0.66 at the highest energy. The $\vec{\gamma}p \to p\pi^+\pi^-$ reaction channel was identified in this kinematically complete experiment by the missing-mass technique, requiring either the detection of all three final-state particles or the detection of two out of the three particles in the CLAS detector.

A total of 3×10^7 $p\pi^+\pi^-$ events were accumulated. Experimental values of the helicity asymmetry were then obtained as

$$I^{\odot}_{\text{exp}} = \frac{1}{\bar{P}_\gamma} \cdot \frac{Y^+ - Y^-}{Y^+ + Y^-}, \tag{2}$$

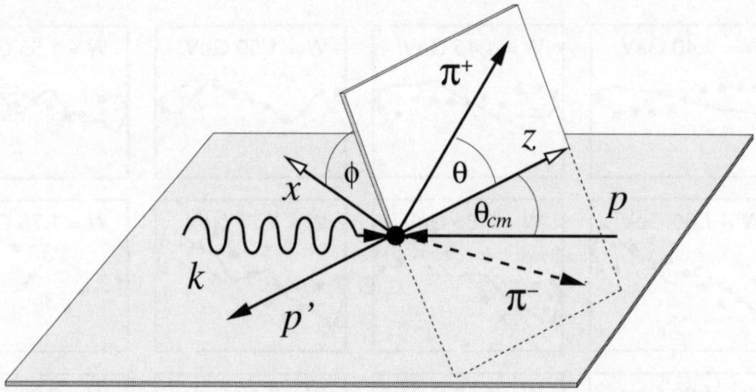

Fig. 1. Angle definitions for the circularly polarized real-photon reaction $\vec{\gamma}p \to p\pi^+\pi^-$; θ_{cm} is defined in the overall center-of-mass frame, and θ and ϕ are defined as the π^+ polar and azimuthal angles in the rest frame of the $\pi^+\pi^-$ system with the z direction along the total momentum of the $\pi^+\pi^-$ system (helicity frame).

where $Y^\pm$ are the experimental yields, corrected for a small electron-beam-charge asymmetry. The experimental asymmetries have not been corrected for the CLAS acceptance to avoid systematic uncertainties. Instead, the data are compared with event-weighted mean values of asymmetries from model calculations. These mean values of asymmetries in a kinematical bin are given by

$$\bar{I}^{\odot}_{\text{model}} = \frac{1}{\bar{P}_\gamma} \cdot \frac{1}{N} \sum_{i=1}^{N} P_{\gamma,i} I^{\odot}_i , \tag{3}$$

where the sum runs over all N events observed in that bin; $P_{\gamma,i}$ and $I^{\odot}_i$ are the degree of circular beam polarization and the model asymmetry, respectively, for the kinematics of each of those events. The only major source of systematic uncertainty is the degree of the beam polarization, which is known to about 3%. The uncertainty from the beam-charge asymmetry is negligible (less than 10^{-3}).

3. Results

Figure 2 shows ϕ-angular distributions of the helicity asymmetry for various selected 50-MeV-wide center-of-mass energy bins between $W = 1.40$ GeV and 2.30 GeV. The data are integrated over the full CLAS acceptance. The analysis shows large asymmetries which change markedly with W up to 1.80 GeV; thereafter they remain rather stable. The asymmetries are odd functions of ϕ and vanish for coplanar kinematics ($\phi = 0$ and 180°), as

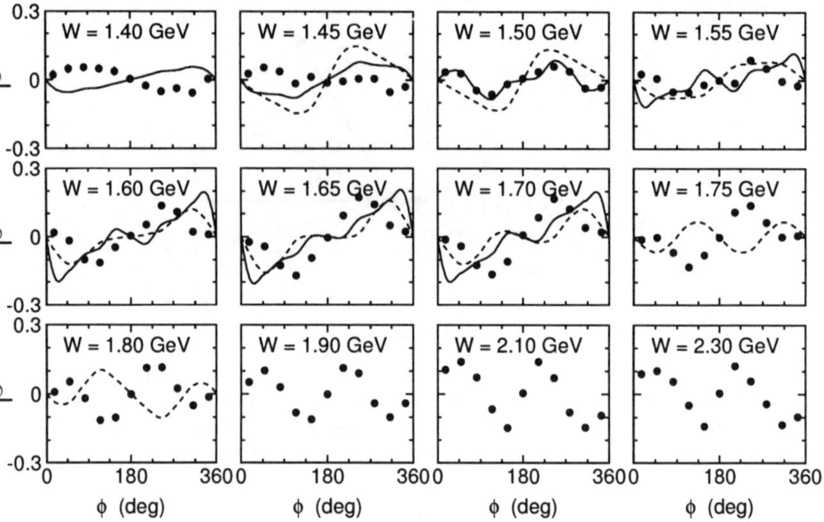

Fig. 2. Angular distributions for selected center-of-mass energy bins (each with $\Delta W =$ 50 MeV) of the beam-helicity asymmetry for the $\vec{\gamma} p \to p\pi^+\pi^-$ reaction. The data are integrated over the detector acceptance. The statistical uncertainties are smaller than the symbol size. The dashed and solid curves are the results from model calculations by Mokeev *et al.*[10] (for 1.45 GeV $\leq W \leq$ 1.80 GeV), and by Fix and Arenhövel[11] (for $W \leq 1.70$ GeV), respectively.

expected from parity conservation.[6] Thus only sine terms contribute to the Fourier expansion of these distributions, $\sum a_k \sin(k\phi)$. If two of the final-state particles are identical, like in the $\pi^0\pi^0 p$ final state, the form of the angular distributions is further restricted and only even-order terms enter the Fourier series. This has indeed been observed by the CLAS collaboration in beam-helicity asymmetry data in the $\vec{\gamma}{}^3\mathrm{He} \to ppn$ reaction, where the lowest order term is $a_2 \sin(2\phi)$.[12]

The data are compared with results of available phenomenological models. In the approach by Mokeev *et al.*[10] (dashed curves), double-charged-pion photo- and electroproduction are described by a set of quasi-two-body mechanisms with unstable particles in the intermediate states: $\pi\Delta$, ρN, $\pi N(1520)$, $\pi N(1680)$, $\pi\Delta(1600)$ and with subsequent decays to the $\pi^+\pi^- p$ final state. Residual direct $\pi^+\pi^- p$ mechanisms are parametrized by exchange diagrams. The model provides a good description of all available CLAS cross-section and world data on double-pion photo- and electroproduction at $W < 1.9$ GeV and $Q^2 < 1.5$ GeV^2. Results have also been obtained by Fix and Arenhövel using an effective Lagrangian approach with

Born and resonance diagrams at the tree level.[11] The corresponding results are shown in Fig. 2 as solid curves.

Although both models previously provided a good description of unpolarized cross sections, neither of the models give a reasonable description of the present beam-asymmetry data over the entire kinematic range covered in this experiment. Even though the model predictions agree remarkably well for certain conditions, for other conditions they are much worse and sometimes even out of phase entirely.

The result of a Fourier analysis of the ϕ distributions is shown in Fig. 3 and compared with model calculations by Fix and Arenhövel.[11] Apart of the $W = 1.4$ GeV region, excellent agreement is achieved in these calculations for the a_1 coefficient; yet the distribution of the a_2 coefficients deviates from the data above $W \approx 1.55$ GeV. It would be interesting to find out if there are specific reaction mechanisms contributing to specific terms in the Fourier decomposition of the angular distributions.

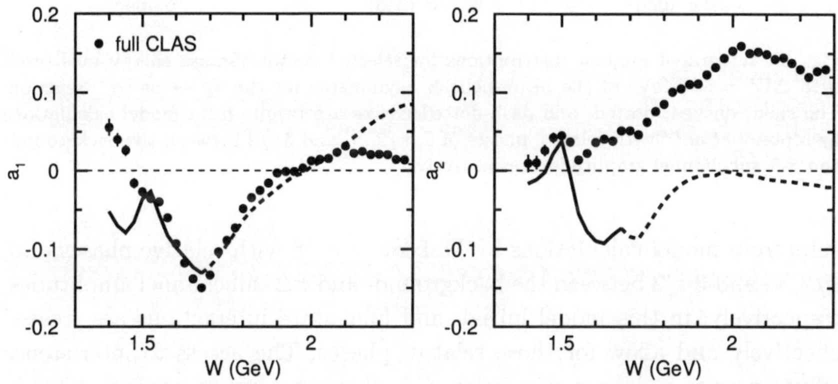

Fig. 3. First (a_1) and second (a_2) order Fourier components of the beam-helicity asymmetry as a function of the γp center-of-mass energy. The data are integrated over the full CLAS acceptance and compared with model calculations by Fix and Arenhövel[11] within (solid) and outside (dashed) the range of validity of that model ($W < 1.7$ GeV).

The main theoretical challenge for double-pion photoproduction lies in the fact that several subprocesses may contribute, even though any given individual contribution may be small. In his recent work Roca[13] studies the beam-helicity asymmetry and our preliminary data[14] in the framework of the Valencia model for double-pion photoproduction. He shows that the shape and strength of the asymmetries depend strongly on the internal mechanisms and interferences among different contributions to the process. In this connection, the polarization measurements should be very helpful in separating the individual terms. The particular sensitivity of the beam asymmetry to interference effects among various amplitudes is illustrated in Fig. 4. The solid, dashed, dotted, and dash-dotted curves are the re-

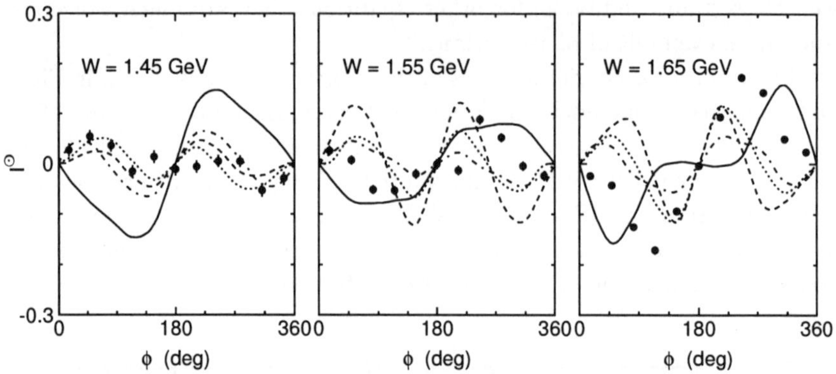

Fig. 4. Integrated angular distributions for selected center-of-mass energy bins (each with $\Delta W = 50$ MeV) of the beam-helicity asymmetry for the $\vec{\gamma}p \to p\pi^+\pi^-$ reaction. The solid, dashed, dotted, and dash-dotted curves are results from model calculations by Mokeev *et al.*[10] with relative phases of 0, $\pi/2$, π, and $3\pi/2$ between the background- and $\pi\Delta$-subchannel amplitudes, respectively.

sults from model calculations by Mokeev *et al.*[10] with relative phases of 0, $\pi/2$, π, and $3\pi/2$ between the background- and $\pi\Delta$-subchannel amplitudes, respectively; in this model initial- and final-state interactions are treated effectively and allow for those relative phases. The access to interference effects permit a cleaner separation of background and resonances. This in turn makes it possible to make more reliable statements about the existence and properties of nucleon resonances.

The large number of observed $\vec{\gamma}p \to p\pi^+\pi^-$ events allows for a confident analysis of the data in selected kinematic regions, making it possible to tune the different parts of the production amplitude independently. Figures 5 and 6 give two examples of more selective distributions of the same data around $W = 1.50$ GeV binned into nine bins in the invariant mass $M(p\pi^+)$ and into nine bins in the invariant mass $M(\pi^+\pi^-)$, respectively.

The double-pion final state allows further the study of sequential decays of nucleon resonances such as $N(1520) \to \pi\Delta \to \pi\pi N$. Figure 7 shows the Fourier coefficients a_1 and a_2 of the ϕ angular distributions as a function

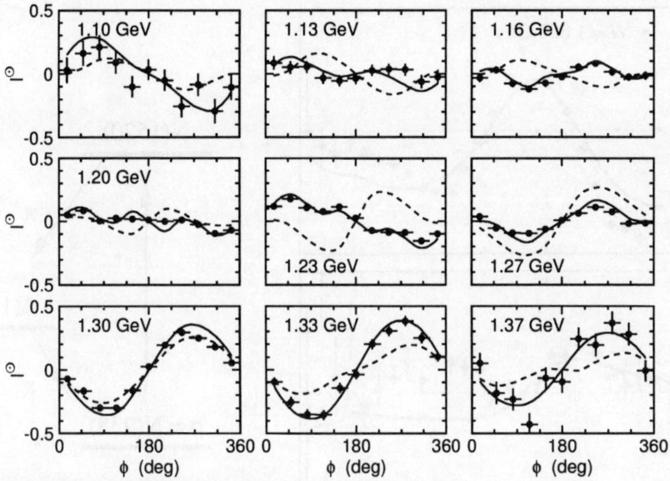

Fig. 5. Helicity asymmetries at $W = 1.50$ GeV for nine bins of the invariant mass $M(p\pi^+)$ with its mean values indicated in each panel. The asymmetries are otherwise integrated over the full CLAS acceptance. The solid curves are the results of Mokeev *et al.*;[10] the dashed curves show results of calculations by Fix and Arenhövel.[11]

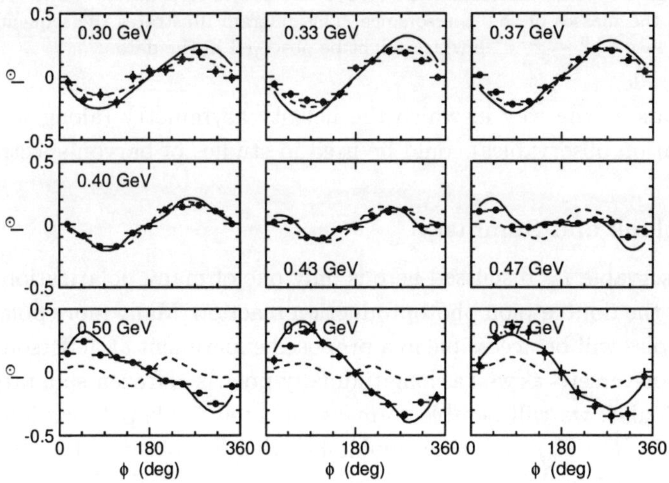

Fig. 6. Same as Fig. 5 for nine bins of the invariant mass $M(\pi^+\pi^-)$.

of the invariant mass $M(p\pi^-)$ for $W = 1.520$ GeV. The most interesting feature of these data is the change that occurs as $M(p\pi^-)$ traverses the $\Delta(1232)$ resonance. A maximum is seen in the region of this resonance.

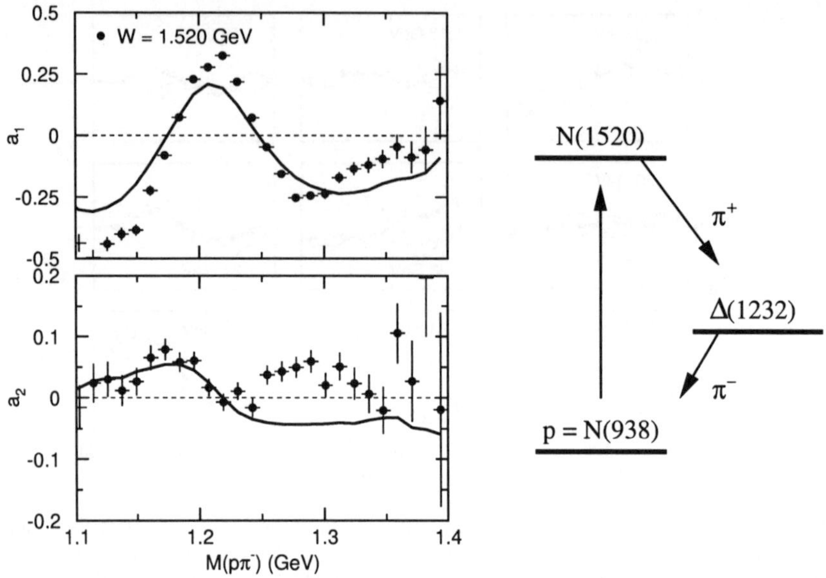

Fig. 7. Fourier coefficients a_1 and a_2 as a function of the invariant mass $M(p\pi^-)$ for $W = 1.520$ GeV. The curves are the results of Fix and Arenhövel.[11] The vertical line indicates the masses of the Δ resonance. The diagram illustrates the sequential decay $N(1520) \rightarrow \pi^+\Delta^0 \rightarrow \pi^+\pi^-p$ which might be observed in the data.

This hints at the way in which the helicity asymmetry (along with other polarization observables) could be used in studies of baryon spectroscopy.

4. Outlook and Summary

The observable $I^{\odot}$ discussed here is only one of many polarization observables in the double-pion photoproduction reaction. Many more polarization observables will be accessible in a proposed experiment at Jefferson Lab using a transversely- as well as longitudinally-polarized frozen-spin target.[15,16] In particular, we will be able to measure three single-polarization observables (P_x, P_y, P_z) and nine double-polarization observables ($P_x^{c,s}$, $P_y^{c,s}$, $P_z^{c,s}$, $P_x^{\odot}$, $P_y^{\odot}$, and $P_z^{\odot}$) in the mass range up to 2 GeV/c^2. Figure 8 shows as an example predictions of the double-polarization observable $P_z^{\odot}$ from the model of Fix and Arenhövel along with expected uncertainties of the proposed data. The data will be able to differentiate between various assumptions in the reaction dynamics; here, the question of the s- and d-wave decay of the $N(1520)$ resonance into $\pi\Delta$.

Additional constraints for models or partial-wave analyses are provided

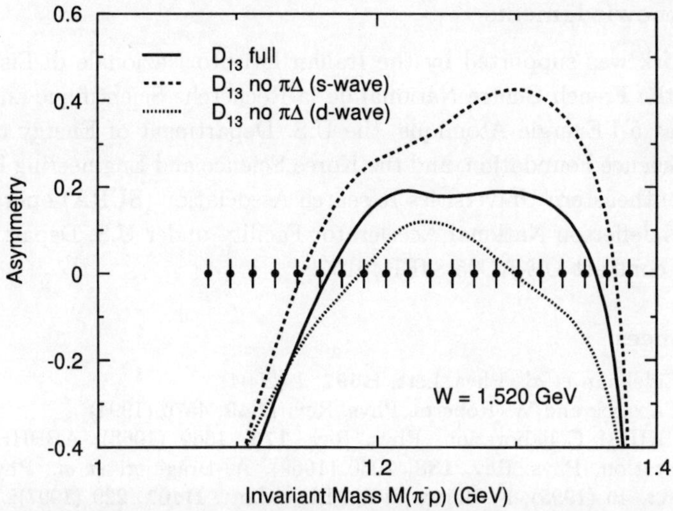

Fig. 8. Predictions of the double-polarization observable $P_z^{\odot}$ in the $\vec{\gamma}\vec{p} \to p\pi^+\pi^-$ reaction with circularly polarized photon beam and longitudinally polarized target at $W = 1.520$ GeV. The curves show results from a model by Fix and Arenhövel[11] for the full model (solid) and assuming no s-wave decay (dashed), or no d-wave decay (dotted) of the $N(1520)$ resonance. The points indicate the uncertainties which could be achieved in a proposed frozen-spin target experiment at CLAS.[15]

by recent studies of the $\gamma p \to \pi^0\pi^0 p$ reaction channel, and are particularly expected from new double-polarization experiments planned at ELSA.[17]

In summary, we have given a brief overview of our $\vec{\gamma}p \to p\pi^+\pi^-$ data, and we have demonstrated, by means of phenomenological models, the sensitivity of this helicity-asymmetry observable to the dynamics of the reaction. Although existing phenomenological models do describe unpolarized cross sections in the double-pion photo- and electroproduction reaction, they have severe shortcomings in the description of the beam-helicity asymmetries. In the region of overlapping nucleon resonances (and model dependent backgrounds), it clearly will be a challenge to any theoretical model to describe this and other new observables from future experiments that depend so sensitively on the interferences between them. Yet, without a proper understanding of the $\pi\pi N$ channel the problem of the "missing" resonances is unlikely to be resolved.

5. Acknowledgments

This work was supported by the Italian Istituto Nazionale di Fisica Nucleare, the French Centre National de la Recherche Scientifique and Commissariat à l'Energie Atomique, the U.S. Department of Energy and National Science Foundation, and the Korea Science and Engineering Foundation. Southeastern Universities Research Association (SURA) operates the Thomas Jefferson National Accelerator Facility under U.S. Department of Energy contract DE-AC05-84ER40150.

References

1. S. Eidelman *et al.*, Phys. Lett. **B592**, 1 (2004).
2. S. Capstick and W. Roberts, Phys. Rev. D **49**, 4570 (1994).
3. ABBHHM Collaboration, Phys. Rev. **175**, 1669 (1968); ABBHHM Collaboration, Phys. Rev. **188**, 2060 (1969); A. Braghieri *et al.*, Phys. Lett. **B363**, 46 (1995); F. Härter *et al.*, Phys. Lett. **B401**, 229 (1997); M. Wolf *et al.*, Eur. Phys. J. **A9**, 5 (2000); Y. Assafiri *et al.*, Phys. Rev. Lett. **90**, 222001 (2003); M. Ripani *et al.*, Phys. Rev. Lett. **91**, 022002 (2003); S. A. Philips, Ph.D. thesis, The George Washington University (2003); M. Bellis, Ph.D. thesis, Rensselaer Polytechnic Institute (2003); W. Langgärtner *et al.*, Phys. Rev. Lett. **87**, 052001 (2001).
4. J. Ballam *et al.*, Phys. Rev. D **5**, 545 (1972); J. Ahrens *et al.* (GDH and A2 Collaborations), Phys. Lett. **B551**, 49 (2003).
5. S. Strauch *et al.* (CLAS Collaboration), Phys. Rev. Lett. **95**, 162003 (2005).
6. W. Roberts and T. Oed, Phys. Rev. **C 71**, 055201 (2005); W. Roberts, these Proceedings.
7. B. A. Mecking *et al.*, Nucl. Instrum. Methods **A503**, 513 (2003).
8. D. I. Sober *et al.*, Nucl. Instrum. Methods **A440**, 263 (2000).
9. H. Olsen and L. C. Maximon, Phys. Rev. **114**, 887 (1959).
10. V. I. Mokeev *et al.*, Yad. Fiz. **64**, 1368 (2001), [Phys. At. Nucl. **64**, 1292 (2001)]; V. Burkert *et al.*, Nucl. Phys. **A737**, S231 (2004); V. I. Mokeev *et al.*, Proc. NSTAR2004 (World Scientific, New Jersey, 2004), p. 321; V. I. Mokeev, these Proceedings.
11. A. Fix and H. Arenhövel, Eur. Phys. J. A **25**, 115 (2005).
12. T.N. Ukwatta *et al.*, submitted to the Proceedings of the "*VI Latin American Symposium on Nuclear Physics and Applications*", Iguazú, Argentina (2005).
13. L. Roca, Nucl. Phys. **A748**, 192 (2005).
14. S. Strauch *et al.*, Proc. NSTAR2004 (World Scientific, New Jersey, 2004), p. 317.
15. Jefferson Lab proposal P06-013, "*Measurement of $\pi^+\pi^-$ Photoproduction in Double-Polarization Experiments using CLAS*", M. Bellis, V. Credé, S. Strauch, spokespeople.
16. F. Klein, these Proceedings.
17. U. Thoma, these Proceedings.

DYNAMICALLY GENERATED BARYON RESONANCES

M. F. M. LUTZ AND J. HOFMANN

Gesellschaft für Schwerionenforschung (GSI)
Planck Str. 1, 64291 Darmstadt, Germany

Identifying a zero-range exchange of vector mesons as the driving force for the s-wave scattering of pseudo-scalar mesons off the baryon ground states, a rich spectrum of molecules is formed. We argue that chiral symmetry and large-N_c considerations determine that part of the interaction which generates the spectrum. We suggest the existence of strongly bound crypto-exotic baryons, which contain a charm-anti-charm pair. Such states are narrow since they can decay only via OZI-violating processes. A narrow nucleon resonance is found at mass 3.52 GeV. It is a coupled-channel bound state of the $(\eta_c\, N), (\bar{D}\, \Sigma_c)$ system, which decays dominantly into the $(\eta' N)$ channel. Furthermore two isospin singlet hyperon states at mass 3.23 GeV and 3.58 GeV are observed as a consequence of coupled-channel interactions of the $(\bar{D}_s\, \Lambda_c), (\bar{D}\, \Xi_c)$ and $(\eta_c\, \Lambda), (\bar{D}\, \Xi_c')$ states. Most striking is the small width of about 1 MeV of the lower state. The upper state may be significantly broader due to a strong coupling to the $(\eta' \Lambda)$ state. The spectrum of crypto-exotic charm-zero states is completed with an isospin triplet state at 3.93 GeV and an isospin doublet state at 3.80 GeV. The dominant decay modes involve again the η' meson.

1. Introduction

The existence of strongly bound crypto-exotic baryon systems with hidden charm would be a striking feature of strong interactions [1–3]. Such states may be narrow since their strong decays are OZI-suppressed [4]. There are experimental hints that such states may indeed be part of nature. A high statistics bubble chamber experiment performed 30 years ago with a K^- beam reported on a possible signal for a hyperon resonance of mass 3.17 GeV of width smaller than 20 MeV [5]. About ten years later a further bubble chamber experiment using a high energy π^- beam suggested a nucleon resonance of mass 3.52 GeV with a narrow width of 7^{+20}_{-7} MeV . In Fig. 1 we recall the measured five body $(p\, K^+\, K^0\, \pi^- \pi^-)$ invariant mass distribution[6] suggesting the existence of a crypto exotic nucleon resonance.

It is the purpose of the present talk to review a study addressing the

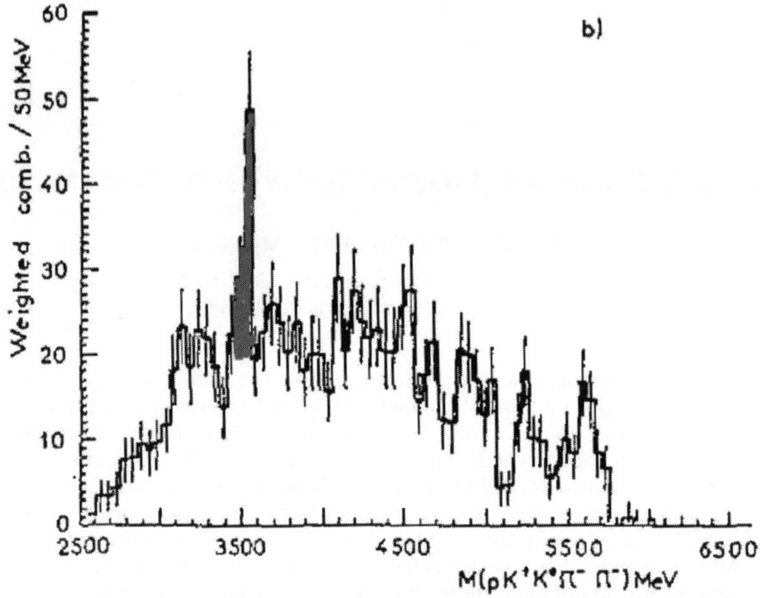

Fig. 1. Measured five-body invariant mass distribution in a 19 GeV π^- beam experiment at CERN. The figure is taken from Ref. [6]

possible existence of crypto-exotic baryon systems[7]. In view of the highly speculative nature of such states it is important to correlate the properties of such states to those firmly established, applying a unified and quantitative framework. We extended previous works [8,9] that performed a coupled-channel study of the s-wave scattering processes where a Goldstone boson hits an open-charm baryon ground state. The spectrum of $J^P = \frac{1}{2}^-$ and $J^P = \frac{3}{2}^-$ molecules obtained in [8,9] is quite compatible with the so far very few observed states. Analogous computations successfully describe the spectrum of open-charm mesons with $J^P = 0^+$ and 1^+ quantum numbers [10,11]. These developments were driven by the hadrogenesis conjecture: meson and baryon resonances that do not belong to the large-N_c ground state of QCD should be viewed as hadronic molecular states [12–16]. Generalizing those computations to include D- and η_c-mesons in the intermediate states offers the possibility to address the formation of crypto-exotic baryon states (see also [17–20]).

The results of [8,9] were based on the leading order chiral Lagrangian, that predicts unambiguously the s-wave interaction strength of Goldstone bosons with open-charm baryon states in terms of the pion decay constant. Including the light vector mesons as explicit degrees of freedom in a chiral Lagrangian gives an interpretation of the leading order interaction in terms of the zero-range t-channel exchange of light vector mesons [21–25]. The latter couple universally to any matter field in this type of approach. Based on the assumption that the interaction strength of D- and η_c-mesons with the baryon ground states is also dominated by the t-channel exchange of the light vector mesons, we performed a coupled-channel study of crypto-exotic baryon resonances[7].

2. Coupled-channel interactions

We consider the interaction of pseudoscalar mesons with the ground-state baryons composed out of u,d,s,c quarks. The pseudoscalar mesons that are considered in this work can be grouped into multiplet fields $\Phi_{[9]}, \Phi_{[3]}$ and $\Phi_{[1]}$ corresponding to the Goldstone bosons, the η', the D-mesons and the η_c. The baryon states are collected into SU(3) multiplet fields $B_{[8]}, B_{[6]}$ and $B_{[\bar{3}]}$ with charm 0,1 and 1. For an explicit representation of the various multiplet fields we refer to [7].

In a first step we construct the interaction of the mesons and baryon fields with the nonet-field $V^\mu_{[9]}$ of light vector mesons. We write down a list of relevant SU(3) invariant 3-point vertices that involve the minimal number of derivatives. Consider first the terms involving pseudo-scalar fields:

$$\begin{aligned}
\mathcal{L}^{\mathrm{SU(3)}}_{\mathrm{int}} &= \tfrac{i}{4}\, h^9_{\bar{3}3}\, \mathrm{tr}\left((\partial_\mu \Phi_{[3]})\, \Phi^\dagger_{[3]}\, V^\mu_{[9]} - \Phi_{[3]}\, (\partial_\mu \Phi^\dagger_{[3]})\, V^\mu_{[9]} \right) \\
&+ \tfrac{i}{4}\, h^1_{\bar{3}3}\, \mathrm{tr}\left((\partial_\mu \Phi_{[3]})\, \Phi^\dagger_{[3]} - \Phi_{[3]}\, (\partial_\mu \Phi^\dagger_{[3]}) \right) \cdot \mathrm{tr}\left(V^\mu_{[9]} \right) \\
&+ \tfrac{i}{4}\, h^9_{99}\, \mathrm{tr}\left((\partial_\mu \Phi_{[9]})\, \Phi_{[9]}\, V^\mu_{[9]} - \Phi_{[9]}\, (\partial_\mu \Phi_{[9]})\, V^\mu_{[9]} \right) . \qquad (1)
\end{aligned}$$

It needs to be emphasized that the terms in (1) are not at odds with the constraints set by chiral symmetry provided the light vector mesons are coupled to matter fields via a gauge principle [21,25]. The latter requires a correlation of the coupling constants h in (1)

$$h^9_{99} = \frac{(m^{(V)}_{[9]})^2}{2\,g\,f^2}\,, \qquad h^9_{\bar{3}3} = 2\,g\,, \qquad (2)$$

with the pion decay constant $f \simeq 92$ MeV. Here the universal vector coupling strength is $g \simeq 6.6$ and the mass of the light vector mesons is $m^{(V)}_{[9]}$.

We continue with the construction of the three-point vertices involving baryon fields. A list of SU(3) invariant terms reads:

$$
\begin{aligned}
\mathcal{L}_{\rm int}^{\rm SU(3)} &= \tfrac{1}{2}\, g^{9}_{\bar{3}3}\, {\rm tr}\left(\bar{B}_{[\bar{3}]}\,\gamma_\mu\, V^\mu_{[9]}\, B_{[\bar{3}]}\right) + \tfrac{1}{2}\, g^{1}_{\bar{3}3}\, {\rm tr}\left(\bar{B}_{[\bar{3}]}\,\gamma_\mu\, B_{[\bar{3}]}\right) {\rm tr}\left(V^\mu_{[9]}\right) \\
&+ \tfrac{1}{2}\, g^{9}_{66}\, {\rm tr}\left(\bar{B}_{[6]}\,\gamma_\mu\, V^\mu_{[9]}\, B_{[6]}\right) + \tfrac{1}{2}\, g^{1}_{66}\, {\rm tr}\left(\bar{B}_{[6]}\,\gamma_\mu\, B_{[6]}\right) {\rm tr}\left(V^\mu_{[9]}\right) \\
&+ \tfrac{1}{2}\, g^{9-}_{88}\, {\rm tr}\left(\bar{B}_{[8]}\,\gamma_\mu \left[V^\mu_{[9]}\,, B_{[8]}\right]_-\right) + \tfrac{1}{2}\, g^{9+}_{88}\, {\rm tr}\left(\bar{B}_{[8]}\,\gamma_\mu \left[V^\mu_{[9]}\,, B_{[8]}\right]_+\right) \\
&+ \tfrac{1}{2}\, g^{1}_{88}\, {\rm tr}\left(\bar{B}_{[8]}\,\gamma_\mu\, B_{[8]}\right) {\rm tr}\left(V^\mu_{[9]}\right) \\
&+ \tfrac{1}{2}\, g^{9}_{\bar{3}6}\, {\rm tr}\left(\bar{B}_{[\bar{3}]}\,\gamma_\mu\, V^\mu_{[9]}\, B_{[6]} + \bar{B}_{[6]}\,\gamma_\mu\, V^\mu_{[9]}\, B_{[\bar{3}]}\right) . \qquad (3)
\end{aligned}
$$

Within the hidden local symmetry model [25] chiral symmetry is recovered with

$$
g^{9}_{\bar{3}3} = g^{9}_{66} = 2\,g\,, \qquad g^{9,-}_{88} = g\,, \qquad g^{9,+}_{88} = 0\,, \qquad g^{9}_{\bar{3}6} = 0\,. \qquad (4)
$$

It is acknowledged that chiral symmetry does not constrain the coupling constants in (1, 3) involving the SU(3) singlet part of the fields. The latter can, however, be constrained by a large-N_c operator analysis [27]. At leading order in the $1/N_c$ expansion the OZI rule [28] is predicted. As a consequence the estimates

$$
h^{1}_{\bar{3}3} = -g\,, \qquad g^{1}_{\bar{3}3} = g^{1}_{66} = 0\,, \qquad g^{1}_{88} = g\,, \qquad (5)
$$

follow. We emphasize that the combination of chiral and large-N_c constraints (2, 4, 5) determine all coupling constants introduced in (1, 3).

We close this section by investigating the coupling of heavy vector mesons, $V^\mu_{[\bar{3}]}$ and $V^\mu_{[1]}$ to the meson and baryon fields. First we construct the most general SU(3) symmetric interaction terms involving the pseudoscalar fields:

$$
\begin{aligned}
\mathcal{L}_{\rm int}^{\rm SU(3)} &= \tfrac{i}{4}\, h^{0}_{\bar{3}3}\, {\rm tr}\left((\partial_\mu \Phi_{[\bar{3}]})\, \Phi^\dagger_{[\bar{3}]} V^\mu_{[1]} - \Phi_{[\bar{3}]}\, (\partial_\mu \Phi^\dagger_{[\bar{3}]})\, V^\mu_{[1]}\right) \\
&+ \tfrac{i}{4}\, h^{\bar{3}}_{9\bar{3}}\, {\rm tr}\left((\partial_\mu \Phi^\dagger_{[\bar{3}]})\, \Phi_{[9]}\, V^\mu_{[\bar{3}]} - \Phi_{[9]}\, (\partial_\mu \Phi_{[\bar{3}]})\, V^{\dagger\mu}_{[\bar{3}]}\right) \\
&+ \tfrac{i}{4}\, h^{\bar{3}}_{\bar{3}9}\, {\rm tr}\left((\partial_\mu \Phi_{[9]})\, \Phi_{[\bar{3}]} V^{\dagger\mu}_{[\bar{3}]} - \Phi^\dagger_{[\bar{3}]}\, (\partial_\mu \Phi_{[9]})\, V^\mu_{[\bar{3}]}\right) \\
&+ \tfrac{i}{4}\, h^{\bar{3}}_{\bar{3}0}\, {\rm tr}\left((\partial_\mu \Phi_{[1]})\, \Phi_{[\bar{3}]} V^{\dagger\mu}_{[\bar{3}]} - \Phi^\dagger_{[\bar{3}]}\, (\partial_\mu \Phi_{[1]})\, V^\mu_{[\bar{3}]}\right) \\
&+ \tfrac{i}{4}\, h^{\bar{3}}_{0\bar{3}}\, {\rm tr}\left((\partial_\mu \Phi^\dagger_{[\bar{3}]})\, \Phi_{[1]} V^\mu_{[\bar{3}]} - \Phi_{[1]}\, (\partial_\mu \Phi_{[\bar{3}]})\, V^{\dagger\mu}_{[\bar{3}]}\right) \\
&+ \tfrac{i}{4}\, h^{\bar{3}}_{\bar{3}1}\, {\rm tr}\left(\Phi^\dagger_{[\bar{3}]}\, V^\mu_{[\bar{3}]} - \Phi_{[\bar{3}]}\, V^{\dagger\mu}_{[\bar{3}]}\right) {\rm tr}\left(\partial_\mu \Phi_{[9]}\right) \\
&+ \tfrac{i}{4}\, h^{\bar{3}}_{1\bar{3}}\, {\rm tr}\left((\partial_\mu \Phi_{[\bar{3}]})\, V^{\dagger\mu}_{[\bar{3}]} - (\partial_\mu \Phi^\dagger_{[\bar{3}]})\, V^\mu_{[\bar{3}]}\right) {\rm tr}\left(\Phi_{[9]}\right) , \qquad (6)
\end{aligned}
$$

where the SU(4) symmetric gerneralization of (1) suggests the identification

$$h^{0}_{3\bar{3}} = \sqrt{2}\,g\,, \quad h^{\bar{3}}_{9\bar{3}} = 2\,g\,, \qquad h^{\bar{3}}_{39} = 2\,g\,,$$
$$h^{\bar{3}}_{30} = \sqrt{2}\,g\,, \quad h^{3}_{0\bar{3}} = \sqrt{2}\,g\,, \quad h^{\bar{3}}_{31} = g\,, \quad h^{3}_{1\bar{3}} = g\,. \tag{7}$$

The prediction of the SU(4)-symmetric coupling constants can be tested against the decay pattern of the D-meson. From the empirical branching ratio [26] we deduce $(h^{\bar{3}}_{39} + h^{3}_{9\bar{3}})/4 = 10.4 \pm 1.4$, which is confronted with the SU(4) estimate $(h^{\bar{3}}_{39} + h^{3}_{9\bar{3}})/4 = g \simeq 6.6$. We observe a moderate SU(4) breaking pattern. Based on this result one may expect the relations (7) to provide magnitudes for the coupling constants reliable within a factor two.

We close this section with the construction of the most general SU(3) symmetric interaction involving baryon fields

$$\begin{aligned}
\mathcal{L}^{\mathrm{SU}(3)}_{\mathrm{int}} &= \tfrac{1}{2}\, g^{0}_{88}\, \mathrm{tr}\left(\bar{B}_{[8]}\,\gamma_\mu\, B_{[8]}\, V^{\mu}_{[1]}\right) + \tfrac{1}{2}\, g^{0}_{66}\, \mathrm{tr}\left(\bar{B}_{[6]}\,\gamma_\mu\, B_{[6]}\, V^{\mu}_{[1]}\right) \\
&+ \tfrac{1}{2}\, g^{0}_{\bar{3}\bar{3}}\, \mathrm{tr}\left(\bar{B}_{[\bar{3}]}\,\gamma_\mu\, B_{[\bar{3}]}\, V^{\mu}_{[1]}\right) \\
&+ \tfrac{1}{2}\, g^{3}_{86}\, \mathrm{tr}\left(\bar{B}_{[8]}\,\gamma_\mu\, B_{[6]}\, V^{\dagger\mu}_{[3]} + \bar{B}_{[6]}\,\gamma_\mu\, B_{[8]}\, V^{\mu}_{[3]}\right) \\
&+ \tfrac{1}{2}\, g^{3}_{8\bar{3}}\, \mathrm{tr}\left(\bar{B}_{[8]}\,\gamma_\mu\, B_{[\bar{3}]}\, V^{\dagger\mu}_{[3]} + \bar{B}_{[\bar{3}]}\,\gamma_\mu\, B_{[8]}\, V^{\mu}_{[3]}\right),
\end{aligned} \tag{8}$$

where a SU(4) symmetric vertex implies

$$g^{0}_{88} = 0\,, \quad g^{0}_{66} = \sqrt{2}\,g\,, \quad g^{0}_{\bar{3}\bar{3}} = \sqrt{2}\,g\,, \quad g^{3}_{86} = \sqrt{2}\,g\,, \quad g^{3}_{8\bar{3}} = -\sqrt{6}\,g\,, \tag{9}$$

Unfortunately there appears to be no way at present to check on the usefulness of the result (9). Eventually simulations of QCD on a lattice may shed some light on this issue. The precise values of the coupling constants (7, 9) do not affect the major results of this study. This holds as long as those coupling constants range in the region suggested by (7, 9) within a factor two to three.

3. S-wave baryon resonances with zero charm

The spectrum of $J^P = \frac{1}{2}^-$ baryon resonances as generated by the t-channel vector-meson exchange interaction via coupled-channel dynamics falls into two types of states. Resonances with masses above 3 GeV couple strongly to mesons with non-zero charm content. In the SU(3) limit those states form an octet and a singlet. All other states have masses below 2 GeV. In the SU(3) limit they group into two degenerate octets and one singlet. The presence of the heavy channels does not affect that part of the spectrum at all. This is reflected in coupling constants of those states to the heavy channels within the typical range of $g \sim 0.1$ (see Tabs. 1-2). We reproduce the success of previous coupled-channel computations [15,16], which predicts the existence of the s-wave resonances $N(1535), \Lambda(1405), \Lambda(1670), \Xi(1690)$ unambiguously with masses and branching ratios quite compatible with empirical information. There are some quantitative differences. This is the consequence of the t-channel vector meson exchange, which, only in the SU(3) limit with degenerate vector meson masses, is equivalent to the Weinberg-Tomozawa interaction the computation in [15,16] was based on.

Most spectacular are the resonances with hidden charm above 3 GeV. The multiplet structure of such states is readily understood. The mesons with $C = -1$ form a triplet which is scattered off the $C = +1$ baryons forming an anti-triplet or sextet. We decompose the products into irreducible tensors

$$3 \otimes \overline{3} = 1 \oplus 8 \,, \qquad 3 \otimes 6 = 8 \oplus 10 \,. \tag{10}$$

The coupled channel interaction is attractive in the singlet for the triplet of baryons. Attraction in the octet sector is provided by the sextet of baryons. The resulting octet of states mixes with the $\eta' \,(N, \Lambda, \Sigma, \Xi)$ and $\eta_c \,(N, \Lambda, \Sigma, \Xi)$ systems. A complicated mixing pattern arises. All together the binding energies of the crypto-exotic states are large. This is in part due to the large masses of the coupled-channel states: the kinetic energy the attractive t-channel force has to overcome is reduced.

The states are narrow as a result of the OZI rule. The mechanism is analogous to the one explaining the long life time of the J/Ψ-meson. We should mention, however, a caveat. It turns out that the width of the crypto-exotic states is quite sensitive to the presence of channels involving the η' meson. This is a natural result since the η' meson is closely related to the $U_A(1)$ anomaly giving it large gluonic components. The latter work against the OZI rule. We emphasize that switching off the t-channel exchange of charm or using the SU(4) estimate for the latter, strongly bound crypto-exotic states are formed. In Tabs. 1-2 the zero-charm spectrum insisting on the SU(4) estimates (7, 9) is shown in the 3rd and 4th column. The mass of the crypto-exotic nucleon resonance comes at 3.33 GeV in this case. Its width of 160 MeV is completely dominated by the $\eta' N$ decay. The properties of that state can be adjusted easily to be consistent with the empirical values claimed in [6]. The η' coupling strength to the open-charm mesons can be turned off by decreasing the magnitude of $h_{\bar{3}1}^{\bar{3}}$ and $h_{1\bar{3}}^{\bar{3}}$ by 33.3 % away from their SU(4) values. As a result the width of the resonance is down to about 1-2 MeV. It is stressed that the masses of the crypto-exotic

Table 1. Spectrum of $J^P = \frac{1}{2}^-$ baryons with charm zero. The 3rd and 4th columns follow with SU(4) symmetric 3-point vertices. In the 5th and 6th columns SU(4) breaking is introduced with $h^1_{\bar{3}\bar{3}} \simeq -1.19\,g$ and $h^{\bar{3}}_{\bar{3}1} = h^{\bar{3}}_{1\bar{3}} \simeq 0.71\,g$. We use $g = 6.6$.

$C=0:\ (I,\ S)$	state	M_R[MeV] Γ_R [MeV]	$\lvert g_R\rvert$	M_R[MeV] Γ_R [MeV]	$\lvert g_R\rvert$
	πN		0.3		0.3
	ηN		2.1		2.1
	$K\Lambda$		1.7		1.6
$(\frac{1}{2},\ 0)$	$K\Sigma$	1535	3.3	1536	3.3
	$\eta' N$	95	0.0	94	0.0
	$\eta_c N$		0.0		0.0
	$\bar{D}\Lambda_c$		0.2		0.2
	$\bar{D}\Sigma_c$		0.2		0.2
	πN		0.1		0.07
	ηN		0.1		0.11
	$K\Lambda$		0.1		0.08
$(\frac{1}{2},\ 0)$	$K\Sigma$	3327	0.1	3520	0.08
	$\eta' N$	156	1.4	7.3	0.22
	$\eta_c N$		0.7		1.0
	$\bar{D}\Lambda_c$		0.5		0.05
	$\bar{D}\Sigma_c$		5.7		5.3
	$\pi\Sigma$		0.7		0.7
	$\bar{K}N$		2.7		2.7
	$\eta\Lambda$		1.1		1.1
	$K\Xi$	1413	0.1	1413	0.1
$(0, -1)$	$\eta'\Lambda$	10	0.0	10	0.0
	$\eta_c\Lambda$		0.0		0.0
	$\bar{D}_s\Lambda_c$		0.2		0.2
	$\bar{D}\Xi_c$		0.0		0.0
	$\bar{D}\Xi'_c$		0.0		0.0
	$\pi\Sigma$		0.2		0.2
	$\bar{K}N$		0.6		0.6
	$\eta\Lambda$		1.1		1.1
	$K\Xi$	1689	3.6	1689	3.6
$(0, -1)$	$\eta'\Lambda$	35	0.0	35	0.0
	$\eta_c\Lambda$		0.0		0.0
	$\bar{D}_s\Lambda_c$		0.1		0.1
	$\bar{D}\Xi_c$		0.1		0.1
	$\bar{D}\Xi'_c$		0.1		0.1
	$\pi\Sigma$		0.04		0.04
	$\bar{K}N$		0.03		0.03
	$\eta\Lambda$		0.03		0.03
	$K\Xi$	3148	0.04	3234	0.04
$(0, -1)$	$\eta'\Lambda$	1.0	0.08	0.57	0.01
	$\eta_c\Lambda$		0.08		0.06
	$\bar{D}_s\Lambda_c$		3.2		3.0
	$\bar{D}\Xi_c$		5.0		5.0
	$\bar{D}\Xi'_c$		0.1		0.01

states are not affected at all. The latter are increased most efficiently by allowing an OZI violating $\phi_\mu D\bar{D}$ vertex. We adjust $h^1_{\bar{3}\bar{3}} \simeq -1.19\,g$ and $h^{\bar{3}}_{\bar{3}1} = h^{\bar{3}}_{1\bar{3}} \simeq 0.71\,g$ as to obtain the nucleon resonance mass and width at 3.52 GeV and 7 MeV. For all other parameters the SU(4) estimates are used. The result of this choice of parameters is shown in the last two rows of Tabs. 1-2. Further crypto-exotic states, members of the aforementioned octet, are predicted at mass 3.58 GeV $(0, -1)$ and 3.93 GeV $(1, -1)$. The

Table 2. Continuation of Tab. 1.

$C=0:\ (I,\ S)$	state	M_R[MeV] Γ_R [MeV]	$\lvert g_R \rvert$	M_R[MeV] Γ_R [MeV]	$\lvert g_R \rvert$
	$\pi\,\Sigma$		0.1		0.06
	$\bar{K}\,N$		0.0		0.01
	$\eta\,\Lambda$		0.0		0.03
	$K\,\Xi$		0.1		0.07
$(0,-1)$	$\eta'\Lambda$	3432 161	1.3	3581 4.9	0.20
	$\eta_c\Lambda$		0.7		0.93
	$\bar{D}_s\Lambda_c$		0.6		0.05
	$\bar{D}\,\Xi_c$		0.1		0.02
	$\bar{D}\,\Xi'_c$		5.6		5.3
	$\pi\,\Lambda$		0.1		0.08
	$\pi\,\Sigma$		0.1		0.04
	$\bar{K}\,N$		0.2		0.12
	$\eta\,\Sigma$		0.1		0.08
	$K\,\Xi$		0.1		0.06
$(1,-1)$	$\eta'\Sigma$	3602 227	1.5	3930 11	0.27
	$\eta_c\Sigma$		1.2		1.8
	$\bar{D}\,\Xi_c$		0.6		0.11
	$\bar{D}_s\Sigma_c$		4.6		3.6
	$\bar{D}\,\Xi'_c$		2.9		2.4
	$\pi\,\Xi$		0.1		0.1
	$\bar{K}\,\Lambda$		0.4		0.4
	$\bar{K}\,\Sigma$		2.8		2.8
	$\eta\,\Xi$		1.3		1.3
$(\frac{1}{2},-2)$	$\eta'\Xi$	1644 3.0	0.0	1644 3.1	0.0
	$\eta_c\Xi$		0.0		0.0
	$\bar{D}_s\Xi_c$		0.2		0.2
	$\bar{D}_s\Xi'_c$		0.1		0.1
	$\bar{D}\,\Omega_c$		0.0		0.0
	$\pi\,\Xi$		0.1		0.08
	$\bar{K}\,\Lambda$		0.1		0.04
	$\bar{K}\,\Sigma$		0.1		0.04
	$\eta\,\Xi$		0.0		0.01
$(\frac{1}{2},-2)$	$\eta'\Xi$	3624 204	1.4	3798 6.0	0.22
	$\eta_c\Xi$		1.0		1.2
	$\bar{D}_s\Xi_c$		0.6		0.10
	$\bar{D}_s\Xi'_c$		3.3		2.9
	$\bar{D}\,\Omega_c$		4.3		4.0

multiplet is completed with a $(\frac{1}{2},-2)$ state at 3.80 GeV. The decay widths of these states center around ~ 7 MeV. This reflects the dominance of their decays into channels involving the η' meson. The coupling constants to the various channels are included in Tabs. 1-2. They confirm the interpretation that the crypto-exotic states discussed above are a consequence of a strongly attractive force between the charmed mesons and the baryon sextet.

We close with a discussion of the crypto-exotic SU(3) singlet state, which is formed due to strong attraction in the $(\bar{D}_s\Lambda_c)$, $(\bar{D}\,\Xi_c)$ system. Its nature is quite different as compared to the one of the octet states. This is because its coupling to the $\eta'\Lambda$ channel is largely suppressed. Indeed its width is independent on the magnitude of $h^{\bar{3}}_{\bar{3}1} = h^{\bar{3}}_{1\bar{3}}$ as demonstrated in Tabs. 1-2. We identify this state with a signal claimed in the K^-p reaction, where a narrow hyperon state with 3.17 GeV mass and width smaller than 20 MeV was seen [5]. Using values for the coupling constants as suggested by SU(4) the state has a mass and width of 3.148 GeV and 1 MeV (see 3rd and 4th column of Tabs. 1-2). Our favored parameter set with $h^{1}_{\bar{3}\bar{3}} \simeq -1.19\, g$ and $h^{\bar{3}}_{\bar{3}1} = h^{\bar{3}}_{1\bar{3}} \simeq 0.71\, g$ predicts a somewhat reduced binding energy.

4. Summary

We reviewed a coupled-channel study of s-wave baryon resonances with charm 0. The interaction is defined by the exchange of light vector mesons in the t-channel. All relevant coupling constants are obtained from chiral and large-N_c properties of QCD. Less relevant vertices related to the t-channel forces induced by the exchange of charmed vector mesons were estimated by applying SU(4) symmetry. Most spectacular is the prediction of narrow crypto-exotic baryons with charm zero forming below 4 GeV. Such states contain a $c\bar{c}$ pair. Their widths parameters are small due to the OZI rule, like it is the case for the J/Ψ meson. We predict an octet of crypto-exotic states which decay dominantly into channels involving an η' meson. An even stronger bound crypto-exotic SU(3) singlet state is predicted to have a decay width of about 1 MeV only. We recover the masses and widths of a crypto-exotic nucleon and hyperon resonance suggested in high statistic bubble chamber experiments [5,6].

References

1. S.J. Brodsky, I. Schmidt and G.F. de Teramond, Phys. Rev. Lett. **64** (1990) 1011.
2. A.B. Kaidalov and P.E. Volkovitsky, Phys. Rev. Lett. **69** (1992) 3155.
3. C. Gobbi, D.O. Riska and N.N. Scoccola, Phys. Lett. **B 296** (1992) 166.
4. L.G. Landsberg, Phys. Atom. Nucl. **57** (1994) 2127.
5. T. Amirzadeh et al., Phys. Lett. **B 89** (1979) 125.
6. V.M. Karnaukhov et al., Phys. Lett. **B 281** (1991) 148.
7. J. Hofmann and M.F.M. Lutz, hep-ph/0507071.
8. M.F.M. Lutz and E.E. Kolomeitsev, Nucl. Phys. **A 730** (2004) 110.
9. M.F.M. Lutz and E. E. Kolomeitsev, Nucl. Phys. **A 755** (2005) 29c.
10. E.E. Kolomeitsev and M.F.M. Lutz, Phys. Lett. **B 585** (2004) 243.
11. J. Hofmann and M.F.M. Lutz, Nucl. Phys. **A 733** (2004) 142.
12. M.F.M. Lutz and E. E. Kolomeitsev, Nucl. Phys. **A 700** (2002) 193.
13. M.F.M. Lutz, Gy. Wolf and B. Friman, Nucl. Phys. **A 706** (2002) 431.
14. M.F.M. Lutz and E.E. Kolomeitsev, Nucl. Phys. **A 730** (2004) 392.
15. C. García-Recio, M.F.M. Lutz and J. Nieves, Phys. Lett. **B 582** (2004) 49.
16. E.E. Kolomeitsev and M.F.M. Lutz, Phys. Lett. **B 585** (2004) 243.
17. Ch. Hong-Mo and H. Högaasen, Z. Phys. **C 7** (1980) 25.
18. M. Rho, D.O. Riska and N.N. Scoccola, Z. Phys. **A 341** (1992) 343.

19. D.-P. Min, Y. Oh, B.-Y. Park and M. Rho, Int. J. Mod. **E 4** (1995) 47.
20. Y. Oh and H. Kim, Phys. Rev. **D 70** (2004) 094022.
21. S. Weinberg, Phys. Rev. **166** (1968) 1568.
22. H.W. Wyld, Phys. Rev. **155** (1967) 1649.
23. R.H. Dalitz, T.C. Wong and G. Rajasekaran, Phys. Rev. **153** (1967) 1617.
24. P.B. Siegel and W. Weise, Phys. Rev. C **38** (1988) 2221.
25. M. Bando et al., Phys. Rev. Lett. **54** (1985) 1215.
26. S. Eidelman et al., Phys. Lett. **B 592** (2004) 1.
27. R. Dashen, E. Jenkins and A.V. Manohar, Phys. Rev. **D 49** (1994) 4713.
28. S. Okubo, Phys. Lett. **5** (1963) 165; G. Zweig, CERN Reports TH-401, TH-412 (1964), unpublished; J. Iiuzuka, Prog. Theor. Part. Sci. **37** (1966) 21.

DESCRIBING THE BARYON SPECTRUM WITH $1/N_c$ QCD

R. F. LEBED
Department of Physics & Astronomy,
Arizona State University,
Tempe, AZ 85287-1504, USA
E-mail: richard.lebed@asu.edu

This talk outlines recent advances using QCD in the $1/N_c$ limit aimed at understanding baryon scattering processes and their embedded short-lived baryon resonances. In this presentation we emphasize developing qualitative physical insight over presenting results of detailed calculations.

1. Introduction

When addressing an audience of baryon resonance experts, it is hardly necessary to emphasize the elusive nature of the N^*s as both experimental and theoretical objects: Owing to their extremely short $O(10^{-23}$ s) lifetimes, they are often barely discernable, lurking in baryon scattering amplitudes like strangers in a fog. My previous talk write-ups on this material[1] have been geared exclusively towards theory audiences, but an N^* conference is attended by a large number of experimentalists as well, who view theory talks with an eye toward picking up new notions of physical understanding for the phenomena that they study, rather than focusing on calculational detail. I therefore wish to focus here on the qualitative description of the motivation behind and the results of my recent work with Tom Cohen on excited baryons.[2–9] The reader who craves more detail is welcomed to peruse Refs. 1 or the original works.

2. Two Physical Pictures for N^*s

The most frequently invoked picture for baryons is that suggested by the constituent quark model, in which the light (masses ~ 5 MeV) fundamental quarks of the QCD Lagrangian somehow agglomerate with the multitude of gluons and virtual quark-antiquark pairs to form constituent (~ 300 MeV) quarks. In order to be discernable as distinct entities, such pseudoparticles must nevertheless remain weakly bound to one another. If this physical picture is valid, then the baryon, originally a complicated many-body object only describable using full quantum field theory, reduces to a simple three-particle quantum-mechanical system interacting through a potential, not unlike a miniature atomic nucleus. In this case the baryon excited states consist of orbital and radial excitations of the three constituents. Inasmuch as the constituent quark masses are larger than the energies that bind them, the baryons fill well-defined multiplets based upon approximate invariances of the state under quark spin flips, quark flavor substitutions, and spatial exchanges, the SU(6)×O(3) symmetry.

Constituent quark models therefore predict numerous excited hadron multiplets, the lowest of which have indeed been observed. For example, the ground states, consisting of the nucleons, the Δ resonances (related to the nucleons by a spin flip), and their strange partners, fill a spin-flavor-space symmetric $(\mathbf{56}, 0^+)$ of SU(6)×O(3), while the lightest excitations appear to fill the orbitally-excited mixed-symmetry multiplet $(\mathbf{70}, 1^-)$ or a radially-excited $(\mathbf{56}, 0^+)$. However, higher in the spectrum the picture becomes much murkier, with numerous partly-filled multiplets as well as predicted multiplets whose members remain unobserved.

Alternately, the chiral soliton picture for baryons, starting directly from a hadronic perspective, recognizes that hadrons rather than quarks are the states observed in nature. *Solitons* are semiclassical finite-energy solutions to a field theory, which is to say that they are non-dissipating "lumps" of energy (such as a lump in a rug placed in a room too small: It can be moved from place to place, but not eliminated). Chiral Lagrangians, which have been so successful in delimiting light meson dynamics, admit solitonic solutions that couple to mesons according to chiral symmetry constraints. Their semiclassical nature is guaranteed if they are heavy compared to the mesons, just as is physically true for the baryons. In the best-studied variant, the Skyrme model, the solitons are shown to carry fermionic statistics.

The basic soliton configuration, called a *hedgehog*, turns out not to possess a single well-defined isospin or spin quantum number, but rather a quantum number that is the magnitude of their vector sum $\mathbf{K} \equiv \mathbf{I} + \mathbf{J}$, sometimes called the *grand spin*. Physical baryon states with particular spin and isospin eigenvalues are then recovered by forming a judicious linear combination of hedgehog states of different K; these "judicious" couplings are none other than Clebsch-Gordan coefficients (CGC). The couplings of mesons to the underlying hedgehog, as arise in scattering processes, also induce spin and isospin CGC.

Excited baryons in chiral soliton models appear as rotational or vibrational excitations of the basic hedgehog configuration. Much of the particular spectrum generated by such excitations depends strongly upon the details of the dynamical "profile" functions multiplying the hedgehog, making predictions of baryon resonance multiplets in chiral soliton models less than robust.

3. Large N_c QCD and the $1/N_c$ Expansion

While both the constituent quark and chiral soliton models warrant attention for incorporating observable features of baryons, they remain just that—models. In both cases, an expansive literature demonstrates that one may refine the models by including subleading effects, but it is not *a priori* obvious which corrections are essential for understanding baryon dynamics. Instead, we prefer to obtain a method directly from QCD that combines the best features of both pictures. *Ab initio* lattice calculations applied to excited baryons hold great promise for the future,[11] but even when completed will provide numerical results rather than definitive dynamical statements.

Large N_c QCD, obtained by supposing that QCD had not 3 but some larger number N_c of color charges, is not a model but rather an extension of the field theory representing strong interactions. It is physically useful if *i*) physical observables have well-defined limits as $N_c \to \infty$ [*i.e.*, with small $O(1/N_c)$ corrections], and *ii*) the values of these observables do not change excessively as N_c is allowed to decrease from a large value down to 3. The key question then becomes whether one can recognize in observables unambiguous signatures of this expansion in powers of $1/3$, and in fact the $(\mathbf{56}, 0^+)$ baryons provide ample evidence[12] in their spectra and couplings.

We first require a few fundamental baryon results. For N_c colors, the baryons contain at least N_c quarks, the number required to form a colorless state. Baryons have $O(N_c^1)$ masses, and meson couplings that are $O(N_c^{1/2})$ (trilinear) and $O(N_c^0)$ (quartic).[13] The latter fact implies that ordinary baryon resonances, since they appear in baryon-meson scattering amplitudes, have masses above the ground states and widths each of $O(N_c^0)$. The baryons themselves, despite having large masses at large N_c, maintain an essentially constant $[O(N_c^0)]$ size, which follows from the suppression of multiple-quark interactions by powers of N_c. Lastly, order-by-order unitarity in N_c powers in baryon-meson scattering processes (called *consistency conditions*[14,15]) require the ground-state multiplet to have not only spin-$\frac{1}{2}$ but spin-$\frac{3}{2}$ members as well, the large N_c analogue to the **56** [for $N_c > 3$ the completely symmetric SU(6) multiplet also contains up to spin-$\frac{N_c}{2}$ states].

Both the quark model and the chiral soliton model have straightforward extensions to arbitrary N_c. Of course, N_c must be odd for baryons to remain fermions. In the quark model case, one may define[16] quantum fields with all the properties of constituent quarks by noting that ground-state baryons carry precisely the quantum numbers of N_c quarks (which remains true for $N_c = 3$; this of course was the original motivation of the quark model), and dividing the baryon into N_c non-overlapping "interpolating fields" that exhaust its wave function. Using this definition for the quarks, the suppression of multiquark operators by powers of $1/N_c$ allows one to conclude that effects carrying the spin-flavor quantum numbers of such operators are also suppressed. If the states are stable against strong decays (as is the case for the ground-state multiplet), one may construct a Hamiltonian for which these baryons are the asymptotic states, and matrix elements are computed by means of the Wigner-Eckart theorem. For example, the nucleon and Δ masses are split only at $O(1/N_c)$ because this is the order of the lowest-order (hyperfine) Hamiltonian operator distinguishing their masses; the exact coefficient remains incalculable unless the strong interactions can be solved from first principles, but if the $1/N_c$ expansion is valid, then it should be a typical hadronic scale (a few hundred MeV) times an $O(1)$ number. Indeed, the observed N-Δ splitting follows this pattern.[17]

One may attempt an extension of this approach to the excited baryons. A large body of literature[18] treats (for example) the lightest negative-parity resonances as filling the analogue to the $(\mathbf{70}, 1^-)$, a symmetrized core of $N_c - 1$ quarks and one excited quark. While this approach has yielded many interesting phenomenological insights, its strict application seems sensible only when *i*) the excited baryons are also asymptotically stable states of a Hamiltonian, and *ii*) can be represented uniquely as 1-quark excitations of a ground state (*i.e.*, *configuration mixing* with states having 2 or more excited quarks but the same overall quantum numbers are ignored).

Chiral soliton models also combine efficiently with the $1/N_c$ expansion. Indeed, much of the interest in such models during the early 1980s centered on the fact that the semiclassical nature of the solitons was consistent with the heaviness of large N_c baryons, in that many of their predictions turned out to be independent of the particular choice of profile function.[19] Subsequent work[20] showed that quark and soliton models for ground-state baryons share common group-theoretical features in the large N_c limit. But these results apply only to the ground-state multiplet, whose members are related by various rotations of the basic hedgehog state.

4. Resonances in the $1/N_c$ Expansion

Since soliton models can be used to study baryon scattering amplitudes, it begs the question whether one can use these models to reach beyond the ground states and obtain definite statements about resonances with a degree of model independence inherited from large N_c. A successful picture for resonances ought not put them in by hand; they are intrinsically excitations in baryon scattering amplitudes and should be generated as complex-valued poles ($z_R = M_R + \frac{i}{2}\Gamma_R$) within them. Work along these lines in the mid-1980s began with Ref. 21 and rapidly progressed to focus upon model-independent group-theoretical features:[22] In particular, from this approach one finds a number of linear relations between distinct partial-wave amplitudes.

The central feature driving these works is the underlying conservation of K-spin. As we have seen, not only the composition of baryon states from the hedgehog, but also the couplings of baryon-meson scattering processes, introduce group-theoretical factors. Carefully combining them yields the full set of baryon partial wave amplitudes written as linear combinations of a smaller set of underlying *reduced amplitudes* labeled by K, while composing the CGC leads to coefficients that are purely group-theoretical $6j$ and $9j$ factors. As a trivial example, for πN scattering one obtains $S_{11} = S_{31}$.

Based upon interesting regularities noted for scattering processes viewed in the t-exchange channel,[23] K-spin conservation (expressed in terms of the usual s-channel quantum numbers) was shown[24] to be equivalent to the t-channel rule $I_t = J_t$. It was not until several years later, however, that the $I_t = J_t$ rule was shown[25] to follow directly from large N_c consistency conditions, completing the ingredients of the proof[2] that underlying K-spin conservation is a direct result of the large N_c limit.

To say that full baryon partial waves are linearly related for large N_c means that a resonant pole occurring in any one of them must appear in at least one of the others, or more fundamentally, in one of the reduced amplitudes. However, since a given reduced amplitude contributes to multiple partial waves, the same resonant pole appears in each one: Large N_c baryon resonances appear in multiplets degenerate in both mass and width.[2]

Large N_c baryon resonances are not the exclusive provenance of soliton models; if one considers the large N_c generalization of the $(\mathbf{70}, 1^-)$ using the Hamiltonian approach described above, one finds[2,9,26] that only 5 distinct mass eigenvalues occur up to $O(N_c^0)$ inclusive, the level at which distinct resonances of the ground states split in mass. When one examines all partial waves in which states carrying these quantum numbers can occur, one finds that all of the states in the multiplet are induced by one pole in each of the reduced amplitudes with $K = 0, \frac{1}{2}, 1, \frac{3}{2}$, and 2 (and only $K = 0, 1, 2$ occur for the nonstrange states). From the point of view of large N_c, the irreducible multiplet $(\mathbf{70}, 1^-)$ of SU(6)×O(3) is therefore actually a reducible collection of 5 *distinct* irreducible multiplets, which are labeled by $K=0, \frac{1}{2}, 1, \frac{3}{2}$, and 2; let us label the masses as m_K. When SU(3) flavor symmetry is invoked, K may also be defined for strange states, where it is simply defined as the magnitude of $\mathbf{I}+\mathbf{J}$ for the nonstrange member of the SU(3) multiplet. A similar pattern, which we call *compatibility*,[3,7] occurs for every SU(6)×O(3) multiplet, each of which decomposes at large N_c into a collection of irreducible multiplets labeled by K: Each quark-model multiplet forms a collection of distinct resonance multiplets. This result generalizes the one discussed above, that the ground-state multiplet in large N_c forms a complete $(\mathbf{56}, 0^+)$ (in this case, only $K=0$ appears).

5. Phenomenological Consequences

The quark and chiral soliton approaches thus find common ground for large N_c by having compatible resonance multiplets. But this is a formal result; to find phenomenological successes, one needs to go no further than examining which reduced amplitudes appear in a given partial wave amplitude.

To illustrate this point, let us consider the lightest $I=\frac{1}{2}$, $J=\frac{1}{2}$ $(N_{1/2})$ negative-parity states. It turns out for any $N_c \geq 3$ that $(\mathbf{70}, 1^-)$ contains precisely 2 $N_{1/2}$ states; for $N_c=3$ these are $N(1535)$ and $N(1650)$. Using only the group theory imposed by the $N_c \to \infty$ limit, ηN states at large N_c allow only $K=0$ amplitudes, while the process $\pi N \to \pi N$ allows only $K=1$. Thus, only the resonance of mass m_0 appears in ηN amplitudes, and only m_1 appears in $\pi N \to \pi N$. As is well known to this audience, $N(1535)$ lies just barely above the ηN threshold and yet decays to it as frequently as to the heavily phase-space favored πN channel. Alternately, the $N(1650)$ has a πN branching ratio many times larger than for ηN despite a much more comparable phase space in these channels.[2] The $N(1535)$ πN and $N(1650)$ ηN couplings thus arise only through subleading corrections of the size expected from the $1/N_c$ expansion.

Results of this sort also appear among the strange resonances.[9] In particular, the $N(1535)$ appears to be just the nonstrange member of an entire $K=0$ octet of resonances, all of which therefore are η-philic and π-phobic. As evidence, note that the $\Lambda(1670)$ lies only 5 MeV above $\eta\Lambda(1116)$ threshold, and yet this channel has a 10–25% branching ratio.

Even stranger selection rules occur when full SU(3) group theory is taken into account.[9] For example, one can prove for N_c arbitrary that resonances in SU(3) multiplets whose highest hypercharge states are nonstrange (**8** and **10**) decay preferentially (by a factor $N_c^{\frac{1}{2}}$) with a π or η, while those whose top states are strange (**1**) prefer $\overline{K}$ decays by $O(N_c^{\frac{1}{2}})$. Evidence for this peculiar prediction is borne out by the $\Lambda(1520)$: Its branching ratios for $\overline{K}N$ and $\Sigma\pi$ are roughly equal, but when the near-threshold p^{2L+1} behavior for this d wave is taken into account, one finds the effective coupling constant ratio $g(\Lambda(1520)\to\overline{K}N)/g(\Lambda(1520)\to\Sigma\pi)\sim 4\text{–}5 = O(N_c)$, as advertised.

$1/N_c$ corrections may be incorporated by noting the demonstration that the $I_t = J_t$ rule is equivalent to the large N_c limit[25] also shows amplitudes with $|I_t - J_t| = n$ to be suppressed by at least $1/N_c^n$. To incorporate all possible $O(1/N_c)$ effects one simply appends to all possible amplitudes with $I_t = J_t$ those with $I_t - J_t = \pm 1$.[6] The number of reduced amplitudes then increases while the number of observable partial waves of course remains the same, making linear relations tougher to obtain; for example, no such $1/N_c$-corrected relations occur among $\pi N \to \pi N$, but $\pi N \to \pi\Delta$ relations do occur, and definitely improve by about a factor of 3 when the $1/N_c$ corrections are taken into account.[6]

We have noted that configuration mixing between different states with the same overall quantum numbers can be a nuisance within specific models by requiring additional assumptions. A true advantage of treating excited baryons as resonances in partial wave amplitudes is that configuration mixing can occur naturally. As an example of this philosophy, if one model predicts an especially narrow excited baryon [say, a width of $O(1/N_c)$], and if there exist broad resonances $[O(N_c^0)]$ in the same mass region with the same overall quantum numbers, then generically the states mix and produce two broad resonances.[4] In the quark picture, for example, this mixing occurs any time one can find a Hamiltonian operator with transition matrix elements of $O(N_c^0)$ between the two states.

The existence of well-defined multiplets of resonances at large N_c is also an aid to searching for exotic states.[5] For example, let us suppose that the pentaquark candidate $\Theta^+(1540)$ were confirmed with hypercharge $+2$, $I=0$, $J=\frac{1}{2}$, and either parity. Then large N_c, independently of any model, mandates that it must have $I = 1$, $J = \frac{1}{2}, \frac{3}{2}$ and $I = 2$, $J = \frac{3}{2}$ partners with the same mass [up to $O(1/N_c)$ corrections, less than about 200 MeV] and the same width [which of course can magnify or shrink in response to nearby thresholds, again indicating $O(1/N_c)$ differences].

Studies of baryon scattering amplitudes are not limited only to couplings with mesons. As long as the quantum numbers and the $1/N_c$ couplings of the field to the baryons is known, precisely the same methods apply. Processes such as photoproduction, electroproduction, real or virtual Compton scattering are then open to scrutiny. In the case of pion photoproduction, the photon carries both isovector and isoscalar quantum numbers, with the former dominating[27] by a factor N_c. Including the leading and first subleading isovector and the leading isoscalar amplitudes then gives linear relations among multipole amplitudes with relative $O(1/N_c^2)$ corrections.[8] Some of the relations obtained this way (*e.g.*, the prediction that isovector amplitude combinations dominate isoscalar ones) agree quite impressively with data. Some, however, do not appear to the eye to fare as well. In those cases, the threshold behaviors still agree quite well, followed by seemingly disparate behavior in the respective resonant regions. Does this mean that the $1/N_c$ expansion is failing? Not so: The disagreements come from resonances in the different partial waves whose masses are split at $O(1/N_c)$, giving critical behavior occurring in different places in distinct partial waves. When this effect is taken into account by extracting couplings *on resonance* (as presented by the Particle Data Group[28]), the linear relations good to $O(1/N_c^2)$ do indeed produce results that agree to within 10–15%.[8]

6. Looking Ahead

A very brief summary tells us where this program is at the current time: We now have at our disposal the correct large N_c method of studying baryon resonances of finite widths model-independently, *i.e.*, in the context of a full quantum field theory. Multiplets of resonances degenerate in masses and widths naturally arise in this approach, and are similar but not identical to old quark-model multiplets. The first phenomenological results have been very encouraging, demonstrating that the $1/N_c$ expansion continues to bear a rich harvest for the excited states. Not only the resonances themselves, but the partial wave amplitudes in which they appear, can be studied using the same methods.

The most important issue yet unsolved in this program is how to treat spurious states, *i.e.*, those that occur only for $N_c > 3$. Indeed, we were loose in our notation when we spoke of, for example, the SU(6) **56** or the SU(3) **8**, which contain (due to quark combinatorics) many more than the given number of states when $N_c > 3$. As commented above, we obtain interesting results for specific states occurring with the same multiplicities for all $N_c \geq 3$, such as negative-parity $N_{1/2}$'s. However, many more high-spin and high-isospin states occur for $N_c > 3$. Which ones survive at $N_c = 3$ and which ones do not? Since this is a difference between $N_c \to \infty$ and

$N_c = 3$, it represents a special kind of $1/N_c$ correction yet to be mastered.

All results thus far obtain from 2-to-2-particle scattering processes. In fact, multiparticle processes such at $\pi N \to \pi\pi N$ are not substantially more difficult in many cases of interest. For example, if the $\pi\pi$ pair is identified by reconstruction as originating from a ρ, then the suppressed width $[O(1/N_c)]$ of ρ allows the process to be studied in factorized form.

The reader should note that physical input within this method has been virtually nil: Only the imposition of an organizing principle, around suppressions in powers of $1/N_c$, has occurred. In this sense, the $1/N_c$ methods employed thus far have the flavor of chiral Lagrangians, which obtain results using only symmetries and a low-momentum expansion. Indeed, one thrust of future work will be the folding of chiral symmetry (*e.g.*, low-energy theorems) into the $1/N_c$ expansion; our preliminary examination suggests this to be a promising direction.

The essential tools thus appear to be in place to disentangle the fundamental features of the N^* spectrum using a systematic approach, much as chiral Lagrangians have done for the light mesons. Given sufficient time and resources, it is a program well within the reach of the N^* community.

Acknowledgments

This work was supported in part by the National Science Foundation under Grant No. PHY-0456520.

References

1. R.F. Lebed, hep-ph/0406236, published in *Continuous Advances in QCD 2004*, edited by T. Gherghetta, World Scientific, Singapore, 2004; hep-ph/0501021, published in *Large N_c QCD 2004*, edited by J.L. Goity *et al.*, World Scientific, Hackensack, NJ, USA (2005); hep-ph/0509020, invited talk at *International Conference on QCD and Hadronic Physics*, Beijing, 16–20 June 2005 (to appear in proceedings).
2. T.D. Cohen and R.F. Lebed, *Phys. Rev. Lett.* **91**, 012001 (2003); *Phys. Rev.* D **67**, 012001 (2003).
3. T.D. Cohen and R.F. Lebed, *Phys. Rev.* D **68**, 056003 (2003).
4. T.D. Cohen, D.C. Dakin, A. Nellore, and R.F. Lebed, *Phys. Rev.* D **69**, 056001 (2004).
5. T.D. Cohen and R.F. Lebed, *Phys. Lett.* B **578**, 150 (2004); *Phys. Lett.* B **619**, 115 (2005).
6. T.D. Cohen, D.C. Dakin, A. Nellore, and R.F. Lebed, *Phys. Rev.* D **70**, 056004 (2004).
7. T.D. Cohen and R.F. Lebed, *Phys. Rev.* D **70**, 096015 (2004).

8. T.D. Cohen, D.C. Dakin, R.F. Lebed, and D.R. Martin, *Phys. Rev.* D **71**, 076010 (2005).
9. T.D. Cohen and R.F. Lebed, *Phys. Rev.* D **72**, 056001 (2005).
10. T.D. Cohen, P.M. Hohler, and R.F. Lebed, *Phys. Rev.* D **72**, 074010 (2005).
11. See C. Morningstar, these proceedings.
12. R.F. Lebed, hep-ph/0301279, in *NStar 2002: Proceedings of the Workshop on the Physics of Excited Nucleons*, edited by S.A. Dytman and E.S. Swanson, World Scientific, Singapore, 2003.
13. E. Witten, *Nucl. Phys.* **B160**, 57 (1979).
14. R.F. Dashen and A.V. Manohar, *Phys. Lett.* B **315**, 425 (1993).
15. R.F. Dashen, E. Jenkins, and A.V. Manohar, *Phys. Rev.* D **49**, 4713 (1994).
16. A.J. Buchmann and R.F. Lebed, *Phys. Rev.* D **62**, 096005 (2000).
17. E. Jenkins, *Phys. Lett.* B **315**, 441 (1993); E. Jenkins and R.F. Lebed, *Phys. Rev.* D **52**, 282 (1995).
18. C.D. Carone, H. Georgi, L. Kaplan, and D. Morin, *Phys. Rev.* D **50**, 5793 (1994); J.L. Goity, *Phys. Lett.* B **414**, 140 (1997); D. Pirjol and T.-M. Yan, *Phys. Rev.* D **57**, 1449 (1998); D **57**, 5434 (1998); C.E. Carlson, C.D. Carone, J.L. Goity, and R.F. Lebed, *Phys. Lett.* B **438**, 327 (1998); *Phys. Rev.* D **59**, 114008 (1999); J.L. Goity, C. Schat, and N. Scoccola, *Phys. Rev. Lett.* **88**, 102002 (2002); *Phys. Rev.* D **66**, 114014 (2002); *Phys. Lett.* B **564**, 83 (2003); J.L. Goity and N.N. Scoccola, *Phys. Rev.* D **72**, 034024 (2005); J.L. Goity, hep-ph/0504121; C.E. Carlson and C.D. Carone, *Phys. Rev.* D **58**, 053005 (1998); *Phys. Lett.* B **441**, 363 (1998); B **484**, 260 (2000); N. Matagne and Fl. Stancu, *Phys. Rev.* D **71**, 014010 (2005); *Phys. Lett.* B **631**, 7 (2005); N. Matagne, hep-ph/0501123; hep-ph/0511247 (these proceedings).
19. E. Witten, *Nucl. Phys.* **B223**, 433 (1983); G.S. Adkins, C.R. Nappi, and E. Witten, *Nucl. Phys.* **B228**, 552 (1983); G.S. Adkins and C.R. Nappi, *Nucl. Phys.* **B249**, 507 (1985).
20. A.V. Manohar, *Nucl. Phys.* **B248**, 19 (1984).
21. A. Hayashi, G. Eckart, G. Holzwarth, H. Walliser, *Phys. Lett.* B **147**, 5 (1984).
22. M.P. Mattis and M. Karliner, *Phys. Rev.* D **31**, 2833 (1985); M.P. Mattis and M.E. Peskin, *Phys. Rev.* D **32**, 58 (1985); M.P. Mattis, *Phys. Rev. Lett.* **56**, 1103 (1986); *Phys. Rev.* D **39**, 994 (1989); *Phys. Rev. Lett.* **63**, 1455 (1989); *Phys. Rev. Lett.* **56**, 1103 (1986).
23. J.T. Donohue, *Phys. Rev. Lett.* **58**, 3 (1987); *Phys. Rev.* D **37**, 631 (1988).
24. M.P. Mattis and M. Mukerjee, *Phys. Rev. Lett.* **61**, 1344 (1988).
25. D.B. Kaplan and M.J. Savage, *Phys. Lett.* B **365**, 244 (1996); D.B. Kaplan and A.V. Manohar, *Phys. Rev.* C **56**, 76 (1997).
26. D. Pirjol and C. Schat, *Phys. Rev.* D **67**, 096009 (2003).
27. E. Jenkins, X. Ji, and A.V. Manohar, *Phys. Rev. Lett.* **89**, 242001 (2002).
28. *Review of Particle Properties* (S. Eidelman *et al.*), *Phys. Lett.* B **592**, 1 (2004).

TOWARDS A DETERMINATION OF THE SPECTRUM OF QCD USING A SPACE-TIME LATTICE

K. J. JUGE, A. LICHTL, AND C. MORNINGSTAR*

Department of Physics, Carnegie Mellon University, Pittsburgh, PA 15213, USA

R. G. EDWARDS AND D. G. RICHARDS

Thomas Jefferson National Accelerator Facility, Newport News, VA 23606, USA

S. BASAK AND S. WALLACE

Department of Physics, University of Maryland, College Park, MD 20742, USA

I. SATO

Lawrence Berkeley Laboratory, 1 Cyclotron Road, Berkeley, CA 94720, USA

G. T. FLEMING

Sloane Physics Laboratory, Yale University, New Haven, CT 06520, USA

Progress by the Lattice Hadron Physics Collaboration in determining the baryon and meson resonance spectrum of QCD using Monte Carlo methods with space-time lattices is described. The extraction of excited-state energies necessitates the evaluation of correlation matrices of sets of operators, and the importance of extended three-quark operators to capture both the radial and orbital structures of baryons is emphasized. The use of both quark-field smearing and link-field smearing in the operators is essential for reducing the couplings of the operators to the high-frequency modes and for reducing statistical noise in the correlators. The extraction of nine energy levels in a given symmetry channel is demonstrated, and identifying the continuum spin quantum numbers of the levels is discussed.

1. Introduction

A charge from the late Nathan Isgur to use Monte Carlo methods to extract the spectrum of hadron resonances and hadronic properties resulted in

---

*Speaker.

the formation of the Lattice Hadron Physics Collaboration (LHPC) in the year 2000. As part of a national collaboration of lattice QCD theorists, the LHPC acquired funding from several sources, including the DOE's Scientific Discovery through Advanced Computing initiative, to build large computing clusters at JLab, Fermilab, and Brookhaven, as well as to develop the software to carry out the needed large-scale computations. The LHPC has several broad goals: to compute the spectrum of QCD from first principles, to investigate hadron structure by computing form factors, structure functions, and other matrix elements, and to study hadron-hadron interactions. This talk focuses solely on our efforts to determine the spectrum of QCD.

Extracting the spectrum of resonances in QCD is a big challenge. The determination of excited-state energies requires the use of correlation matrices of sets of operators. The masses and widths of unstable hadrons (resonances) must be deduced from the energies of single-particle and multi-hadron stationary states in a finite-sized box. This necessitates the use of multi-hadron operators in the correlation matrices, and the computations must be performed in full QCD at realistically-light quark masses. For these reasons, the computation of the QCD spectrum is indeed a long-term project. This talk is a brief status report of our efforts.

I first describe in Sec. 2 how excited-state energies are extracted in our Monte Carlo calculations, and discuss issues related to unstable resonances. Our construction of baryon operators is then outlined in Sec. 3 (see Ref. 1 for further details), and in Sec. 4, the importance of using smeared quark and gluon fields is demonstrated. (see also Ref. 2). The first-time extraction of *nine* nucleon energy levels is also presented in this section. Concluding remarks are made and future work is outlined in the concluding Sec. 5.

2. Excited states and resonances

In the path integral formulation of quantum field theory with imaginary time, stationary-state energies are extracted from the asymptotic decay rates of temporal correlations of the field operators. If $\Phi(t)$ is a Heisenberg-picture operator which annihilates the hadron of interest at time t, then its evolution $\Phi(t) = e^{Ht}\Phi(0)e^{-Ht}$, where H is the Hamiltonian, can be used to show that in a finite-sized box,

$$C(t) \equiv \langle 0|\Phi(t)\Phi^\dagger(0)|0\rangle = \sum_n \langle 0|e^{Ht}\Phi(0)e^{-Ht}|n\rangle\langle n|\Phi^\dagger(0)|0\rangle, \quad (1)$$

$$= \sum_n |\langle 0|\Phi(0)|n\rangle|^2 e^{-(E_n-E_0)t} = \sum_n A_n e^{-(E_n-E_0)t}, \quad (2)$$

where $\{|n\rangle\}$ is the complete set of discrete eigenvectors of H. We assume the existence of a transfer matrix, and temporal boundary conditions have been ignored for illustrative purposes. One can then extract A_1 and $E_1 - E_0$ as $t \to \infty$, assuming $\langle 0|\Phi(0)|0\rangle = 0$ and $\langle 0|\Phi(0)|1\rangle \neq 0$.

A convenient visual tool for demonstrating energy-level extraction is the so-called "effective mass" defined by $m_{\rm eff}(t) = \ln[C(t)/C(t+a_t)]$, where t is time and a_t is the temporal lattice spacing. The effective mass tends to the actual mass (or energy) of the ground state as t becomes large, signalled by a plateau in the effective mass. At smaller times before this plateau is observed, the effective mass varies due to contributions from other states in the spectrum. The correlation $C(t)$ is estimated with some statistical uncertainty since the Monte Carlo method is used, and usually, the ratio of the noise to the signal increases with temporal separation t. Hence, judiciously-chosen operators having reduced couplings with contaminating higher-lying states are important in order to observe a plateau in the effective mass before noise swamps the signal. Key ingredients in constructing such operators are the use of smeared quark and gluon fields, the incorporation of spatially-extended assemblages of the fields, and the use of sets of different operators to exploit improvements from variational methods.

Methods of extracting excited-state energies are well known[3,4]. For a given $N \times N$ Hermitian matrix of correlations $C_{\alpha\beta}(t) = \langle 0|\Phi_\alpha(t)\Phi_\beta^\dagger(0)|0\rangle$, the N *principal correlators* $\lambda_\alpha(t, t_0)$ are defined as the eigenvalues of the matrix $C(t_0)^{-1/2}C(t)C(t_0)^{-1/2}$, where t_0 is some reference time (typically small), and one can show that

$$\lim_{t\to\infty} \lambda_\alpha(t, t_0) = e^{-(t-t_0)E_\alpha}(1+O(e^{-t\Delta E_\alpha})), \qquad \Delta E_\alpha = \min_{\beta\neq\alpha} |E_\beta - E_\alpha|, \quad (3)$$

assuming $E_0 = 0$ and $\lambda_0 \geq \lambda_1 \geq \lambda_2 \cdots$. The N principal effective masses $m_\alpha^{\rm eff}(t) = \ln[\lambda_\alpha(t, t_0)/\lambda_\alpha(t+a_t, t_0)]$ now tend (plateau) to the N lowest-lying stationary-state energies which couple to the N operators. The associated eigenvectors are orthogonal, and a knowledge of the eigenvectors can yield information about the partonic structure of the states.

Many of the hadron states we wish to study are unstable resonances. Our computations are done out of necessity in a box of finite volume with periodic boundary conditions. Hence, the momenta of the particles we study are quantized, so all states are discrete in our computations. Thus, we can only determine the discrete energy spectrum of stationary states in a periodic box, which are admixtures of single hadrons and multi-hadron states. Resonance masses and widths must somehow be deduced from the finite-box spectrum[5–8]. Once the masses of the stable single particle states

Table 1. Continuum limit spin identification: the number n^J_Λ of times that the Λ irrep. of the octahedral point group O_h occurs in the (reducible) subduction of the J irrep. of $SU(2)$. The numbers for G_{1u}, G_{2u}, H_u are the same as for G_{1g}, G_{2g}, H_g, respectively.

J	$n^J_{G_{1g}}$	$n^J_{G_{2g}}$	$n^J_{H_g}$	J	$n^J_{G_{1g}}$	$n^J_{G_{2g}}$	$n^J_{H_g}$
$\frac{1}{2}$	1	0	0	$\frac{9}{2}$	1	0	2
$\frac{3}{2}$	0	0	1	$\frac{11}{2}$	1	1	2
$\frac{5}{2}$	0	1	1	$\frac{13}{2}$	1	2	2
$\frac{7}{2}$	1	1	1	$\frac{15}{2}$	1	1	3

have been determined, the placement and pattern of their scattering states are known approximately, and the dependences of their energies on the volume are roughly known. Resonances show up as extra states with little volume dependence. Our initial goal is simply to ferret out these resonances, not to pin down their properties to high precision.

3. Operator construction

Our approach to constructing hadron operators is to directly combine the physical characteristics of hadrons with the symmetries of the lattice regularization of QCD used in simulations. For baryons at rest, our operators are formed using group-theoretical projections onto the irreducible representations (irreps) of the O_h symmetry group of a three-dimensional cubic lattice. There are four two-dimensional irreps G_{1g}, G_{1u}, G_{2g}, G_{2u} and two four-dimensional representations H_g and H_u. The continuum-limit spins J of our states must be deduced by examining degeneracy patterns across the different O_h irreps (see Table 1). For example, a $J^P = \frac{1}{2}^+$ state will show up in the G_{1g} channel without degenerate partners in the other channels, and a $J^P = \frac{3}{2}^+$ state will show up in the H_g channel without degenerate partners in the other channels. Four of the six polarizations of a $J^P = \frac{5}{2}^+$ state show up as a level in the H_g channel, and the other two will occur as a degenerate partner in the G_{2g} channel, whereas three degenerate levels, one in each of the three G_{1g}, G_{2g}, H_g channels, may indicate a single $J^P = \frac{7}{2}^+$ state (or the accidental degeneracy of a spin-$\frac{1}{2}$ and a $\frac{5}{2}$ state).

Baryons are expected to be rather large objects, so local operators will not suffice. Our approach to constructing spatially-extended operators is to use covariant displacements of the quark fields along the links of the lattice. Displacements in different directions are used to build up the appropriate

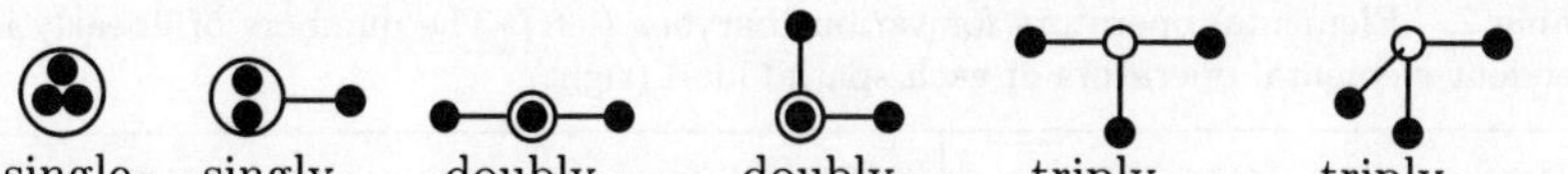

Fig. 1. The spatial arrangments of the extended three-quark baryon operators used. Smeared quark fields are shown by solid circles, line segments indicate gauge-covariant displacements, and each hollow circle indicates the location of a Levi-Civita color coupling. For simplicity, all displacements have the same length in an operator.

orbital structure, and displacements of different lengths can build up the needed radial structure. All our three-quark baryon operators are superpositions of gauge-invariant, translationally-invariant terms of the form

$$\Phi^{ABC}_{\alpha\beta\gamma,ijk}(t) = \sum_{\vec{x}} \varepsilon_{abc}\, (\tilde{D}^{(p)}_i\tilde{\psi}(\vec{x},t))^A_{a\alpha}\ (\tilde{D}^{(p)}_j\tilde{\psi}(\vec{x},t))^B_{b\beta}\ (\tilde{D}^{(p)}_k\tilde{\psi}(\vec{x},t))^C_{c\gamma}, \quad (4)$$

where A, B, C indicate quark flavor, a, b, c are color indices, α, β, γ are Dirac spin indices, $\tilde{\psi}$ indicates a smeared quark field, and $\tilde{D}^{(p)}_j$ denotes the smeared p-link covariant displacement operator in the j-th direction. The smearing of the quark and gauge field will be discussed later. There are six different spatial orientations that we use, shown in Fig 1. The singly-displaced operators are meant to mock up a diquark-quark coupling, and the doubly-displaced and triply-displaced operators are chosen since they favor the Δ-flux and Y-flux configurations, respectively.

Next, the $\Phi^{ABC}_{\alpha\beta\gamma,ijk}$ are combined into *elemental* operators $B^F_a(t)$ having the appropriate flavor structure characterized by isospin, strangeness, *etc.* We work in the $m_u = m_d$ (equal u and d quark masses) approximation, and thus, require that the elemental operators have definite isospin, that is, they satisfy appropriate commutation relations with the isospin operators τ_3, τ_+, τ_-. Since we plan to compute full correlation matrices, we need not be concerned with forming operators according to an $SU(3)$ flavor symmetry. Maple code which manipulates Grassmann fields was used to identify maximal sets of linearly independent elemental operators. The operators used are shown in Table 2, and the numbers of such independent operators are listed in this table as well.

The final step in our operator construction is to apply group-theoretical projections to obtain operators which transform irreducibly under all lattice rotation and reflection symmetries:

$$B^{\Lambda\lambda F}_a(t) = \frac{d_\Lambda}{g_{O_h}} \sum_{R\in O_h} \Gamma^{(\Lambda)}_{\lambda\lambda}(R)\ U_R\ B^F_a(t)\ U^\dagger_R, \quad (5)$$

Table 2. Elemental operators for various baryons (left). The numbers of linearly independent elemental operators of each spatial kind (right).

Baryon	Operator	Spatial Type	Δ, Ω	N	Σ, Ξ	Λ
Δ^{++}	$\Phi^{uuu}_{\alpha\beta\gamma,ijk}$	single-site	20	20	40	24
Σ^{+}	$\Phi^{uus}_{\alpha\beta\gamma,ijk}$	singly-displaced	240	384	624	528
N^{+}	$\Phi^{uud}_{\alpha\beta\gamma,ijk} - \Phi^{duu}_{\alpha\beta\gamma,ijk}$	doubly-displaced-I	192	384	576	576
Ξ^{0}	$\Phi^{ssu}_{\alpha\beta\gamma,ijk}$	doubly-displaced-L	768	1536	2304	2304
Λ^{0}	$\Phi^{uds}_{\alpha\beta\gamma,ijk} - \Phi^{dus}_{\alpha\beta\gamma,ijk}$	triply-displaced-T	768	1536	2304	2304
Ω^{-}	$\Phi^{sss}_{\alpha\beta\gamma,ijk}$	triply-displaced-O	512	1024	1536	1536

where Λ refers to an O_h irrep, λ is the irrep row, g_{O_h} is the number of elements in O_h, d_Λ is the dimension of the Λ irrep, $\Gamma^{(\Lambda)}_{mn}(R)$ is a Λ representation matrix corresponding to group element R, and U_R is the quantum operator which implements the symmetry operations. The projections in Eq. (5) are carried out using computer software written in the Maple[9] symbolic manipulation language.

4. Field smearing and operator pruning

For single-site (local) hadron operators, it is well known that the use of spatially-smeared quark fields is crucial. For extended baryon operators, one expects quark-field smearing to be equally important, but the relevance and interplay of link-field smearing is less well known. Thus, we decided that a systematic study of both quark-field and link-variable smearing was warranted.

The link variables were smeared $U \to \tilde{U}$ using the analytic stout link method of Ref. 10. There are two tunable parameters, the number of iterations n_ρ and the staple weight ρ. For the quark-field, we employed gauge-covariant Gaussian smearing

$$\tilde{\Psi}(x) = \left(1 + \frac{\sigma_s^2}{4n_\sigma}\tilde{\Delta}\right)^{n_\sigma} \Psi(x), \tag{6}$$

$$\tilde{\Delta}\Psi(x) = \sum_{k=\pm1,\pm2,\pm3} \left(\tilde{U}_k(x)\Psi(x+\hat{k}) - \Psi(x)\right), \tag{7}$$

where $\tilde{\Delta}$ denotes the smeared three-dimensional gauge-covariant Laplacian. The two parameters to tune in this smearing procedure are the smearing radius σ_s and the integer number of iterations n_σ.

For our tests of the efficacy of quark-field and gauge-link smearing,

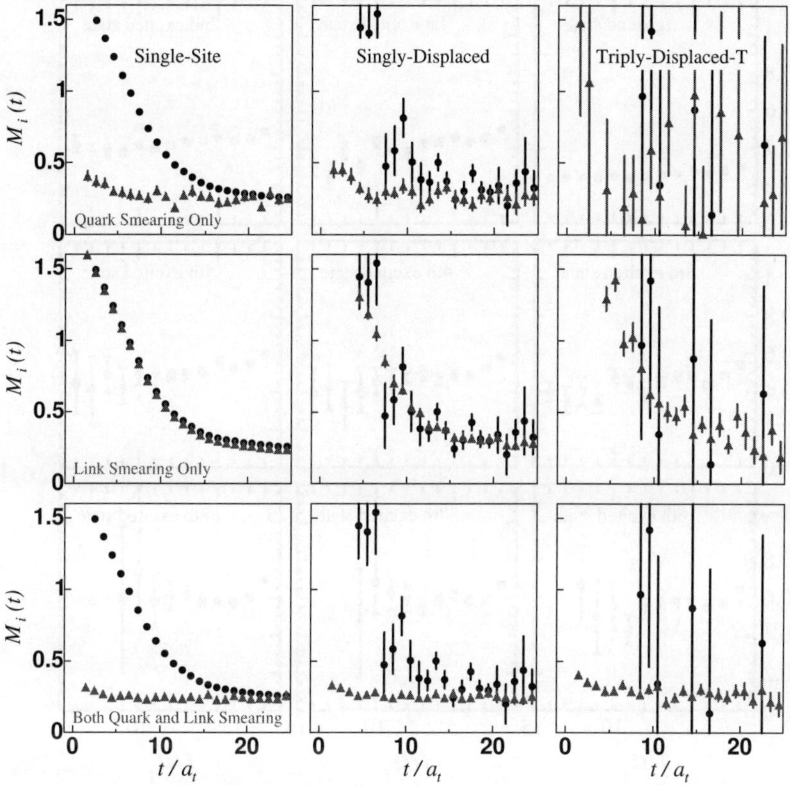

Fig. 2. Effective masses $M(t)$ for unsmeared (circles) and smeared (triangles) operators O_{SS}, O_{SD}, O_{TDT}. Top row: only quark-field smearing $n_\sigma = 32$, $\sigma_s = 4.0$ is used. Middle row: only link-variable smearing $n_\rho = 16$, $n_\rho \rho = 2.5$ is applied. Bottom row: both quark and link smearing $n_\sigma = 32$, $\sigma_s = 4.0$, $n_\rho = 16$, $n_\rho \rho = 2.5$ are used, dramatically improving the signal for all three operators. Results are based on 50 quenched configurations on a $12^3 \times 48$ anisotropic lattice using the Wilson action with $a_s \sim 0.1$ fm, $a_s/a_t \sim 3.0$. The quark mass was chosen so that $m_\pi \simeq 700$ MeV.

correlators were computed for three particular nucleon operators: a single-site operator O_{SS} in the G_{1g} irreducible representation of the cubic point group, a singly-displaced operator O_{SD} with a particular choice of each Dirac index, and a triply-displaced-T operator O_{TDT} with a specific choice of each Dirac index. Our findings are summarized in Fig. 2. The top row shows that applying only quark-field smearing to the three selected nucleon operators significantly reduces couplings to higher-lying states, but the displaced operators remain excessively noisy. The second row illustrates

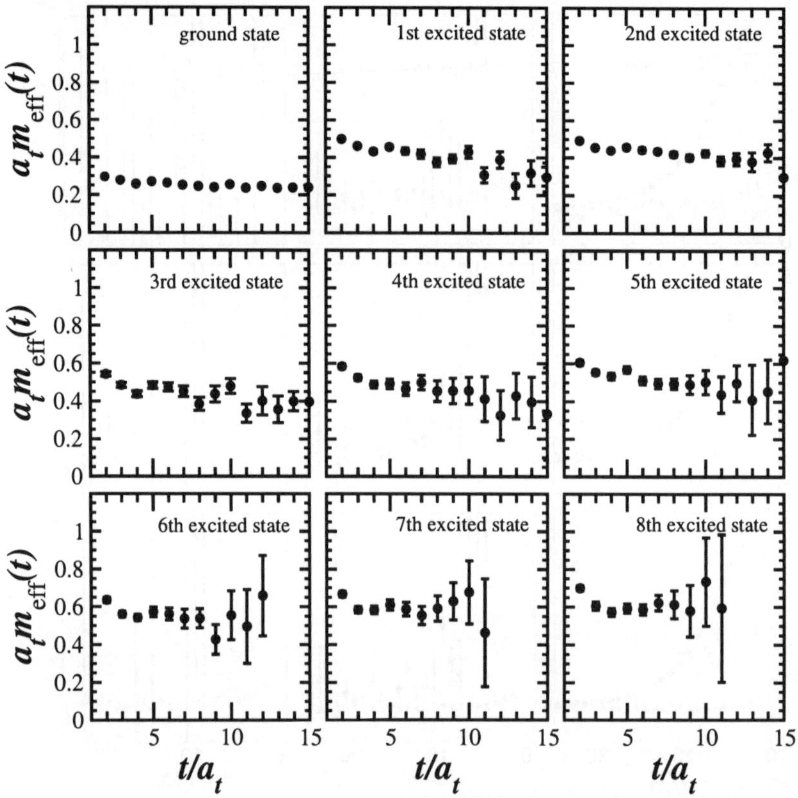

Fig. 3. Principal effective masses for the lowest-lying *nine* states in the G_{1g} nucleon channel. Results are based on 100 quenched configurations on a $12^3 \times 48$ anisotropic lattice using the Wilson action with $a_s \sim 0.1$ fm, $a_s/a_t \sim 3.0$. The quark mass was chosen so that $m_\pi \simeq 700$ MeV. Quark-field smearing $n_\sigma = 32$, $\sigma_s = 3.0$ and link-smearing $n_\rho = 16$, $n_\rho\rho = 2.5$ are used. These principal effective masses come from a 10×10 correlation matrix, where 2 singly-displaced, 5 doubly-displaced-I, 2 doubly-displaced-L, and 1 triply-displaced-T operators were chosen.

that including only link-field smearing substantially reduces the noise, but does not appreciably alter the effective masses themselves. The bottom row shows dramatic improvement from reduced couplings to excited states and dramatically reduced noise when both quark-field and link-field smearing is applied, especially for the extended operators. The "best" quark-field smearing parameters n_σ and σ were determined by requiring that the effective mass for the three chosen operators reach a plateau as close to the source as possible. The gauge-link smearing parameters were tuned so as to minimize the noise in the effective masses. One interesting point we also learned was that the preferred link-smearing parameters determined from the static quark-antiquark potential produced the smallest error in the extended baryon operators.

The computation of correlation matrices using hundreds of operators is not feasible, so it is necessary to "prune" out unnecessary operators. The first step in this pruning is to examine the effective masses of the diagonal elements of the correlation matrices to identify and eliminate noisy operators. Keeping only operators with small statistical uncertainties yields a set of about forty to fifty operators in each symmetry channel. We then computed the correlation matrix of this reduced set of operators, examining whether further reductions to the operator sets could be made without increased contamination in the principal effective masses. These computations are still ongoing, but preliminary results are shown in Fig. 3. This figure shows that it is possible to extract at least *nine* levels in a given symmetry channel, a feat which has never before been accomplished. Demonstrating that this number of energy levels can be reliably extracted is an important milestone in our long-term project.

5. Conclusion

We have outlined a program to study the resonance spectrum in lattice QCD. The use of the variational method and the need to isolate several energy levels in each channel require a sufficiently broad basis of operators. Having developed suitable group-theory methods to project operators onto the irreducible representations of the cubic group, and having examined the efficacy of both quark- and gauge-link-smearing, we are now identifying a more limited set of operators that we will employ in a large-scale study of the hadron spectrum. Our methods are applicable not only to baryons, but also to mesons, tetra-quark and pentaquark systems, and to states with excited glue. Only by performing such a program can we hope to identify the states of QCD, and in particular their spins and parities, in the continuum limit. Ultimately, when quark loops are included at realistically light quark masses, multi-hadron (baryon-meson) operators must be included in our correlation matrices, and finite-volume techniques will need to be employed to ferret out the baryon resonances from uninteresting scattering states. We are currently exploring different ways of building such operators.

This work was supported by the U.S. National Science Foundation through grants PHY-0354982, PHY-0510020, and PHY-0300065, and by the U.S. Department of Energy under contracts DE-AC05-84ER40150 and DE-FG02-93ER-40762. Computations were performed using the *Chroma* software package[11].

References

1. S. Basak, R. Edwards, G. Fleming, U. Heller, C. Morningstar, D. Richards, I. Sato, S. Wallace, *Phys. Rev. D* **72**, 094506 (2005); **72**, 074501 (2005).
2. S. Basak, R. Edwards, G. Fleming, U. Heller, A. Lichtl, C. Morningstar, D. Richards, I. Sato, and S. Wallace, *Proc. Sci. LAT2005:* **076** (2005).
3. C. Michael, *Nucl. Phys. B* **259**, 58 (1985).
4. M. Lüscher and U. Wolff, *Nucl. Phys. B* **339**, 222 (1990).
5. B. DeWitt, *Phys. Rev.* **103**, 1565 (1956).
6. U. Wiese, *Nucl. Phys. B (Proc. Suppl.)* **9**, 609 (1989).
7. M. Lüscher, *Nucl. Phys. B* **364**, 237 (1991).
8. K. Rummukainen and S. Gottlieb, *Nucl. Phys. B* **450**, 397 (1995).
9. Visit the web site `http://www.maplesoft.com`.
10. C. Morningstar and M. Peardon, *Phys. Rev. D* **69**, 054501 (2004).
11. R. Edwards and B. Joo, *Nucl. Phys. B (Proc. Suppl.)* **140**, 832 (2005).

DYNAMICAL GENERATION OF $J^P = \frac{3}{2}^-$ RESONANCES AND THE $\Lambda(1520)$ RESONANCE

S. SARKAR, L. ROCA, E. OSET, V. K. MAGAS AND M. J. VICENTE VACAS

Departamento de Física Teórica and IFIC,
Centro Mixto Universidad de Valencia-CSIC,
Investigación de Paterna, Aptdo. 22085, 46071 Valencia, Spain

The lowest order chiral Lagrangian is used to study s-wave interactions of the baryon decuplet with the octet of pseudoscalar mesons. The coupled channel Bethe Salpeter equation is used to obtain dynamically generated $\frac{3}{2}^-$ resonances in the partial wave amplitudes which provide a reasonable description to a number of 3 and 4-star resonances like $\Delta(1700)$, $\Lambda(1520)$, $\Sigma(1670)$, $\Sigma(1940)$, $\Xi(1820)$, $\Omega(2250)$ etc. The phenomenological introduction of d-wave channels in the coupled channel scheme along with the existing s-wave channels is shown to provide a much improved description of the $\Lambda(1520)$ resonance. The prediction of the absolute strength of the cross section in the reactions $K^-p \to \Lambda\pi^0\pi^0$ and $K^-p \to \Lambda\pi^+\pi^-$ provided by this scheme shows a good agreement with existing data.

1. Introduction

The introduction of unitary techniques in a chiral dynamical treatment of the meson baryon interaction has been very successful. It has lead to good reproduction of meson baryon data with a minimum amount of free parameters, and has led to the dynamical generation of many low lying resonances which qualify as quasibound meson baryon states [1,2]. In particular, the application of these techniques to the s-wave scattering of the baryon octet and the pseudoscalar meson octet have led to the successful description of many $J^P = \frac{1}{2}^-$ resonances like the $N^*(1535)$, the $\Lambda(1405)$, the $\Lambda(1670)$ and the $\Sigma(1620)$ [3–6]. Naively one may expect that this scheme is not suitable for studying d-wave resonances due to a large number of unknown parameters in the corresponding chiral Lagrangian. However d-wave resonances could be studied in s-wave interactions of the meson octet with the baryon decuplet [7–9], in which case chiral dynamics is quite predictive. Applying unitary techniques to the lowest order chiral Lagrangian we have been successful in generating a number of $\frac{3}{2}^-$ resonances. From the information of the pole positions and couplings to the channels involved we could associate many of these resonances to the $N^*(1520)$, $\Delta(1700)$, $\Lambda(1520)$, $\Sigma(1670)$, $\Sigma(1940)$, $\Xi(1820)$ resonances tabulated by the Particle Data Group (PDG).

We then consider the dynamics of one of the above resonances, namely the $\Lambda(1520)$, which is capturing renewed attention, particularly since it appears invariably in searches of pentaquarks in photonuclear reactions like $\gamma p \to K^+K^-p$ and $\gamma d \to K^+K^-np$. In the simple picture [8] mentioned above, this resonance couples to the $\pi\Sigma^*(1385)$ and $K\Xi^*(1530)$ channels, particularly to the former one. With the $\pi^+\Sigma^{*-}$, $\pi^-\Sigma^{*+}$, $\pi^0\Sigma^{*0}$ masses 7 MeV above, 2 MeV above and 1 MeV below the nominal $\Lambda(1520)$ mass and the strong coupling of the resonance to $\pi\Sigma^*$, this state could qualify as a loosely bound $\pi\Sigma^*$ state. However, the lack of other relevant channels which couple to the quantum numbers of the resonance makes the treatment of [8] only semiquantitative. In particular, the $\Lambda(1520)$ appears at higher energy than the nominal one and with a large width of about 130 MeV, nearly ten times larger than the physical width. This large width is a necessary consequence of the large coupling to the $\pi\Sigma^*$ channel and the fact that the pole appears at energies above the $\pi\Sigma^*$ threshold. On the other hand, if we modify the subtraction constants of the meson baryon loop function to bring the pole below the $\pi\Sigma^*$ threshold, then the pole appears without imaginary part. Since the width of the $\Lambda(1520)$ resonance comes basically from the decay into the $\bar{K}N$ and $\pi\Sigma(1193)$, the introduction of these channels is mandatory to reproduce the shape of the $\Lambda(1520)$ resonance.

In [10,11] we phenomenologically include the $\bar{K}N$ and $\pi\Sigma$ channels into the set of coupled channels which build up the $\Lambda(1520)$. The novelty with respect to the other channels already accounted for [8], which couple in s-wave, is that the new channels couple in d-waves. The transition matrix elements for the d-wave channels $\bar{K}N$ and $\pi\Sigma$ to the s-wave channel $\pi\Sigma^*$ were parametrized in terms of unknown constants which were fitted to experimental values for the real and imaginary parts of the partial wave amplitudes for the reactions $\bar{K}N \to \bar{K}N$ and $\bar{K}N \to \pi\Sigma$. The coupling of the $\Lambda(1520)$ to the $\pi\Sigma^*$ channel is then a prediction of this scheme and we use this to study the reaction $K^-p \to \pi^0\pi^0\Lambda$ and $K^-p \to \pi^+\pi^-\Lambda$ which are closely related to the strength of this coupling. The prediction of the absolute strength of the cross sections in the above two reactions close to and above the $\Lambda(1520)$ energy find a good agreement with data.

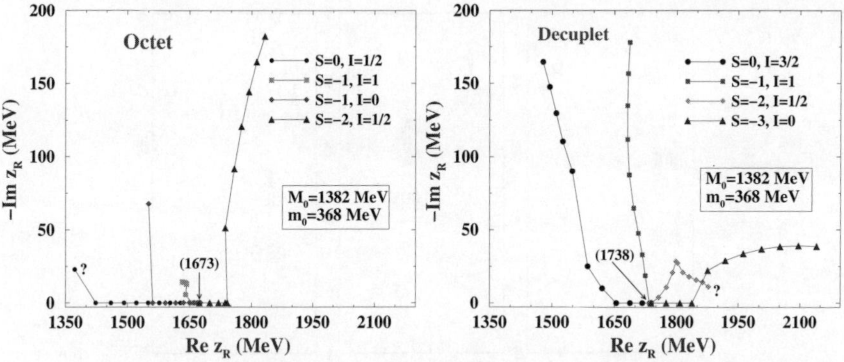

Fig. 1. Trajectories of the poles in the scattering amplitudes obtained by increasing the value of the $SU(3)$ breaking parameter in steps of 0.1 (symbols) from zero, which is the $SU(3)$ symmetric situation, up to 1 which corresponds to the physical masses.

2. Dynamical generation of spin 3/2 baryon resonances

The tree-level scattering amplitude involving the baryon decuplet and the pseudoscalar octet is obtained from the dominant lowest order chiral Lagrangian which accounts for the Weinberg Tomozawa term. The matrix containing these amplitudes, V is used as the kernel of a coupled channel Bethe Salpeter equation given by

$$T = V + VGT$$

to obtain the transition matrix fulfilling exact unitarity in coupled channels. The factor G in the above equation is the meson baryon loop function. We have looked in detail [8,9] at the $\frac{3}{2}^-$ resonances which are generated dynamically by this interaction, by searching for poles of the transition matrix in the complex plane in different Riemann sheets. We start from a $SU(3)$ symmetric situation where the masses of the baryons are made equal and the same is done with the masses of the mesons. In this case we found attraction in the octet, decuplet and the 27 representations, while the interaction was repulsive in the 35 representation. In the $SU(3)$ symmetric case all states of the $SU(3)$ multiplet are degenerate and the resonances appear as bound states with no width. As we gradually break $SU(3)$ symmetry by changing the masses, the degeneracy is broken and the states with different strangeness and isospin split apart generating pole trajectories in the complex plane which lead us to the physical situation in the final point, as seen in fig. 1. This systematic search allows us to trace the poles to their $SU(3)$ symmetric origin, although there is a mixing of representations when the

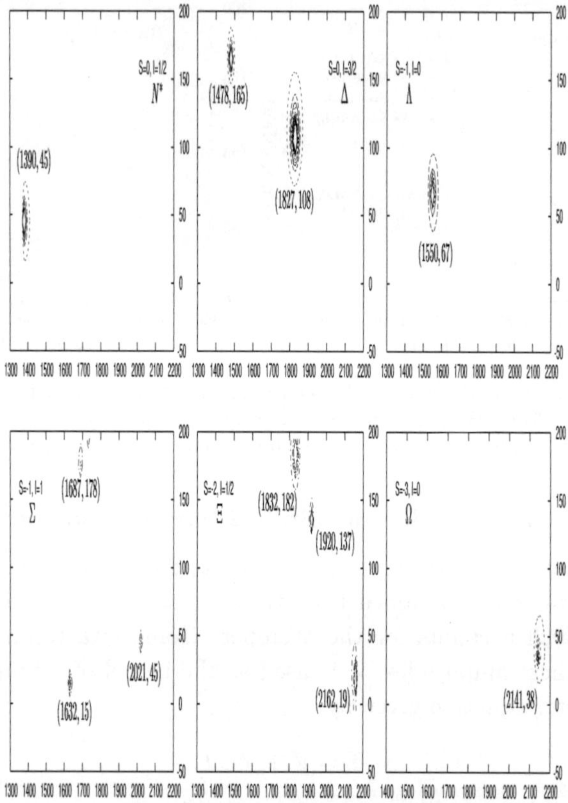

Fig. 2. Contour plots of the scattering amplitudes showing poles in the unphysical Riemann sheets. The x and y axes denote the real and imaginary parts of the CM energy.

symmetry is broken.

In fig. 2 we show contour plots of the partial wave amplitudes as a function of the complex CM energy. The projection on the real (energy) axis is shown in fig. 3. We have also evaluated the residues of the poles from where the couplings of the resonances to the different coupled channels were found and this allowed us to make predictions for partial decay widths into a decuplet baryon and a meson. There is very limited experimental information on these decay channels but, even then, it represents an extra check of consistency of the results which allowed us to more easily identify the resonances found with some resonances known, or state that the resonance should correspond to a new resonance not yet reported by the PDG. In particular, in view of the information of the pole positions and couplings to

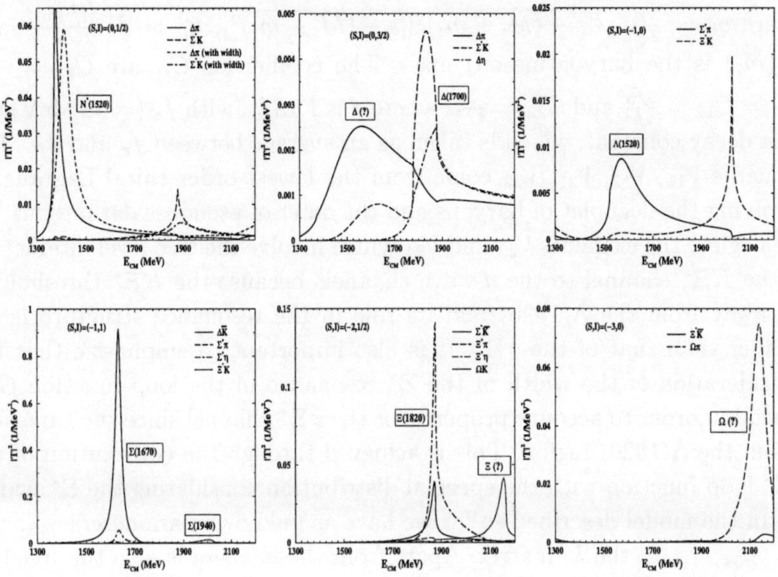

Fig. 3. Scattering amplitudes as a function of CM energy showing $3/2^-$ resonances generated dynamically in decuplet baryon - pseudoscalar octet interactions.

channels we could associate some of the resonances found to the $N^*(1520)$, $\Delta(1700)$, $\Lambda(1520)$, $\Sigma(1670)$, $\Sigma(1940)$ and $\Xi(1820)$.

3. Improved description of the $\Lambda(1520)$ resonance

In addition to the transition amplitudes involving the s-wave channels $\pi\Sigma^*$ and $K\Xi^*$ which were used to dynamically generate the $\Lambda(1520)$ as discussed in the previous section we introduce phenomenologically tree level transition potentials involving the d-wave channels $\bar{K}N$ and $\pi\Sigma$. As discussed in [10], we use for the vertices $\bar{K}N \to \bar{K}N$, $\bar{K}N \to \pi\Sigma$ and $\pi\Sigma \to \pi\Sigma$ effective transition potentials which are proportional to the incoming and outgoing momentum squared. Denoting $\pi\Sigma^*$, $K\Xi^*$, $\bar{K}N$ and $\pi\Sigma$ channels by 1, 2, 3 and 4 respectively, the matrix containing the tree level amplitudes is written as:

$$V = \begin{pmatrix} C_{11}(k_1^0 + k_1^0) & C_{12}(k_1^0 + k_2^0) & \gamma_{13}\, q_3^2 & \gamma_{14}\, q_4^2 \\ C_{21}(k_2^0 + k_1^0) & C_{22}(k_2^0 + k_2^0) & 0 & 0 \\ \gamma_{13}\, q_3^2 & 0 & \gamma_{33}\, q_3^4 & \gamma_{34}\, q_3^2\, q_4^2 \\ \gamma_{14}\, q_4^2 & 0 & \gamma_{34}\, q_3^2\, q_4^2 & \gamma_{44}\, q_4^4 \end{pmatrix},$$

where $q_i = \frac{1}{2\sqrt{s}}\sqrt{[s-(M_i+m_i)^2][s-(M_i-m_i)^2]}$, $k_i^0 = \frac{s-M_i^2+m_i^2}{2\sqrt{s}}$ and $M_i(m_i)$ is the baryon(meson) mass. The coefficients C_{ij} are $C_{11} = \frac{-1}{f^2}$, $C_{21} = C_{12} = \frac{\sqrt{6}}{4f^2}$ and $C_{22} = \frac{-3}{4f^2}$, where f is $1.15 f_\pi$, with f_π (= 93 MeV) the pion decay constant, which is taken as an average between f_π and f_K. The elements V_{11}, V_{12}, V_{21}, V_{22} come from the lowest order chiral Lagrangian involving the decuplet of baryons and the octet of pseudoscalar mesons [8,7]. We neglect the elements V_{23} and V_{24} which involve the tree level interaction of the $K\Xi^*$ channel to the d-wave channels because the $K\Xi^*$ threshold is far away from the $\Lambda(1520)$ and its role in the resonance structure is far smaller than that of the $\pi\Sigma^*$. It is also important to emphasize that the consideration of the width of the Σ^* resonance in the loop function G is crucial in order to account properly for the $\pi\Sigma^*$ channel since the threshold lies in the $\Lambda(1520)$ region. This is achieved through the convolution of the $\pi\Sigma^*$ loop function with the spectral distribution considering the Σ^* width.

In the model described so far we have as unknown parameters γ_{13}, γ_{14}, γ_{33}, γ_{34}, γ_{44} in the V matrix. Apart from these, there is also the freedom in the value of the subtraction constants in the loop functions. We will consider one subtraction constant for the s-wave channels (a_0) and one for the d-wave ones (a_2). Despite the apparent large number of free parameters in the V matrix, it is worth emphasizing that the largest matrix elements are V_{11}, V_{12} and V_{22} [8] which come from a chiral Lagrangian without any free parameters. Due to the d-wave behavior the other ones are expected to be smaller, as we will see below. In order to obtain these parameters we fit the partial wave amplitudes obtained by using the V matrix given above as the kernel in the Bethe-Salpeter equation to the experimental results on the $\bar{K}N$ and $\pi\Sigma$ scattering amplitudes in d-wave and $I = 0$. We use experimental data from [12,13] where $\bar{K}N \to \bar{K}N$ and $\bar{K}N \to \pi\Sigma$ amplitudes are provided from partial wave analysis. These experimental amplitudes are related to the amplitudes of Eq. (2) through $\tilde{T}_{ij}(\sqrt{s}) = -\sqrt{\frac{M_i q_i}{4\pi\sqrt{s}}}\sqrt{\frac{M_j q_j}{4\pi\sqrt{s}}}\,T_{ij}(\sqrt{s})$, where M and q are the baryon mass and the on-shell C.M. momentum of the channel respectively. Note that in this normalization the branching ratio is simply given by the strength of the imaginary part as $B_i = \Gamma_i/\Gamma = Im\tilde{T}_{ii}(\sqrt{s} = M_R)$.

From the fit we obtain the subtraction constants $a_0 = -1.8$ for the s-wave channels and $a_2 = -8.1$ for the d-wave channels. The unknown constants in the V matrix are given by $\gamma_{13} = 0.98$ and $\gamma_{14} = 1.1$ in units of 10^{-7} MeV^{-3} and $\gamma_{33} = -1.7$, $\gamma_{44} = -0.7$ and $\gamma_{34} = -1.1$ in units of 10^{-12} MeV^{-5}. We can see that the value obtained for the subtraction

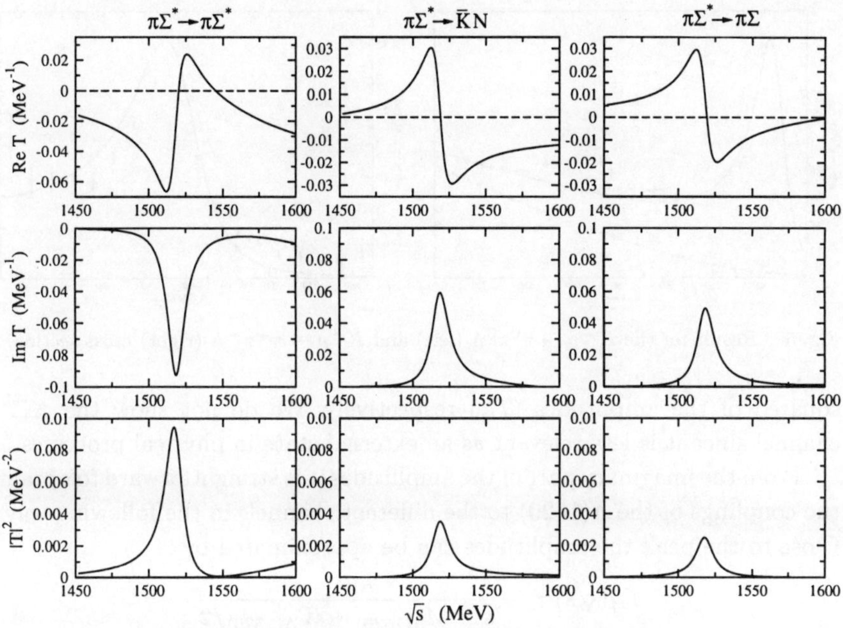

Fig. 4. Unitary amplitudes involving the $\pi\Sigma^*$ channel. From left to right: $\pi\Sigma^* \to \pi\Sigma^*$, $\pi\Sigma^* \to \bar{K}N$ and $\pi\Sigma^* \to \pi\Sigma$.

constant for the s-wave channels is of natural size (~ -2) since it agrees with the result obtained with the cutoff method using a cutoff of about 500 MeV (at $\sqrt{s} \simeq 1520$ MeV). On the other hand, regarding the d-wave loops, the large value obtained for a_2 can be understood comparing also to the cutoff method. If one keeps the momentum dependence of the d-wave vertices inside the loop integral (i.e., one does not use the on-shell approximation mentioned above) and evaluates the integral with the cutoff method, then also a cutoff of about 500 MeV gives the same result as the dimensional regularization with on-shell factorization and $a_2 \sim -8$. In summary, the use of the dimensional regularization method along with the on-shell factorization for both the s and d-wave loops, correspond to the result obtained with the cutoff method without on-shell factorization using the same cutoff of about 500 MeV.

We now show in fig. 4, the prediction for the unitarized amplitudes for the different channels involving the $\pi\Sigma^*$. From left to right the columns represent the $\pi\Sigma^* \to \pi\Sigma^*$, $\pi\Sigma^* \to \bar{K}N$ and $\pi\Sigma^* \to \pi\Sigma$ channels. The rows denote from top to bottom the real part, imaginary part and modulus

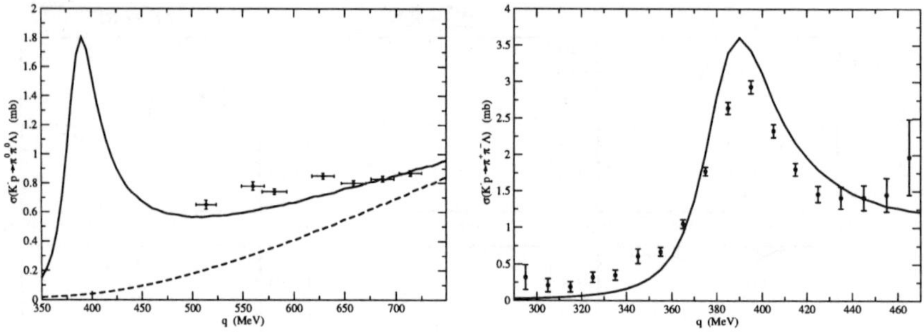

Fig. 5. Result for the $K^-p \to \pi^0\pi^0\Lambda$ (left) and $K^-p \to \pi^+\pi^-\Lambda$ (right) cross section.

squared of the amplitudes (T_{ij}) respectively. We do not show the $K\Xi^*$ channel since it is less relevant as an external state in physical processes.

From the imaginary part of the amplitudes it is straightforward to obtain the couplings of the $\Lambda(1520)$ to the different channels in the following way. Close to the peak the amplitudes can be approximated by

$$T_{ij}(\sqrt{s}) = \frac{g_i g_j}{\sqrt{s} - M_{\Lambda(1520)} + i\Gamma_{\Lambda(1520)}/2}$$

from where we have

$$g_i g_j = -\frac{\Gamma_{\Lambda(1520)}}{2} \frac{|T_{ij}(M_{\Lambda(1520)})|^2}{Im[T_{ij}(M_{\Lambda(1520)})]},$$

where $M_{\Lambda(1520)}$ is the position of the peak in $|T_{ij}|^2$ and $\Gamma_{\Lambda(1520)} = 15.6$ MeV. Up to a global sign of one of the couplings (we choose g_1 to be positive), we obtain: $g_1 = 0.91$, $g_2 = -0.29$, $g_3 = -0.54$ and $g_4 = -0.45$ for the coupling of the $\Lambda(1520)$ to the $\pi\Sigma^*$, $K\Xi^*$, $\bar{K}N$ and $\pi\Sigma$ channels respectively. We can see from the values that the $\Lambda(1520)$ resonance couples most strongly to the $\pi\Sigma^*$ channel. The fact that we are able to predict the value of this coupling is a non trivial consequence of the unitarization procedure that we employ. With the value for g_1 obtained above, we can evaluate the partial decay width of the $\Lambda(1520)$ into $\pi\pi\Lambda$ assuming that this process is dominated by the $\pi\Sigma^*$ channel, and this leads us to a branching ratio of 0.14. All these branching ratios (the branching ratios obtained for $\bar{K}N$ and $\pi\Sigma$ are 0.45 and 0.43 respectively) essentially sum up to unity considering the uncertainties in the calculations.

The prediction of the amplitudes involving $\pi\Sigma^*$ channels can be checked in particular reactions where this channel could play an important role. We evaluate the cross section for $K^-p \to \pi\pi\Lambda$ in the lines of [10] but using the new coupled channel formalism. The mechanisms and the expressions for

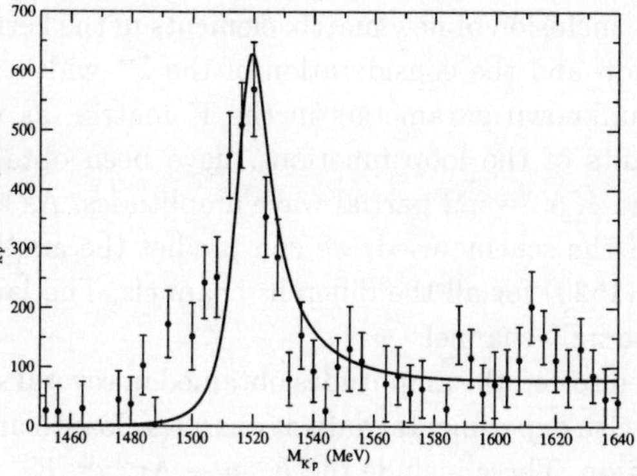

Fig. 6. K^-p invariant mass distribution for the $\gamma p \to K^+K^-p$ reaction with photons in the range $E_\gamma = 2.8 - 4.8$ GeV. Experimental data from [17].

the amplitudes and the cross sections can be found in [10] where, apart from the coupled channel unitarized amplitude, other mechanisms of relevance above the $\Lambda(1520)$ peak were also included. In fig. 5 we show our results for $K^-p \to \pi^0\pi^0\Lambda$ and $K^-p \to \pi^+\pi^-\Lambda$ cross section on the left and right panels respectively along with experimental data from refs. [15] and [16] respectively. The dashed line in the left figure represents the contribution from mechanisms other than the unitarized coupled channels, and the solid line gives the coherent sum of all the processes. These cross sections depend essentially on the $T_{\bar{K}N\to\pi\Sigma^*}$ amplitude which is obtained from our coupled channel analysis. It is a non-trivial prediction of the theory since this amplitude has not been included in the fit.

We have also evaluated the K^-p invariant mass distribution for the $\gamma p \to K^+K^-p$ reaction. In [14] the basic phenomenological model is explained but there only the $\pi\Sigma^*$ and $K\Xi^*$ channels were considered. The result in the present model is shown in Fig. 6. The normalization is arbitrary in the experimental data as well as in our calculation. For the purpose of the present work the shape of the distribution is the most important part and we can see that the agreement is quite fair.

4. Conclusions

We have done a coupled channel analysis of the $\Lambda(1520)$ resonance using the $\pi\Sigma^*$, $K\Xi^*$, $\bar{K}N$ and $\pi\Sigma$ channels. We have used the Bethe-Salpeter equation to implement unitarity in the evaluation of the different amplitudes. The main novelty from previous coupled channel approaches to this resonance is the inclusion of new matrix elements in the kernel of the Bethe-Salpeter equation and the consideration of the Σ^* width in the $\pi\Sigma^*$ loop function. The unknown parameters in the V matrix, as well as the subtraction constants of the loop functions, have been obtained by a fit to $\bar{K}N \to \bar{K}N$ and $\bar{K}N \to \pi\Sigma$ partial wave amplitudes. As a consequence of the unitarity of the scheme used, we can predict the amplitudes and couplings of the $\Lambda(1520)$ for all the different channels. The largest coupling is obtained for the $\pi\Sigma^*$ channel.

We have then tested the amplitudes obtained in several specific reactions and compared with experimental data at energies close to and slightly above the $\Lambda(1520)$ region. These include the $K^-p \to \Lambda\pi^0\pi^0$, $K^-p \to \Lambda\pi^+\pi^-$ and $\gamma p \to K^+K^-p$ reactions. We have obtained a reasonable agreement with the experimental results that allows us to be confident in the procedure followed to describe the nature of the $\Lambda(1520)$ resonance.

Acknowledgments

This work is partly supported by DGICYT contract number BFM2003-00856, and the E.U. EURIDICE network contract no. HPRN-CT-2002-00311. This research is part of the EU Integrated Infrastructure Initiative Hadron Physics Project under contract number RII3-CT-2004-506078.

References

1. N. Kaiser, P. B. Siegel and W. Weise, Nucl. Phys. A **594** (1995) 325; N. Kaiser, T. Waas and W. Weise, Nucl. Phys. A **612** (1997) 297.
2. E. Oset and A. Ramos, Nucl. Phys. A **635** (1998) 99.
3. T. Inoue, E. Oset and M. J. Vicente Vacas, Phys. Rev. C **65** (2002) 035204.
4. D. Jido, J. A. Oller, E. Oset, A. Ramos and U. G. Meissner, Nucl. Phys. A **725** (2003) 181.
5. E. Oset, A. Ramos and C. Bennhold, Phys. Lett. B **527** (2002) 99 [Erratum-ibid. B **530** (2002) 260];A. Ramos, E. Oset and C. Bennhold, Phys. Rev. Lett. **89** (2002) 252001.
6. C. Garcia-Recio, J. Nieves, E. Ruiz Arriola and M. J. Vicente Vacas, Phys. Rev. D **67** (2003) 076009.
7. E. E. Kolomeitsev and M. F. M. Lutz, Phys. Lett. B **585** (2004) 243; M. F. M. Lutz and E. E. Kolomeitsev, Nucl. Phys. A **755** (2005) 29.
8. S. Sarkar, E. Oset and M. J. Vicente Vacas, Nucl. Phys. A **750** (2005) 294; Eur. Phys. J. A **24** (2005) 287.
9. M. J. Vicente Vacas, E. Oset and S. Sarkar, Int. J. Mod. Phys. A **20** (2005) 1826; S. Sarkar, E. Oset and M. J. Vicente Vacas, Nucl. Phys. A **755** (2005) 665 ; E. Oset, S. Sarkar, M. J. Vicente Vacas, A. Ramos, D. Jido, J. A. Oller and U. G. Meissner, Int. J. Mod. Phys. A **20** (2005) 1619.
10. S. Sarkar, E. Oset and M. J. Vicente Vacas, Phys. Rev. C **72** (2005) 015206.

11. L. Roca, S. Sarkar, V. K. Magas and E. Oset, (in preparation).
12. G. P. Gopal *et al.*, Nucl. Phys. B **119** (1977) 362.
13. M. Alston-Garnjost *et al.*, Phys. Rev. D **18** (1978) 182.
14. L. Roca, E. Oset and H. Toki, arXiv:hep-ph/0411155.
15. S. Prakhov *et al.*, Phys. Rev. C **69** (2004) 042202.
16. T. S. Mast *et al.*, Phys. Rev. D **7** (1973) 5.
17. D. P. Barber *et al.*, Z. Phys. C **7** (1980) 17.

COUPLED-CHANNEL FIT TO PION-NUCLEON ELASTIC AND ETA PRODUCTION DATA

W. BRISCOE, R. ARNDT, I. STRAKOVSKY, AND R. WORKMAN

Center for Nuclear Studies, Department of Physics
The George Washington University, Washington, D.C. 20052

1. Motivation

This study was motivated by the results of two BNL experiments (E909[1] and E913/E914[2]) measuring differential cross sections for the reaction $\pi^- p \to \eta n$. The most recent measurements were obtained using the Crystal Ball detector (now at MAMI) and have provided precise angular distributions extending from threshold up to 750 MeV/c, clearly showing the onset of D-wave contributions to an S-wave dominated process. This behavior is summarized in Fig. 1, where we have plotted the coefficients of a fit

$$\frac{d\sigma}{d\Omega} = a_0 + a_1 P_1(\cos\theta^*) + a_2 P_2(\cos\theta^*) \tag{1}$$

to the differential cross sections from E913/E914. The interference of S- and D-wave components has allowed a determination of N(1520) properties, even though this resonance has a very small ηN coupling.

The dominance of S-wave contributions (near threshold) is also evident in the integrated cross section, when plotted as a function of the CM momentum p_η^*. The recent BNL data are plotted against this variable and compared to other near-threshold measurements in Fig. 2. A linear fit provides a lower limit for the imaginary part of the ηN scattering length through the optical theorem. This and a more complete K-matrix analysis has allowed us to determine the scattering length and study its model dependence [5].

2. Results

We have performed a coupled-channel fit to the low-energy $\pi^- p \to \eta n$ data base and the full πN elastic scattering database, including the new BNL measurements. In the S- and D-waves, our fit was modified to use a simple K-matrix representation, from which we have directly determined widths and branching fractions for the N(1535) and N(1520) resonances. This fit was then extrapolated to threshold in order to determine the ηN scattering length. Results for the extracted resonance properties are listed in Table 1, wherein the fits A and C have

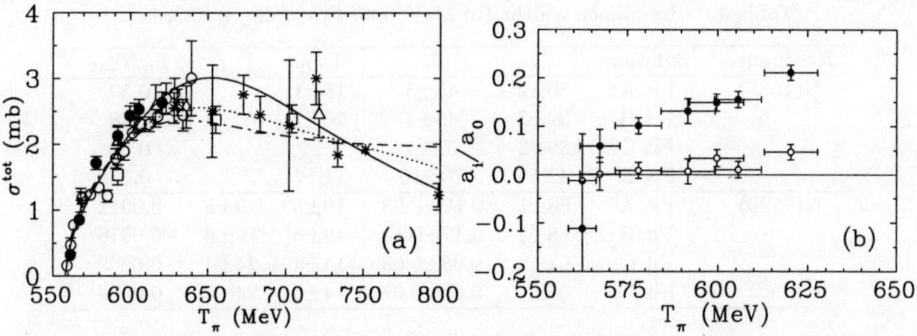

Fig. 1. Energy dependence of total cross sections and the ratio of fit coefficients. (a) Total cross section. Experimental data are from E909[1] (open circles) and E913/E914[2] (filled circles). Other previous measurements are from the SAID database[3]. FA02[4] (E913/E914 data not included), G380, and Fit A shown as solid, dash-dotted, and dotted lines, respectively. (b) Ratio of the coefficients a_1/a_0 (open symbol) and a_2/a_0 (filled symbol). Experimental data are from E913/E914[2]. The horizontal errors include the ± 2.5 MeV/c beam momentum uncertainty and the beam momentum spread.

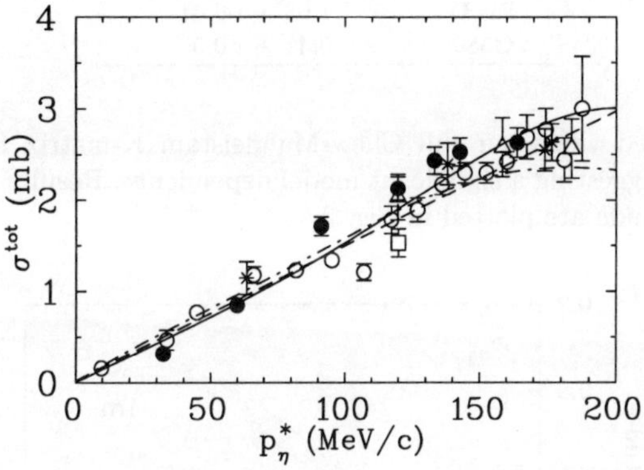

Fig. 2. p^*_η dependence of $\sigma^{tot}(\pi^- p \to \eta n)$. Data and notation given in Fig. 1(a). Dashed line shows a linear fit to E909[1] (open circles) data.

included the Crystal Ball data (B and D have not). Fits A and B have been constructed to obtain approximately equal $\pi\Delta$ and ρN couplings (to the N(1520)), whereas in fits C and D, the effect of a larger ρN coupling is studied. The ηN branching fraction for N(1520) was found to be sensitive to this choice, though values are reasonably consistent with previous determinations.

Results for the ηN scattering length are compiled in Table 2. Within the simple K-matrix approximation, these results were found to be rather insensitive to details of the database and couplings to the ρN and $\pi\Delta$ channels. The values listed for fits A to D are also reasonably consistent with previous determinations by Penner and Mosel[6] and by Green and Wycech[7]. However, a very different

Table 1. Resonance widths (in MeV) and branching fractions.

Resonance	Solution	Γ_π	Γ_η	$\Gamma_{\pi\Delta}$	$\Gamma_{\rho N}$	Γ_η/Γ_{tot}
N(1535)	Fit A	30±2	45±3	15±1		0.50
	Fit B	32±3	45±4	16±1		0.48
	Fit C	39±3	67±4	9±2		0.58
	Fit D	42±6	70±10	11±2		0.57
N(1520)	Fit A	68±1	0.12±0.03	19±5	19±5	0.0012
	Fit B	68±1	0.17±0.12	19±6	19±6	0.0016
	Fit C	67±1	0.08±0.03	14±4	24±4	0.0008
	Fit D	67±1	0.09±0.07	14±5	24±5	0.0009

Table 2. ηN scattering lengths from K-matrix fits (resonance plus background, see text) and the global energy-dependent fit (G380).

Solution	Scattering Length (fm)
Fit A	1.14 + i 0.31
Fit B	1.10 + i 0.30
Fit C	1.12 + i 0.39
Fit D	1.03 + i 0.41
G380	0.41 + i 0.56

result is obtained when our full Chew-Mandelstam K-matrix fit (fit G380) is extrapolated, suggesting a significant model dependence. Results for the deduced $\eta n \to \eta n$ amplitude are plotted in Fig. 3.

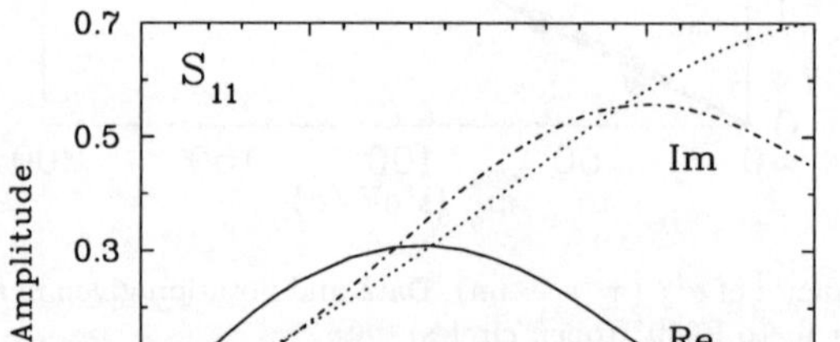

Fig. 3. p^*_η dependence of the S_{11} amplitude for the reaction $\eta n \to \eta n$. Dash-dotted (dotted) curves give the real (imaginary) parts of amplitudes corresponding to the solution G380. Solid (short-dash-dotted) lines represent the real (imaginary) parts of amplitudes Fit A. All amplitudes are dimensionless.

We note that these new measurements have only extended to the peak of the N(1535) resonance, leaving considerable uncertainty above this energy. Further progress will require data covering a broader energy range with precision comparable to the Crystal Ball measurements.

Acknowledgments

This work was supported in part by a U. S. Department of Energy Grant No. DE–FG02–99ER41110. We also acknowledge a contract from Jefferson Lab under which part of this work was done. Jefferson Lab is operated by the Southeastern Universities Research Association under DOE contract DE–AC05–84ER40150.

References

1. T. W. Morrison *et al.*, Bull. Am. Phys. Soc. **45**, 58 (2000);
   T. W. Morrison, Ph. D. Thesis, The George Washington University, Dec. 1999.
2. S. Prakhov *et al.* [Crystal Ball Collaboration], Phys. Rev. C **72**, 015203 (2005).
3. The full database and numerous PWAs can be accessed at the website http://gwdac.phys.gwu.edu.
4. R. A. Arndt, W. J. Briscoe, I. I. Strakovsky, R. L. Workman, and M. M. Pavan, Phys. Rev. C **69**, 035213 (2004).
5. R. A. Arndt, W. J. Briscoe, T. W. Morrison, I. I. Strakovsky, R. L. Workman, and A. B. Gridnev, Phys. Rev. C **72**, 045202 (2005).
6. G. Penner and U. Mosel, Phys. Rev. C **66**, 055211 (2002).
7. A. M. Green and S. Wycech, Phys. Rev. C **71**, 014001 (2005).

THE CARNEGIE MELLON UNIVERSITY PROGRAM FOR STUDYING BARYON RESONANCE PHOTOPRODUCTION USING PARTIAL WAVE ANALYSIS

M. WILLIAMS, D. APPLEGATE, M. BELLIS,
C. A. MEYER AND Z. KRAHN

Carnegie Mellon University,
5000 Forbes Ave,
Pittsburgh, PA 15213, USA

We are performing partial wave analyses on the photoproduction of $p\pi^0$, $n\pi^+$, $p\pi^+\pi^-$, $K\Sigma$, $K\Lambda$, $p\omega$, $p\eta$ and $p\eta'$ final states. The goal is to extract information on baryon resonances produced and to identify any missing baryons which couple to these final states. We have adopted a covariant tensor formalism for our amplitude calculation. We are using an event-based unbinned maximum likelihood method for our fitting procedure. This note gives a brief overview of our program and a forecast of the future of this project.

1. Program Goals

Constituent quark models do an excellent job of predicting the hadron spectrum[1–3]. However, many of the baryon states predicted by these models are not observed experimentally[4,5]. This prompted Lichtenberg[6] to propose the di-quark model, where two of the three quarks become tightly bound. This constraint leads to a spectrum devoid of the missing states of the full model. It should be noted that there is nothing in QCD which would imply any sort of di-quark interaction. Later calculations[1–3] suggest that these missing states may couple more strongly to final states other than $N\pi$, which is where most of the experimental data lies.

We are currently studying the baryon spectrum by performing partial wave analyses on the photoproduction of $p\pi^0$, $n\pi^+$, $p\pi^+\pi^-$, $K\Sigma$, $K\Lambda$, $p\omega$, $p\eta$ and $p\eta'$ final states. Our goal is to extract information about the baryon resonances produced and to identify any missing baryons which may couple to these final states.

Table (1) shows the approximate yields for each final state, along with the CLAS experiment used to acquire the data. The incident photon energy ranges for the two CLAS data sets involved in this analysis are $0.5-2.3 GeV$ for g1c and $0.8-3.8 GeV$ for g11a.

2. Formalism and Fitting Procedure

In the first phase of the analysis, each channel listed in section (1) will be fit separately in a mass independent fit. In the mass independent approach, the data is binned in $\sqrt{s}$, which allows us to omit any mass dependence from $s-$channel baryon resonance propagators. Each $\sqrt{s}$ bin is fit independently to obtain the contributions of the different J^P's. The advantage to this approach is that it allows us to not bias the fit by imposing a Breit-Wigner or K-matrix. At this stage, we simply let the fit decide how much each J^P contributes in each $\sqrt{s}$ bin.

Once a good understanding of each of the individual channels is obtained, a coupled channel analysis will be performed. This could proceed in two different ways. Multiple channels could be fit simultaneously in a mass independent procedure, or the output from each single channel analysis could be used as input to a mass dependent approach. We plan on implementing both of these techniques in at least some of the channels involved in this analysis.

We have adopted a covariant tensor approach, based on the work of Rarita and Schwinger[7], for calculating our partial wave amplitudes. The formalism is very similar to that of Anisovich, et al[8], which has been used to perform a coupled channel partial wave analysis on CB-ELSA and Mainz-TAPS data[9,10]. Amplitudes are constructed to obey Lorentz invariance, gauge invariance and, where applicable, parity conservation.

The fitting procedure used is an event-based unbinned maximum likelihood method. We do not fit to cross sections. Instead, we calculate amplitudes for each event and use them to calculate the likelihood, which is maximized by the fit. This method is similar to that used by Chung[11].

The large statistics of our datasets (see Table (1)) give rise to some technical difficulties. Consider, for example, the $p\omega$ final state which has approximately 10M data events. To implement this procedure, we must first calculate every partial wave amplitude for each event (and also for the Monte Carlo). In the covariant tensor formalism, these amplitudes are built using tensors (which can be up to rank 8 if a spin-$\frac{9}{2}$ resonance is involved) and Dirac spinors and matrices. This can make the calculations tedious. We have developed a comprehensive software package, written in C++, that allows us to calculate these amplitudes in a very general way. It is now trivial to program almost any amplitude. We have also developed a highly optimized fitting program to handle the likelihood maximization process. Documentation on all of this software can be found at http://www-meg.phys.cmu.edu/pwa.

3. Program Outlook

We have now completed two very important preliminary stages of this analysis. First, we have developed a software package that is able to handle the technical difficulties associated with this type of project (see Section(2)). Second, we have put a large amount of effort into understanding the CLAS detector acceptance. When performing a partial wave analysis, it is extremely important that the detector acceptance is as well known as possible. Small discrepancies near the edges of acceptance can, in some cases, cause the fit to miscalculate the contributions of certain partial waves. We have now reached a level of knowledge about our detector that gives us a high degree of confidence in the results of our fits.

We have also completed numerous preliminary fits on a number of the channels listed in Section(1). In these fits, we have developed methods for handling $t-$ and $u-$channel contributions along with systematic techniques for analyzing the quality of the fits. All of this work has lead to where we are now. We are now running mass independent fits on our final datasets for the $p\pi^+\pi^-$, $p\eta$, $p\eta'$ and $p\omega$ final states. We will soon combine the $p\eta$ and $p\eta'$ channels into a coupled channel mass independent fit. By the end of 2006, we hope to be running mass independent fits on the rest of the channels list in Section(1). In this same time frame, we also hope to expand the number of coupled channel fits we're running along with beginning to run some mass dependent fits.

Acknowledgments

References

1. Simon Capstick and Nathan Isgur. Baryons in a relativized quark model with chromodynamics. *Phys. Rev.* **D34** (1986).
2. Simon Capstick and Winston Roberts. $N\pi$ decays of baryons in a relativized model. *Phys. Rev* **D47** (1993).
3. Simon Capstick and Winston Roberts. Quasi two-body decays of non-strange baryons. *Phys. Rev.* **D49** (1994).
4. David Fairman and Archibald W. Hendry. Harmonic oscillator model for baryons. *Phys. Rev.* **173** (1968).
5. David Fairman and Archibald W. Hendry. Harmonic oscillator model for baryons. *Phys. Rev.* **180** (1969).
6. D. B. Lichtenberg. Baryon supermultiplets of su(6) x o(3) in a quark-diquark model. *Phys.Rev.* **178** 2197-2200 (1969).
7. William Rarita and Julian Schwinger. On a theory of particles with half integral spin. *Phys. Rev.* **60:61** (1941).
8. A. Anisovich, E. Klempt, A. Saranstev, and U. Thoma. Partial wave decomposition of pion and photoproduction amplitudes. *Eur. Phys. J.*, **A24**, 111-128 (2005).
9. A.V. Anisovich et al. Photoproduction of baryons decaying into $N\pi$ and $N\eta$. *Eur. Phys. J.* **A25** (2005)
10. A. V. Saranstev et al. Decays of baryon resonances into ΛK^+, $\Sigma^0 K^+$ and $\Sigma^+ K^0$. *Eur. Phys. J.* **A25** (2005)
11. S. U. Chung. Formulas for Partial Wave Analysis Version II. *BNL preprint* **BNL-QGS-93-05** (1999).

Table 1. Approximate yields for each channel in our analysis.

Final State	CLAS Data Set	Events
$p\pi^0$	g1c	$> 10,000,000$
$n\pi^+$	g1c	$> 10,000,000$
$p\pi^+\pi^-$	g1c	$> 10,000,000$
$p\omega$	g11a	$\sim 10,000,000$
$p\eta$	g11a	$\sim 2,000,000$
$p\eta'$	g11a	$\sim 500,000$
ΛK^+	g11a	$\sim 1,000,000$
ΣK	g11a	TBD

MULTICHANNEL PARTIAL-WAVE ANALYSIS OF $\overline{K}N$ SCATTERING

H. Y. ZHANG, D. M. MANLEY, AND J. TULPAN

Department of Physics,
Kent State University,
Kent, OH 44240, USA
E-mail: manley@kent.edu

Recently the Crystal Ball Collaboration measured precise new data for several important $\overline{K}N$ reactions at the BNL-AGS. These data have motivated a new multichannel partial-wave analysis of all available data for $\overline{K}N$, $\pi\Lambda$, and $\pi\Sigma$ final states. The main goal is to obtain better information about the properties of Λ^* and Σ^* resonances. This contribution briefly describes some problems with prior analyses and presents preliminary results of the present analysis.

1. Introduction

In comparison with nonstrange baryon resonances, our knowledge of the properties of hyperon resonances is relatively poor. Furthermore, unlike for nonstrange baryon resonances, our knowledge about strangeness -1 hyperons is derived almost entirely from energy-dependent partial-wave analyses (PWAs)[1]. Most prior energy-dependent PWAs of $\overline{K}N$ reactions used a simple parametrization for the partial-wave amplitudes, which introduced a model-dependent bias and (unless a K-matrix procedure was used) resulted in a violation of two-body unitarity of the S-matrix. For example, typical PWAs assumed that partial-wave T-matrix amplitudes could be written in the form $T = T_B + T_R$, where T_B was a background term and T_R was a sum of Breit-Wigner resonance terms[2–5].

One objective of our work is to reduce this bias as much as possible by carrying out an energy-independent partial-wave analysis. Such analyses are also called single-energy PWAs. In order to do so, it is necessary to introduce a number of constraints to account for the facts that (i) the experimental database is sparse compared with that for πN scattering and (ii) unlike for πN scattering, there exist known hyperon resonances that are so narrow that conventional single-energy PWAs are not possible.

2. Fitting Procedure and Results

Table 1 summarizes the available world database for the K^-p scattering reactions of interest here. Column 1 gives the final-state channel, column 2 gives the laboratory momentum range, column 3 gives the available number of differential cross-section measurements, column 4 gives the corresponding number of recoil polarization measurements, and column 5 gives the number of integrated cross-section measurements.

Table 1. Summary of database for selected $\overline{K}N$ scattering reactions.

channel	P_{lab} (MeV/c)	$d\sigma/d\Omega$	P	σ
K^-p	281–1815	3987	585	170
$\overline{K}^0 n$	281–1434	2913		213
$\pi^0\Lambda$	436–1843	2265	128	182
$\pi^+\Sigma^-$	436–1842	1867		141
$\pi^0\Sigma^0$	436–1730	501	72	94
$\pi^-\Sigma^+$	436–1842	1876		131
Total	15025	13409	785	831

Because the database is somewhat sparse, various constraints were implemented in order to obtain partial-wave amplitudes with a reasonably smooth energy dependence. The first step consisted of performing a global, unitary multichannel fit of the best available partial-wave amplitudes of the $\overline{K}N$ reactions involving the final-state channels $\overline{K}N$, $\pi\Lambda$, $\pi\Sigma$, $\pi\Lambda(1520)$, $\pi\Sigma(1385)$, $\overline{K}\Delta$, and $\overline{K}^*N$. This first step is the dissertation project of Kent State student, John Tulpan[6,7]. Such an approach promises to be the best chance of success for determining the resonance parameters of states with small elasticity or for states immersed in large nonresonant backgrounds.

Once suitable unitary amplitudes were obtained from the global fit, we were able to begin our single-energy partial-wave analyses. We found it adequate to select energy bins with widths of about 30 MeV. The partial-wave amplitudes that resulted from selecting narrower energy bins were unacceptably noisy. In order to decrease the number of free parameters to be searched, we also held fixed the very small T-matrix amplitudes (those with $|T| < 0.05$). The energy dependence of these amplitudes was taken from the initial global fit. This constraint is believed to introduce only a small bias to the final partial-wave solution. Finally, we parametrized the energy dependence of each amplitude in a given energy bin by

$$T(E) = T(E_0) + T'(E_0)(E - E_0) ,$$

where $T(E_0)$ is the complex T-matrix amplitude at c.m. energy E_0, which was taken as the central energy of the bin. As an additional constraint, we calculated the slope parameters $T'(E_0)$ from the initial global unitary fits, and held them fixed in subsequent single-energy fits. Thus, the free parameters of the fit were real and imaginary parts of $T(E_0)$. Finally, we held the D_{03} amplitudes fixed at the values from the global unitary fit in the vicinity of the narrow $\Lambda(1520)$ resonance, which has a width of only about 16 MeV.

We have obtained very good fits of all available data by this procedure. Figure 1 shows typical results of our single-energy partial-wave analysis. The data shown are $d\sigma/d\Omega$ and the recoil Λ polarization for $K^- p \to \pi^0 \Lambda$ at 750 MeV/c as measured by the Crystal Ball Collaboration[8]. The solid curves are predictions derived from the prior PWA of Gopal *et al.*,[2] and the open boxes are the results of our single-energy solution.

The key remaining stage of this project involves incorporating the resulting single-energy solutions into a global multichannel unitary fit, which will permit the extraction of resonance parameters. We will check that the resulting energy-dependent solution still provides a good description of the $\overline{K}N$ observables, and if necessary, we will iterate the entire procedure.

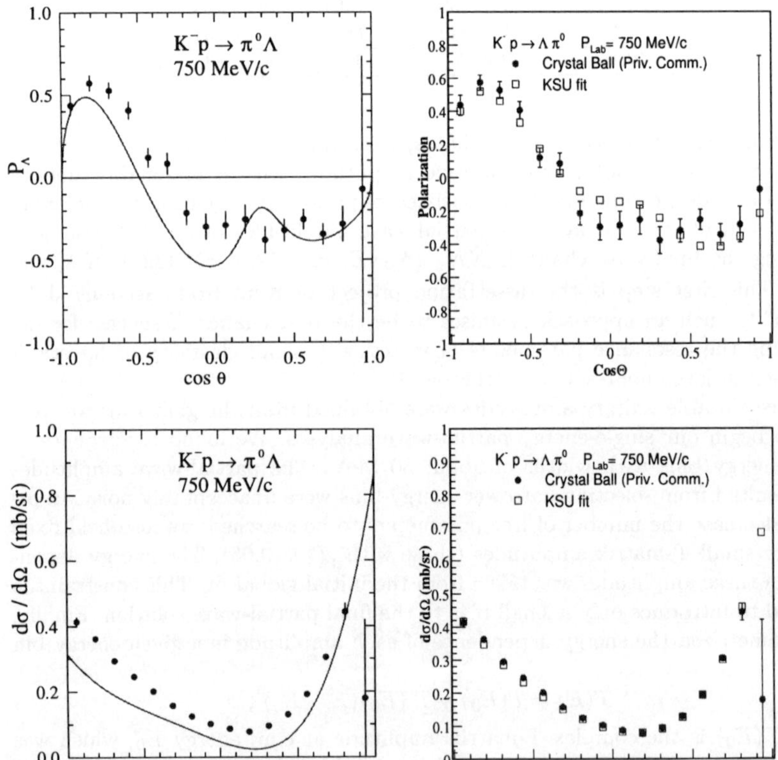

Fig. 1. Differential cross section and Λ recoil polarization for $K^- p \to \pi^0 \Lambda$ at 750 MeV/c. (See text for details.)

3. Summary

The available world database of differential cross sections, recoil polarizations, and integrated cross sections up to a c.m. energy of ~ 2 GeV, has been compiled for the reactions $\overline{K}N \rightarrow \overline{K}N$, $\overline{K}N \rightarrow \pi\Lambda$, and $\overline{K}N \rightarrow \pi\Sigma$. Constrained single-energy fits were carried out successfully to provide partial-wave amplitudes at c.m. energy bins of 30-MeV width. These amplitudes will be incorporated into a new global fit to extract resonance parameters that are reliably determined and consistent with S-matrix unitarity.

Acknowledgments

This work was supported in part by grant DE-FG02-01ER41194 from the U.S. Department of Energy. We are grateful to the Crystal Ball Collaboration for permission to show unpublished results for $K^-p \rightarrow \pi^0\Lambda$ at 750 MeV/c.

References

1. Particle Data Group, S. Eidelman *et al.*, *Phys. Lett.* **B592**, 1 (2004).
2. G. P. Gopal *et al.*, *Nucl. Phys.* **B119**, 362 (1977).
3. W. Cameron *et al.*, *Nucl. Phys.* **B146**, 327 (1978).
4. W. A. Morris *et al.*, *Phys. Rev.* **D17**, 55 (1978).
5. H. Koiso, F. Sai, S. S. Yamamoto, and R. R. Kofler, *Nucl. Phys.* **A433**, 619 (1985).
6. J. Tulpan, *Ph.D. Dissertation*, Kent State University (2006).
7. D. M. Manley and J. Tulpan, *NSTAR 2004: Proceedings of the Workshop on the Physics of Excited Baryons*, edited by J.-P. Bocquet, V. Kuznetsov, and D. Rebreyand, pp. 341-344, (World Scientific, 2004).
8. S. Prakhov *et al.*, *Crystal Ball Collaboration* (private communication), (2005).

HELICITY AMPLITUDES AND ELECTROMAGNETIC DECAYS OF STRANGE BARYON RESONANCES

T. VAN CAUTEREN* AND J. RYCKEBUSCH

Ghent University, B-9000 Gent, Belgium

B. C. METSCH AND H. R. PETRY

Helmholtz-Institut für Strahlen- und Kernphysik
Bonn University, D-53115 Bonn, Germany

We present results for the helicity amplitudes of the lowest-lying hyperon resonances Y^*, computed within the framework of the Bonn constituent-quark model, which is based on the Bethe-Salpeter approach [1–3]. The seven parameters entering the model are fitted against the best known baryon masses [4]. Accordingly, the results for the helicity amplitudes are genuine predictions. Some hyperon resonances are seen to couple more strongly to a virtual photon with finite Q^2 than to a real photon. Other Y^*'s, such as the $S_{01}(1670)$ Λ resonance or the $S_{11}(1620)$ Σ resonance, have large electromagnetic decay widths and couple very strongly to real photons. The negatively-charged and neutral members of a Σ^* triplet may couple only moderately to the $\Sigma(1193)$, while the positively-charged member of the same Σ^* triplet displays a relatively large coupling to the $\Sigma^+(1193)$ state. This illustrates the necessity of investigating all isospin channels in order to obtain a complete picture of the hyperon spectrum.

1. Introduction

The implementation of the electromagnetic (EM) couplings to a hadron resonance in isobar models constitutes one of its major sources of uncertainty. This is particularly the case for models including hyperon resonances, for which little experimental information is available. These Y^*'s can play an important role in the background of kaon electroproduction processes, photo- and electroproduction of the elusive $\Theta(1540)$ resonance, and radiative kaon capture reactions. In this work, the Bonn constituent-quark (CQ) model, based on the Lorentz-covariant Bethe-Salpeter approach, is used to compute the electromagnetic helicity amplitudes (HA's) of the lowest-lying hyperon resonances. The results may be used in isobar models where a $\gamma^{(*)}YY^*$ coupling gets introduced.

*e-mail:Tim.VanCauteren@UGent.be

2. Helicity Amplitudes in the Bethe-Salpeter Approach

The expression for the current matrix elements in the framework of the Bonn CQ model can be found in Refs. 1, 2, 3 and 5. In the rest frame of the incoming baryon resonance, the EM transition matrix element reads :

$$\langle \overline{P} \,|\, j^{\mu} \,|\, \overline{M}^{*} \rangle \simeq -3 \iint \frac{\mathrm{d}^4 \left[\frac{1}{2}(p_1 - p_2)\right]}{(2\pi)^4} \frac{\mathrm{d}^4 \left[\frac{1}{3}(p_1 + p_2 - 2p_3)\right]}{(2\pi)^4} \times \overline{\Gamma}_{\overline{P}}^{\Lambda} \, S_F^1(p_1) \otimes S_F^2(p_2) \otimes \left[S_F^3(p_3 + q) \, \hat{q}\gamma^{\mu} S_F^3(p_3) \right] \Gamma_{\overline{M}^*}^{\Lambda} , \quad (1)$$

where Γ and $\overline{\Gamma}$ are the amputated BS amplitude and its adjoint of the incoming and outgoing baryon respectively, and S_F^i is the i'th CQ propagator. In the above matrix element, the operator $\hat{q}\gamma^{\mu}$ describes how the photon couples to a pointlike CQ with charge $\hat{q}$.

The electromagnetic properties of baryon resonances are usually presented in terms of helicity amplitudes, which are functions of the squared fourmomentum of the photon. The HA's are directly proportional to the current matrix element with the appropriate spin projections for incoming and outgoing hyperon :

$$A_{1/2}(B^* \to B) = \mathcal{D} \left\langle B, \overline{P}, \frac{1}{2} \right| j^1(0) + i j^2(0) \left| B^*, \overline{P}^*, -\frac{1}{2} \right\rangle , \quad (2a)$$

$$A_{3/2}(B^* \to B) = \mathcal{D} \left\langle B, \overline{P}, -\frac{1}{2} \right| j^1(0) + i j^2(0) \left| B^*, \overline{P}^*, -\frac{3}{2} \right\rangle , \quad (2b)$$

$$C_{1/2}(B^* \to B) = \mathcal{D} \left\langle B, \overline{P}, \frac{1}{2} \right| j^0(0) \left| B^*, \overline{P}^*, \frac{1}{2} \right\rangle , \quad (2c)$$

with $\mathcal{D} = \sqrt{\frac{\pi\alpha}{2M(M^{*2}-M^2)}}$, where α is the fine-structure constant and M^* (M) the mass of the incoming (outgoing) hyperon.

3. Results and Conclusions

In Fig. 1 we present our predictions for the HA's of the lowest-lying spin $J = 3/2$ Λ resonances, decaying to the $\Lambda(1116)$. One notices that the $A_{1/2}$ and $C_{1/2}$ of the lightest resonance with either parity show a maximum at finite Q^2. This feature also occurs in some of the other HA's we have computed within the Bonn CQ model [3]. As a consequence, u-channel contributions to $p(e, e'K)Y$ processes, can be expected to exhibit a peculiar Q^2 dependence.

In Table 1, the EM decay widths of the hyperon resonances with masses around 1660 MeV are given. This is the energy region investigated by the Crystal Ball Collaboration in Brookhaven for the $\overline{K}^- p \to \gamma Y^0$ processes [6]. Our results indicate that the $S_{01}(1670)$ and $D_{03}(1690)$ Λ-resonances could be important if $Y^0 = \Sigma^0(1193)$. If on the other hand $Y^0 = \Lambda(1116)$, the $P_{11}(1660)$, $S_{11}(1620)$ and

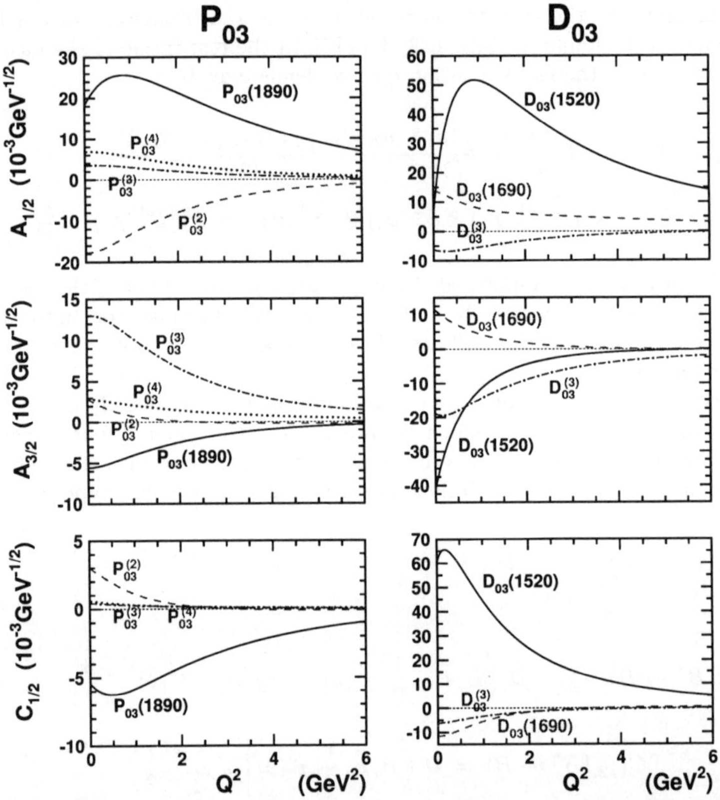

Fig. 1. The Q^2 dependence for the $\Lambda^* + \gamma^* \to \Lambda$ decays for spin $J = 3/2$ resonances : left (right) panels show the results for the positive (negative) parity Λ^* resonances.

$D_{13}(1670)$ Σ-resonances seem to be more appropriate candidates for governing the reaction dynamics.

Another distinct feature illustrated by Table 1 is the dependence of the EM decay widths of the Σ resonances on the isospin-3 component. For positively-charged Σ^*'s, the reported decay widths can be an order of magnitude larger than for the negatively-charged or neutral members of the isospin triplet. These resonances could therefore contribute to the background of the $p(\gamma, K^0)\Sigma^+$ process significantly, while being marginal for the $p(\gamma, K^+)\Sigma^0$ reaction.

References

1. D. Merten, U. Löring, K. Kretzschmar, B. Metsch and H.-R. Petry, *Eur. Phys. J.* **A14**, 477 (2002).

Table 1. Electromagnetic decay widths of the hyperon resonances around 1660 MeV to the different ground-state hyperons in units of MeV.

Y^*	$\Gamma_{Y^* \to \Lambda(1116)}$	$\Gamma_{Y^{*0} \to \Sigma^0(1193)}$	$\Gamma_{Y^{*+} \to \Sigma^+(1193)}$	$\Gamma_{Y^{*-} \to \Sigma^-(1116)}$
$S_{01}(1670)$	0.159×10^{-3}	3.827	—	—
$D_{03}(1690)$	0.0815	1.049	—	—
$P_{11}(1660)$	0.451	0.0578	0.733	0.141
$S_{11}(1620)$	1.551	0.688	5.955	0.613
$D_{13}(1670)$	1.457	0.0214	0.440	0.184

2. T. Van Cauteren, D. Merten, J. Ryckebusch, T. Corthals, S. Janssen, B. Metsch and H.-R. Petry, *Eur. Phys. J.* **A20**, 283 (2004).
3. T. Van Cauteren, J. Ryckebusch, B. Metsch and H.-R. Petry, submitted to *Eur. Phys. J.* **A**, *arXiv:nucl-th/0509047*.
4. U. Löring, *A Covariant Quark Model of Baryons with Instanton-induced Forces*, PhD. Thesis, Rheinische Friedrich-Wilhelms-Universität Bonn, Germany (2001).
5. D. Merten, *Hadron Form Factors and Decays*, PhD. Thesis, Rheinische Friedrich-Wilhelms-Universität Bonn, Germany (2002).
6. N. Phaisangittisakul, *First Measurement of the Radiative Process* $K^- p \to \Lambda\gamma$ *at Beam Momenta* 520-750 *MeV/c Using the Crystal Ball Detector*, PhD. Thesis, UCLA, USA (2001); S. Prakhov, *Progress on Study of* K^-*-Proton Reactions*, Technical Report CB-01-010, UCLA, USA, March 2001, http://bmkn8.physics.ucla.edu/Crystalball/Docs/.

PROGRESS REPORT FOR A NEW KARLSRUHE-HELSINKI TYPE PION-NUCLEON PARTIAL WAVE ANALYSIS*

S. WATSON AND M. E. SADLER

Abilene Christian University, Abilene, TX 79699, USA

J. STAHOV

University of Tuzla, 35000 Tuzla, Bosnia and Herzegovina

A new phase shift analysis has been started by groups from Abilene and Tuzla using formalism and methods of the Karlsruhe-Helsinki partial wave analysis. The first step is an amplitude analysis where invariant amplitudes consistent with fixed-t analyticity are obtained. Next, a similar amplitude analysis is performed along interior hyperbolas. As a final step, a single energy partial wave analysis is performed using results of amplitude analyses as constraints. In an iterative procedure, pion-nucleon partial waves consistent with Mandelstam analyticity are obtained. The resulting partial waves will be used for determination of resonance parameters. Pion-nucleon partial waves having analytic properties required by Mandelstam hypothesis are a valuable part of input for multichannel analyses.

1. Introduction

The goal of this Karlsruhe-Helsinki (KH) type partial wave analysis (PWA) effort is to obtain a single energy (SE) partial wave solution that provides a good description of available data while complying with theoretical and analytic constraints. To accomplish this, an iterative method is used where the constraints are applied in different components of the analysis. These different components should be constrained to agree with one another numerically, so that the final solution satisfies all applied constraints.

2. Single Energy PWA

The first component of this analysis is a standard single energy or fixed energy PWA. In this analysis, experimental data are binned and kinematically shifted to fixed values of lab momentum. Partial wave amplitudes at fixed values of lab

*This work is supported by USDOE grant DE-FG03-94ER40860 and the Federal Ministry of Education and Science, Bosnia and Herzegovina

momentum are determined by fitting Legendre polynomial based expansions to the experimental observables. Unitarity of the PW amplitudes is imposed as a strong constraint in the SE analysis. One serious issue related to SE PWA is the consistency of solutions as a function of momentum. Gaps in available data as well as inconsistencies present between data sets are sources of this issue. A second issue, which is even more serious, is ambiguities of PW amplitudes obtained in the SE PWA. Without external constraints, a PW amplitudes from SE analysis may have both discrete and continuum ambiguities. Since uniqueness of the solution is critical, dispersion relation based analytic constraints are used to resolve ambiguities.

3. Invariant Amplitude Analyses

In order to implement dispersion relation constraints in this amplitude analysis, Pietarinen's expansion method is used.[1,2] This method is based on analytic parameterizations of invariant scattering amplitudes. The parameterization is model independent and flexible. This method is model independent in the sense that the expansion coefficients are not physically significant and are not related to resonance parameters. The forms of the amplitude expansions are chosen to require the relationship between the real and imaginary parts of amplitudes to be analytic as required by the Mandelstam hypothesis. The B^+, B^-, C^+, and C^- invariant amplitudes are used for the following analyses. Each amplitude is expanded separately, and observables are calculated from combinations of the expansions. A major advantage of this type of analysis over explicit dispersion integrals is the possibility to fit analytic representations of invariant amplitudes to experimental observables. Here, the amplitude expansions allow flexible interpolation between measurements at finite points while obeying Mandelstam analyticity. A χ^2 function is minimized to determine the values of the expansion coefficients.

3.1. *Fixed-t Amplitude Analysis*

Fixed-t dispersion relations (FTDR) are a strong analytic constraint used in the KH78 and KH80 analyses. The KH analyses used fixed-t amplitude analyses to obtain invariant amplitudes consistent with FTDR.[3] The process for performing invariant amplitude analyses at fixed values of the Mandelstam variable t has been redeveloped for this analysis. The goal of this amplitude analysis is to acquire invariant amplitudes that satisfy crossing symmetry and fixed-t dispersion relations. Here, experimental observables are binned and shifted to fixed values of t which correspond to fixed values of momentum transfer. Then, invariant amplitude expansions are fit to the observables by χ^2 minimization. This analysis provides full angular coverage at low energy and covers very high energies in the near forward region as shown in Figure 1. Experimental observables for this analysis are available up to lab momentum greater than 300 GeV/c.

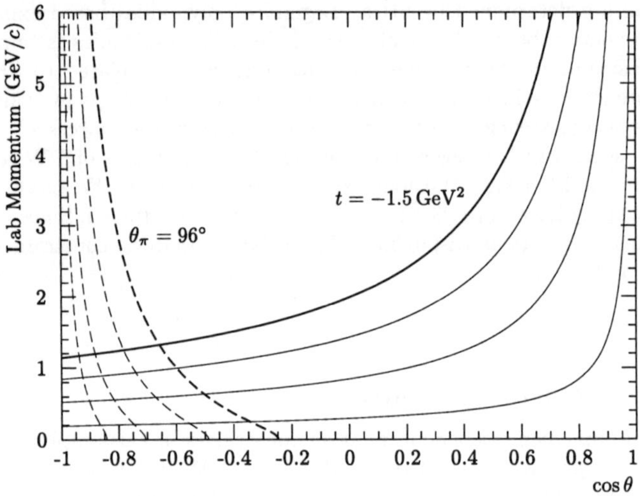

Fig. 1. Kinematic coverage of dispersion relation analyses in the $\cos\theta_{\mathrm{cms}}$–$p_{\mathrm{lab}}$ plane. Solid curves denote fixed-t, and dashed curves denote fixed-θ_π. The bold curves show bounds of the FT and IDR analyses. The nonbold curves correspond to a representative subset of values for each variable ($\theta_\pi = 111°, 128°, 142°$ and $t = -0.1, -0.5, -1.0\,\mathrm{GeV}^2$).

3.2. *Interior Hyperbola Amplitude Analysis*

An interior dispersion relation (IDR) analysis is currently being developed.[4] IDR analysis of this form was not used in the KH analyses. This is an amplitude analysis with similarities to the fixed-t analysis. This analysis is performed along interior hyperbolas in the Mandelstam plane. In the physical region, interior hyperbolas correspond to fixed lab angle (θ_π). The hyperbolas suitable for this analysis cover the backward region ($180° \geq \theta_\pi > 95°$) at intermediate to high energies as shown in Figure 1. The paths along interior hyperbolas corresponding to $\theta_\pi < 95°$ enter the double spectral region, which limits the extent of IDR analysis.[3] Experimental observables for this analysis are available up to lab momentum greater than 25 GeV/c.

4. Conclusions

The two invariant amplitude analyses should complement one another well to stabilize the SE PW solution and resolve ambiguities. At lower energies, these analytic constraints cover the full angular range and overlap in the backward region. At higher energies, the amplitudes are constrained in the forward region by the FT analysis and in the backward region by the analysis along interior hyperbolas. Additional constraints on PW amplitudes, such as PW dispersion relations, are desired and should be applied as a next step.

References

1. E. Pietarinen, *Nuovo Cimento* **12A**, 522 (1972).
2. E. Pietarinen, *Nucl. Phys.* **B107**, 21 (1976).
3. G. Hoehler, Landolt-Börnstein, **Vol. I/9b2**, *Pion-nucleon Scattering*, Springer, Berlin (1983).
4. G. E. Hite, F. Steiner, *Phys. Rev.*, **D8**, 3205 (1973).

BARYON EXCITATION THROUGH MESON HADRO- AND PHOTOPRODUCTION IN A COUPLED-CHANNELS FRAMEWORK: CHIRAL-SYMMETRY-INSPIRED MODEL

A. WALUYO AND C. BENNHOLD

Physics Department, The George Washington University
725 21st Street NW,
Washington, DC 20052, USA
E-mail: waluyoab@gwu.edu, bennhold@gwu.edu

We report the results of an investigation of baryon excited states up to an energy of $W = 2$ GeV. A newly developed Chiral-Symmetry-Inspired model is employed in this investigation to improve the background description. We extract resonance properties and search for missing states. The status of the "missing" $D_{13}(1900)$ state is discussed.

1. Introduction/Motivation

Considerable theoretical and experimental efforts are expended to understand the baryon excited state spectra in the region of $1 < W < 3$ GeV. The Particle Data Group[1] registers 44 excited nucleon states of isospin $\frac{1}{2}$ and $\frac{3}{2}$. QCD in this region becomes non-perturbative, thus difficult to solve. Lattice QCD and effective field theories have begun to make inroads in the study of baryon resonance spectra.

Most of our knowledge about baryon excited states have been obtained within a phenomenological field-theoretical framework. In this paper, we report the results on the investigation of baryon resonance properties in the region of W up to 2 GeV. The investigation is performed using a newly developed Chiral-Symmetry-Inspired (CSI) coupled-channels model where the channels $\pi N, 2\pi N, \eta N, K\Lambda$, and $K\Sigma$ hadro- and photoproduction are calculated simultaneously. 25 resonance states are identified and investigated. The resonance properties are extracted through the interpretation of high- quality $\eta N, K\Lambda$, and $K\Sigma$ photoproduction data recently provided by the CLAS collaboration[2].

2. Chiral-Symmetry-Inspired (CSI) Model

Our Chiral-Symmetry-Inspired model is based on the work of Feuster and Mosel[3] that solves the Bethe-Salpeter equation in the $K-$matrix approximation. The driving terms consist of the conventional background and resonance contributions, in addition to new contact terms consistent with $SU(3)$ chiral dynamics. The new background amplitudes are thus constructed from the standard $s-, u-$ and $t-$ Born tems, scalar and vector meson resonances, and higher order chiral contact terms. At the same time, resonance contributions have been updated using the modern gauge-invariant resonance Lagrangians derived by Pascalutsa[6]. This fixes the problem of the wrong spin degrees of freedom and discards the ambiguous off-shell terms inherent in the traditional Rarita-Schwinger types of Lagrangians. Spin$-\frac{1}{2}$, $\frac{3}{2}$, and $\frac{5}{2}$ baryon resonances are included.

3. Results and Discussion

Baryon resonances are identified according to their quantum numbers. Naturally, these numbers point to partial waves where resonance states are excited. According to their appearance in each partial wave, baryon resonance states are categorized as first-tier, second-tier, and third-tier, i.e., in the D_{13} partial wave we find the$D_{13}(1520)$, $D_{13}(1700)$, and $D_{13}(1900)$ states, respectively.

Table 1. Extracted baryon resonance properties in the first and second resonance region from the present calculation (1st line) in comparison to the values taken PDG (2nd line). The mass and total width are given in MeV, the decay ratio β is in %. ^{a} the decay is given in 0.01%. ng: not given.

$L_{2I,2S}$	Mass	Γ_{tot}	$\beta_{\pi N}$	$\beta_{2\pi N}$	$\beta_{\eta N}$	$\beta_{K\Lambda}$	$\beta_{K\Sigma}$
$D_{13}(1520)$	1505	93	55	45	0.01^a	–	–
	$1515 - 1530$	$110 - 135$	$50 - 60$	$40 - 50$	–	–	–
$D_{13}(1700)$	1692	299	2.0	86	0.35	12	0.57
	$1650 - 1750$	$50 - 150$	$5 - 15$	$85 - 95$	$0.0 - 0.1$	< 3	–
$D_{13}(1900)$	1961	313	7.0	48	0.5	15	31
	ng	–	–	–	–	–	–

In most cases, first-tier resonances are clearly isolated with typical Breit-Wigner behavior and small background. Only the $S_{31}(1620)$ and $P_{13}(1720)$ deviate from this situation with large and unusual background contributions. Since the first-tier resonance masses and decay branching ratios are in agreement within $20-30\%$ between each other, their extractions is an important test that needs to be satisfied by the CSI model. The successful extraction of the first-tier $D_{13}(1520)$ properties is shown in Table 1.

For the second-tier resonances, rough agreement is found only for the resonance mass, however, the extracted total and partial widths vary widely in most cases. This is reflected by the extracted properties of the $D_{13}(1700)$ shown in Table 1. Finally, the very existence of third-tier states in each partial wave is tenuous at best, examples are the 1- and 2-star states $S_{11}(2090)$, $P_{11}(2100)$ and $D_{13}(2080)$. The only exception here is the $P_{33}(1920)$ which appears well established and happens to be the only third-tier state with three stars.

In order to study these lesser-known states, we focus our effort on the interpretation of high-quality $\eta N, K\Lambda$ and $K\Sigma$ photoproduction data recently provided by the CLAS collaboration[2].

In search of missing resonances, we have investigated four possible partial waves: the S_{11}, P_{11}, P_{13}, and D_{13} in the energy region of 1.8 - 2.0 GeV New resonances would correspond to third-tier states in each partial wave. Only in the D_{13} partial wave is a new state found with mass of 1961 MeV and a total width of 313 MeV . As shown in Table 1, branching ratios of 15% and 31% are found into the $K\Lambda$ and $K\Sigma$ decay channels, respectively. Other possible missing states came out with total widths of less than 40 MeV and were thus deemed unphysical. The quality of the fit of our calculation to the $K\Lambda$ and $K\Sigma$ photoproduction differential cross section data[2] at high energies are shown in Figure 1 and 2.

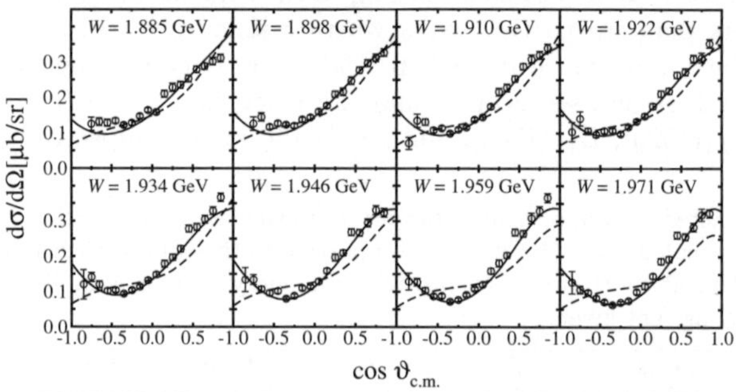

Fig. 1. Differential cross section for $\gamma N \to K\Lambda$ at high energies. The solid line shows our full results, while the dashed, dotted, and dot-dashed lines show our calculations without the $D_{13}(1900)$.

4. Summary

In this talk, we have discussed the results of our investigation on 25 N^* and Δ states with $W < 2.0$ GeV through meson hadro- and photoproduction in a coupled-channels framework using a newly developed CSI model. The properties of resonance states are successfully extracted. Most 2-star resonances are found but most 1-star states are not. The "missing" $D_{13}(1900)$ resonance state is identified. Using the CSI model, we are able to describe observables of $\pi N \to \pi N, \pi N \to 2\pi N, \gamma N \to \pi N, \gamma N \to \eta N, \gamma p \to K\Lambda$, and $\gamma p \to K^0\Sigma^+$

References

1. L. Alvarez-Gaumeé et al., Phys. Lett. B **592**, 1-1109 (2004).
2. M. Dugger *et. al.*, CLAS Collaboration Phys. Rev. Lett. **89**, 222002 (2002); Mc Nabb *et. al* CLAS 2003, collaboration; Mc Nabb, Ph.D. dissertation, Carnegie-Mellon University, Pittsburg USA, 2003.
3. T. Feuster and U. Mosel Phys. Rev. C**58** 457 (1998), Phys. Rev. C**59** 460 (1999).
4. Agung B. Waluyo, Ph.D. dissertation, the George Washington University, Washington DC USA, 2005.

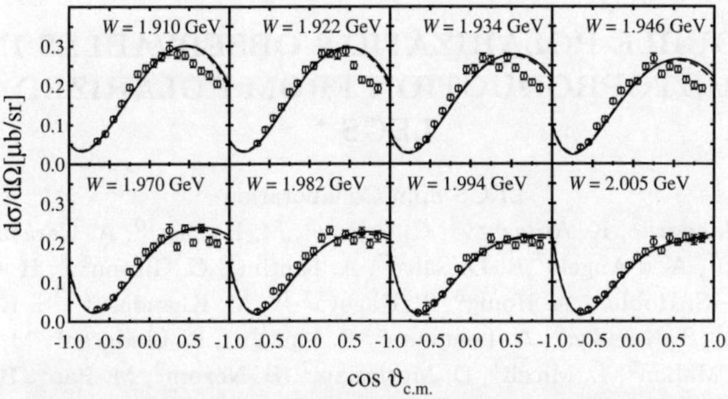

Fig. 2. Differential cross section for $\gamma N \to K^+\Sigma^0$ at high energies. Notation as in Figure 1

5. G. Penner and U. Mosel Phys. Rev. C**66**, 055211 (2002); **66**, 055212 (2002) .
6. V. Pascalutsa Phys. Lett. **B503** 85 (2001) and private communication.

DOUBLE-POLARIZATION OBSERVABLES IN PION-PHOTOPRODUCTION FROM POLARIZED HD AT LEGS *

LEGS Spin Collaboration

A. M. Sandorfi[1], K. Ardashev[8], C. Bade[1,5], M. Blecher[10], A. Caracappa[1], A. D'Angelo[7], A. d'Angelo[7],R. Di Salvo[7], A. Fantini[7], C. Gibson[1,8], H. Glückler[2], K. Hicks[5], S. Hoblit[1], A. Honig[6], T. Kageya[1,10], M. Khandaker[4], S. Kizilgul[5], O. Kistner[1], S. Kucuker[9], A. Lehmann[8], F. Lincoln[1], R. Lindgren[9], M. Lowry[1], M. Lucas[5], J. Mahon[5], L. Miceli[1], D. Moricciani[7], B. Norum[9], M. Pap[2], B. Preedom[8], C. Schaerf[7], H. Seyfarth[2], H. Ströher[2], C. Thorn[1], K. Wang[9], X. Wei[1], C. Whisnant[3]

(1) Brookhaven National Lab., (2) Forschungszentrum Jülich GmbH, (3) James Madison Univ., (4) Norfolk State Univ., (5) Ohio Univ., (6) Syracuse Univ., (7) Univ. di Roma, (8) Univ. of South Calolina, (9) Univ. of Virginia, (10) Virginia Polytechnic Institute and State Univ.

Frozen-spin HD targets have been used to measure beam-target double-polarization observables in pion production at the Laser Electron Gamma Source (LEGS). The targets are pure; the only unpolarizable nucleons are associated with the cell and their small contributions are subtracted with conventional empty-cell measurements. H and D polarizations of 58% and 29% were achieved with in-beam lifetimes of about a year. Exclusive measurements of $D(\gamma, \pi^0 n)$ are discussed. Requirements on the π^0-n azimuthal angle can be used to emphasize quasi-free neutron reactions. Some preliminary results are presented.

For the past eight years, the LEGS Spin Collaboration has been developing a new class of frozen-spin polarized target for nucleon spin structure measurements [1]. This adventure has recently culminated in two highly successful data collection periods. In this paper we provide a brief overview of the physics of the new target, report on its in-beam characteristics and provide a sampling of some preliminary π^0 photo-production results.

An H_2 molecule which consists of two identical protons has two spin states, with spins either antiparallel (para-H_2) or parallel (ortho-H_2). Ortho-H_2, which accounts for 3/4 of room temperature hydrogen, is readily polarized at high magnetic field and low temperature while para-H_2 is not. At a low temperature, ortho-H_2 decays to the para-H_2 state with a half life of about six days. Because

---

*Supported by the U.S. Department of Energy under contract DE-AC02-98CH10886, the U.S. National Science Foundation and the Istituto Nazionale di Ficisa Nucleare

the two protons in H_2 are identical particles, para-H_2 is the lowest energy state of the molecule, so that pure H_2 cannot reach a frozen-spin state.

The HD molecule has no limitations of symmetry. Its lowest energy state is L=0 and S=3/2 with proton and deuteron spins parallel. With no phonons to couple to the crystal lattice in solid HD, spin-relaxation times are extremely long. A small concentration of polarized ortho-H_2 (on order of 10^{-4}) is used to polarize HD, using a spin-spin coupling between an H in H_2 and an H in HD. After a significant number of ortho-H_2 -to- para-H_2 half-lives (typically, about three months) almost all of the H_2 has decayed to the magnetically inert para-H_2 and the target has reached a frozen-spin state. The D in HD can be polarized in a similar way, although to a lesser degree due to its smaller magnetic moment.

At LEGS, one mole HD targets are condensed in a 2K/4K dewar, where equilibrium NMR measurements are made for polarization calibration, and then transferred to a dilution refrigerator for polarization at 15 tesla and 12 mK. (Thin aluminum cooling wires imbedded in the target cell are used to conduct away the heat generated by the ortho-H_2 -to- para-H_2 conversion and these limit the HD temperature to about 12 mK.) Typically, two targets are polarized simultaneously, although the system accommodates three. After about 3 months, a target is moved to a 0.25K In-Beam Cryostat (IBC) with a thin 1.0 tesla solenoid and rolled into the 4π calorimeter for data collection. Polarizations are monitored frequently with a coil positioned around the target in the IBC. Spin differences from reactions on H and on D can be separated using data taken with different target spins. For that reason, the H spin is flipped with an allowed RF transition about half way through the run period. At the end of the run period, the target is transferred back to the original 2K/4K dewar for cross calibration of NMR polarization monitoring [2].

The first production run in November, 2004, focused on H polarization, starting with P(H)=58% and P(D)= 8%. The second production run in April, 2005, focused on D polarization. At the start of the second run, an RF-modulated saturated forbidden transition was used to transfer polarization from H to D, resulting in P(H)=29% and P(D)=29%. For both run periods, the in-beam lifetimes of the H and D polarizations were about a year. The polarization profile during the April, 2005, running period is shown in Fig. 1. The H polarization was flipped after 20 days into the run. (The efficiency of this allowed transition was limited by instabilities in a power supply.) The polarization relaxation times continued to grow as the run progressed, due to the continued decay of the para-D_2 and ortho-H_2 impurities.

At LEGS, circular and linear polarized gamma-ray beams are produced by laser backscattering from 2.8 GeV electrons. Data are collected in six beam polarization states simultaneously, right and left circular, 0° and 90° linear and $+45^0$ and -45^0 linear. A small component of unpolarized bremsstrahlung (1%) is also periodically sampled.

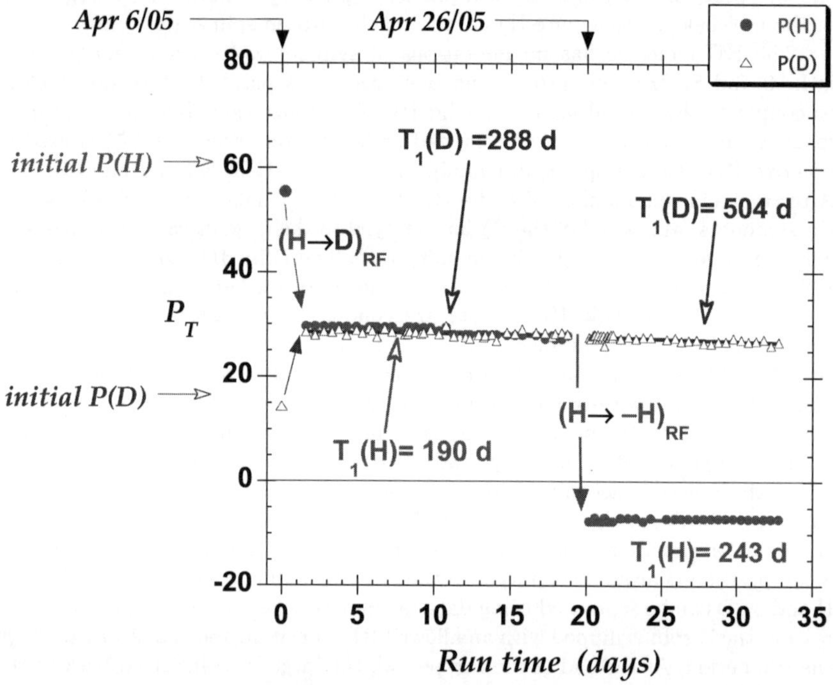

Fig. 1. Target polarizations during the April, 2005, run period. At the start of the run, polarization was transferred from H to D using an RF saturated forbidden transition. The H spin was flipped on day 20.

The cross section for reactions of polarized photons with a deuteron whose vector polarization is oriented along the beam can be written as,

$$\frac{d\sigma}{d\Omega}(\Theta,\phi,E_\gamma)$$
$$= \frac{d\sigma_0}{d\Omega}(\Theta,E_\gamma)\cdot\left\{\begin{array}{l} 1+P_\gamma^L\cdot\left[\tilde{\Sigma}(\Theta;E_\gamma)+\frac{1}{\sqrt{2}}P_D^T\cdot T_{20}^L(\Theta;E_\gamma)\right]\cdot cos2\phi \\ \quad +P_\gamma^L\cdot P_D^V\cdot\tilde{G}(\Theta;E_\gamma)\cdot sin2\phi \\ \quad -P_\gamma^C\cdot P_D^V\cdot\tilde{E}(\Theta;E_\gamma)+\frac{1}{\sqrt{2}}P_D^T\cdot T_{20}^0(\Theta;E_\gamma) \end{array}\right\}.$$

Here, P_D^V and P_D^T are the degrees of vector and tensor polarization of the deuteron. This expression is similar to the corresponding cross section for reactions on a longitudinally polarized proton, except for the addition of the T_{20}^L and T_{20}^0 tensor observables [4]. By combining data runs with different gamma-ray polarizations and different target polarizations, and fitting the different azimuthal angular dependences, all five polarization observables can be extracted. This is now underway.

Missing energy spectra for the $\pi^0 n$ channel near the peak of the $P_{33}\Delta$ resonance are plotted in Fig. 2 for a portion of the April, 2005, run. The Kel-F

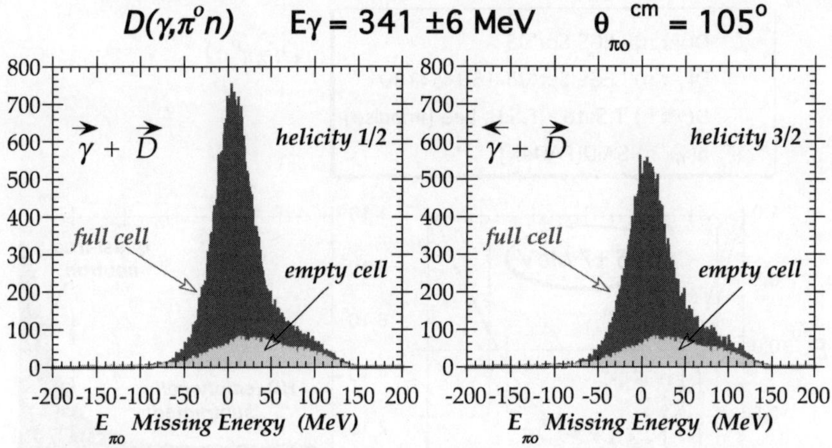

Fig. 2. π^0 missing energy distributions for the $D(\gamma, \pi^0 n)$ reaction with beam and target spins parallel (left) and antiparallel (right). The flux-weighted empty cell yields are indicated.

target cell walls and aluminum cooling wires are the only sources of unpolarizable nucleons and these are sampled in conventional empty-target runs. Their flux-weighted contributions (about 20%) are indicated. The comparison between yields with parallel γ and D spins (helicity 1/2) and antiparallel spins (helicity 3/2) shows a pronounced asymmetry.

The ratio of the difference in the yields of Fig. 2 to their sum is the E Asymmetry, constructed from circularly polarized photons and longitudinally polarized nucleons. The spin-difference in the numerator of this asymmetry enters the Gerasimov-Drell-Hearn sum rule. Theoretical calculations have shown that many of the nuclear corrections are small in this observable and as such it is expected to be particularly effective in helping to constrain fits to the underlying "free" neutron π-production amplitude. Data from the April, 2005, run at 400 MeV are shown as the solid circles in the left panel of Fig. 3. The solid curve is an impulse-level calculation by Lee and Sato for the $D(\gamma, \pi^0 n)p$ reaction [3]. The dotted curve is a SAID predictions for π^0 photo-production from the "free" neutron, which appears shifted to background angles.

A measurement of the azimuthal angle between the recoil neutron and the $< \gamma - \pi^0 >$ plane provides an indication of the degree of nuclear binding. For a completely free neutron, this would be 180°. The distribution of this angle is shown in the right panel of Fig. 3. As expected, a peak from quasi-free neutrons at 180° is apparent, though far from dominant. Using a selection of events within 30^0 of this quasi-free peak results in the triangles shown in the left panel of Fig. 3. These appear shifted from the circles to larger angles and are in fact much closer to the SAID predictions (the dotted curve) for a "free" neutron target.

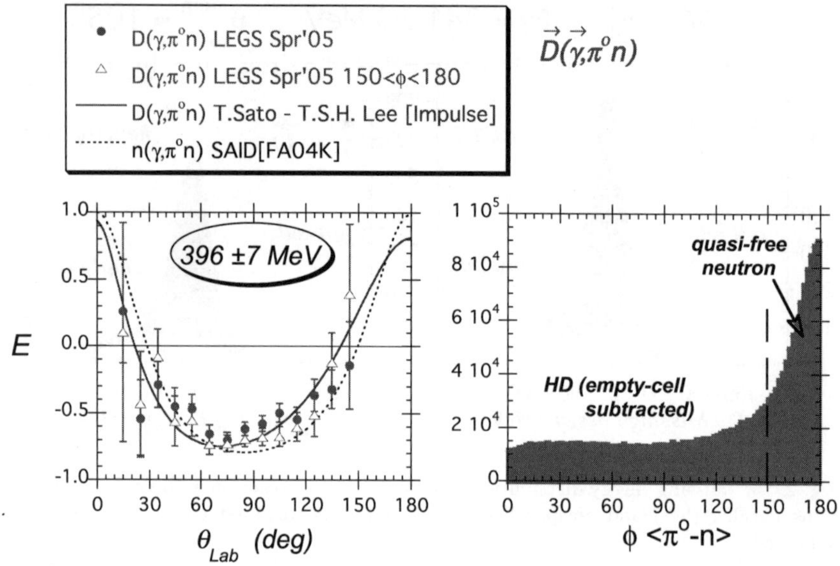

Fig. 3. E asymmetries in the $D(\gamma, \pi^0 n)$ channel, left panel; azimuthal angle between the detected π^0 and n, right panel. The solid curve is an impulse model calculation; the dashed line is the prediction from a multipole analysis for a "free" neutron target. A 180 ± 30 degree cut on the latter brings the E asymmetries close to the "free" neutron expectation.

A significant degree of freedom presently exists in predictions of "neutron" observables, due chiefly to the sparce neutron data. The LEGS data sets are presently under analysis and there is every expectation that constraints provided by the new results with polarized HD will be very effective in removing this unwanted freedom.

References

1. X. Wei, C.M. Bade, A. Caracappa, T. Kageya, F.C. Lincoln, M.M. Lowry, J.C. Mahon, A.M. Sandorfi, C.E. Thorn, C.S. Whisnant, NIM A526 (2005) 157, and references therein.
2. A. Caracappa and C. Thorn, *Proceedings of 15th International Spin Symposium, Brookhaven National Lab, Upton, New York.* AIP 675, 867 (2002).
3. T.S.-H. Lee and T. Sato, private communication. T.S.-H. Lee, A. Matshuyama and T. Sato, Proc. NSTAR 2004, workshop on Phys. of Excited Nucleons, Grenoble, World Scientific, NJ (2004) 104.
4. A. Fix and H. Arenhoevel, nucl-th/0506018.

ELECTROEXCITATION OF THE $P_{11}(1440)$, $D_{13}(1520)$, $S_{11}(1535)$, AND $F_{15}(1680)$ UP TO 4 $(GeV/c)^2$ FROM CLAS DATA

I. G. AZNAURYAN

Thomas Jefferson National Accelerator Facility
Newport News, VA 23606 USA
Yerevan Physics Institute, Yerevan, 375036 Armenia
E-mail: aznaury@jlab.orG

We present the helicity amplitudes for the electroexcitation of the resonances $P_{11}(1440)$, $D_{13}(1520)$, $S_{11}(1535)$, and $F_{15}(1680)$ on protons extracted from CLAS-JLab data on the π, 2π, and η electroproduction. The obtained results extend over wide region of Q^2, thus providing strict test for the approaches for the descriptrion of the nucleon and nucleon resonances. None of the existing approaches gives satisfactory description of the obtained results.

1. Introduction

We are currently at a very encouraging stage of the study of hadron structure as experiments on electroproduction of mesons off protons in wide Q^2 region have been carried out at JLab. These experiments allow to measure the Q^2 dependence of the electroexcitation of the nucleon resonances, and so to probe simultaneously the structure of the nucleon and its excited states. The knowledge of the electroexcitation amplitudes in wide Q^2 region put the approaches of hadron physics in the nonperturbative region to stringent test and by this reason is very important both for tesing the existing models and for developing new approches.

In this report, we present the results on the helicity amplitudes of the transitions $\gamma^* p \to P_{11}(1440)$, $D_{13}(15200)$, $S_{11}(1535)$, $F_{15}(1680)$ extracted from the analysis of the following data:

(a) $Q^2 = 0.4$ $(GeV/c)^2$: the data used in our analysis [1] at this value of Q^2 are the results of CLAS measurements of π^0 ($W = 1.1 - 1.68$ GeV) and π^+ ($W = 1.1 - 1.55$ GeV) differential cross sections [2,3], and polarized beam asymmetry in π^0 and π^+ electroproduction ($W = 1.1 - 1.66$ GeV) [4,5].

(b) $Q^2 = 0.65$ $(GeV/c)^2$: here we have used CLAS measurements of π^0 electroproduction cross sections ($W = 1.1 - 1.52$ GeV, $E_e = 1.645$ GeV and $W = 1.1 - 1.68$ GeV, $E_e = 2.445$ GeV) [2], and polarized beam asymmetry in π^0 and π^+ electroproduction ($W = 1.1 - 1.66$ GeV) [4,5]. We have also used CLAS data [3] on π^+ differential cross sections at $W = 1.1 - 1.41$ GeV and $Q^2 = 0.6$ $(GeV/c)^2$.

As the values of Q^2 in [3] and the main data set are different, and the data on π^+ differential cross sections extend over more restricted range in W, we have complemented this data set by the older DESY and NINA data on π^0 and π^+ differential cross sections at $W = 1.4 - 1.7\ GeV$ and $Q^2 = 0.6 - 0.64\ (GeV/c)^2$ [6–10].

(b′) At $Q^2 = 0.65\ (GeV/c)^2$, we have also performed a combined analysis [11] of single- and double-pion electroproduction data. These data were successfully described in the second and third resonance regions with common N^* photocouplings, thus providing a confirmation of the relability of the background description and resonance/background separation in both channels.

(c) It is known that η photo- and electroproduction provide a unique opportunity to study the electroexcitation of the $S_{11}(1535)$, because the contributions of nearby resonances in these reactions are strongly suppressed in comparison with the $S_{11}(1535)$ contribution. In order to obtain additional information on the electroexcitation of the $S_{11}(1535)$, we have performed an analysis [1] of the CLAS data on η electroproduction cross sections at $Q^2 = 0.375,\ 0.75\ (GeV/c)^2$ $(W = 1.5 - 1.62\ GeV)$ [12].

(d) We present also the results of the analysis of the preliminary CLAS data [13] on π^+ differential cross sections and polarized beam asymmetries at $W = 1.1 - 1.79\ GeV$ and $Q^2 = 1.72,\ 2.05,\ 2.44,\ 2.91,\ 3.48,\ 4.16\ (GeV/c)^2$.

The analyses of data sets at relatively small Q^2 ((a),(b), and (c)) were carried out using two approaches: dispersion relations and isobar model. With increasing Q^2, the applicability of dispersion relations at energies above first resonance region becomes questionable; by this reason the analysis of data set (d) was carried out only using isobar model.

2. Results and conclusion

The obtained results for the helicity amplitudes of the transitions $\gamma^* p \to P_{11}(1440)$, $S_{11}(1535)$, $D_{13}(15200)$, $F_{15}(1680)$ are presented in Figs. 1-4 in comparison with quark model predictions [15–20].

In Figs. 1-4 we have also presented the results which follow from older data. These are bands for the $S_{11}(1535)$ and $D_{13}(1520)$ obtained in Ref. [23] on the basis of single quark transition model, and curves obtained in the global fit [21] of older data and more recent JLab data on π^0 electroproduction at $W < 1.39\ GeV$ and $Q^2 = 2.8,\ 4\ (GeV/c)^2$ [24].

On the basis of the presented results we can conclude:

- Our results obtained from π, 2π, and η electroproduction data, and those extracted at different Q^2 are in good agreement with each other.

- For the first time transverse and longitudinal amplitudes for the transition $\gamma p \to P_{11}(1440)$ are obtained in wide Q^2 region: $Q^2 = 0 \div 4\ (GeV/c)^2$. These results are based on the measurements performed in wide W region which includes the $P_{11}(1440)$ resonance. In contrast with this, the results of Ref. [21] (thin solid curves) at high Q^2 are based on π^0 electroproduction data [24] at $Q^2 = 2.8,\ 4\ (GeV/c)^2$ and small W:

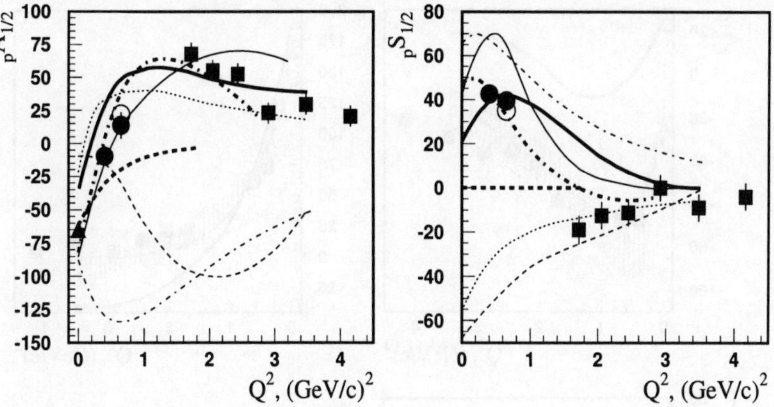

Fig. 1. Helicity amplitudes for the $\gamma^* p \to P_{11}(1440)$ transition (in 10^{-3} $GeV^{-1/2}$ units). Full circles are the results obtained in the analysis of π electroproduction data [1]. Open circles are the results obtained in the combined analysis of single- and double-pion electroproduction data [11]. Full boxes are the results obtained in the analysis of preliminary π^+ electroproduction data [13]. Full triangle at $Q^2 = 0$ is the PDG estimate [14]. Bold solid and dashed-dotted curves and thin dashed, dashed-dotted and dotted curves correspond to quark models of Refs. [15–19], respectively. Bold dashed curves are the predictions obtained assuming that the $P_{11}(1440)$ is a q^3G hybrid state [20]. Thin solid curves correspond to the results obtained in the analysis of older data in Ref. [21].

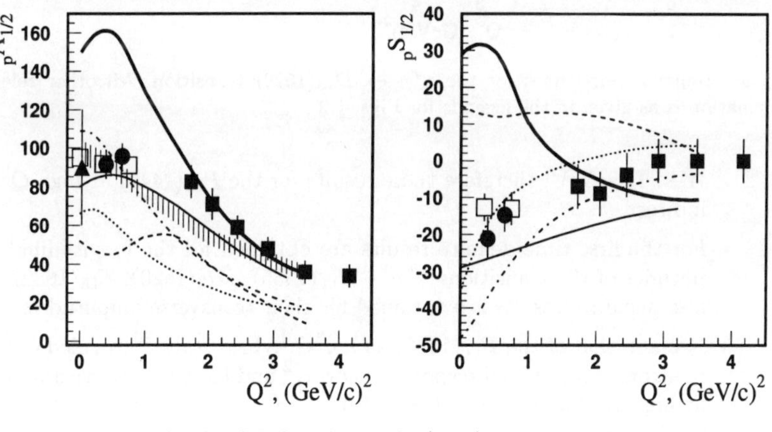

Fig. 2. Helicity amplitudes for the $\gamma^* p \to S_{11}(1535)$ transition. Open boxes are the results obtained in the analysis of η photo and electroproduction data in Refs. [1,22]. Shadowed area [23] corresponds to the results obtained from the existing Bonn, DESY and NINA data, and from JLab measurements of η electroproduction [12]. All other relevant information is as given in the legend for Fig. 1.

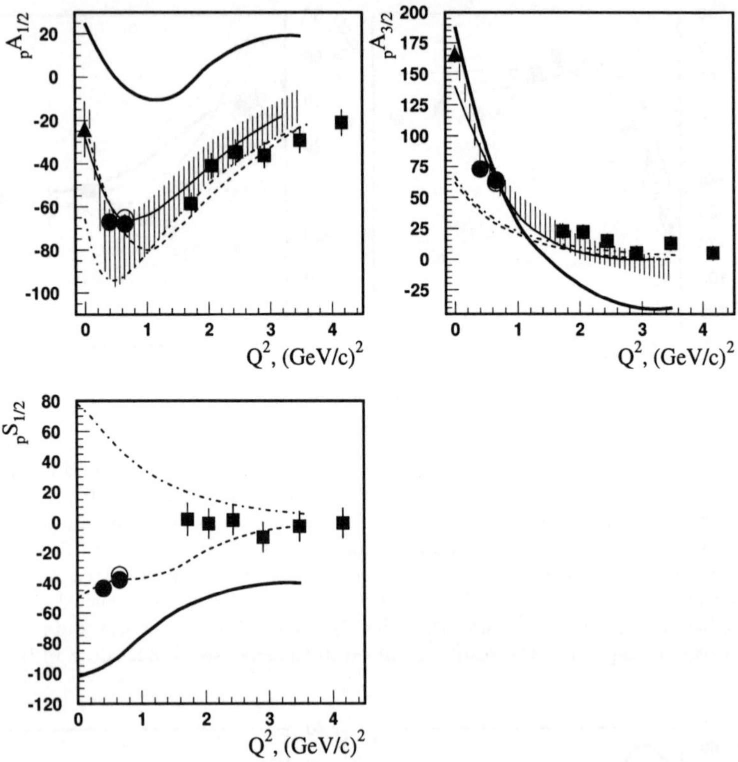

Fig. 3. Helicity amplitudes for the $\gamma^* p \rightarrow D_{13}(1520)$ transition. All other relevant information is as given in the legends for Figs. 1,2.

$W < 1.39$ GeV; therefore these results for the $P_{11}(1440)$ at high Q^2 are indirect.

- For the first time definite results are obtained for the longitudinal amplitudes of the transitions $\gamma^* p \rightarrow S_{11}(1535)$, $D_{13}(1520)$, $F_{15}(1680)$, and also accurate results are obtained for their transverse amplitudes.
- Presentation of the $P_{11}(1440)$ as a q^3G hybrid state [20] is ruled out due to strong longitudinal response at low Q^2 and large transverse amplitude at high Q^2 for the $\gamma p \rightarrow P_{11}(1440)$.
- Comparison of our results with the existing quark model predictions show that none simultaneously gives a good description of the $\gamma^* p \rightarrow P_{11}(1440)$, $S_{11}(1535)$, $D_{13}(1520)$ transition amplitudes.

References

1. I. G. Aznauryan, V. D. Burkert et al., *Phys. Rev.* **C71**, 015201 (2005).

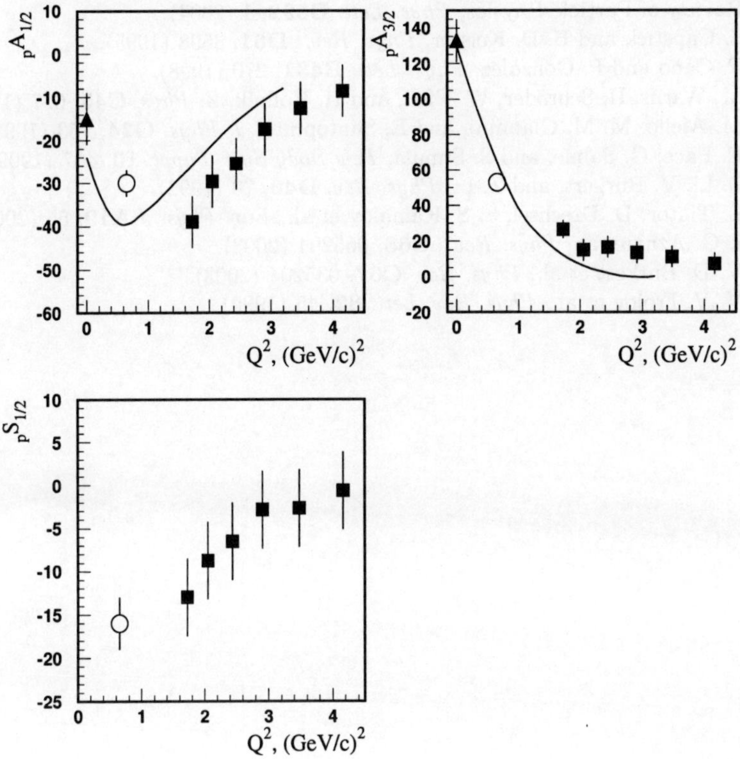

Fig. 4. Helicity amplitudes for the $\gamma^* p \to F_{15}(1680)$ transition. All other relevant information is as given in the legend for Fig. 1.

2. K. Joo, L. C. Smith et al., CLAS Collaboration, *Phys. Rev. Lett.* **88**, 12201 (2002).
3. K. Joo, L. C. Smith et al., CLAS Collaboration, *Phys. Rev.* **C70**, 042201 (2004).
4. K. Joo, L. C. Smith et al., CLAS Collaboration, *Phys. Rev.* **C68**, 032201 (2003).
5. H. K. Egiyan et al., CLAS Collaboration, to be submitted to Phys.Rev.C; L. C. Smith , hep-ph/0306199.
6. J. C. Alder et al., *Nucl. Phys.* **B105**, 253 (1976).
7. J. C. Alder et al., *Nucl. Phys.* **B99**, 1 (1975).
8. Ch. Gerhardt et al., Preprint DESY-F21-73/3 (1971).
9. Ch. Gerhardt et al., Preprint DESY-F21-79/02 (1979).
10. W. J. Shuttleworth et al., *Nucl. Phys.* **B45**, 428 (1972).
11. I. G. Aznauryan, V. D. Burkert et al., *Phys. Rev.* **C72**, 045201 (2005).
12. R. Thompson et al., CLAS Collaboration, *Phys. Rev. Lett.* **86**, 1702 (2001).
13. K. Park et al., CLAS Collaboration, to be submitted to Phys.Rev.C.

14. Review of Particle Physics, *Phys. Lett.* **B592**, 1 (2004).
15. S. Capstick and B. D. Keister, *Phys. Rev.* **D51**, 3598 (1995).
16. F. Cano and P. Gonzàlez, *Phys. Lett.* **B431**, 270 (1998).
17. M. Warns, H. Schröder, W. Pfeil, and H. Rollnik, *Z. Phys.* **C45**, 627 (1990).
18. M. Aiello, M. M. Giannini, and E. Santopinto, *J. Phys.* **G24**, 753 (1998).
19. E. Pace, G. Salme, and S. Simula, *Few Body Syst. Suppl.* **10**, 407 (1999).
20. Z. Li, V. Burkert, and Z. Li, *Phys. Rev.* **D46**, 70 (1992).
21. L. Tiator, D. Drechsel, S. S. Kamalov et al., *Eur. Phys. J* **A19**, 55 (2004).
22. I. G. Aznauryan, *Phys. Rev.* **C68**, 065204 (2003).
23. V. D. Burkert et al., *Phys. Rev.* **C67**, 035204 (2003).
24. V. V. Frolov et al., *Phys. Rev. Lett.* **82**, 45 (1999).

A GENETIC ALGORITHM ANALYSIS OF N* RESONANCES

D. G. IRELAND
Department of Physics and Astronomy,
University of Glasgow,
Glasgow, G12 8QQ, UK
E-mail: d.ireland@physics.gla.ac.uk

Extracting information on N* resonances by comparing models to data requires an efficient fitting strategy. A genetic algorithm is one of a range of tools for optimization, but it has not yet been applied in this field. The fitting of $p(\gamma, K^+)\Lambda$ data with a typical isobar model will serve to illustrate the method developed which incorporates a genetic algorithm, together with a traditional optimizer. Within the limits of this tree-level analysis, we find the data are not sufficient to resolve ambiguities. The main outcomes of this work are to show that an extensive search of parameter space is mandatory in analyses of this kind, and to outline a scheme for making a quantitative comparison of different models.

1. Introduction

The problem of establishing how many N* resonances actually exist has persisted for some time. A detailed knowledge of the baryonic spectrum is crucial to determine the properties of QCD at non-perturbative energies. Study of the $p(\gamma, K^+)\Lambda$ reaction channel has been proposed[1] as a means of discovering new N* resonances, since undiscovered resonances may couple more strongly to this channel than pion channels.

The total cross section for the reaction, as measured at SAPHIR[2] showed a distinct bump at an invariant mass of 1.9 GeV. A preliminary analysis by Mart and Bennhold[3] showed that the shape of the total cross section could be better described with the inclusion of an extra D_{13} resonance. However, a different analysis[4] appeared to show that a tuning of background parameters removed the need for a new resonance. The lack of channel coupling also cast doubt on whether there was any evidence for a new resonance.

Against this backdrop, the Gent group developed a tree-level calculation[5–7] which required the fitting of data to extract values for coupling constants (a typical calculation requires 20 - 30 free parameters). The fitting process involves running several genetic algorithms, whose results are used to provide starting values for Minuit[8] optimization.

2. Using a Genetic Algorithm

Genetics algorithms (GAs) have been described extensively[9,10], so only a brief description is presented here. A number of trial solutions (i.e. sets of values of free parameters) are generated randomly. Each solution is assigned a fitness, depending on the goodness of fit of the calculation to data. The "population" of solutions is then evolved in a manner analogous to biological evolution, using both "crossover" and "mutation" operators to try new solutions. In this way, the population migrates to one, or more, better regions of parameter space.

The efficiency of using a genetic algorithm can be seen in figure 1. In this example the progress of 10 GAs is compared to 10 Minuit calculations. The GA points represent the χ^2 of the best-of-generation solution. The Minuit calculations start from the best solution from the initial GA populations. It is clear that the improvement of the GA outstrips Minuit in the first 3000 calculations by more than two orders of magnitude. As the GAs progress, their rate of improvement d

t]

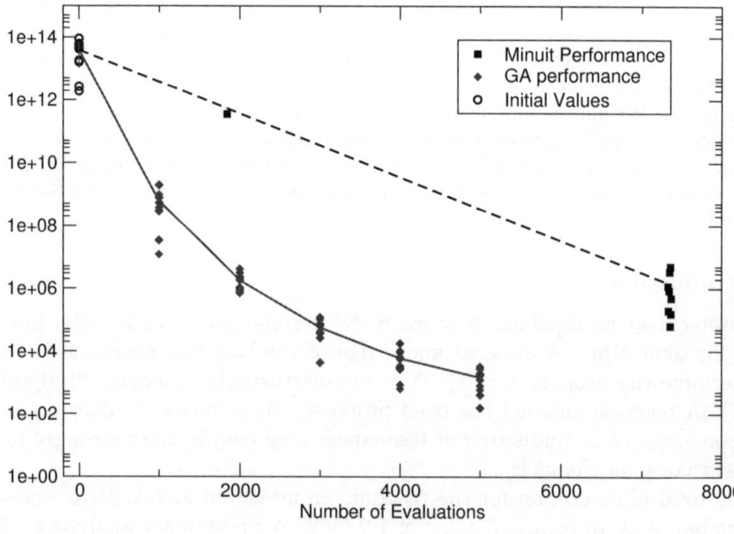

Fig. 1. Comparison of a GA and Minuit.

3. First Results

Applying the GA-Minuit strategy and the Gent calculations to the SAPHIR data set[2] led to some interesting conclusions. In this analysis, an additional $D_{13}(1895)$ resonance was included, together with the known resonances. The calculations of total cross-sections using several sets of parameters which gave similar χ^2 values are shown in figure 2.

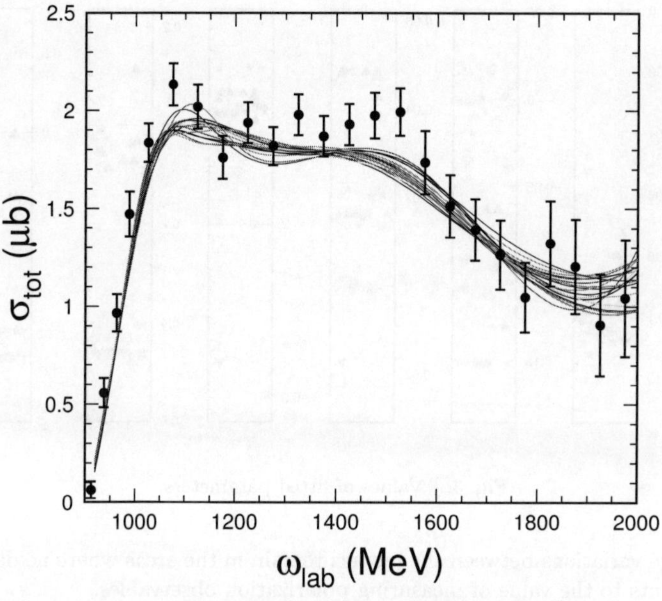

Fig. 2. First calculations with additional D_{13}.

By examining the spread of resonance parameters shown in figure 3, one sees that for each of the parameters, a range of values is permissible. The range of values is typically larger that the uncertainty calculated from Minuit. Looking in detail at the values for the $D_{13}(1895)$ coupling constant $G^{(1)}$ (fifth plot in figure 3), two clusters of solutions appear: one is centred around zero (indicating that the resonance is not there), and the other is clear non-zero (indicating that it is there). This ambiguity clearly highlights the need for analyses to study parameter space extensively before making firm conclusions.

4. Systematic Study

Following the publication of results of photon beam asymmetry measurements from SPring-8[11], and including some electroproduction results from Jlab[12]. A systematic study was undertaken to look for evidence of additional resonances[13]. Variants of calculations were used, each of which included a core set of known resonances, but also included one of an additional S_{11}, P_{11}, P_{13} or D_{13} at a mass of 1895 MeV.

Figure 4 shows a comparison of each of these model variants with the photon beam asymmetry data. The inclusion of the photon beam asymmetry in the fit greatly reduced the ambiguities within each model variant, but it can be seen

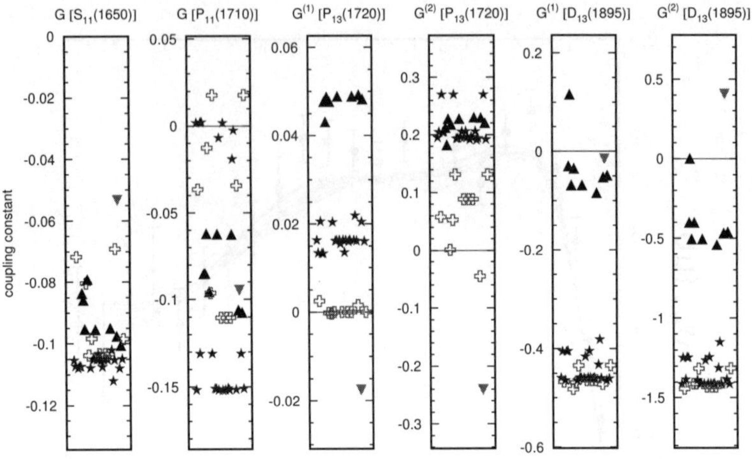

Fig. 3. Values of fitted parameters.

that huge variations between the models remain in the areas where no data exist. This points to the value of measuring polarization observables.

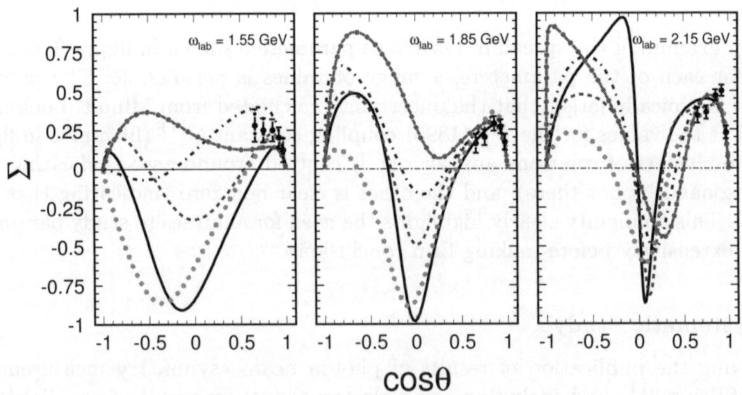

Fig. 4. Calculations compared to SPring-8 data. The model variant calculations are: core set (solid line), S_{11} (dashed line), P_{11} (small circles), P_{13} (dot-dashed line) and D_{13} (solid plus circles).

However, in establishing which of the model variants leads to the best description of the data, we need to take into account not only the value of χ^2, but the different number of free parameters used for in each case. Details of this are given in [13], but in short, models with more parameters are penalised by a so-called "Occam factor"[14]. This is a product of factors for each parameter which, roughly, are the ratios of the fitted errors on the parameter to the ranges of search space over which they can vary.

Some evidence was found to indicate that physics beyond the known set of resonances is at work, but given the lack of channel coupling in these calculations, little more information of substance can be extracted from the present analysis. The main conclusion, which essentially boils down to common sense, is that a more extensive set of measurements of polarization observables (single and double) will be required.

In this preliminary investigation the calculation of the Occam factors was performed in an approximate manner to illustrate how the method might be applied to a quantitative comparison between different models trying to describe the same data. The full calculation of these factors will be a much more complicated process since it involves the integration of likelihood functions over the whole parameter space. Techniques such as Markov Chain Monte Carlo methods, used for inference in other scientific disciplines, will need to be deployed.

5. Conclusions

This contribution has used an analysis of the $p(\gamma, K^+)\Lambda$ reaction with a tree-level model to illustrate problems associated with the extract of information from reactions involving baryon resonances. The conclusions can be summarised as follows:

- Genetic algorithms have been shown (at least in this case) to be a powerful additional tool, to be harnessed to a fitting problem in combination with well known optimization methods.
- An extensive study of parameter space is essential to establish whether two or more islands of ambiguity exist. Unfortunately, this calls for many (unfortunately CPU intensive) sets of fitting calculations to be performed, irrespective of fitting strategy.
- Some effort is required to develop a method for quantitative comparison among models. It is likely that this will also involve a great deal of computing power, but possibly no more so than is presently utilised by the lattice QCD community.

References

1. S. Capstick and W. Roberts. *Phys. Rev. D*, 58:074011, 1998.
2. M.Q. Tran *et al. Phys. Lett. B*, 445:20, 1998.
3. T. Mart and C. Bennhold. *Phys. Rev. C*, 61:(R)012201, 2000.
4. B. Saghai. *nucl-th/0105001*, 2001.
5. S. Janssen, J. Ryckebusch, W. Van Nespen, D. Debruyne, and T. Van Cauteren. *Eur. Phys. J. A*, 11:105, 2001.

6. S. Janssen, J. Ryckebusch, D. Debruyne, and T. Van Cauteren. *Phys. Rev. C*, 65:015201, 2002.
7. S. Janssen, J. Ryckebusch, and T. Van Cauteren. *Phys. Rev. C*, 67:052201(R), 2003.
8. CERN. *MINUIT 95.03*, cern library d506 edition, 1995.
9. D.E. Goldberg. *Genetic Algorithms in Search, Optimization and Machine Learning*. Addison Wesley, Reading, MA, 1989.
10. L. Davis. *Handbook of Genetic Algorithms*. Van Nostrand Reinhold, New York, 1991.
11. R.G.T. Zegers *et al. Phys. Rev. Lett.*, 91(9):092001–1, 2003.
12. R.M. Mohring *et al. Phys. Rev. C*, 67:055205, 2003.
13. D.G. Ireland, S. Janssen, and J. Ryckebusch. *Nucl. Phys. A*, 740:147, 2004.
14. D.S. Sivia. *Data Analysis - A Bayesian Tutorial*. Oxford University Press, Oxford, UK, 1996.

MEASUREMENT OF THE $N \to \Delta^+(1232)$ TRANSITION AT HIGH MOMENTUM TRANSFER BY π^0 ELECTROPRODUCTION

MAURIZIO UNGARO, KYUNGSEON JOO

University of Connecticut,
2152 Hillside Rd.,
Storrs, CT 06269
E-mail: ungaro@jlab.org

PAUL STOLER

Rensselaer Polytechnic Institute,
Troy, New York 12180-3590

The measurement of the differential cross section of the exclusive electroproduction reaction $\gamma^* p \to \pi^0 p$ in W region of the $\Delta^+(1232)$ resonance is reported. The magnetic form factor (G^*_M) and ratios of electric to magnetic and scaler to magnetic multipole amplitudes $R_{EM} = E_{1+}/M_{1+}$ and $R_{SM} = S_{1+}/M_{1+}$ for the $\gamma^* p \to \Delta^+(1232)$ transition are extracted in the framework of a unitary isobar model.

1. Introduction

The $\Delta(1232)$ resonance is the lowest and most prominent baryon excitation. For electromagnetic excitations in which the Δ decays into a pion and nucleon, the transition amplitudes are most often expressed in terms of multipoles, which for the $N \to \Delta$ transition are denoted [1] M_{1+}, E_{1+} and S_{1+} . Here, the subscript 1 corresponds to the decay pion orbital angular momentum $l = 1$ and the subscript + corresponds to a total angular momentum $j = 3/2$. Alternatively, the $N \to \Delta$ transition is frequently expressed in terms of form factors [2] G^*_M, G^*_E, G^*_C.

In this report, differential cross sections for exclusive π^0 electroproduction have been obtained in W from threshold to 1.4 GeV, in Q^2 from 2.5 to 5.5 GeV^2, and solid angle 4π in the center of mass. The quantities G^*_M , $R_{EM} = Re(E_{1+}/M_{1+})$ and $R_{SM} = Re(S_{1+}/M_{1+})$ have been extracted from the measured cross sections through the application of an isobar model [3].

2. The experiment

In the present experiment an electron beam of energy of 5.75 GeV was incident on a 5.0 cm long liquid hydrogen target. The CLAS spectrometer [4] was used to detect the scattered electrons and final state protons.

A Monte Carlo simulation based on GEANT3 [5] was used to determine the acceptance and efficiency of the CLAS detector.

To account for non-elastic radiative processes a program *Exclurad* [6] developed at JLAB was used.

Systematic errors were estimated by taking into account variations in the cross sections obtained by varying the kinematic cuts such as missing mass, detector acceptance, and particle identification. Estimated uncertainties in the radiative and bin averaging corrections are also included.

3. JANR Analysis

The measured unpolarized differential cross sections consist, in addition to the Δ, of non-resonant π^0 production and of tails of higher mass resonances. In order to extract the Δ amplitudes G^*_M , R_{EM} and R_{SM} these contributions need to be modeled. For the present case, the JANR model, which has been developed at JLab, has been used to extract the amplitudes. The JANR model incorporates the unitary isobar approach as in Ref. [7]. The non-resonant background consists of the pion Born terms and the t-channel ρ and ω contributions. To calculate the Born terms the latest available values of the nucleon and pion form factors are used. Underlying tails from higher resonances such as the $P_{11}(1440)$, $D_{15}(1520)$ and $S_{11}(1535)$ which are modeled as Breit-Wigner shapes are also incorporated.

4. Results, conclusions

Figure 1 shows the extracted $G^*_M/3G_D$, where $G_D = (1+Q^2/0.71)^{-2}$, is plotted as a function of Q^2. Also shown are the results of selected earlier published experiments. The most notable feature is that G^*_M continues to decrease with Q^2 compared with the elastic magnetic form factor and other prominent resonant form factors such as that for the $S_{11}(1535)$, and also faster than the underlying non-resonant background. Figure 2 shows the extracted ratios R_{EM} and R_{SM}. The ratio R_{EM} remains small over the entire range of measured Q^2, and the ratio R_{SM} continues to be negative, with increasing magnitude as a function of Q^2.

The results suggest that the region of Q^2 where pQCD processes would be expected to be valid is higher than currently accessible.

Extensions of effective Lagrangian or isobar based models to higher Q^2 continue to be driven by these data, and continue to provide insights into the underlying physics in terms of traditional baryon and meson degrees of freedom. Calculations of these and other form factors in terms QCD based or inspired models such as Lattice QCD, GPDs or LCSR involving soft processes are at an early stage, and greater theoretical progress will be necessary before good quantitative agreement reaches these higher Q^2 results.

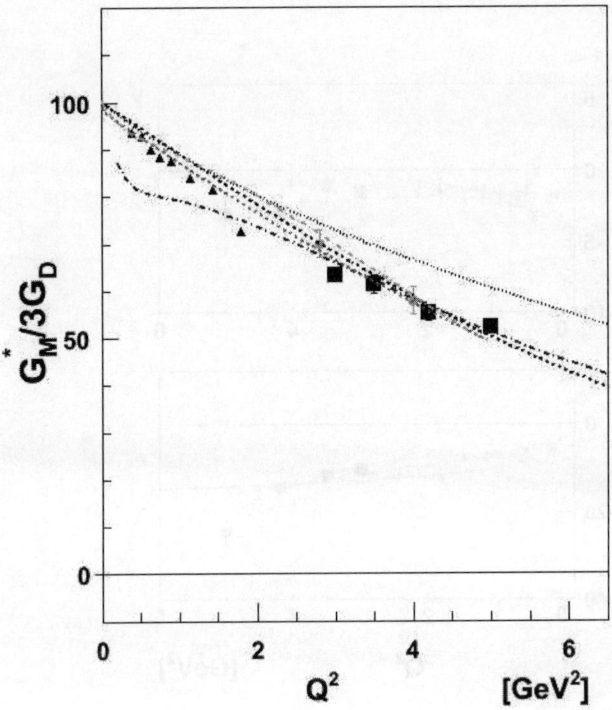

Fig. 1. The extracted form factor $G^*_M/3G_D$, where G_D is the dipole form $G_D = (1 + Q^2/0.71)^{-2}$. The filled squares are from the current CLAS experiment utilizing the JANR method. The filled triangles at lower Q^2 are the previously reported CLAS results. The filled circles are from an earlier JLab Hall C experiment. The curves are as follows. Black dots: Sato-Lee. Dashes: MAID-2003. Dot-dash: DMT.

References

1. G.F. Chew, M.L. Goldberger, F.E. Low, Y. Nambu, Phys. Rev. **106**, (1957).
2. Jones-Scadron, Annals Phys. **81**:1 (1973).
3. I. Aznauryan , Phys. Rev. C **68**, 065204, (2003).
4. B. Mecking *et al.*, Nucl. Inst. Meth. **A503**, 513, (2003).
5. S. Agostinelli *et al.*, Nucl. Inst. Meth. **A 506**, 250, (2003).
6. A. Afanasev *et al.* , PiN Newslett. **16**, 343, (2002).
7. D. Drechsel, *et al.*, Nucl. Phys. **A645**, 145, (1999).

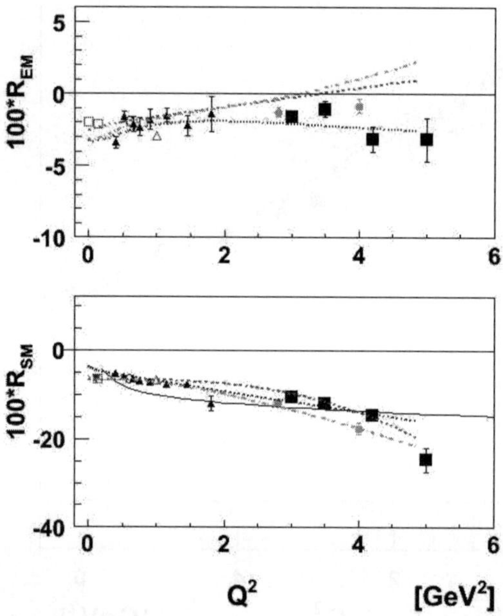

Fig. 2. the extracted ratios R_{EM} (upper panel) and and R_{SM} (lower panel). The filled squares are from the JANR model fit to the current CLAS experiment. The filled triangles at lower Q^2 are the previously reported CLAS results. The filled circles are from an earlier JLab Hall C experiment. The various the open symbols at lower Q^2 are results of measurements by other laboratories as Bates, Mainz, Bonn. The curves are as follows. Black dots: Sato-Lee. Dashes: MAID-2003. Dot-dash: DMT.

MEASUREMENT OF CROSS SECTION AND ELECTRON ASYMMETRY OF THE $p(\vec{e}, e'\pi^+)n$ REACTION IN THE $\Delta(1232)$ AND HIGHER RESONANCES FOR $Q^2 \leq 4.9$ $(GeV/c)^2$

K. PARK, I. AZNAURYAN, V. BURKERT, W. KIM
FOR THE CLAS COLLABORATION

Thomas Jefferson National Accelerator Facility
12000 Jefferson Ave., Newport News, VA 23606, USA

The cross section and beam asymmetry were measured in channel of $\vec{e}p \to e'\pi^+n$ using 5.754 GeV electron beam with CEBAF Large Acceptance Spectrometer(CLAS). This measurement covers 4 π angular coverage and high Q^2 upto 4.9 GeV^2 under various resonance mass regions. The structure functions $\sigma_T + \epsilon_L\sigma_L$, σ_{TT}, σ_{LT} and $\sigma_{LT'}$ were extracted from fit angular distribution of cross section and asymmetry.

1. Introduction

Due to the small mass of the pion, the single pion-nucleon decay is the favorite channel for many resonances, and not surprisingly, single pion electroproduction is being extensively exploited to understand the structure of nucleon. In this analysis we study the transition amplitudes from protons into the higher mass resonances by observing the $n\pi^+$ final state. Especially, The nature of $P_{11}(1440)$ is not understood in the frame of the constituent quark model(CQM), and there are suggestions that the Roper may be a hybrid state (q^3G) [18], or a small quark core with a large vector meson cloud [17]. The Q^2 evolution of the $A_{1/2}$ photon coupling amplitude is predicted to be different for 3-quark and hybrid states for $P_{11}(1440)$. Another state of interest in the second resonance region is the $S_{11}(1535)$. This state was found to have an unusually hard transition form factor, i.e. the Q^2 evolution shows a slow fall-off. This state is often studied in the $p\eta$ channel which shows a strong s-wave resonance near the η-threshold with very little non-resonant background. Older data show some discrepancies as to the total width and photocoupling amplitude. In particular, analysis of pion photoproduction data disagree with the analysis of the η-photoproduction data by a wide margin.

The unpolarized cross sections of the single π^+ electronproduction in one photon exchange approximation can be written [2] as:

$$\frac{d\sigma}{d\Omega_\pi{}^*} = \sigma_T + \epsilon_L\sigma_L + \epsilon\sigma_{TT}cos2\phi_\pi{}^* + \sqrt{2\epsilon_L(1+\epsilon)}\sigma_{LT}cos\phi_\pi{}^* \tag{1}$$

The ϵ is the virtual photon polarization parameter and $\frac{d\sigma}{d\Omega_\pi^*}$ is the photoabsorption cross section. The σ_T, σ_L, σ_{TT} and σ_{TL} are the structure functions, that depend on Q^2, W, $\theta_\pi{}^*$ and $\phi_\pi{}^*$. σ_T, σ_L are the photoabsorption cross section of the unpolarized transverse/longitudinal virtual photons. The σ_{TT} is the interference between the two transverse linear photon polarizations, while σ_{TL} is due to the interference between transverse and longitudinal polarizations of the virtual photons.

An antisymmetric part arises with unpolarized targets when an emitted particle is detected in coincidence with the scattered electron. The electron beam asymmetry is directly proportional to the polarized longitudinal-transverse structure function ($\sigma_{LT'}$) in Eq. 2

$$A_{LT'} = \frac{d^2\sigma^+ - d^2\sigma^-}{d^2\sigma^+ + d^2\sigma^-} = \frac{\sqrt{2\epsilon_L(1-\epsilon)}\,\sigma_{LT'}\,sin\theta_\pi^*\,sin\phi_\pi^*}{\sigma_0} \tag{2}$$

where, $d^2\sigma^\pm$ is the differential cross section for each helicity state. The superscripts($\pm$) of $d^2\sigma$ correspond to electron helicity states. $A_{LT'}$ can be measured by scaling the measured number of π^+ events with both positive and negative helicities with considering electron beam polarization. The cross section(σ_0) is the unpolarized cross section. The fifth structure function ($\sigma_{LT'}$) is intrinsically different from the other four structure functions of the unpolarized cross section. The fifth structure function is produced by the imaginary part of terms involving the interference between longitudinal and transverse components of the hadronic and leptonic current [11].

2. Results

The CEBAF Large Acceptance Spectrometer(CLAS) detector has 4π detecting charged particle acceptance range. The continuous beam provided to CEBAF is well suited for carrying out experiments which require two and more particles in coincidence in the final state with a very small accidental background to signal ratio of $< 10^{-3}$ over a large angular range in the laboratory frame at the luminosities up to $10^{34}cm^{-2}sec^{-1}$. This experiment was performed at Hall B using incident beam had 7 nA of intensity and 5.754 GeV of energy with 70% beam polarization on unpolarized proton target.

Radiative corrections were applied with the procedure recently developed by A. Afanasev et al. specifiedly for exclusive single pion production. The MAID00 was used for the actual RC correction and the MAID03 model was used for systematic uncertainty studies. Model dependence is less than 1%. The differential cross section and beam asymmetry had been compared with various physics models such as Dubna-Mainz-Taipei(DMT) [12], SL [10], SL04, MAID98, MAID00 [8], MAID03.

The most interesting quantities that can be extracted from the single π^+ electroproduction cross section data are the photocoupling amplitudes for the

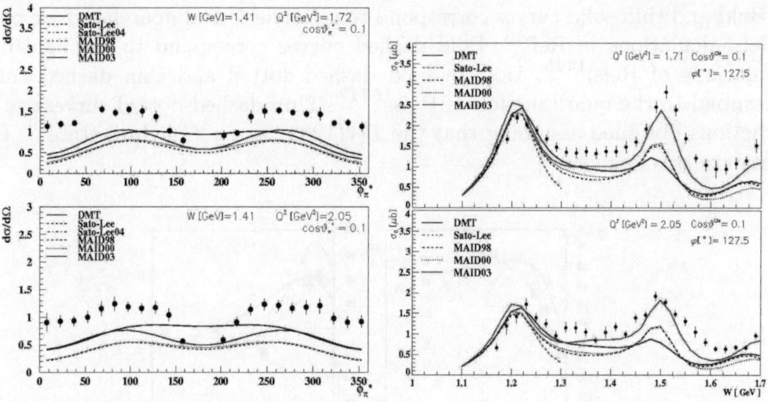

Fig. 1. Preliminary results of the differential cross section of ϕ_π^* dependence with various physical models after applying 2.5% mininum acceptance cut at $W = 1.41\ GeV$, $Q^2 = 1.72,\ 2.05\ GeV^2$, $cos\ \theta_\pi^* = 0.1$.(Left) W dependent differential cross section at $Q^2 = 1.71,\ 2.05\ GeV^2$, $cos\ \theta_\pi^* = 0.1$ and $\phi_\pi^* = 127.5^o$ (Right)

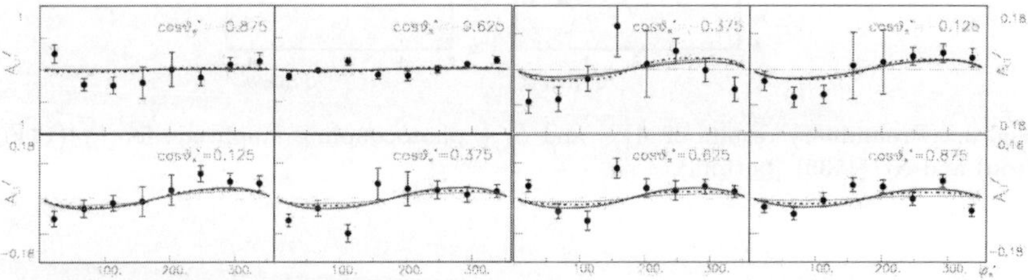

Fig. 2. Preliminary results of the Electron Beam Asymmetry as a function $\phi_\pi{}^*$ in different $cos\theta_\pi{}^*$ bin at $W = 1.41\ GeV$, $Q^2 = 2.05\ GeV^2$ with comparison to different physics models : MAID00(thin solid), MAID03(bold solid), DMT(dash-dot), and SL(bold dot)

resonance transition. The π^+ data from this analysis were fitted using the JANR program [4]. The multipole amplitudes for $P_{11}(1440)$, $S_{11}(1535)$, $D_{13}(1520)$ and $F_{15}(1680)$ were extracted. Figure 3 shows the $A_{1/2}$ and $S_{1/2}$ photocoupling amplitude for $P_{11}(1440)$ with previous data.

The $A_{1/2}$ amplitude for Roper clearly changes sign and becomes large and positive with slow fall off at higher Q^2. The longitudinal amplitude $S_{1/2}$ drops rapidly with Q^2. Both features are consistant with the Roperusing a small quad core and a large vector meson cloud. The new data rule out the Roper resonance as a gluonichybride state. The S_{11} $A_{1/2}$ amplitude confines the resuts from η production, and for the foot times gives definite results for the $S_{1/2}$ amplitude.

Bold and thin solid curves correspond to relativistic and nonrelativistic quark model calculations in Ref.[13]. Bold dashed curves correspond to the light-front calculations of Refs,[1415]. Dotted, bold dashed-dotted and thin dashed curves correspond to the quark models of Refs[16717]. Thin dashed-dotted curves are the predictions obtained assuming that the $P_{11}(1232)$ is a q^3G hybrid state[18]. Blue points are this analysis

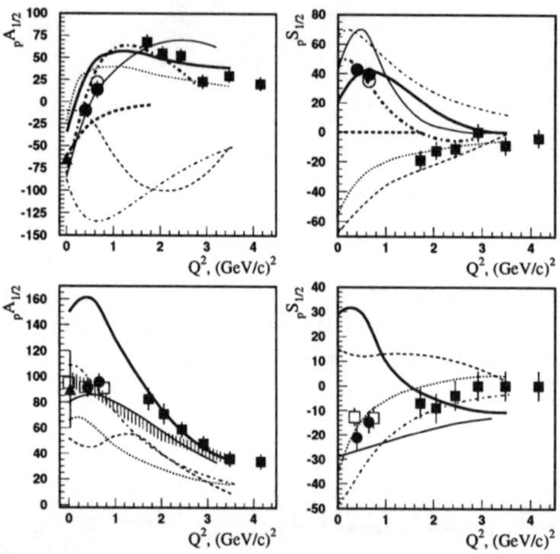

Fig. 3. Preliminary results of $A_{1/2}$ and $S_{1/2}$ photocoupling amplitude for $P_{11}(1440)$ (top) and $S_{11}(1535)$ (bottom).

References

1. V. D. Burkert *et al.*, Phys. Rev. **C67**, 035204 (2003).
2. F. Foster, G. Hughes, Rept. Prog. Phys. **46**, 1445, (1983).
3. S. Eidelman *et al.*, Phys. Lett. B **592**, 1 (2004).
4. I.G. Aznauryan, Phys. Rev. C**67**, 015209, (2003).
5. I.G. Aznauryan, Phys. Rev. C**68**, 065204, (2003).
6. I.G. Aznauryan, Phys. Rev. C**71**, 015201, (2005).
7. M. Aiello *et al.*, J. Phys. **G24**, 753, (1998).
8. D. Drechsel *et al.*, Nucl. Phys. **A645**,145,(1999)
9. R. A. Arndt, I.I. Strakovski and R.L. Workman, Phys. Rev. C **53**, 430, (1996).
10. T. Sato and T.-S. Lee, Phys. Rev. C **54**, 2660 (1996).
11. S. Boffi, C. Guisti, F. D. Pacati, and M. Radici, Oxford science publications, 1996.
12. S. S. Kamalov and S. N. Yang, Phys. Rev. Lett. **83**, 4494 (1999).
13. S. Capstick and B. D. Keister, Phys. Rev. D **51**, 3598 (1995).
14. E. Pace, G. Salme and S. Simula, Few Body Syst. Suppl. **10**, 407 (1999).

15. S. Simula, Proceedings of the Workshop on The Physics of Excited Nucleons, NSTAR 2001, p.135.
16. M. Warns, H. Schroder, W. Pfeil, and H. Rollnik, Z. Phys. C-Particles and Fields, **45**, 627 (1990).
17. F. Cano and P. Gonzalez, Phys. Lett. B **431**, 270 (1998).
18. Z. Li and V. Burkert, Phys. Rev. D **46**, 70 (1992).

PION-NUCLEON CHARGE EXCHANGE IN THE N*(1440) RESONANCE REGION*

M. E. SADLER†

Abilene Christian University
320B Foster Science Building,
Abilene, TX 79699, USA
E-mail: sadler@physics.acu.edu

Results for pion-nucleon charge exchange scattering at eight laboratory momenta from 355 to 659 MeV/c are reported from the Crystal Ball Collaboration at Brookhaven National Laboratory. The results agree well with the GWU FA02 partial wave analysis but differences exist at the backward angles at higher momenta. These data complement previously reported measurements made by the Crystal Ball Collaboration of all-neutral final states in the first and second resonance regions (charge exchange near the $\Delta(1232)$ and near eta threshold, eta production, multiple pi-zero final states, and inverse photoproduction). Complete sets of measurements for pion-nucleon scattering (differential cross sections, analyzing powers and spin rotation parameters) have been measured in the N*(1440) resonance region since 1980, when the last Karlsruhe-Helsinki and Carnegie-Mellon-Berkeley partial wave analyses were reported. The accepted resonance parameters for the N*(1440) should be reevaluated in a new partial wave analysis.

*This work is supported by USDOE, NSF, NSERC (Canada), Russian Ministry of Sciences, Croatian Ministry of Science and Technology, and Volkswagen Stiftung.

†The Crystal Ball Collaboration at BNL consists of A. Barker(deceased), C. Bircher, C. Carter, M. Daugherity, B. Draper, S. Hayden, J. Huddleston, D. Isenhower, M. Jerkins, M. Joy, C. Robinson, M. Sadler and S. Watson, *Abilene Christian University*, C. Allgower, R. Cadman and H. Spinka, *Argonne National Laboratory*, J. Comfort, K. Craig and A. Ramirez, *Arizona State University*, T. Kycia (deceased), *Brookhaven National Laboratory*, M. Clajus, A. Marusic, S. McDonald, B. M. K. Nefkens, N. Phaisangittisakul, S. Prakhov, J. Price, A. Starostin and W. B. Tippens, *University of California at Los Angeles*, J. Peterson, *University of Colorado*, W. Briscoe, A. Shafi and I. Strakovsky, *George Washington University*, H. Staudenmaier, *Universität Karlsruhe*, D. M. Manley and J. Olmsted, *Kent State University*, D. Peaslee, *University of Maryland*, V. Abaev, V. Bekrenev, N. Kozlenko, S. Kruglov, A. Kulbardis, and I. Lopatin, *Petersburg Nuclear Physics Institute*, N. Knecht, G. Lolos and Z. Papandreou, *University of Regina*, I. Supek and D. Mekterovic, *Rudjer Boskovic Institute* and D. Grosnick, D. D. Koetke, R. Manweiler and S. Stanislaus, *Valparaiso University*.

1. Introduction

The motivation of AGS experiment E913, measurements of $\pi^- p \to$ Neutrals, was to improve the determination of the masses, widths and decay modes of N* and Δ* resonances, and to determine the ηn scattering length. Differential cross sections for pion-nucleon charge exchange, $\pi^- p \to \pi^0 n$, have been published in the region of the Δ(1232) resonance[1] and in the vicinity of the threshold for η production, $\pi^- p \to \eta n$[2]. The results presented here are from 355 to 659 MeV/c, the region of the N*(1440) resonance.

2. The Experiment

The Crystal Ball (CB) detector is a highly-segmented, total-energy electromagnetic calorimeter and spectrometer that covers ≈93% of 4π steradians. Details of the CB and its installation in the C6 beam line of the AGS at BNL are given in the literature[3]. The $\pi^- p \to \pi^0 n$ reaction was identified by measuring the energy and direction of the two photons from $\pi^0 \to \gamma\gamma$ decay. The acceptance of the Crystal Ball for detecting π^0's from $\pi^- p \to \pi^0 n$ was calculated using a Monte Carlo program based on GEANT[4]. The data analysis and the treatment of systematic effects are discussed in our previous paper.[1]

The last run with the CB at BNL in 2002 was to determine more accurately the muon and electron contamination in the pion beam at momenta above 350 MeV/c, data taken in 1998. At lower momenta, the lepton fractions were determined by TOF. Crossed-field DC separators in the C6 beam line were effective in minimizing the contamination. The separators become less effective as the momentum is increased and produced fluctuations at higher momenta that were not well understood. They were not used for the 2002 data. A gas Cherenkov counter provided independent measurements of the electron fractions. It was utilized in both the 1998 and 2002 runs, but was moved much closer to the target in 2002.

Above 500 MeV/c the TOF distributions for muons and pions are too close to disentangle reliably. Special runs were completed in 2002 with beam tunes that had been used in previous experiments that had total muon contaminations of ~4%. A 3-mm lead sheet was placed in the beam to reduce the electron contamination from 30% to ~3%, which was monitored by the Cerenkov counter. Data were taken with a solid 2-cm CH_2 target, allowing the placement of a small beam defining counter 30 cm upstream. This counter produced background that must be subtracted using a target empty run. The carbon background was measured using a solid C target and also subtracted. The shapes of the angular distributions agreed very well with the 1998 data but the statistical uncertainties were larger due to the background subtractions. Therefore, the 1998 data are used for the shape but the normalization is obtained from the 2002 data.

3. Results

Differential cross sections for $\pi^- p \to \pi^0 n$ are plotted in Figure 1 and Figure 2 together with the results of the FA02 partial-wave analysis (PWA) of the George Washington group.[5] The statistical errors are typically 2-4% except for the most forward angles which is at the edge of the acceptance and for the backward-angle data where the cross section dips near 540 MeV/c. The present results exhibit excellent agreement with the FA02 PWA.

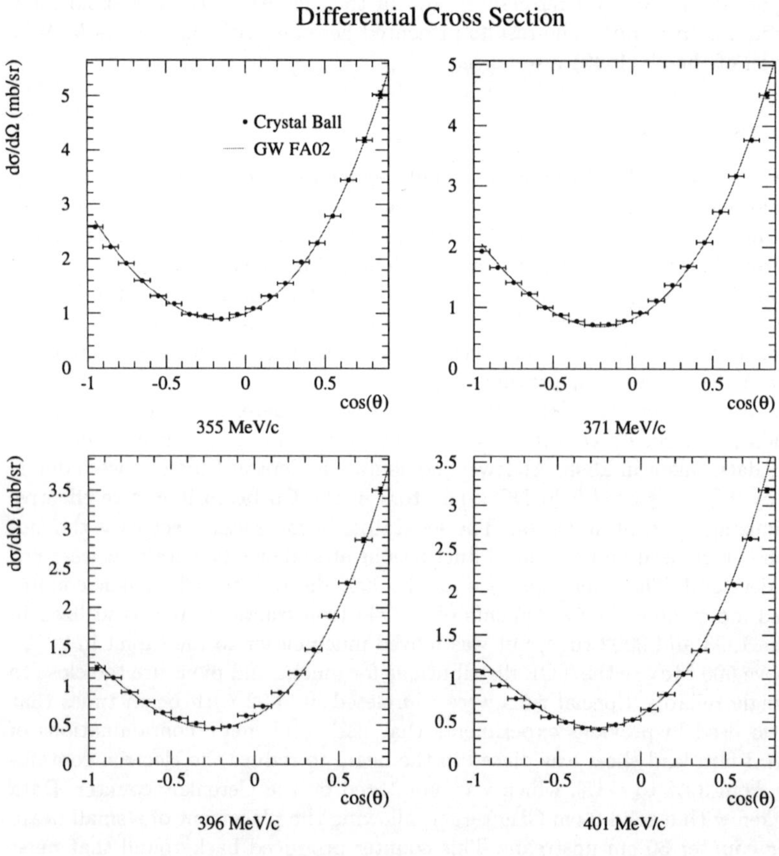

Fig. 1. Differential cross sections of reaction $\pi^- p \to \pi^0 n$ at 355, 371, 396 and 401 MeV/c. Black circles are the values obtained in this experiment. The curves show the results of the FA02 partial-wave analysis of the George Washington group based on earlier experiments.

These data complete an experimental program of systematic measurements of pion-nucleon scattering observables in the P_{11} resonance region that was started at LAMPF in 1978. The program has included differential cross sections $(d\sigma/d\Omega)$[6], analyzing powers (A_N)[7,8], and spin rotation parameters (A and R)[9] for $\pi^{\pm} p \to \pi^{\pm} p$, and $d\sigma/d\Omega$[1,2] and A_N[10,11] for $\pi^- p \to \pi^0 n$. Similar measure-

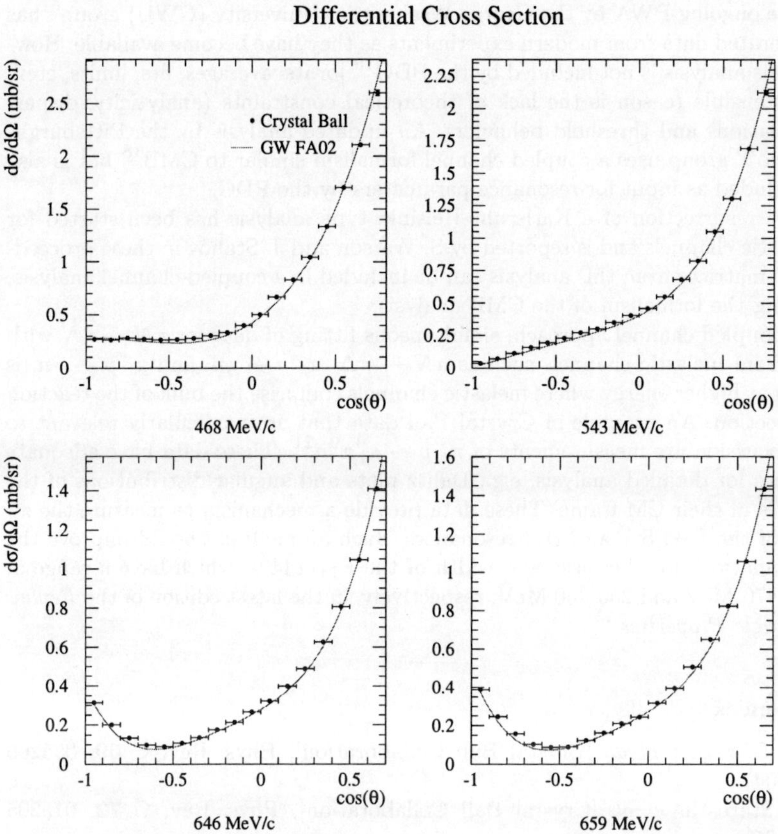

Fig. 2. Differential cross sections of reaction $\pi^- p \to \pi^0 n$ at 468, 543, 646 and 659 MeV/c.

ments were made at PNPI for the charged final states.[12–14] There is general agreement where the data sets overlap, resulting in a robust database available for partial wave analysis (PWA).

Existing determinations of the masses, widths and decay modes of low-lying excited states of the nucleon, as compiled by the Particle Data Group (PDG)[15], are determined from energy-independent partial wave analyses of pion-nucleon scattering data. For the P_{11}, also known as the N*(1440) or Roper resonance, the analyses cited are the Karlsruhe-Helsinki[16,17] Carnegie Mellon-Berkeley (CMB)[18], and Kent State[19] analyses, the latter of which used the elastic amplitudes from the other two. The data included in these analyses were published before 1980. The complete sets of measurements that have been published in the P_{11} resonance region since 1980 are not included.

The ongoing PWA by the George Washington University (GWU) group[5] has incorporated data from modern experiments as they have become available. However, this analysis is not included by the PDG[15] for its 'averages, fits, limits, etc.'. The ostensible reason is the lack of theoretical constraints (analyticity, dispersion relations and threshold behavior). An updated analysis by the Pittsburgh-Argonne[20] group uses a coupled channel formalism similar to CMB[18] but is also not included as input for resonance parameters by the PDG.

The resurrection of a Karlsruhe-Helsinki type analysis has been started for the elastic channels and is reported by S. Watson and J. Stahov in these proceedings. T-matrices from this analysis can be included in a coupled-channel analysis, following the formalism of the CMB analysis.

A coupled channel approach, simultaneous fitting of data for $\pi N \to \pi N$ with those from inelastic channels such as $\pi N \to \pi\Delta$, $\pi^- p \to \eta n$, and $\pi^- p \to \gamma n$, is needed at higher energy where inelastic channels comprise the bulk of the reaction cross section. An example of Crystal Ball data that are particularly relevant to this discussion are measurements of $\pi^- p \to \pi^0\pi^0 n$[21]. These data have adequate statistics for detailed analysis, e.g., Dalitz plots and angular distributions of the two π^0's in their CM frame. These data provide a mechanism to measure the π^0 decay of the P_{11}, S_{11} and D_{13} resonances. Such an analysis should improve the determinations for the mass and width of the $P_{11}(1440)$, which have a range of 1430-1470 MeV and 250-450 MeV, respectively, in the latest edition of the *Review of Particle Properties.*[15]

References

1. M. E. Sadler *et al.* [Crystal Ball Collaboration], Phys. Rev. C **69**, 055206 (2004).
2. A. Starostin *et al.* [Crystal Ball Collaboration], Phys. Rev. C **72**, 015205 (2005).
3. A. Starostin *et al.* [Crystal Ball Collaboration], Phys. Rev. C **64**, 055205 (2001).
4. GEANT 3.21 CERN Program Library Long Writeup W5013, CERN, Geneva, Switzerland.
5. R. A. Arndt, W. J. Briscoe, I. I. Strakovsky, R. L. Workman, and M. M. Pavan, Phys. Rev. C **69**, 035213 (2004).
6. M. E. Sadler, W. J. Briscoe, D. H. Fitzgerald, B. M. K. Nefkens and C. J. Seftor, Phys. Rev. **D35** 2718 (1987).
7. A. Mokhtari, A. D. Eichon, G. J. Kim, B. M. K. Nefkens, J. A. Wightman, D. H. Fitzgerald, W. J. Briscoe, and M. E. Sadler, Phys. Rev. **D35**, 810 (1987).
8. C. J. Seftor *et al.* Phys. Rev. **D39** 2457 (1989).
9. I. Supek *et al.* Phys. Rev. **D47** 1762 (1993).
10. J. A. Wightman *et al.* Phys. Rev. **D38** 3365 (1988).
11. G. J. Kim *et al.* Phys. Rev. **D41** 733 (1990).
12. V. A. Gordeev*et al.* Nucl. Phys. **A364** 408 (1981).
13. V. S. Bekrenev*et al.* Nucl. Phys. **A364** 515 (1981).

14. V. S. Bekrenev*et al.* J. Phys. **G13** L19 (1987).
15. S. Eidelman *et al.* (Particle Data Group) Review of Particle Physics, Particle Data Group, Phys. Lett. **B 592**, 1 (2004).
16. G. Hoehler, F. Kaiser,R. Koch, and E. Pietarinen, *Handbook of Pion-Nucleon Scattering* Fachinformationszentrum, Karlsruhe (1979).
17. G. Hoehler, *Pion-Nucleon Scattering*, Springer, Berlin (1983).
18. R. E. Cutkosky *et al.*Phys. Rev. **D20** 2804 (1979).
19. D. M. Manley and E. M. Saleski, Phys. Rev. **D45** 4002 (1992).
20. T. P. Vrana, S. A. Dytman and T. S. H. Lee, Phys. Rep. bf 328 181 (2000).
21. S. Prakhov *et al.* [Crystal Ball Collaboration], Phys. Rev. C **69**, 045202 (2004).

NUCLEON RESONANCE DECAY BY THE $K^0\Sigma^+$ CHANNEL

R. CASTELIJNS, J. BACELAR, H. LÖHNER,
J. G. M. MESSCHENDORP, AND S. SHENDE

KVI Groningen,
Zernikelaan 25,
9747 AA, Groningen, the Netherlands
For the CBELSA/TAPS Collaboration

At the tagged photon beam of the ELSA electron synchrotron at the University of Bonn in Germany the Crystal Barrel and TAPS photon spectrometers have been combined to provide a 4π detector for multi-neutral-particle final states from photonuclear reactions. In a series of experiments on single and multiple neutral meson emission we have concentrated on the hyperon production off the proton, and in particular on the $K^0\Sigma^+$ channel. High-quality excitation function, recoil polarizations, and angular distributions from the $K\Sigma$ threshold up to 2.3 GeV c.m. energy were obtained. Particular care was taken to establish the cross section normalization. The experimental results are compared with predictions aof a recent coupled-channels calculation within the K-matrix formalism by A. Usov and O. Scholten[1].

1. Introduction

Recent predictions in a quark-pair creation model[2] or a collective string-like three-quark model[3] revealed substantial decay branches of baryon resonances into the $K\Lambda$ and $K\Sigma$ channels. Kaon production experiments will therefore be an important tool to establish or disprove "missing" resonances and thus to determine the relevant degrees of freedom of quark models. It has been proposed that the observed η decay branches of S_{11} baryon resonances find an explanation in mixing with quasi-bound a S_{11} state near the $K\Lambda$ or $K\Sigma$ threshold. Furthermore, the coupled-channel analysis of hyperon production data[1] requires accurate data in the $K^0\Sigma^+$ channel. Here we present data for the neutral decay branch of this channel:

$$\gamma p \rightarrow K^0\Sigma^+ \rightarrow (\pi^0\pi^0)(\pi^0 p) \rightarrow 6\gamma p \qquad (1)$$

2. Experimental setup

The measurement was performed using the combination of TAPS and Crystal Barrel, which has been set up at the bremsstrahlung-tagger photon beam facility at the ELSA electron accelerator in Bonn. This facility provides tagged photons in the energy range from 0.5 to 2.9 MeV. The Crystal Barrel is a calorimeter consisting of 1290 CsI crystals. It covers all polar angles from 30° to 168°, with full azimuthal coverage. The forward angles, from 5° to 30°, are covered by TAPS, a calorimeter consisting of 528 hexagonal BaF_2 crystals arranged in a wall configuration. Together the two calorimeters have a geometrical acceptance of almost 4π. The high granularity and good acceptance of this system make it very well suited to measure reactions with a high number of photons in the final state.

A plastic scintillator mounted in front of each TAPS crystal is used to identify charged particles online. Offline, an analysis of the pulse shape and time of flight methods are used to identify the particles detected in TAPS. The Crystal Barrel can identify charged particles via a cylindrical scintillating fiber detector mounted around the target. In the center of the Crystal Barrel a 5 cm long target cell was mounted, filled with liquid hydrogen.

A first and second level trigger were defined for this experiment. The TAPS crystals are read out with photo-multipliers giving fast signals suitable for a first level trigger. To produce a trigger on this level at least two hits in TAPS depositing more than 50 MeV or at least a single hit depositing more than 100 MeV was required.

The signals from the Crystal Barrel crystals, which are read out via photodiodes, are fed through an online cluster encoder, which supplies a second level trigger. On this level at least one cluster was required whenever the first level trigger was produced by a single hit in TAPS.

3. Analysis

3.1. *Event selection*

From the complete measured data set, events containing $3\pi^0$'s and an additional hit were selected. To improve the invariant mass resolutions each event was kinematically fitted. The constraints used in the kinematic fit are the conservation of momentum and energy (4 constraints) and the π^0 invariant mass (3 constraints). It is important to note that neither the K^0 nor the Σ^+ mass was used as a constraint. The fitted values are the energies deposited by the six photons and the two angles of their trajectories. For the proton only the angles of its trajectory are fitted. Its energy is calculated from the other observables since protons with a kinetic energy in excess of 400 MeV will result in a significant leakage of the shower. The resulting fit is 6 fold overconstraint. A confidence level cut is applied at 10%.

The kinematic fit improves the K^0 and Σ^+ invariant-mass resolutions by a factor 3. This is particularly important since, due to the conservation of strangeness in the strong decay, both particles have to decay weakly and have a relatively long lifetime. Both particles decay after traveling on average 2 cm, lowering the measured invariant mass resolution of the K^0 approximately by a factor 2.

After the fit, the events produced by the background reaction:

$$\gamma p \to \eta p \to 3\pi^0 p \to 6\gamma p \tag{2}$$

are removed from the data set by selecting the η peak in the $3\pi^0$ invariant mass spectrum. At this point the remaining background is caused mainly by $3\pi^0$ sequential resonance decay and, to a lesser degree, combinatorics. It is further reduced by dropping all events that do not contain a Σ^+ hyperon, by selecting the region outside the Σ^+ invariant mass peak in the $p\pi^0$ invariant mass spectrum. The $\pi^0\pi^0$ invariant mass of the remaining events are plotted in figure 1(left). In total a number of 10000 K^0's were found in the 1400 hours of beam time that have been analyzed.

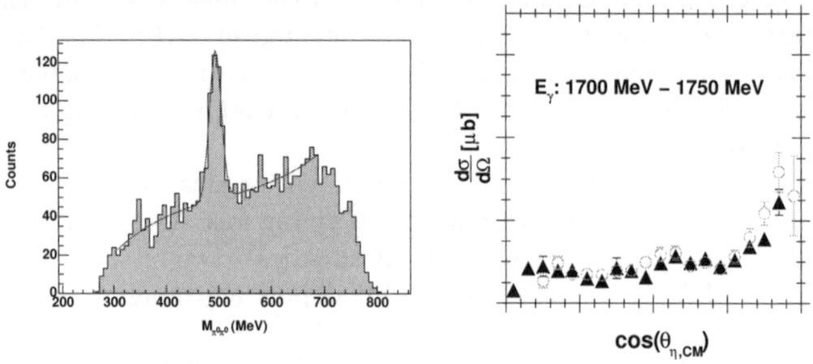

Fig. 1. Left: The K^0 peak in the $\pi^0\pi^0$ invariant mass spectrum after the kinematic fit. Right: The $\gamma p \to \eta p$ differential cross section measured by this experiment (solid triangles), compared against the results published by the Crystal Barrel collaboration[4] (open circles).

3.2. *Acceptance and normalization*

The acceptance for the process $\gamma p \to K^0\Sigma^+ \to 3\pi^0 p \to 6\gamma p$ was determined using a Monte Carlo simulation. The resulting acceptance varies between 11% and 14%, dependent of the incoming photon energy and the center of mass angle. It covers all center of mass angles. Using the same acceptance calculation the experimental cross section for the process $\gamma p \to \eta p$ was obtained from our data. The differential cross section was then compared against previously published data[4], as shown in figure 1(right). The agreement between both datasets gives confidence in the acceptance calculation.

The $\gamma p \to \eta p$ differential cross sections were also used to obtain a normalization for the cross sections for $\gamma p \to K^0\Sigma^+$. By comparing the differential cross sections for the reaction $\gamma p \to \eta p$ measured in this experiment to the published results the photon flux is extracted. It is verified that the energy dependence of the extracted photon flux agrees with the theoretical description of the bremsstrahlung spectrum[6].

4. Results

The excitation function measured in this experiment is shown in figure 2 compared to the data measured by the SAPHIR[7] and CLAS[5] collaborations. In addition, the differential cross sections and the recoil polarization have been measured in energy bins of 200 MeV, covering all center of mass angles with 6 bins.

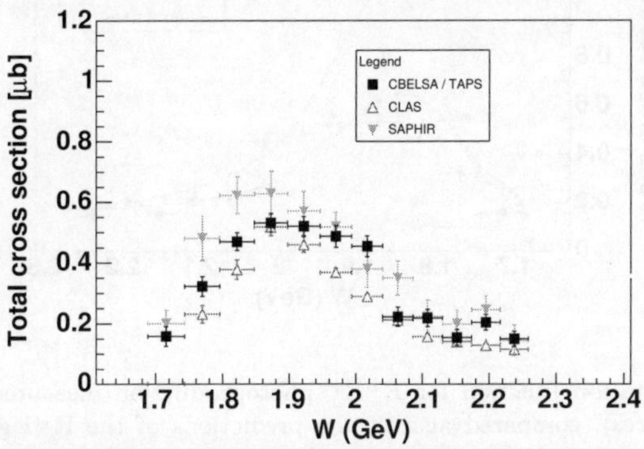

Fig. 2. The excitation function for $K^0\Sigma^+$ photoproduction measured by this experiment (solid squares), compared to the results of CLAS[5] (open triangles) and SAPHIR[7] (upside-down triangles).

A comparison of the $K^0\Sigma^+$ cross section measured in the present work to the two published results by SAPHIR[8,7] generally confirms the result published in the second reference. For the bins near threshold, our result does not show the peak at forward angles in the angular distribution shown by the SAPHIR result, and therefore the excitation function is somewhat lower in that region. The recoil polarization measured in this experiment matches that of SAPHIR well, both in sign and strength. The excitation function and the differential cross sections also agree with the result from CLAS[5]; the forward angles of the differential cross sections were not measured in that experiment.

In figure 3 the excitation function is compared to the K-matrix calculations of the KVI group[1]. Taking all known resonances into account, the peak in the total cross section is predicted too high by almost a factor 2. When taking an additional $P_{13}(1830)$ into account, much better agreement is obtained. This is caused by destructive interference effects between the new resonance and the $P_{33}(1855)$ already present in the model.

To investigate the influence of a third S_11 at 1729 MeV as proposed by Saghai[9] the data have been compared with the predictions obtained from the K-matrix calculations[1], using two sets of input parameters. The sets were obtained by fitting the η photoproduction cross section from the Crystal Barrel[4] once with and once without the proposed third S_{11}. The parameters describing the $S_{11}(1535)$ and the $S_{11}(1650)$ were not refitted as they are determined by the π sector. Both obtained sets of parameters describe the η photoproduction data well, but the sensitivity of the $K^0\Sigma^+$ photoproduction cross section is far below the accuracy

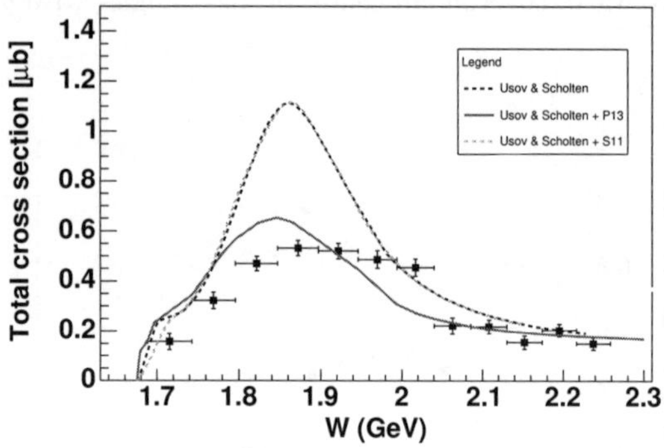

Fig. 3. The excitation function for $K^0\Sigma^+$ photoproduction measured by this experiment (solid squares), compared against the predictions of the KVI group[1], using all known resonances (black dashed line), an additional $P_{13}(1830)$ (solid gray line), and an additional $S_{11}(1729)$ (dashed gray line).

of the data points obtained from the experiment.

5. Conclusions

This analysis of the reaction $\gamma p \rightarrow K^0\Sigma^+$ produced differential cross sections, Σ^+ recoil polarizations, and the excitation function in the region between threshold (1050 MeV) and 2250 MeV in incoming photon energy.

The statistical accuracy of our measurement is comparable to the two analyses (CLAS and SAPHIR) published during the course of this work. For the most part, the three analyses agree, both in shape and in absolute magnitude.

The present data are sensitive to inclusions of particular additional resonances. The total cross section predicted by the K-matrix calculations of the KVI group[1] is too high. Inclusion of an additional $P_{13}(1830)$ resonance improves the agreement between theory and measurement for the differential cross sections and the excitation function considerably. Inclusion of the $S_{11}(1729)$ does not improve the agreement between prediction and measurement, neither for the differential cross sections, nor for the Σ^+ recoil polarization.

Acknowledgments

We would like to gratefully acknowledge the support of the ELSA accelerator crew in providing a high quality beam.

References

1. A. Usov and O. Scholten, *Phys. Rev.* **C72**, 025205 (2005).

2. S. Capstick and W. Roberts, *Phys. Rev.* **D58**, 074011 (1998).
3. R. Bijker, F. Iachello, and A. Leviathan, *Phys. Rev.* **D55**, 2862 (1997).
4. V. Crede *et al.*, *Phys. Rev. Lett.* **94**, 012004 (2005).
5. B. Carnahan, PhD thesis, Catholic University of America, Washington, (2003).
6. Y.-S. Tsai, *Rev. Mod. Phys.* **46**, 815 (1974).
7. R. Lawall *et al.*, *Eur. Phys. J.* **A24**, 251 (2004).
8. S. Goers *et al.*, *Phys. Lett.* **B464**, 331 (1999).
9. B. Saghai and Z. Li, *Eur. Phys. J.* **A11**, 217 (2001).

COUPLED CHANNEL STUDY OF $K^+\Lambda$ PHOTOPRODUCTION

B. JULIA-DIAZ, B. SAGHAI

Département d'Astrophysique, de Physique des Particules, de Physique Nucléaire et de l'Instrumentation Associée, DSM, CEA/Saclay, 91191 Gif-sur-Yvette, France
E-mail: bruno.julia-diaz@cea.fr, bijan.saghai@cea.fr

T.-S. H. LEE*

Physics Division, Argonne National Laboratory, Argonne, IL 60439, USA
E-mail: lee@phy.anl.gov

F. TABAKIN†

Department of Physics and Astronomy,
University of Pittsburgh, PA 15260, USA
E-mail: tabakin@pitt.edu

A coupled channel model with γN, KY and πN channels has been used to analyze the recent data of $\gamma p \to K^+\Lambda$. The non-resonant interactions within the subspace $KY \oplus \pi N$ are derived from effective Lagrangians using a unitary transformation method. The direct photoproduction reaction is obtained from a chiral constituent quark model with $SU(6) \otimes O(3)$ breaking. Missing baryon resonances issues are briefly discussed.

1. Introduction

Our knowledge of associated strangeness photoproduction processes has been greatly improved in recent years thanks to measurements performed at several facilities, JLAB [1,2], ELSA [3,4] and Spring-8 [5]. This database should serve to shed light on the properties of known, poorly known, and missing resonances.

Our main aim in this contribution is to analyze the reaction $\gamma p \to K^+\Lambda$ making use of a coupled channel formalism. This work is an extension of Refs. [6, 7, 8] with improvements in the derivation of the meson-baryon intermediate states ($\pi N \to \pi N$, $\pi N \to KY$, and $KY \to KY$) and making use of the quark model of Refs. [9, 10] for the direct and resonance photoproduction of $K^+\Lambda$. A more detailed description of this investigation will be reported elsewhere [11].

*U.S. Department of Energy, Office of Nuclear Physics, Contract No. W-31-109-ENG-38.
†National Science Foundation, grant No. 0244526 at the University of Pittsburgh.

The main physics under scrutiny are the properties of resonances. This analysis should improve our knowledge of the properties of more or less well known resonances and also possible manifestations of missing resonances, predicted by QCD-inspired approaches [12]. In our study a 3rd S_{11}, 3rd P_{13} and 3rd D_{13} resonances are considered and their relevance is examined.

2. Theoretical model

A simple glimpse at the cross sections for meson photoproduction reveals that the pion photoproduction process is orders of magnitude larger than for strange photoproduction [13]. Thus, it is obvious that part of the strange production flux will come from first producing a πN intermediate state which subsequently decays into the KY system. A suitable method to account for these processes, as well for the final state interactions, is to consider a coupled channel formalism which includes the most relevant opened channels in the considered regime.

Studying the effects of coupled channels is however an involved task due to the many intermediate and final state channels which are active. Thus, we need to have reasonable models for the following mechanisms: $\gamma N \to \pi N$, $\pi N \to \pi N$, $\pi N \to KY$ and $KY \to KY$ in the considered total center-of-mass energy regime, $W \approx 1.6 - 2.7$ GeV. The main sources utilized in this work are Ref. [14] for πN and Ref. [7] with a number of improvements in the formulation for the KY hadronic channels.

The procedure employed [7] to fit the parameters involved in the model has been to first fix the meson-baryon model parameters performing a χ^2 fit to the available $\pi N \to KY$ data. Then the photoproduction process has been studied keeping the meson-baryon part unaltered.

The direct $K\Lambda$ photoproduction process is handled using the quark model of Refs. [9, 10]. That approach allows one to include all 3 and 4 star resonances and contains one adjustable parameter per resonance with masses below 2 GeV, due to the $SU(6) \otimes O(3)$ symmetry breaking.

A detailed description of the formalism will be published elsewhere [11].

3. Results

The present experimental status is the following: there are two quite complete measurements of the differential cross section performed at CLAS [1] and SAPHIR [3]. The most recent CLAS data [1] for differential cross sections shows a closer agreement to SAPHIR data than those released [2] previously. Secondly there are measured data for recoil polarization asymmetry from CLAS [2] and also polarized photon asymmetry measured at LEPS [5].

We have performed a thorough study of the compelete database [11]. Fitting separately SAPHIR and CLAS data, leads to reduced χ^2 of 1.3 and 2.1, respectively. Then, fitting *simultaneously* all cross section and polarization asymmetry data, we get $\chi^2_{d.o.f} \approx 3$.

In fig. 1 the total cross section obtained with this model is compared to the data. To study the role played by new resonances introduced, we also show results

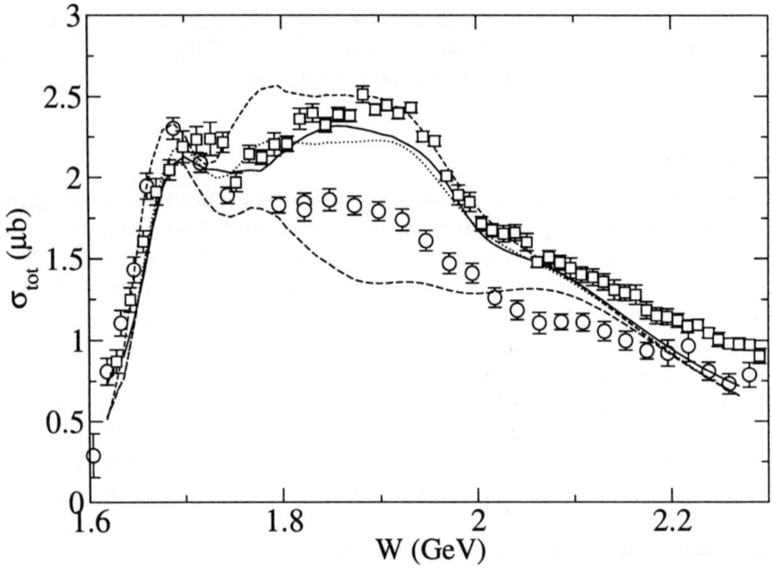

Fig. 1. Total cross section as a function of the centre of mass energy. Solid curve corresponds to the full model, dotted, dashed and long dashed correspond to switching off the $3^{rd}P_{13}$, $3^{rd}S_{11}$ and $3^{rd}D13$ resonances, respectively. Circle symbols, are SAPHIR data, Ref. 3. Square symbols correspond to the data from CLAS [1].

obtained by switching off those resonances one by one, without further fittings. Here, we emphasize that discrepencies between the two data sets are much smaller in the differential cross sections than in the total cross sections shown here. Larger discrepencies in the latter case are very likely due to the angular range covered by each data set and extrapolation methods used to extract the total from the differential cross sections.

The figure summarizes our main findings which are to be presented in a longer discussion [11]: the full model allows for a good reproduction of the data, the possible influence of the $3^{rd}P_{13}$ is very minor, and finally the role played by $3^{rd}S_{11}$ and $3^{rd}D_{13}$ are sizable. The mass and width of these resonances in this model are : S_{11} [M=1.84 GeV, Γ=283 MeV] and D_{13}[M=1.93 GeV, Γ=252 MeV]. These findings are in line with other studied with respect to the manifestations of a new S_{11}, Ref. [10], and a new D_{13}, Ref. [15], resonances.

References

1. R. Bradford *et al.*, arXiv:nucl-ex/0509033; R. Schumacher, private communication (2005).
2. J.W.C. McNabb *et al.*, Phys. Rev. C **69**, 042201 (2004); J.W.C. McNabb, PhD

Thesis, CMU (2002); R. Schumacher, private communication (2003).
3. K.H. Glander *et al.*, Eur. Phys. J. A **19**, 251 (2004).
4. R. Lawall *et al.*, Eur. Phys. J. A **24**, 275 (2005).
5. R.G.T. Zegers *et al.*, Phys. Rev. Lett. **91**, 092001 (2003).
6. W.-T. Chiang, F. Tabakin, T.-S. H. Lee, B. Saghai, Phys. Lett. B **517**, 101 (2001).
7. W.-T. Chiang, B. Saghai, F. Tabakin, T.-S. H. Lee, Phys. Rev. C **69**, 065208 (2004).
8. B. Juliá-Díaz, *et al.*, Nucl. Phys. A **755**, 463 (2005).
9. Z. Li, Phys. Rev. C **52**, 1648 (1995).
10. B. Saghai and Z. Li, Eur. Phys. J. A **11**, 217 (2001).
11. B. Juliá-Díaz, *et al.*, to be submitted to Phys. Rev. C.
12. See e.g. S. Capstick, W. Roberts, Prog. Part. Nucl. Phys. **45**, 5241 (2000).
13. V. D. Burkert and T. S. H. Lee, Int. J. Mod. Phys. E **13**, 1035 (2004).
14. T. Sato and T.-S. H. Lee, Phys. Rev. C **54**, 2660 (1996); Phys. Rev. C**63**, 055201 (2001).
15. A.V. Sarantsev *et al.*, Eur. Phys. J. A **25**, 441 (2005); T. Corthals *et al.*, arXiv:nucl-th/0510056.

MEASUREMENTS OF C_z, C_x FOR $K^+\Lambda$ AND $K^+\Sigma^\circ$ PHOTOPRODUCTION

R. BRADFORD

Department of Physics and Astronomy
University of Rochester
500 Wilson Boulevard
Rochester, NY 14627-0171, USA
E-mail: bradford@pas.rochester.edu

R. SCHUMACHER

Dept of Physics
Carnegie Mellon University
5000 Forbes Avenue
Pittsburgh, PA 15213
E-mail: schumacher@cmu.edu

FOR THE CLAS COLLABORATION

The CLAS collaboration has recently completed first measurements of the double polarization observables C_x and C_z for the reactions $\gamma p \to K^+\Lambda$ and $\gamma p \to K^+\Sigma^\circ$. C_x and C_z are the beam-recoil polarization asymmetries measuring the polarization transfer from incoming circularly polarized photons to outgoing hyperons along two directions in the reaction's production plane. The Λ is found to nearly maximally polarized along the direction of incident photon's polarization for forward-going kaons. Polarization transfer to the Σ° is different from the Λ case.

1. Introduction

Measurement of polarization observables have long been recognized as key to unraveling baryon production mechanisms. This work represents the first measurement of C_x and C_z for $\gamma p \to K^+\Lambda$ and $\gamma p \to K^+\Sigma^\circ$.

2. C_x and C_z

C_x and C_z measure the polarization transfer from a circularly polarized incident photon beam to the recoiling Λ or Σ° baryons along two orthogonal directions in the production plane of the K^+-hyperon system. The $\hat{x}$ and $\hat{z}$ directions are defined in the CM frame, with $\hat{z}$ lying along the directions of the incident photon beam's polarization. In this analysis, the polarization transfer was measured through the beam helicity asymmetry according to

$$A_{x/z}(\cos\theta_p) = \frac{N^+(\cos\theta_p) - N^-(\cos\theta_p)}{N^+(\cos\theta_p) + N^-(\cos\theta_p)} = \alpha_{eff}\eta C_{x/z}\cos\theta_p \tag{1}$$

where $\cos\theta_p$ is the direction of the proton from the hyperon's decay measured in the hyperon rest frame with respect to the $\hat{x}$ or $\hat{z}$ axis. η is the polarization of the incident photon beam, $N^+(\cos\theta_p)$ and $N^-(\cos\theta_p)$ are the beam helicity dependent hyperon yields in a given $\cos\theta_p$ bin. α_{eff} is the effective weak decay asymmetry parameter, and has a value of 0.642 for $K^+\Lambda$ and -0.165 for $K^+\Sigma^\circ$. The value of α_{eff} for the Σ° decay arises from our technique of measuring the proton distribution in the rest frame of the Σ°, not the Λ; this dilutes its value to less than the nominal -0.642/3.

3. Experimental Setup and Analysis

The data were taken using the CLAS spectrometer in Hall B at Jefferson Lab. The experiment used a circularly polarized photon beam incident on a liquid hydrogen target. Data were taken with endpoint photon energies of 2.4 and 2.9 GeV. From this dataset, we also measured differential cross sections, which are currently available in preprint[1].

All analyzed events were required to have explicit detection of the K^+ and proton. The Λ or Σ° hyperons were identified in the $p(\gamma, K^+)Y$ missing mass. The data were binned in beam helicity, the cosine of the kaon angle in the CM frame $(\cos(\theta_{KCM}))$, the cosine of the proton angle $(\cos(\theta_p))$ and photon energy (E_γ). Within kinematic each bin, yields were extracted by fitting a Gaussian peak to each hyperon in the missing mass spectrum. Backgrounds were modeled with a polynomial. The beam helicity asymmetry was plotted against $\cos(\theta_p)$ and the slope of this distribution was extracted with a linear fit. Complete analysis details are available elsewhere[2].

4. Results and Discussion

Sample results are presented in Figures 1 and 2. The data are plotted with predictions from the Kaon-MAID[4] (solid line) and Janssen[3] (dashed line) isobar models. The data plotted are for only a few representative bins in $\cos(\theta_{KCM})$. The full results include nine bins in $\cos(\theta_{KCM})$ for $K^+\Lambda$ and six bins for $K^+\Sigma^\circ$.

The Λ results show some W-dependent structure at backward kaon angles and then stabilize at more forward-going kaon angles. For this hyperon, C_z is near one over most of the forward hemisphere of the kaon angle while C_z is near zero for the same range.

The $K^+\Sigma^\circ$ results show no preferred direction for the polarization over the kaon angle range. The precision of these results here appears worse due to the small value of α_{eff}.

Of the models shown, the Janssen does a good job of following the data. The MAID curve does not fair as well. This model has the oddity of predicting that C_z saturates at -1 in $K^+\Lambda$ for forward-going kaons.

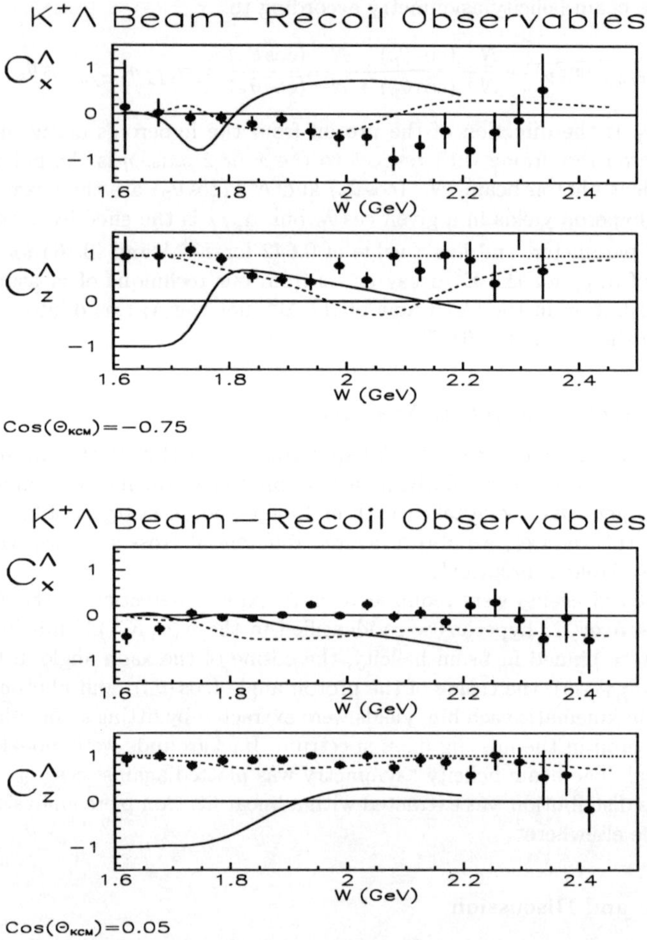

Fig. 1. C_x and C_z for $K^+\Lambda$ in two different kaon-angle bins. Top: $\cos(\theta_K) = -0.75$, bottom: $\cos(\theta_K) = 0.25$. The data are a subset of the 2005 CLAS results. The curves are predictions from the Kaon-MAID (solid line, [4]) and Janssen (dashed line, [3]) isobar models.

References

1. R. Bradford, *et al.* Preprint available at: nucl-ex/0509033. Submitted to *Phys. Rev. C.*
2. R. Bradford, *Measurement of differential cross sections and C_x and C_z for $\gamma p \to K^+\Lambda$ and $\gamma p \to K^+\Sigma^\circ$ using CLAS at Jefferson Lab.* Ph.D. the-

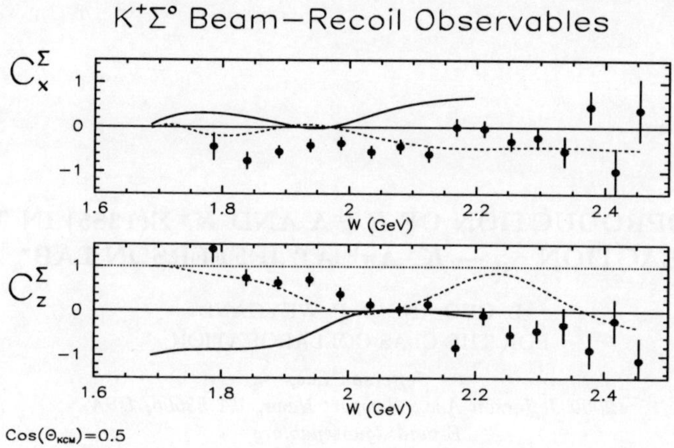

Fig. 2. C_x and C_z for $K^+\Sigma^o$ for $\cos(\theta_K) = 0.5$. The data are a subset the 2005 CLAS results. The curves are predictions from the Kaon-MAID (solid line, [4]) and Janssen (dashed line, [3]) isobar models.

sis, Carnegie Mellon University, 2005, to be published. Available on-line at: http://www.jlab.org/Hall-B/general/thesis/bradford/index.html

3. S. Janssen, *Strangeness production on the nucleon.* Ph.D. thesis, University of Gent, 2002.
4. F.X. Lee, *et al.*, *Nucl. Phys. A* **695**, 237 (2001).

PHOTOPRODUCTION OF $K^{*+}\Lambda$ AND $K^{+}\Sigma(1385)$ IN THE REACTION $\gamma p \rightarrow K^{+}\Lambda\pi^{0}$ AT JEFFERSON LAB*

L. GUO AND D. P. WEYGAND
FOR THE CLAS COLLABORATION

Jefferson Lab,
12000 Jefferson Ave., Newport News, VA 23606, USA
E-mail: lguo@jlab.org

The search for missing nucleon resonances using coupled channel analysis has mostly been concentrated on $N\pi$ and KY channels, while the contributions of $K^{*}Y$ and KY^{*} channels have not been investigated thoroughly mostly due to the lack of data. With an integrated luminosity of about 75 pb^{-1}, the photoproduction data using a proton target recently collected by the CLAS Collaboration at Jefferson Lab with a photon energy range of 1.5-3.8 GeV provided large statistics for the study of light hyperon photoproduction through exclusive reactions. The reaction $\gamma p \rightarrow K^{+}\Lambda\pi^{0}$ has been investigated. Preliminary results of the $K^{*+}\Lambda$ and $K^{+}\Sigma(1385)$ cross sections are not negligible compared with the KY channels. The $\Lambda\pi^{0}$ invariant mass spectrum is dominated by the $\Sigma(1385)$ signal and no significant structure was found around the $\Sigma(1480)$ region.

1. Introduction

Presently, coupled channel analysis has been mostly performed for N^{*} physics by including pion, eta, and kaon production. However, it should be pointed out that $K^{*}Y$ and KY^{*} production cannot be ignored. A recent study of $N^{*} \rightarrow K^{*}Y/KY^{*}$ by Capstick and Roberts [1], using a quark-pair creation model, has predicted that a few low-lying negative-parity states could couple strongly to the $K^{*}\Lambda$ channel. The well established $N[\frac{7}{2}^{-}]_{1}(2190)$ from pion production data should also be clearly visible in photon reactions, while the two star state $N[\frac{3}{2}^{-}]_{1}(2080)$ and the weakly established state $N[\frac{1}{2}^{-}]_{1}(2090)$ are predicted to couple to $K\Lambda$ and $K^{*}\Lambda$ with similar strength. As for the KY^{*} channel , in particular the $K\Sigma(1385)$, the nominal threshold is lower, and several N^{*} states are predicted to couple strongly to $K\Sigma(1385)$. These states include $N[\frac{1}{2}^{-}]_{5}(2070)$ (established in pion production), $\Delta[\frac{3}{2}^{-}]_{3}(2145)$, and the relatively lighter predicted states such as

*This work is supported by DOE under contract DE-AC05-84ER40150

$N[\frac{5}{2}^+]_2(1980)$ and $\Delta[\frac{3}{2}^+]_3(1985)$. The new CLAS data (g11) represents a total luminosity of about 75 pb^{-1} with a tagged photon energy range of 1.5-3.8 GeV [2,3], making it possible to study the photoproduction of K^*Y and KY^* in many channels which previously lacked statistics.

2. Basic data features

The reaction of $\gamma p \to K^+\Lambda\pi^0$ has been investigated by selecting events with the $K^+\pi^-$ and proton being detected by the CLAS spectrometer. The Λ is identified from the invariant mass of the $p\pi^-$ system, while the π^0 is reconstructed from the $K^+\Lambda$ missing four momentum (Fig. 1, top). Overall, the data is dominated by $K^*\Lambda$ and $K^+\Sigma(1385)$ production (Fig. 1, bottom). Recently, there has been some renewed interest in the possible $\Sigma(1480)$ state, with the latest observation reported by the COSY collaboration in the reaction $pp \to pK^+Y^{0*}$ [4]. No significant signal was found in the $\Lambda\pi^0$ invariant mass spectrum for any range of photon energy and other kinematic variables. However, direct comparison is difficult due to the different production mechanisms. Various kinematic requirements were applied to suppress background processes such as $\gamma p \to K^+\Lambda/\Sigma^0$ and $\gamma p \to K^+\Lambda^*, \Lambda^* \to \Lambda\pi^0\gamma$. The contribution from the former can be removed by requiring a minimum missing momentum. A total of about 250 k events are identified as $K^+\Lambda\pi^0$ events. The latter process, however, can not be totally eliminated because of the kinematic overlapping with the reaction of interest. In Fig. 2 (left), the K^+ missing mass shows a clear $\Lambda(1405)$ and $\Lambda(1520)$ contribution when the events between the high limit of the π^0 peak and the 2 π threshold are selected, indicating the $\Sigma^0\pi^0$ decay of the Λ^*'s.

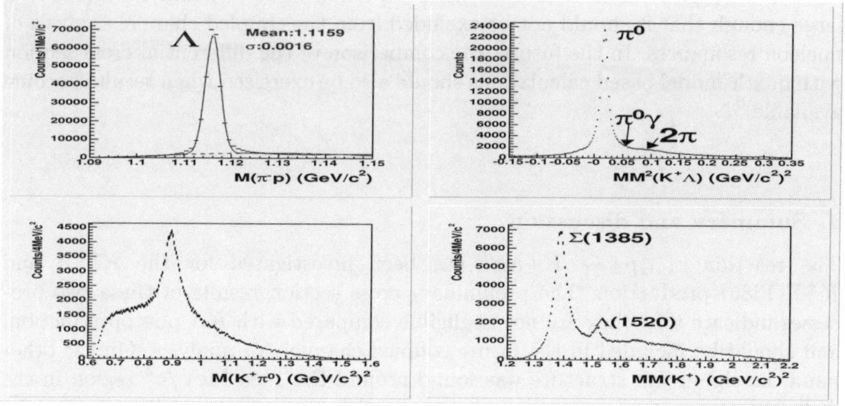

Fig. 1. Top Left: Invariant mass spectrum of $p\pi^-$ system. Top right: Missing mass squared spectrum off of the $K^+\Lambda$ system. Bottom left: Invariant mass spectrum of the $K^+\pi^0$ system. Bottom right: K^+ missing mass spectrum.

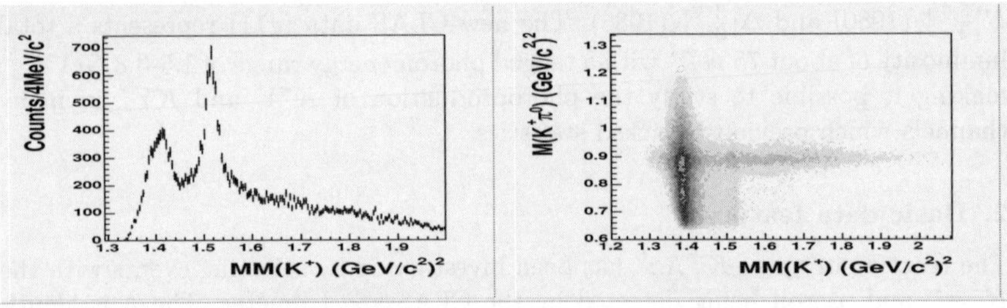

Fig. 2. Left: K^+ missing mass spectrum for the $\pi^0\gamma$ region indicated by Fig. 1(Top right); Right: Invariant mass of the $K^+\pi^0$ system vs. K^+ missing mass.

3. Cross section results

The two processes of $\gamma p \rightarrow K^{*+}\Lambda$ and $\gamma p \rightarrow K^+\Sigma(1385)$ overlap kinematically, as shown in Fig. 2(right), for events above the nominal K^{*+} threshold. Extensive simulations were conducted for these two reactions assuming a t-channel process. The differential cross section results were used as the new input of the next iteration of simulation. The $\Sigma(1385)$ yield was obtained using a $p-$wave Breit-Wigner function with background shape obtained from the simulation of $K^{*+}\Lambda$ events. The K^{*+} yield was also obtained similarly. Due to the preliminary status of the results and the limit of the proceedings, only total cross section results of $K^{*+}\Lambda$ ($\sigma_{total} = \int \frac{d\sigma}{dcos^*\theta_\Lambda}$) and $K^+\Sigma(1385)$ ($\sigma_{total} = \int \frac{d\sigma}{dcos^*\theta_{K^+}}$) are included in this paper†. Compared with the most recent CLAS results of $K^+\Lambda$ total cross section [5] (Fig. 3), it is clear that the production of $K^{*+}\Lambda$ and $K^+\Sigma(1385)$ is large enough that it should not be excluded from the coupled channel analysis of nucleon resonances. In the future, the comparison of the differential cross section with quark-model based calculations should also be exercised when results become available [6].

4. Summary and discussion

The reaction of $\gamma p \rightarrow K^+\Lambda\pi^0$ has been investigated for the $K^{*+}\Lambda$ and $K^+\Sigma(1385)$ production. The preliminary cross section results of these two processes indicate that they are not negligible compared with $K\Lambda$ photoproduction, and should be included in the future coupled channel N^* analysis. On the other hand, no significant structure was found around the 1480 MeV/c^2 region in the $\Lambda\pi^0$ invariant mass spectrum, in contrast with the recent results reported by COSY [4].

†The cross sections are preliminary, and 20% systematic errors are expected

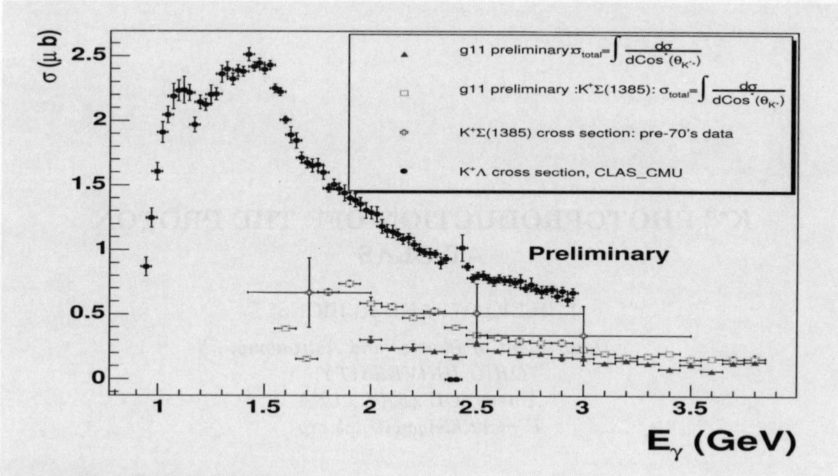

Fig. 3. Total cross sections of $K^*\Lambda$ and $K^+\Sigma(1385)$ compared with recent CLAS results of $K^+\Lambda$ photoproduction and earlier measurements of $K^+\Sigma(1385)$ photoproduction.

5. Acknowledgment

We wish to thank all of the CLAS collaborators, the extraordinary efforts of the CEBAF staff, and particularly the g11 running group.

References

1. S. Capstick and W. Roberts, *Phys. Rev.* **D58**, 074011 (1998).
2. B.A. Mecking *et al.*, *Nucl. Instrum. Methods* **A503**, 513 (2003);
3. D. Sober *et al.*, *Nucl. Instrum. Methods* **A440**, 263 (2000);
4. I. Zychor *et al.*, nucl-ex/0506014. Submitted to *Phys. Rev. Lett.* (2005).
5. R. Bradford *et al.*, nucl-ex/0509033. Submitted to *Phys. Rev.* **C** (2005).
6. Q. Zhao, J.S. Al-Khalili, and C. Bennhold, *Phys. Rev.* **C64**, 052201 (2001).

K*0 PHOTOPRODUCTION OFF THE PROTON AT CLAS

I. HLEIQAWI AND K. HICKS*

Department of Physics and Astronomy,
OHIO UNIVERSITY,
Athens, OH 45701., USA
E-mail: hleiqawi@jlab.org

We report here differential cross sections for the reaction $\gamma p \to K^{*0}\Sigma^+$ for the first time with high statistics. The measurements were done using the CEBAF Large Acceptance Spectrometer (CLAS) at Jefferson Lab. Data were taken at 3.115 GeV electron beam energy using tagged photons incident upon a liquid hydrogen target. In addition, our analysis provides information about nucleon resonances and their couplings to K^* decay. Our data will be used to test predictions of theoretical models of K* production, and to provide constraints on the $K^*\Sigma N$ coupling constant used in these models.

1. Introduction

Some Quark Models predict yet undetected baryon resonances [1]. These "*missing*" resonances have not been observed via strong interactions and could be observed via the photoproduction method. Of special interest are the nucleon resonances N* which could couple strongly to $K\Lambda$ and $K\Sigma$ [2]. Moreover, higher mass nucleon resonances could favor decaying into $K^*\Sigma$, near threshold. Vector meson electro- and photoproduction near threshold might provide good knowledge about these resonances, their internal structure, and their couplings to vector mesons. This has been the main motivation for studying strange mesons photoproduction off the proton, γp → KY, where Y denotes the hyperon,

$$\gamma \mathrm{p} \to K^+\Lambda, \qquad \gamma \mathrm{p} \to K^+\Sigma^0, \qquad \gamma \mathrm{p} \to K^{*0}\Sigma^+. \tag{1}$$

Analyses and results on the first two reactions have been published [3, 4]. However, the third channel has not been studied due to its small cross section and the difficulty of detecting kaons. The availability of the high intensity electron facility and the CEBAF Large Acceptance Spectrometer (CLAS), at Jefferson Lab, has made it possible to study this channel.

*For CLAS Collaboration.

Theoretical Model: a theoretical quark model [5] has been developed to study nucleon resonances and to present quark model predictions for the K^{*0} production. In addition to using common quark model parameters, this model uses two free parameters: the vector and tensor couplings, a and b, for the quark-K^* interaction. These are the basic parameters in the model and are related to the $K^*\Sigma N^*$ couplings that appear in the quark model symmetry limit. The SU(3)-flavor-blind assumption of non-perturbative QCD is adopted, which suggests the above two parameters should have values close to those used in the ω and ρ photoproduction. That is, the model predicts cross sections for the K^{*0} production using couplings extracted from non-strange production. Our data will provide a good test of this model.

K^* production is related to other strangeness production in Eq. (1). At the *hadronic level*, (γ, K^*) and (γ, K) are related to each other since one reaction contains meson production in the other process as the t-channel exchanged particle, and therefore constrains the range of available couplings. This allows both reactions to use the same observables. At the *quark level*, both reactions involve the creation of $s\bar{s}$ pair from the vacuum. Hence, these reactions are related also through the quark model.

2. Experiment and Analysis

The K^{*0} photoproduction data were extracted from the "g1c" data set (4.5 billion triggers) using the CLAS detector [6], at Jefferson Lab's Hall B. Data for our analysis were taken at 3.115 GeV electron beam energies.

The photon beam was produced in Hall B photon tagger [7], by directing the electron beam toward the tagger's radiator where the beam strikes an Al foil creating bremsstrahlung photons. Each photon was tagged by measuring energies of the recoiling electrons in the tagger spectrometer, up to 95% of incident electron beam energy. The photon beam was then directed to hit the target at the heart of CLAS, through a pair of collimators which trim the beam halo. The target, a liquid hydrogen of length 17.85 cm, was surrounded by a start counter which provides prompt timing measurements once photons interact with the target. A coincidence between the tagger and the start counter provided the trigger. See Ref. [6] for details on CLAS spectrometer.

When the γ strikes the p, in addition to K^{*0} other background channels are produced from K^+ production,

$$\gamma \mathrm{p} \rightarrow K^+\Sigma^0, \qquad \gamma \mathrm{p} \rightarrow K^+\Lambda/\mathrm{Y}^* \tag{2}$$

In this experiment, K^{*0} decays to $K^+\pi^-$, while both ground states $\Sigma^0(1193)$ and $\Lambda(1116)$, in the background channels Eq. (2), decay (with 64% probability) to $\pi^-\mathrm{p}$, and the $\Lambda(1520)$ decays (with 14% probability) into $\pi^-\Sigma^+$. In addition, we have other backgrounds, from higher excited states of Λ. Fig. (1) shows diagrams of the K^{*0} (signal) and $\Lambda(1520)$ production.

The only detected particles are K^+ and π^-. The K^{*0} meson is identified from its decay products, $K^+\pi^-$, while Σ^+ hyperon is reconstructed from the missing

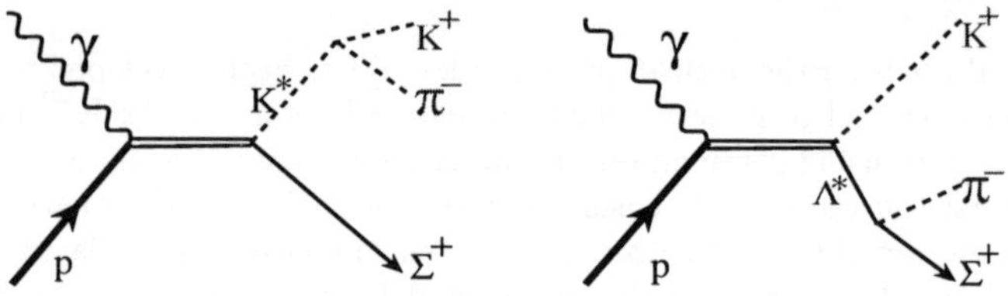

Fig. 1. The reaction of interest (left) and a background channel, Λ(1520), Eq. (2).

mass of K^{*0}. We have two major background contributions: (i) from misidentified π^+ as a K^+, and (ii) from the K^+ production, Eq. (2). Contributions from $\Sigma^0(1193)$ and $\Lambda(1116)$ hyperons were removed by applying a cut on the K^+ missing mass, $MM(K^+)$. To remove contributions from Y^* states and the misidentified pions, and to identify Σ^+ as well as to extract the yields from it, we applied two cuts on K^{*0} mass spectrum: a peak cut centered at 0.892 GeV (the K^{*0} cut) and side-band cuts (the Y^* cut) on either side of K^{*0} peak. Using the peak cut, our signal (the Σ^+ yield) is reconstructed from the missing mass, $MM(K^+\pi^-)$, and is fit with a gaussian plus a polynomial fit to the background. On the other hand, using the side-band cut, the Y^* *physics* background is calculated from the $MM(K^+\pi^-)$, obtaining another Σ^+ peak that is fit in the same way. From these fits we obtained two Σ^+ yields, one from K^* cut and the other from Y^* cut. By subtracting the latter from the former one we obtained the final $K^{*0}\Sigma^+$ yields. The measured yields were then normalized by the real photon flux and corrected for the CLAS detector acceptance (from Monte Carlo).

3. Results and conclusions

Fig. (2) shows the differential cross sections for the K^{*0} photoproduction. Although not shown in this figure, the cross sections are in good agreement with the theoretical model of Zhao [5], after a small modification of the K^*-quark vector and tensor coupling constants (see Section 1). More details will be presented in an upcoming paper. This suggests that the theoretical model of Zhao for vector meson production within the quark model has some predictive power.

At small angles, our data show the cross section is dominated by t-channel exchange except at the highest photon energy bin. At higher energies, the forward-peaking moves out of our detector acceptance, as CLAS can only measure the production angle up to $\cos(\Theta^{c.m.}_{K^{*0}}) < 0.9$ due to the beam-pipe hole through the center of CLAS.

Acknowledgments

We would like to thank A. Tkabladze for his help in this project, and also to C. Bennhold and Q. Zhao for their help and for valuable discussions.

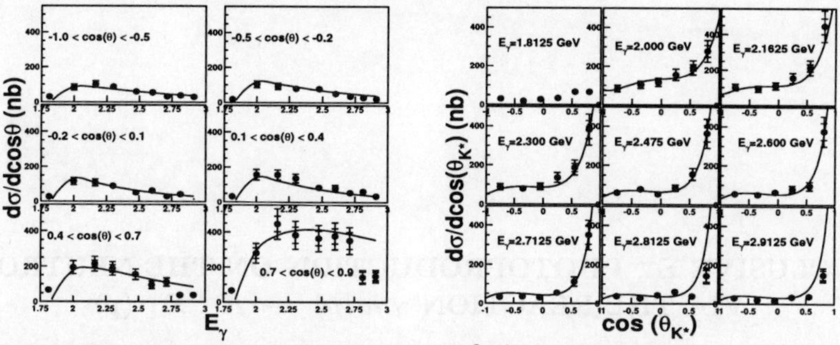

Fig. 2. Differential cross sections for $\gamma p \to K^{*0}\Sigma^{+}$. Energy evolution for six angular bins (left) and angular distribution for nine photon energy bins (right).

References

1. N. Isgur and G. Karl, *Phys. Lett.* **B 72**, 109 (1977);
   N. Isgur and G. Karl, *Phys. Rev.* **D 23**, 817 (1981).
2. S. Capstick and W. Roberts, *Phys. Rev.* **D 58** 74011 (1988).
3. J. McNabb, Ph.D. thesis, Carnegie Mellon University, 2002.
4. R. Bradford *et al.*, nucl-ex/0509033.
5. Q. Zhao, J. Al-Khalili, and C. Bennhold, *Phys. Rev.* **C 64**, 052210 (2001).
6. B. Mecking *et al.*, *Nucl. Instrum. Metheds.* **A 503**, 513 (2003).
7. D.J. Sober *et al.*, *Nucl. Instrum. Metheds.* **A 440**, 263 (2000).

INCLUSIVE Σ^- PHOTOPRODUCTION ON THE NEUTRON VIA THE REACTION $\gamma\, n\, (p) \rightarrow K^+\, \Sigma^-\, (p)$

J. LANGHEINRICH, A. LIMA, AND B. BERMAN
FOR THE CLAS COLLABORATION

The analysis described here is part of a comprehensive survey of the elementary strangeness photoproduction cross sections on the nucleon. The six elementary strangeness reactions are:

$\gamma n \rightarrow K^o\, \Lambda$ and $\gamma p \rightarrow K^+ \Lambda$
$\gamma n \rightarrow K^o\, \Sigma^o$ and $\gamma p \rightarrow K^+ \Sigma^o$
$\gamma n \rightarrow K^+ \Sigma^-$ and $\gamma p \rightarrow K^o\, \Sigma^+$

Although there are γp data from ABBHHM [1], SAPHIR [2], and CLAS [34], there are none for the γn channels; our results from the CLAS g2 experiment are the first values to be reported.

Theoretical model predictions for the strangeness photoproduction cross sections have been made by Bennhold and collaborators [6 7]. Depending on the data set fitted for the proton channels, their predictions for the neutron channels, particularly the $K^o\Lambda$ and $K^+\Sigma^-$ channels, differ widely. Therefore, this work is motivated in part by the certainty that our results will provide a stringent constraint for the theory.

There are two different strategies to extract a differential cross section for the reaction $\gamma n(p) \rightarrow K^+\Sigma^-(p)$ from the CLAS[8] data sample, referred to as "inclusive" and "exclusive". In the exclusive analysis [5] the detection of a K^+, π^-, and neutron is required, and the Σ^- four-momentum vector is reconstructed from the latter two particles. In the inclusive analysis only the K^+ is required, and the Σ is detected using missing mass techniques. The advantage of the exclusive analysis is, that one can correct for Fermi motion and measure the Σ^- yield directly. However, the inclusive analysis which will be described more in detail here has significant advantages when it comes to statistical fluctuations and acceptance corrections. The reconstruction efficiency is at least an order of magnitude better, and the acceptance obtained from Monte-Carlo simulation is not dependent on the input distribution from the event generator. The acceptance is mostly flat and much better understood than the more complicated CLAS multi-particle and neutron acceptance.

The first step is to identify K^+-mesons by their momentum and time of flight. Special attention is given to the fact that the width of the K^+-mass peak is not constant in this kinematic range (see Fig. 1). After dividing the whole statistic in small p_{K^+} bins and evaluating all these bins separately, we found

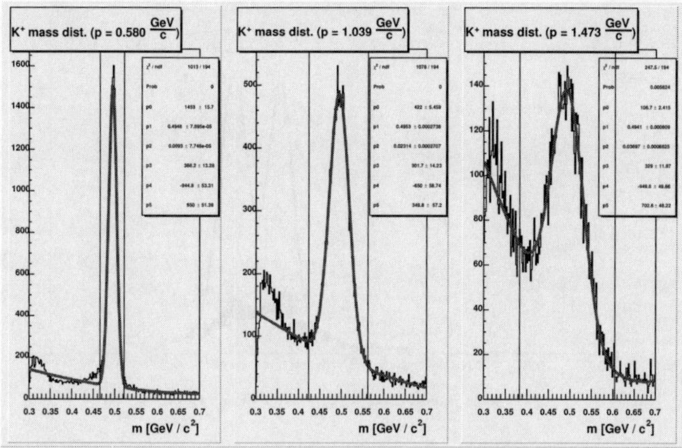

Fig. 1. K^+ mass calculated from momentum and time-of-flight for several K^+ momentum bins.

that the peak width can be parameterized using a third degree polynomial. It has to be emphasized that these K^+-mass distributions can be reproduced with high precision from a Monte Carlo simulation, so that identical cuts and analysis routines can be used for real and simulated ata.

To calculate the four momentum vector of the hyperon from the missing mass once the K^+ has been identified, we assume that the deuteron is very loosely bound so that we deal with a quasi-free proton or neutron target. This assumption can be tested when plotting the missing mass against the dynamic properties of the K^+, like the angle ϑ_K in the laboratory system. We observe that neither the position nor the width of the Λ and Σ hyperon mass peaks change as a function of ϑ_K. Although there is a notable smearing of the peaks by Fermi motion, the Λ and Σ can be clearly separated, and the contents of the mass peaks can be obtained from a Gaussian fit.

A critical part of this analysis is the thorough understanding of the background underneath the hyperon mass distribution. The background distribution can be obtained after rebinning the hyperon mass plot so that there is a reasonable number of events in each bin. For each of these bins a fit to the K^+ mass distribution reveals the signal and background yields (See Fig. 2). Surprisingly, the Λ and Σ hyperon peaks can be clearly seen in the background distribution as well. This seems to indicate that much of the background tracks are a μ^+ resulting from the K^+ decay rather than a misidentified π^+. Investigating the upper tail of the π^+ mass distribution from exclusive $\gamma p \to n\pi^+$ events verifies this hypothesis.

When it comes to the efficiency calculation the single particle analysis pays off. The probability ε_K that the K^+ can be reconstructed from the track momentum and the time-of-flight measured in CLAS is a function of the K^+ momentum and the K^+ spatial angle. Integrating over ϕ_K we obtain the function $\varepsilon_K(\vartheta_K, p_K)$

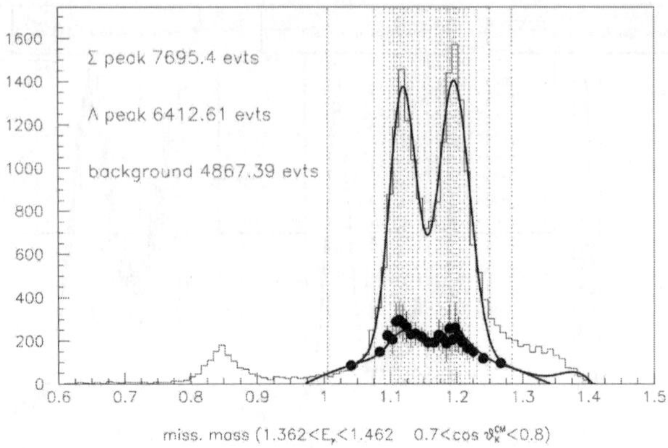

Fig. 2. The missing mass distribution with fits to the Λ and Σ peaks and to the background.

from the Monte Carlo simulation. It is worth mentioning that, within the limits of statistical uncertainties, $\varepsilon_K(\vartheta_K, p_K)$ is independent of the input distribution chosen for the Monte Carlo simulation.

The final step is to separate the Σ^- from the Σ^o channel. The latter one is produced by the reaction $\gamma p \to K^+\Sigma^o$, and has been measured independently on a hydrogen target by the CLAS g1c experiment. However, doubts whether the bound proton in the deuteron might behave slightly different than the proton in a hydrogen target and the ongoing discussion about systematic uncertainties in the CLAS photon flux led us to the decision not to subtract the Σ^o cross section directly, but to use the Λ/Σ^o ratio instead. This ratio is obtained by fitting Legendre polynomials to the published Σ^o and Λ data points.

The acceptance corrected yield of the reaction $\gamma p \to K^+\Lambda$ provides a tool to check agreement of the results from the bound and from the free proton. These results are, for most of the E_γ photon energy bins, in good agreement. However, the results for the unfolded Σ^- yield are a surprise. Not only is the yield below the expectations from the model predictions; there is a minimum around $\cos\theta_{CM} = 0.45$ in the $\cos\theta_{CM}$ distribution which is neither predicted nor visible in the Σ^o channel.

It might be premature to jump to conclusions. However, the statement that the Σ^- yield is significant lower than model expectations has been confirmed by the exclusive analysis as well. The results seem to be robust when tested for systematic uncertainties. The analysis is currently scrutinized by the mandatory CLAS analysis review process, and a publication providing cross sections is expected in the near future.

References

1. ABBHHM Collaboration, Phys. Rev. 188, 2060 (1969).

2. R. Lawall *et al*, Measurement of the reaction $\gamma p \rightarrow K^0$ at photon energies up to 2.6 GeV, Eur. Phys. J. A24, 275 (2005).
3. B. Carnahan, Strangeness photoproduction in the p $K^{0,+}$ reaction, UMI-31-09682, Ph.D. thesis, Catholic University of America (2003).
4. J. W. C. McNabb *et al*, Phys. Rev. C 69, 042201(R) (2004).
5. I. Niculescu, G. Niculescu, CLAS Note (2005).
6. F. X. Lee, T. Mart, C. Bennhold, L. E. Wright, **nucl-th/9907119** (1999).
7. H. Yamamura *et al*, **nucl-th/9907029** (1999).
8. V. Burkert, H. Egiyan, L. Elouadrhiri, S. Stepanyan, CLAS Note, **98-008** (1998).

$S = 0$ PSEUDOSCALAR MESON PHOTOPRODUCTION FROM THE PROTON

M. DUGGER* AND THE CLAS COLLABORATION

Arizona State University,
Tempe, AZ 85287-1504, USA
E-mail: dugger@jlab.org

Many measurements of pseudoscalar mesons with $S = 0$ photoproduced on the proton have been made recently. These new data are particularly useful in theoretical investigations of nucleon resonances. How the new data from various labs complement each other and help fill in the gaps in the world data set is disscussed, with a glance at measurements to be made in the near future. Some theoretical techniques used to explain the data are briefly described.

1. Motivation

The $S = 0$ pseudoscalar mesons include the π^0, π^+, η, and η'. The pions, as the lightest mesons, are copiously produced in the strong interaction. Pion photoproduction data have been vital to gain a first glimpse of the nucleon resonance spectrum. Whereas the pions have isospin 1, the η and η' mesons have isospin 0, so resonances decaying by emitting η or η' mesons can only have isospin $\frac{1}{2}$. Thus, η and η' mesons act as "isospin filters" for the nucleon resonance spectrum. Further, as the only isosinglet, η' can be used to indirectly probe gluonic coupling to the proton through the flavor-singlet Goldberger-Treiman relation[1]:

$$g_A^0 = \frac{3}{4}\frac{F_0}{m_N}(g_{\eta' NN} - g_{GNN}), \tag{1}$$

where m_N is the mass of the nucleon, $g_{\eta' NN}$ is the η'-nucleon-nucleon coupling, g_{GNN} is the gluon-nucleon-nucleon coupling, and F_0 is a renormalization constant. The flavor singlet axial charge of the nucleon (g_A^0) has been measured[2] with a value of $g_A^0 = 0.20 \pm 0.35$; however, the quark and gluon components have yet to be specifically determined.

In addition to these theoretical motivations for studying photoproduction of $S = 0$ pseudoscalar mesons from the proton, there are practical motivations. Electromagnetic interactions are well understood and real photons are particularly simple (only two polarization states), so data from real photon beams are

*Work supported by the National Science Foundation.

easier to analyze than data from other probes. Moreover, the $S = 0$ two-body final states for $\gamma p \to pX$ (X is a meson), and $\gamma p \to n\pi^+$ have the benefit of having an outgoing proton or pion easy to identify and with relatively little contamination; $S \neq 0$ final states require identification of a kaon usually contaminated with both pions and protons. Thus, photoproduction of $S = 0$ mesons offers a relatively simple experimental means to explore nucleon resonances.

2. Resent results:

Four experimental facilities are providing new data on meson photoproduction from the proton: GRAAL[3], SAPHIR[4], CB-ELSA[5], and CLAS[6]. Their contributions to $S = 0$ pseudoscalar meson photoproductions data will be summarized here.

a) $\gamma p \to p\pi^0$: Prior to 2005, the world data set (compiled in SAID[7]) for $\gamma p \to p\pi^0$ differential cross sections had good coverage in incident energy (E_γ) and angle only up to $E_\gamma \approx 1.5$ GeV. In 2005, CB-ELSA published results[8] which extended coverage in E_γ up to 3.0 GeV. While the CB-ELSA data greatly enhance the coverage of $d\sigma/d\Omega$ for π^0 photoproduction from the proton, the systematic errors of the absolute normalization are estimated to be $\sim 15\%$ for E_γ above 1.3 GeV. New preliminary data from CLAS cover E_γ up to 2.125 GeV with an estimated systematic uncertainty in the absolute normalization to be $<$ 5%; a sample of this data is shown in Fig. 1. An additional data set has come from GRAAL[9], which measured differential cross sections and beam asymmetry Σ for E_γ up to 1.496 GeV.

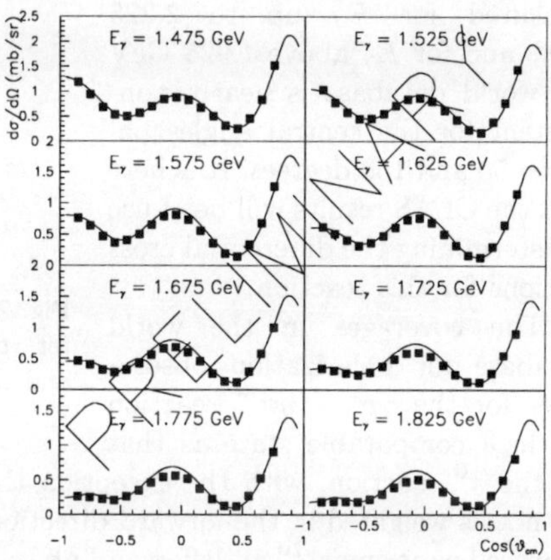

Fig. 1. A sample of differential cross sections for $\gamma p \to p\pi^0$. The data points are from CLAS and the lines are from SAID.

Prior to the GRAAL measurements, angular coverage for Σ was heavily biased in the forward direction. The new GRAAL data for Σ populate the angular range much more uniformly for E_γ up to 1.496 GeV. The rest of the polarization observables in the database are rather sparse. In the near future, a new generation of experiments specifically dedicated to polarization measurements should significantly expand our knowledge of polarization observables.

Thus, the world database for $\gamma p \to p\pi^0$ differential cross sections is becoming quite thorough for E_γ up to ~ 3 GeV, and with coverage by more than one data set up to 2.125 GeV. The new GRAAL results for beam polarization extend the database for Σ to E_γ up to 1.496 GeV. All other polarization observables are very sparsely covered in energy and angle. Future experiments are expected to enhance our knowledge of the polarization observables. In particular, an approved experiment[10] at Jefferson Lab could start taking data for double polarization observables (beam and target) as soon as Fall 2006.

b) $\gamma p \rightarrow n\pi^+$: Of the four collaborations mentioned above, only the CLAS Collaboration has an analysis effort underway for the reaction $\gamma p \rightarrow n\pi^+$ which is in a very preliminary state. Preliminary differential cross section results for $\gamma p \rightarrow n\pi^+$ agree well with the SAID parameterization for E_γ up to 1.925 GeV (see Fig. 2).

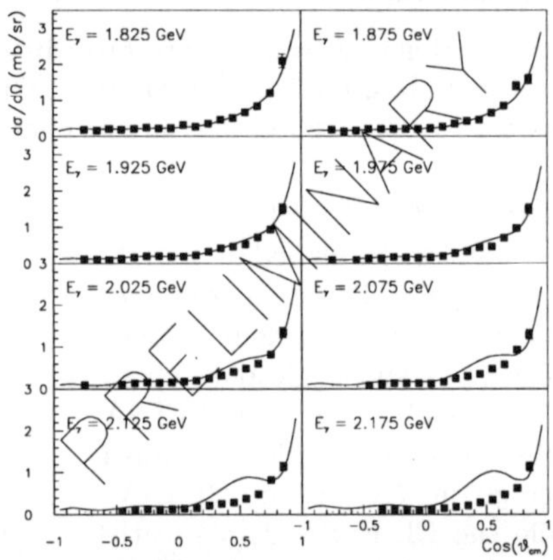

Fig. 2. A sample of Differential cross sections for $\gamma p \rightarrow n\pi^+$. The data points are from CLAS and the lines are from SAID.

The CLAS cross sections were measured for E_γ up to 2.225 GeV, and for E_γ above 1.925 GeV the world database is nearly nonexistent for the central angles between 50 and 150 degrees. It is here that the CLAS results will be of use in determining the differential cross sections for this reaction.

The coverages in the world database for polarization observables for the $\gamma p \rightarrow n\pi^+$ reaction are in a comparable state as that for the π^0 reaction, with the exception that the beam polarization observables are not as weighted in the forward direction. As with the π^0 reaction, there is an approved experiment[10] at Jefferson Lab to obtain double polarization observables that could start taking data as soon as Fall 2006.

c) $\gamma p \rightarrow p\eta$: Before 2002 the world database for $\gamma p \rightarrow p\eta$ differential cross sections was only well covered for E_γ from threshold (0.707 GeV) up to 0.8 GeV. In 2002 GRAAL published results[11] on $d\sigma/d\Omega$ for E_γ up to 1.1 GeV, and CLAS published[12] $d\sigma/d\Omega$ for E_γ up to 1.95 GeV. More recently (2005), CB-ELSA published[13] $d\sigma/d\Omega$ results for E_γ up to 3 GeV.

In 1998, GRAAL published[14] $\gamma p \rightarrow p\eta$ beam polarization results for E_γ up to 1.45 GeV. Polarization observables for $\gamma p \rightarrow p\pi^+$ remain very sparsely populated. This past summer, data were taken at CLAS for beam polarization that should allow extraction of that observable for η photoproduction for E_γ up to 2.1 GeV. As with the pions, an approved experiment[15] at Jefferson Lab should help fill in the database for single and double polarization observables in η photoproduction from the proton.

d) $\gamma p \to p\eta'$: Prior to 1998, only 18 η' photoproduction events had been measured (11 events from the ABBHHM bubble chamber experiment[16], and 7 events from the AHHM streamer chamber experiment[17]). In 1998, the SAPHIR collaboration published results[18] extracted from an additional 250 η' exclusive events. By contrast, the new (unpublished) CLAS results have over 2×10^5 η' photoproduction events detected and used to extract differential cross sections shown in Fig. 3. These CLAS results span E_γ from 1.527 to 2.227 GeV. No polarization observables have been measured for this reaction. Therefore, HHH the differential cross sections provide the only experimental data for the reaction $\gamma p \to p\eta'$.

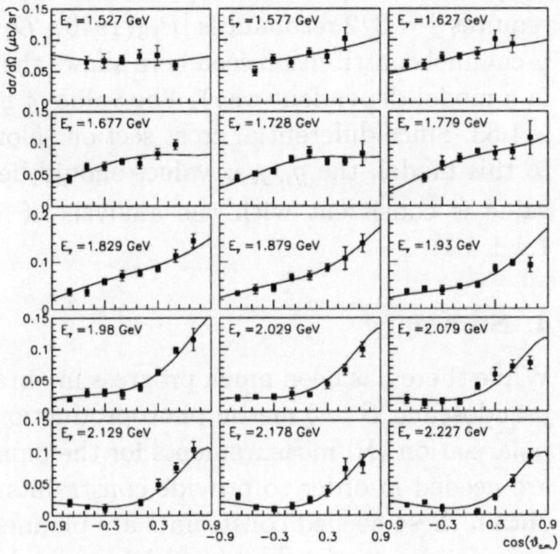

Fig. 3. Differential cross sections for $\gamma p \to p\eta'$. The data points are from CLAS and the lines are from the HK model described in the text.

3. Theoretical Results

As noted above, there are many new differential cross section data for the reactions discussed here. However, these alone are not sufficient to constrain theoretical models to the extent that resonances can be uniquely determined. More data on the polarization observables are desperately needed, and a coupled channel approach is required, in order to constrain the contributions of various resonances.

One step in this direction comes from a model[8,13,19] developed by A. V. Anisovich, E. Klempt, A.Sarantsev, and U.Thoma (AKST model) that couples the reactions $\gamma p \to p\pi^0$, $n\pi^+$, and $p\eta$. AKST included published differential cross sections, as well as the recent GRAAL beam polarization observables. The model uses a K matrix approach for the $S_{11}(1535)$ and the $S_{11}(1650)$ resonances. The remaining resonances are described by Breit-Wigner amplitudes. The model also includes reggeized $u-$ and $t-$channel contributions. Results from their analysis[8] find evidence for a previously unseen $D_{15}(2070)$ resonance, and indications for a new $P_{13}(2200)$ resonance.

One model that considers the η' exclusively comes from K. Nakayama and H. Haberzettl (NH). This model is based upon a relativistic meson-exchange model of hadronic interactions. Allowed processes include s-, t-, and u-channel contributions. The intermediate mesons in the $t-$channel exchanges are the ω and ρ^0. The NH model includes the $S_{11}(1535)$ and $P_{11}(1710)$ resonances ($j = 1/2$), which are known to decay strongly to the ηN channel[21], and also includes two additional S_{11} and two additional P_{11} resonances, albeit with relatively small couplings. The present adaptation of the NH model to the CLAS data now also requires $j = 3/2$ resonances [$P_{13}(1940)$, $D_{13}(1780)$, and $D_{13}(2090)$]. The observed u-channel contribution seen here allows the $g_{\eta' NN}$ coupling to be extracted (albeit in a model-dependent way). The value of $g_{\eta' NN}$ found from the particular NH fit is 1.33. Since differential cross sections alone do not provide sufficient constraints to this model, the $g_{\eta' NN}$ values should be taken with caution. Nonetheless, this value is consistent with the analysis of T. Feldmann[23] which gives $g_{\eta' NN} = 1.4 \pm 1.1$.

4. Summary

While there has been much progress in obtaining differential cross section data for pseudoscalar $S = 0$ meson photoproduction from the proton, and some new beam polarization (Σ) measurements for the pions and η, more polarization observables are needed in order to provide constraints to theoretical models. Experiments to obtain these needed constraints are planned for the near future. A comprehensive program for single, double and even triple polarization measurements in photoproduction is in preparation at Jefferson Lab. From the data already taken, there appears to be evidence for a new D_{15} resonance at 2.09 GeV and indications of a new P_{13} at 2.20 GeV. In addition to improving our knowledge about resonaces and their parameters by fitting the $\gamma p \to p\pi^0$, $n\pi^+$, and $p\eta$ data, the new CLAS η' data also has been analyzed to suggest a value of $g_{\eta' NN} \approx 1.3$, consistent with theoretical predictions. When this value can be determined with a high degree of confidence, it can be used to indirectly determine the gluonic coupling to the proton through the flavor-singlet Goldberger-Treiman relation.

References

1. S.D. Bass, Phys. Lett. B **463**, 286 (1999).
2. J. Ashman, *et al.*, Phys. Lett. B **328**, 1 (1989).
3. F. Ghio, *et al.*, Nucl. Instrum. Meth. A **404**, 71 (1998).
4. W. J. Schwille, *et al.*, Nucl. Instrum. Meth. A **344**, 470 (1994).
5. E. Aker, *et al.*, Nucl. Instrum. Meth. A **321**, 69 (1992).
6. B. Mecking, *et al.*, Nucl. Instrum. Meth. A **503/3**, 513 (2003).
7. R. A. Arndt, W. J. Briscoe, I. I. Strakovsky, and R. L. Workman, Phys. Rev. C **66**, 055213 (2002).
8. O. Bartholomy *et al.*, Phys. Rev. Lett. **94**, 012003 (2005).
9. D. Rebreyend, C. Schaerf, A. Dangelo , private communication.
10. Jefferson Lab Proposal E03-105, "Pion Photoproduction from a Polarized Target" Spokespersons: S. Strauch (contact) *et al.*, (2005).
11. F. Renard, *et al.*, Phys. Lett. B **528**, 215 (2002).
12. M. Dugger, *et al.*, Phys. Rev. Lett. **89**, 222002 (2002).
13. V. Crede, *et al.*, Phys. Rev. Lett. **94**, 012004 (2005).
14. J. Ajaka, *et al.*, Phys. Rev. Lett. **81**, 1797 (1998).
15. Jefferson Lab Proposal E05-012, "Measurement of Polarization Observables in η-photoproduction with CLAS," Spokespersons: E. Pasyuk (contact), M. Dugger (2005).

16. ABBHHM Collaboration, Phys. Rev. **175**, 1669 (1968).
17. AHHM Collaboration, Nucl. Phys. B **108**, 45 (1976).
18. R. Plötzke *et al.*, Phys. Lett. B **444**, 555 (1998).
19. A.V.Anisovich, E.Klempt, A.Sarantsev, U.Thoma, hep-ph/0407211.
20. K. Nakayama, and H. Haberzettl, private communication. The results shown here are based on an extended version of the model given in [22].
21. S. Eidelman *et al.*, Phys. Lett. B **592**, 1 (2004).
22. K. Nakayama and H. Haberzettl, Phys. Rev. C **69**, 065212 (2004).
23. T. Feldmann, Int. J. Mod. Phys. A **15** 159 (2000); hep-ph/9907491.

PHOTOPRODUCTION OF NEUTRAL PION PAIRS OFF THE PROTON WITH THE CRYSTAL BARREL DETECTOR AT ELSA

M. FUCHS

Helmholtz Institut für Strahlen- und Kernphysik
Nußallee 14–16,
53115 Bonn, Germany,
E-mail: Michael.Fuchs@hiskp.uni-bonn.de

The photoproduction of two neutral pions off of the proton was investigated with the Crystal Barrel detector at the ELectron Stretcher Accelerator (ELSA) in Bonn, Germany. Data shows clear resonance structures and hints of cascading decays of baryon resonances with the $\Delta(1232)P_{33}$, $N(1520)D_{13}$ and $X(1660)$ as intermediate states. A partial wave analysis (PWA) was accomplished on the selected events to determine contributing resonances and properties thereof. Additionally results of the PWA were used in a new method of acceptance correction which takes the correct dynamics of the reaction into account. With this method the total and the differential cross sections were calculated for photon energies up to 3 GeV. The total cross section shows two clear peaks at 700 and 1100 MeV. In the differential cross section for $m(p\pi^0)$ contributions of the $\Delta(1232)P_{33}$, $N(1520)D_{13}$ and $X(1660)$ are revealed as well as enhancements from the $f_0(980)$ in the differential cross section for $m(\pi^0\pi^0)$.

1. Introduction

Recent quark model calculations predict many more resonances than have been experimentally discovered so far [1]. The expected small coupling [2] of these missing states to the πN channel is one possible explanation for these 'missing resonances'. Most of the current data stems from πN scattering experiments. Calculations showed that these states have strong couplings to $\Delta\pi$, Nρ and a non-vanishing coupling to γN and hence photoproduction experiments have a good chance to discover these states.

The analysis discussed in this paper is based on the same data set and reconstruction methods as [3] and [4]. Differences appear within the data selection and in the determination of the acceptance.

2. Experiment

Fig. 1 shows the experimental setup at the **EL**ectron **S**tretcher **A**ccelerator (ELSA) in Bonn. The tagging system consisted of 14 scintillators and two multiwire proportial chambers for improved energy resolution. Data was taken at two different electron energies of 1.4 and 3.2 GeV resulting in photon energies between 300 MeV and 3.0 GeV.

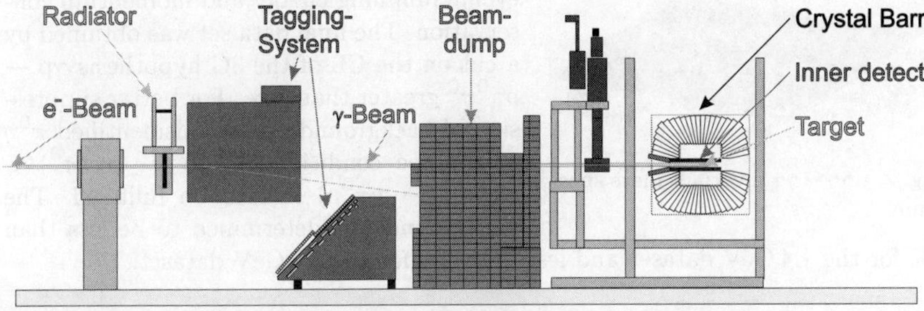

Fig. 1. Experimental setup at ELSA in Bonn

The tagged photon beam hit a liquid H_2 target of 5 cm length and 3 cm diameter. A three–layer scintillating fiber (scifi) detector [5] encircling the target identified charged particles in a polar angle range between 15° and 165°. The Crystal Barrel calorimeter [6] surrounding the target was primarily used to detect photons. It was comprised of 1380 CsI(Tl) crystals with photodiode readout covering 98% of 4π.

3. Reconstruction and Selection

Data selection started by selecting only events with 4 and 5 clusters in the Crystal Barrel detector. By comparing the intersection points from the scifi with the clusters in the calorimeter proton candidates were found. The identified proton was handled as a 'missing' particle, and was not used in the further analysis.

The remaining uncharged clusters were considered as photons. Energy and momentum conservation was assured by performing a kinematic fit and applying a cut of 1% on the confidence level (CL) of the 1C hypothesis $p(\gamma, 4\gamma)p_{\text{missing}}$. Fig. 2 shows the reconstructed $\gamma\gamma$ versus $\gamma\gamma$ invariant mass of these events.

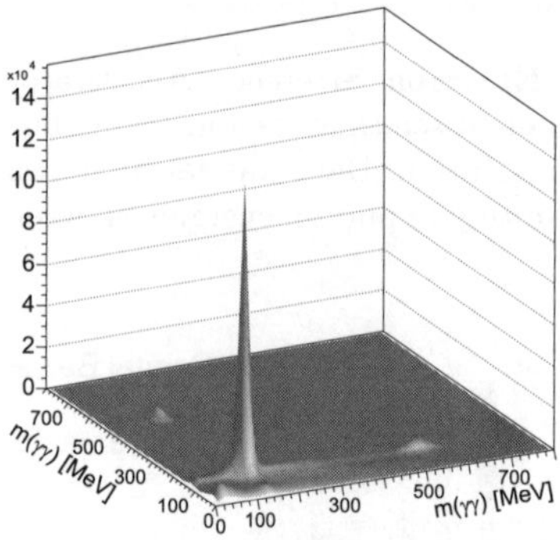

Fig. 2. $\gamma\gamma - \gamma\gamma$ invariant mass spectrum.

A dominant $\pi^0\pi^0$ peak and two small bumps due to $\pi^0\eta$ events can be seen with almost no background visible. Neutral pions were identified by a cut on the invariant mass of two γ in an interval of $m(\pi^0) \pm 2\sigma$. The standard deviation $\sigma = 8\,\text{MeV}$ was determined as the width of the invariant $\gamma\gamma$ mass peak for events fullfilling energy and momentum conservation. The final data set was obtained by a cut on the CL of the 3C hypothesis $\gamma\text{p} \rightarrow \text{p}\pi^0\pi^0$ greater than 10%. For better suppression of background due to misidentified $\pi^0\eta$ events the condition $\text{CL}(\gamma\text{p} \rightarrow \text{p}\pi^0\pi^0) > \text{CL}(\gamma\text{p} \rightarrow \text{p}\pi^0\eta)$ had to be fullfilled. The background was determined to be less than 1% for the 1.4 GeV dataset and less than 2% for the 3.2 GeV dataset.

4. Acceptance

A reaction with three particles in the final state (e.g. $\gamma\text{p} \rightarrow \text{p}\pi^0\pi^0$) depends on five independent variables. Calculating the acceptance of such a reaction in only one dimension implies an integration over the other four variables. If the acceptance in at least one of these four variables changes strongly the calculation for a one dimensional acceptance will lead to incorrect results if the dynamics of the process is not properly included in the MC simulation. This would e.g. be the case if phase space MC is used for acceptance correction.

The solution for this problem is to calculate the acceptance not based on phasespace (PS) distributed Monte Carlo (MC) events but rather on a modified set of MC events. This set was created by using a Partial Wave Analysis (PWA) to calculate a weight factor for each of the PS generated MC events. After normalizing these weight factors to the maximum value a new MC set was created by comparing the normalized weight factor with a generated random number. An event was copied to the new set if its weight factor was greater than the generated random number.

The acceptance correction used for all results in this paper are based on this newly created set of Monte Carlo events. This set now includes the correct dynamics of the process to the extent the PWA describes the process correctly.

5. Results

The absolute normalization of the cross section was derived in two different ways. In the low energy range up to 1.3 GeV the angular distributions [3] for $\gamma\text{p} \rightarrow \text{p}\pi^0$ were compared to theoretical predictions from SAID* resulting in the normalization. To the 3.2 GeV dataset an absolute normalization with a scaling factor was

*Virginia Tech Partial-Wave Facility

applied. The error was determined to be 5% for photon energies below 1.3 GeV and 15 % above 1.3 GeV.

Data, acceptance and flux normalization were determined for each wire of the proportional wire chamber as the smallest experimentally given energy unit resulting in a cross section for each wire. For data presentation, wires were combined and a weighted average cross section for these energy intervals was calculated. The error of the average was calculated by error propagation.

5.1. *Total cross section*

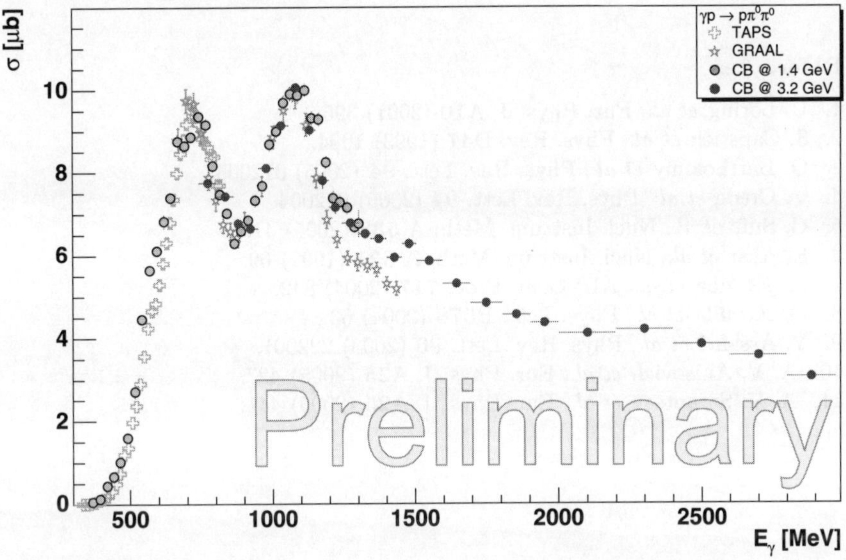

Fig. 3. Total cross section with statistical errors only; systematic errors: flux: 5% for $E_\gamma < 1.3\,\mathrm{GeV}$ and 15% otherwise, acceptance: 6%, reconstruction: 5%

Fig. 3 shows the total cross section of the reaction $\gamma p \to p\pi^0\pi^0$ for the two analyzed energies in comparison to published data [7–9]. Our data confirms the cross section in the energy range between 700 and 1200 MeV and extends the measured energy range up to 3.0 GeV. Deviations are visible between 500 and 700 MeV and at energies above 1200 MeV.

5.2. *Differential cross section*

The differential cross sections $d\sigma/dm(p\pi^0)$ (Figs. 4a–4d) and $d\sigma/dm(\pi^0\pi^0)$ (Figs. 4e–4h) were calculated the same way as the total cross section.

In the energy range between 1350 and 1570 MeV in $\sqrt{s}$ (Fig. 4a) the reaction is dominated by $\gamma p \to X \to \Delta(1232)\pi^0$. With increasing energy a shoulder builds up in Fig. 4b and can be identified as N(1520)D_{13} in Fig. 4c. With even higher energies in Fig. 4d an indication for an additional resonance at 1660 MeV is seen.

The differential cross section $d\sigma/dm(\pi^0\pi^0)$ does not show clear structures in the lower energy intervals (Figs. 4e–4g). A peak only appears in the highest energy interval (Fig. 4h) and it can be identified with the production of $f_0(980)$ mesons.

Acknowledgments

This work is supported by the Deutsche Forschungsgemeinschaft (DFG) within the SFB/TR16.

References

1. U. Löring *et al.*, Eur. Phys. J. A**10** (2001) 395.
2. S. Capstick *et al.*, Phys. Rev. D**47** (1993) 1994.
3. O. Bartholomy *et al.*, Phys. Rev. Lett. **94** (2005) 012003.
4. V. Credé *et al.*, Phys. Rev. Lett. **94** (2005) 012004.
5. G. Suft *et al.*, Nucl. Instrum. Meth. A **538** (2005) 416.
6. E. Aker *et al.*, Nucl. Instrum. Meth. A **321** (1992) 69.
7. M. Kotulla *et al.*, AIP Conf. Proc. **717** (2004) 842.
8. M. Kotulla *et al.*, Phys. Lett. B**578** (2004) 63.
9. Y. Assafiri *et al.*, Phys. Rev. Lett. **90** (2003) 222001.
10. A. V. Anisovich *et al.*, Eur. Phys. J. A**25** (2005) 427.
11. A. V. Sarantsev *et al.*, Eur. Phys. J. A**25** (2005) 441.

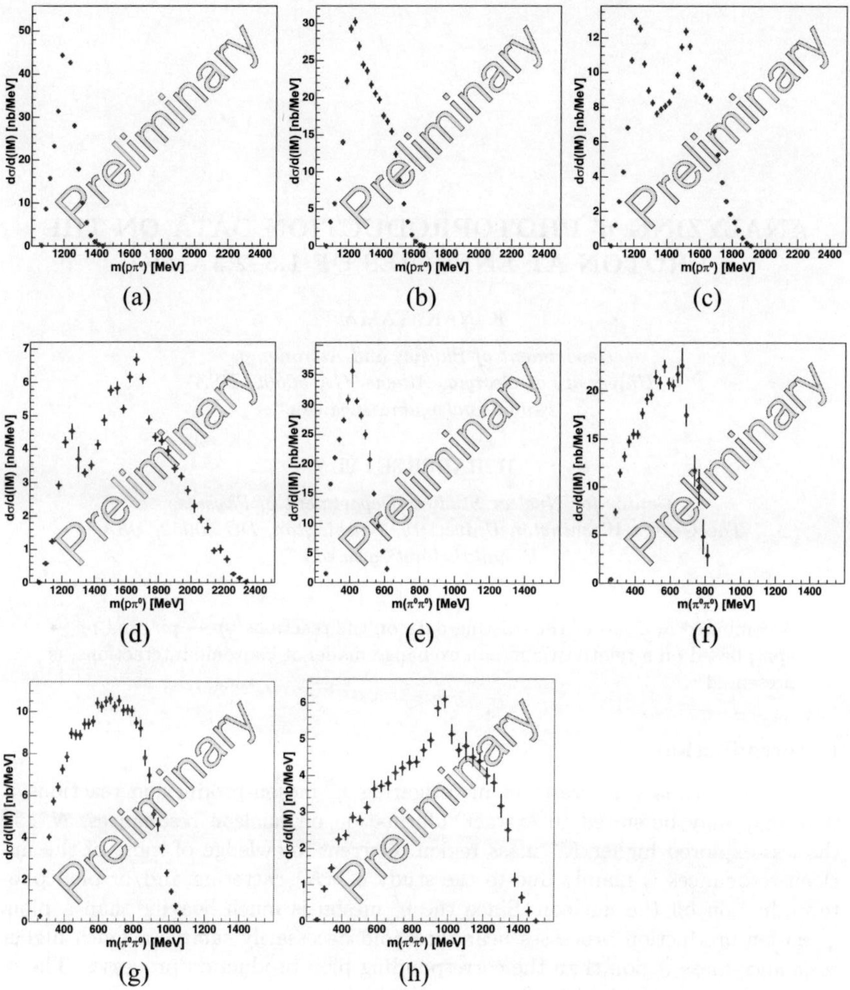

Fig. 4. Differential cross section $\mathrm{d}\sigma/\mathrm{d}m(\mathrm{p}\pi^0)$ (a–d) and $\mathrm{d}\sigma/\mathrm{d}m(\pi^0\pi^0)$ (e–h) with statistical errors only; systematic errors: flux: 5% for $E_\gamma < 1.3\,\mathrm{GeV}$ and 15% otherwise, acceptance: 6%, reconstruction: 5%. The following energy intervals were used (a),(e) : $1350\,\mathrm{MeV} < \sqrt{s} < 1570\,\mathrm{MeV}$, (b),(f) : $1570\,\mathrm{MeV} < \sqrt{s} < 1800\,\mathrm{MeV}$ (c),(g) : $1800\,\mathrm{MeV} < \sqrt{s} < 2060\,\mathrm{MeV}$ and (d),(h) : $2060\,\mathrm{MeV} < \sqrt{s} < 2550\,\mathrm{MeV}$

ANALYZING η' PHOTOPRODUCTION DATA ON THE PROTON AT ENERGIES OF 1.5–2.3 GeV

K. NAKAYAMA

Department of Physics and Astronomy,
University of Georgia, Athens, GA 30602, USA
E-mail: nakayama@uga.edu

H. HABERZETTL

Center for Nuclear Studies, Department of Physics,
The George Washington University, Washington, DC 20052, USA
E-mail: helmut@gwu.edu

A combined analysis of the existing data on the reactions $\gamma p \to p\eta'$ and $pp \to pp\eta'$, based on a relativistic meson exchange model of hadronic interactions, is presented.

1. Introduction

One of the primary interests in investigating η' meson-production reactions is that they may be suited to extract information on nucleon resonances, N^*, in the less-explored higher N^* mass region. Current knowledge of most of the nucleon resonances is mainly due to the study of πN scattering and/or pion photoproduction off the nucleon. Since the η' meson is much heavier than a pion, η' meson-production processes near threshold necessarily sample a much higher resonance-mass region than the corresponding pion production processes. Therefore, they are well-suited for investigating high-mass resonances in low partial-wave states. Furthermore, these reactions provide opportunities to study those resonances that couple only weakly to pions, in particular, those referred to as "missing resonances".

Another special interest in η' photoproduction is the possibility to impose a more stringent constraint on its yet poorly known coupling strength to the nucleon. This has attracted much attention in connection with the so-called "nucleon-spin crisis" in polarized deep inelastic lepton scattering[1] in that the $NN\eta'$ coupling constant can be related to the quark contribution to the "spin" of the nucleon.[2] Reaction processes where the η' meson is produced directly off a nucleon may offer a unique opportunity to extract this coupling constant.

The major purpose of the present contribution is to present the photoproduction part of a combined analysis of the $\gamma p \to p\eta'$ and $pp \to pp\eta'$ reactions within a

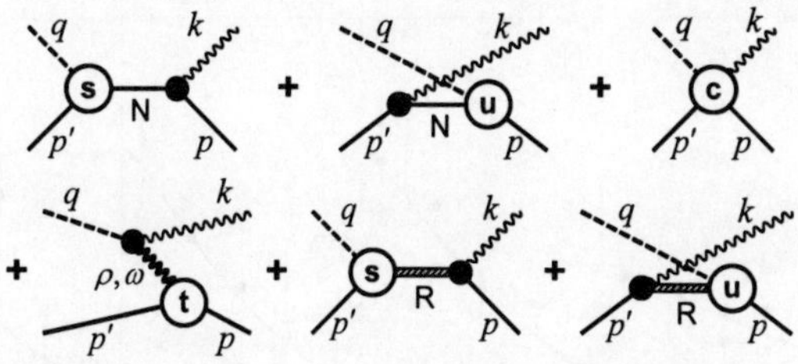

Fig. 1. Diagrams contributing to $\gamma p \to \eta' p$. The intermediate baryon states are denoted N for the nucleon, and R for the nucleon resonances. The total current is made gauge-invariant by an appropriate choice of the contact current depicted in the top-right diagram. The nucleonic current (nuc) referred to in the text corresponds to the top line of diagrams; the mesonic current (mec) and resonance current contributions correspond, respectively, to the leftmost diagram and the two diagrams on the right of the bottom line of diagrams.

relativistic meson-exchange model of hadronic interactions. The photoproduction reaction is described in the tree-level approximation, where a phenomenological contact term is introduced in order to guarantee the gauge-invariance of the full amplitude.[3] The latter consists of nucleonic, mesonic and (nucleon) resonance currents as depicted in Fig. 1. The hadro-production reaction part of our analysis as well as further details can be found in Ref. 3.

2. Analysis of the SAPHIR data on $\gamma p \to p\eta'$

The objectives of analyzing the SAPHIR data[8] on η' photoproduction are:

1) To shed light on the conflicting conclusions of the existing model calculations for these data. These contradictions are: (a) The origin of the shape of the observed angular distribution. Zhao[4] has emphasized that this is due to the interference among the resonance currents, while Chiang et al.[5] have concluded that it is due to the interference between the resonance and t-channel mesonic currents. Yet, Sibirtsev et al.[6] have claimed that the mesonic current is responsible for the observed angular distribution. The latter authors use a t-dependent exponential form factor in their mesonic current. (b) t-channel meson exchange versus Regge trajectory. Chiang et al.[5] have emphasized that the SAPHIR data can be described only if the Regge trajectory is used in the t-channel mesonic current, while other authors[6,7] have used ordinary vector meson exchanges.
2) Can we constrain the $NN\eta'$ coupling constant from the photoproduction reaction?
3) To which extent are we able to identify the nucleon resonances from the differential cross section data, i.e., can, for example, the mass of the resonance be pinned down from the existing cross-section data?
4) Provide inputs for the $NN \to NN\eta'$ reaction.

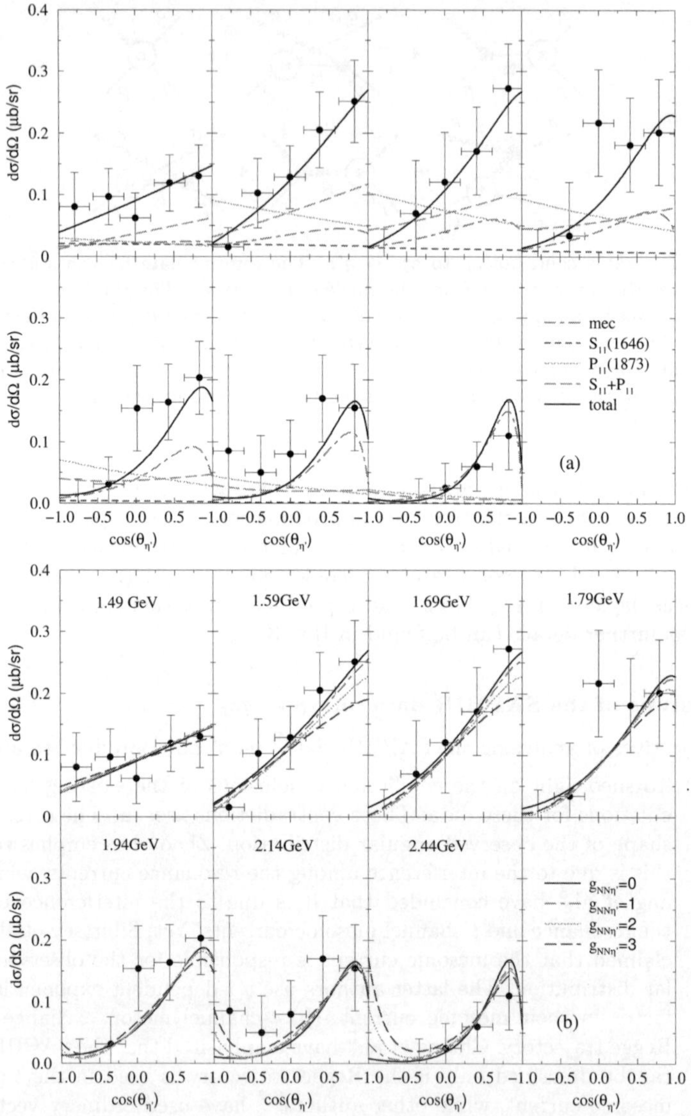

Fig. 2. Differential cross section for $\gamma p \to p\eta'$ according to the mechanisms shown in Fig. 1a. Panel (a) includes the meson-exchange current (mec), the S_{11} and P_{11} resonances. In (b), successively stronger (as indicated by the values of the $g_{NN\eta'}$ coupling constant) nucleonic current (nuc) contributions are added to the results shown in panel (a). In each case, the model parameters are determined by best fits. The meaning of the corresponding lines is indicated in the panels. The data are from Ref. 8.

Our results for the angular distribution at various energies are shown in Fig. 2, together with the SAPHIR data.[8] First of all, the mesonic current is responsible for the measured forward peaked angular shape at higher energies. The P_{11} resonance contribution is much larger than that due to the S_{11} resonance. Also, note the interference effects among different currents; the P_{11} resonance gives raise to a larger backward angle cross sections, while the total resonance current, S_{11} plus P_{11}, yields a larger forward-angle cross sections. Adding the mesonic current leads to a further enhancement of the forward cross sections. We mention that the inclusion of the mesonic and S_{11} resonance currents only is not sufficient to describe the strong angular dependence at lower energies.[3] Also, the nucleonic current, through its interference with the mesonic and S_{11} resonance currents, makes the angular distribution more pronounced,[3] but not as pronounced as the addition of the P_{11} resonance shown in Fig. 2a. It is clear, therefore, that the observed angular distribution is a result of the rather non-trivial interference among different currents.

Figure 2b displays our results for various values of the $NN\eta'$ coupling constant, $g_{NN\eta'}$, in the nucleonic current. Here, we also include the mesonic and the S_{11} and P_{11} resonance currents. Note that the nucleonic current becomes pronounced at backward angles as the energy increases which is due to the u-channel diagram. We see that $g_{NN\eta'}$ cannot be much larger than 3. Naively, we would expect that more accurate data at higher energies will enable us to reduce this upper limit. See, however, the analysis of the high-precision CLAS data in Sec. 3.

Next, we address the issue of the ordinary meson exchange versus Regge trajectory in the t-channel mesonic current. The mesonic current based on the ordinary vector-meson exchange contains an extra form factor at the electromagnetic vertex, while that based on the Regge trajectory contains no such form factor. We verified that both models describe the data equally well, although there are some differences in detail. However, one important point to be noted here is that the resulting resonance parameters are quite different. This is quite disturbing, for it reveals a clear model dependence in the extracted resonance parameters. Further investigation of this important issue is required in future works.

We have also verified the sensitivity/insensitivity of the differential cross section data to the mass value of the nucleon resonance. In particular, we found that the results with the S_{11} mass values which differ by about 100 MeV are hardly distinguishable from each other in the differential cross sections. This gives a rough idea about the uncertainty one should expect on the extracted resonance mass values based only on the differential cross section data.

3. Analysis of the (preliminary) CLAS data on $\gamma p \to p\eta'$ reaction

Our model results for the (preliminary) CLAS data on η' photoproduction[9] are shown in Fig. 3a. First of all, compared to the SAPHIR data[8] analyzed in Sec. 2, the new CLAS data are much more accurate and, as such, may reveal features that were not seen in the analysis of the SAPHIR data. In Fig. 3a, different curves correspond to different sets of fit parameters which yield comparable χ^2 values.

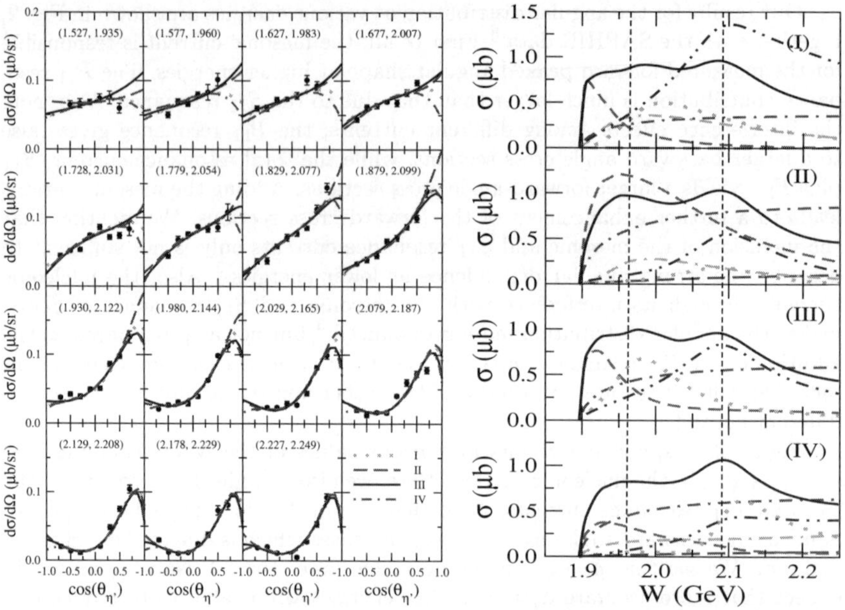

Fig. 3. Left figure (a): Same as Fig. 2 for the CLAS data.[9] The curves correspond to different fit results which yield comparable χ^2 values. The numbers (T_γ, W) in parentheses are the incident photon energy T_γ and the corresponding s-channel energy $W = \sqrt{s}$, respectively, in GeV. Right figure (b): Total cross section for $\gamma p \to p\eta'$ as a function of W. As indicated in the legend, the panels correspond to the fit results shown in the left figure. The overall total cross sections (solid lines) are broken down according to their dynamical contributions. The dash-dotted curves correspond to the mesonic current contribution; the dashed curves to the S_{11} resonance current and the dotted curves to the P_{11} resonance. The dot–double-dashed curves correspond to the P_{13} resonance current while the dash–double-dotted curves show the D_{13} resonance contribution. The nucleonic current contribution (long-dashed curves) are negligible and cannot be seen on the present scale. The two dashed vertical lines are placed to guide the eye through the two bump positions in all panels.

Unlike the case of the SAPHIR data, here one requires not only the spin-1/2 resonances, but also the spin-3/2 resonances in order to reproduce the data. We found that the required spin-1/2 and -3/2 resonances are consistent with those quoted by the PDG.[10]

Although the different parameter sets yield practically the same differential cross sections (except for very forward and backward angles where no data exist), the corresponding dynamical content is very different from each other over the entire angular range. This shows, in particular, that cross sections alone are unable to fix the resonance parameters unambiguously, and that more exclusive observables, such as the beam and target asymmetries, are necessary in order to extract information on nucleon resonances.

Our predictions for the total cross section, as displayed in Fig. 3b, have been obtained by integrating the differential cross section results of Fig. 3a. A common feature present in all of these results is the bump structure around $W = 2.09$ GeV. If this is confirmed, the $D_{13}(2080)$ [and possibly $P_{11}(2100)$] resonance is likely to be responsible for the structure. The PDG[10] quotes $D_{13}(2080)$ and $P_{13}(2100)$ as two and one star resonances, respectively. Another feature we see in Fig. 3b is the sharp rise of the total cross section near threshold which is caused by the S_{11} resonance.

Contrary to the expectation in Sec. 2, the $NN\eta'$ coupling constant cannot be determined even with the high-precision CLAS data. The reason is that even at higher energies, the resonance contribution may be large, especially that of the D_{13} resonance. In fact, two of the results shown in Fig. 3b correspond to practically vanishing $g_{NN\eta'}$. However, we are able to give a more stringent upper limit of $g_{NN\eta'} < 2$. In order to pin down further this coupling constant, one needs to either have more exclusive data than the cross section or go beyond the resonance energy region.

4. Summary

The study of η' production processes is still in its early stage of development. Our study reveals that in order to extract relevant physics, one needs more exclusive data than the cross sections. In addition, measurements of η' production using the neutron/deuteron target are also required. On the other hand, our theoretical model also needs to be improved; in particular, coupled channel effects should be investigated. In this connection, unfortunately, there is no realistic model for the $N\eta'$ final state interaction available at present.

Acknowledgments

This work is partly supported by the Forschungszentrum Jülich, COSY Grant No. 41445282 (COSY-58).

References

1. J. Ashman et al., *Phys. Lett.* **B206**, 364 (1988).
2. G. M. Shore and G. Veneziano, *Nucl. Phys.* **B381**, 23 (1992).
3. K. Nakayama and H. Haberzettl, *Phys. Rev.* **C69**, 065212 (2004); nucl-th/0507044.
4. Q. Zhao, *Phys. Rev.* **C63**, 035205 (2001).
5. W. T. Chiang et al., *Phys. Rev.* **C68**, 045202 (2004).
6. A. Sibirtsev, Ch. Elster, S. Krewald and J. Speth, *AIP Conf. Proc.* **717**, 837 (2004); nucl-th/0303044.
7. B. Borasoy, *Eur. Phys. J.* **A9**, 95 (2000).
8. R. Plötzke et al., *Phys. Lett.* **B444**, 555 (1998).
9. M. Dugger and B. Ritchie, *private communication.*
10. S. Eidelman et al. (Particle Data Group), *Phys. Lett.* **B592**, 1 (2004).

ETA PHOTOPRODUCTION OFF THE NEUTRON AT GRAAL

V. KUZNETSOV[1*], O. Bartalini[2], V. Bellini[3], M. Castoldi[4], A. D'Angelo[2], J-P. Didelez[5], R. Di Salvo[2], A. Fantini[2], D. Franco[2], G. Gervino[6], F. Ghio[7], B. Girolami[7], A. Giusa[3], M. Guidal[5], E. Hourany[5], R. Kunne[5], A. Lapik[1], P. Levi Sandri[8], D. Moricciani[2], L. Nicoletti[3], C. Randieri[3], N. Rudnev[9], G. Russo[3], C. Schaerf[2], M.-L. Sperduto[3], M.-C. Sutera[3], A. Turinge[10].

[1]*Institute for Nuclear Research, 117312 Moscow, Russia*
[2]*INFN sezione di Roma II and Università di Roma "Tor Vergata", 00133 Roma, Italy*
[3]*INFN Laboratori Nazionali del Sud and Università di Catania, 95123 Catania, Italy*
[4]*INFN Genova and Università di Genova, 16146 Genova, Italy*
[5]*IN2P3, Institut de Physique Nucléaire, 91406 Orsay, France*
[6]*INFN sezione di Torino and Università di Torino, 10125 Torino, Italy*
[7]*INFN sezione Sanità and Istituto Superiore di Sanità, 00161 Roma, Italy*
[8]*INFN Laboratori Nazionali di Frascati, 00044 Frascati, Italy*
[9]*Institute of Theoretical and Experimental Physics, Moscow, Russia*
[10]*RRC "Kurchatov Institute", Moscow, Russia*

The $\gamma n \to \eta n$ quasi-free cross section reveals a resonant structure at $W \sim 1.675$ GeV. This structure may be a manifestation of a baryon resonance. *A priori* its propreties, the possibly narrow width and the strong photocoupling to the neutron, look surprising. This structure may also signal the existence of a narrow state.

Meson photoproduction off the neutron offers an attractive tool to study certain baryon resonances. A single-quark transition model[1] predicts only weak photoexcitation of the $D_{15}(1675)$ resonance from the proton target. Photocouplings to the neutron calculated in the framework of this approach, are not small. An isobar model for η photo- and electroproduction η-MAID[2] also suggests significant contribution of the $D_{15}(1675)$ to η photoproduction on the neutron.

Possible photoexcitation of the non-strange pentaquark is of high interest as well. A benchmark signature of this particle (if it exists) is its photoproduction off the nucleon. The chiral soliton model predicts that photoexcitation of the non-strange pentaquark has to be suppressed on the proton and should mainly

*E-mail Slava@cpc.inr.ac.ru, SlavaK@jlab.org

occur on the neutron [4]. Estimates of the chiral soliton approach ranges its mass to 1.65 – 1.7 GeV[5]. Modified parial wave analysis of πN scattering suggests two possible candidates, at 1.68 and/or 1.73 GeV, with the total width about 10 MeV. Among various reactions, η photoproduction has been considered as particularly sensitive to the signal of this particle [4–7].

Up to now, η photoproduction off the neutron was explored mostly in the region of the $S_{11}(1535)$ resonance from threshold up to W$\sim$1.6 GeV [8]. The ratio of the $\gamma n \rightarrow \eta n / \gamma p \rightarrow \eta p$ cross sections was extracted and found almost constant near $\sim$0.67. At higher energies, the GRAAL Collaboration reported the sharp rise of this ratio[9] and the evidence for a resonant structure in the cross section on the neutron at W$\sim$1.675 GeV [10]. Recently the CB/TAPS Collaboration reported similar observation[11].

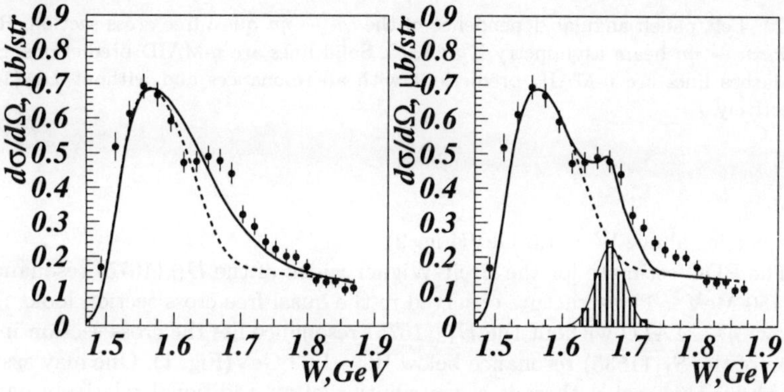

Fig. 1. $\gamma n \rightarrow \eta n$ quasi-free differential cross-section at 137°. Left panel: Soild line is the η-MAID prediction folded with Fermi motion. Dashed line is the same prediction without the $D_{15}(1675)$ resonance. Right panel: Dashed area shows the contribution of a narrow state. Solid line is the sum of the η-MAID cross section without $D_{15}(1675)$, folded with Fermi motion, and the narrow state. Dashed line is the same as in the left panel.

Quasi-free ηn differential photoproduction cross section is shown in Fig. 1. The cross section clearly reveals a resonant structure near $W \sim 1.675$ GeV. We compared this cross section with an isobar model for η photo- and electroproduction $\eta - MAID^2$. The model includes 8 main resonances and suggests the dominance of the $S_{11}(1535)$ and $D_{15}(1675)$ resonances in η photoproduction off the neutron below $W \sim 1.7$ GeV[†]. The model predicts a bump-like structure near $W \sim 1.675$ GeV in the total η photoproduction cross section on the neutron[12]. This structure is caused by the $D_{15}(1675)$ resonance. The η - MAID differential cross section at 137° is smooth (Fig. 1, left panel). $\eta - MAID$ reasonably reproduces the angular dependence of the cross section while predicts larger beam

[†]We refer to the recent update of $\eta - MAID$ which includes the corrected helicity amplitude ${}^nA_{\frac{1}{2}}$ of the $D_{15}(1675)$ resonance.

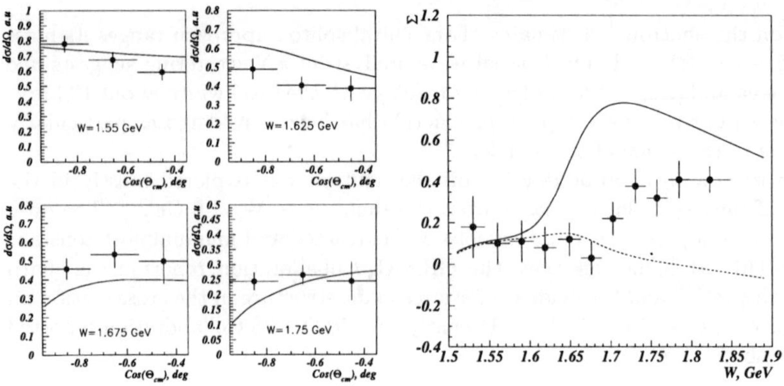

Fig. 2. Left panel: angular dependence of the $\gamma n \to \eta n$ quasi-free cross section. Right panel: $\gamma n \to \eta n$ beam asymmetry Σ at 137°. Solid lines are η-MAID predictions. Solid and dashes lines are η-MAID predictions with all resonances and without $D_{15}(1675)$ respectively.

asymmetries above $W \sim 1.6$ GeV(Fig. 2).

The PDG estimate for the Breit-Wigner width of the $D_{15}(1675)$ resonance is $\Gamma \sim 150$ MeV[3]. The structure observed in the quasi-free cross section looks more narrow. $\eta - MAID$ without the $D_{15}(1675)$ resonance fits the cross section in the region of the $S_{11}(1535)$ resonance below $W \sim 1.62$ GeV(Fig. 1). One may assume that above this region there is a contribution of an additional relatively narrow resonance.

In the right panel of Fig. 1, a simulated narrow ($M = 1675$ GeV, $\Gamma = 10$ MeV) state is shown. This state appears as a bump in the quasi-free cross section due to Fermi motion of the target neutron bound in a deuteron target. The sum of the $\eta - MAID$ without $D_{15}(1675)$ and the narrow state well reproduces the cross section up to $W \sim 1.7$ GeV. At higher energies, the increasing contribution of higher-lying resonances is expected[13]. Thus, the structure in the $\gamma n \to \eta n$ cross section may signal the existence of a relatively narrow state. If confirmed, such state coincide with the expectation of the Chiral Soluton Model[4,5] and modified PWA[6] for the non-strange pentaquark. On the other hand, the manifectation of one of usual resonances is not ruled out. *Apriori* its properties, the possibly narrow width and the strong photocoupling to the neutron, look surprising. More data and detailed partial wave analysis are needed to identify the nature of the observed structure. New programs to study η photorpoduction off the neutron with polarized targets, which are now launched at modern photon factories such as the upgraded MamiC Facility (Mainz, Germany), aim to provide experimental data at new level of quality.

I wish to thank Bill Briscoe and Igor Strakovsky for assistance in preparation and delivery of this talk, Berndt Krusche for fruitful discussions, and Lothar Tiator for providing new η - MAID predictions.

References

1. V. Burkert *et al.*, *Phys. Rev. C* **67**, 035205 (2003).
2. W.-T.Chiang, S.-N.Yang,L.Tiator, and D.Drechsel, *Nucl. Phys* A **700**, 426 (2002), [hep-0110034], http://www.kph.uni-mainz.de
3. S.Eidelman *et al.*, *Phys. Lett.* B**592**, 1 (2004).
4. M. Polyakov and A. Rathke *Eur. Phys. J A* **18**, 691 (2003) [hep-ph/0303138].
5. D. Diakonov and V. Petrov, *Phys. Rev. D* **69**, 094011 (2004) [hep-ph/0310212].
6. R. Arndt *et al.*, *Phys. Rev. C* **69**, 0352008 (2004) [nucl-th/0312126].
7. R. Jaffe and F. Wilczek, *Phys. Rev. Lett.* **91**, 232003 (2003) [hep-ph/0307341].
8. B. Krusche *et al.*, *Phys. Lett. B* **358**, 40 (1995); V. Heiny *et al.*, *Eur. Phys. J. A* **6**, 83 (2000); J. Weiß *et al.*, *Eur. Phys. J. A* **16**, 275 (2003) [nucl-ex/0210003]; P. Hoffman-Rothe *et al.*, *Phys. Rev. Lett.* **78**, 4697 (1997).
9. V.Kouznetsov *et al.*, Proceedings of Workshop on the Physiscs of Excited Nucleons NSTAR2002, Pittsburgh, USA, October 9 - 12 2002, Ed. E.Swanson, World Scientific, 2003, pg.267-270.
10. V. Kuznetsov *et al.*, Proceedings of Workshop on the Physics of Excited Nucleons NSTAR2004, March, 2004, Grenoble, France, Eds. J.-P.Bocquet, V.Kuznetsov, D.Rebreyend, World Scientific, pg.197 - 203, [hep-ex/0409032].
11. J. Jeagle, Contribution to this conference.
12. L.Tiator, Contribution to this conference.
13. F. Renard *et al.*, *Phys. Lett. B* **528**, 215 (2002); M. Dugger *et al.*, *Phys. Rev. Lett.* **89**, 222002 (2002); V. Crede *et al.*, *Phys. Rev. Lett.* **94**, 012004 (2005) [hep-ex/0311045].

η PHOTOPRODUCTION OFF DEUTERIUM

I. JAEGLE FOR THE CBELSA-TAPS COLLABORATION

Institut für Physik
Universität Basel,
Klingelbergstrasse 82, CH-4056 Basel
E-mail: igal.jaegle@unibas.ch

Photoproduction of η-mesons off deuterium has been used as tool for the study of the $\gamma n \rightarrow n\eta$ reaction. The experiment was done at the Bonn ELSA accelerator with the Crystal Barrel and TAPS detectors combined to an almost 4π electromagnetic calorimeter. The η-mesons were detected in coincidence with the recoil nucleons, which allows a direct comparison of the $\gamma n \rightarrow n\eta$ and $\gamma p \rightarrow p\eta$ reaction for nucleons bound in the deuteron, while a comparison to the free proton cross section allows to estimate nuclear effects. Differential cross sections and photon beam asymmetries have been measured. Preliminary results are discussed, which show a bump-like structure in the neutron cross section around incident photon around 1 GeV, which is not observed for the proton.

1. Introduction

The excitation spectrum of baryons is intimately connected to the properties of QCD in the low-energy regime, where it cannot be treated by perturbative approaches. Originally, most experimental information on the baryon spectrum was obtained from hadron induced reactions, in particular from pion-nucleon scattering. More recently, photon induced meson production reactions off the proton have largely contributed, but so far much less effort has gone into the investigation of the corresponding reactions on the neutron. Photoproduction off the neutron serves two purposes. First, it is the only possibility to disentangle the isospin structure of the electromagnetic resonance excitations. Secondly, in some cases the electromagnetic coupling of resonances can be stronger for the neutron than for the proton so that they can be studied in more detail on the neutron.

Quasifree and coherent photoproduction of η mesons off the deuteron, off ^{3}He, and off ^{4}He in the excitation range of the $S_{11}(1535)$ resonance has been studied during the last years in quite some detail [1–7] at different levels of sophistication. The measured cross sections were compared to the free proton data [8] and could be consistently understood in the framework of simple PWIA models, which took into account the Fermi motion of the bound nucleons, when an approximately energy independent ratio $\sigma_n/\sigma_p \approx 2/3$ was assumed (see [9] for an overview). At higher incident photon energy (E_γ > 900 MeV) models of η-photoproduction like the η-MAID model [10] predict a strong rise of the neutron - proton cross section ratio due to the contribution of other nucleon resonances. In the MAID model, the relative rise of he neutron cross section compared to the proton cross section which peaks at incident photon energies around 1 GeV, is mainly due to the excitation of the $D_{15}(1675)$ four star resonance, which has a much stronger photon coupling to the neutron than to the proton. It is interesting to note, that also in the framework of the chiral soliton model [11] a state with similar properties is predicted in this energy range. This is the nucleon-like member of the proposed anti-decuplet of pentaquarks, which would be a P_{11} state. On the experimental side, recently the GRAAL collaboration [12] has reported preliminary results for quasifree η photoproduction off protons and off neutrons bound in the deuteron. The excitation function of quasifree η production measured in coincidence with recoil protons shows a smooth behavior and agrees as expected with the results for the free proton for incident photon energies not too close to the kinematical production threshold. However, on the neutron a relatively narrow structure is observed at incident photon energies around 1 GeV corresponding to $W \approx 1675$ MeV. At the same time the photon beam asymmetry on the neutron also shows a different behavior from the proton.

The present experiment aims at precise measurement of excitation functions, angular distributions and photon beam asymmetries for quasifree η-photoproduction from the neutron.

2. Experiment and Analysis

In the experiment the $\eta \rightarrow 3\pi^o \rightarrow 6\gamma$ decay channel was used. The decay photons were identified in the Crystal Barrel and in the TAPS forward wall with the standard analysis methods, using the inner detector of the barrel, the veto detector of TAPS, time-of-flight and pulse-shape techniques. Events with six photon clusters where than selected for further analysis. The six photons were combined into all disjunct combinations of thee pairs, and the invariant mass of the photon pairs was compared to the pion mass. Cuts were made on the invariant masses of the photon pairs and the kinematical parameters of the three pions were used to calculate the invariant mass of the six-photon state. Typical invariant mass spectra are shown in the upper row of fig. 1 for two regions of incident photon energies and for reactions with coincident recoil protons and recoil neutrons. They are almost background free at low incident photon energies, at higher energies background from triple π^o production via different channels appears. Finally, single η events were selected by a missing mass analysis of the reaction (see fig. 1, lower part). Again no background is visible at low incident photon energies, but at higher energies substantial contributions from $\eta\pi$ final states must be removed with a cut in the missing mass spectra.

3. Preliminary results

As a first check of the data, η-photoproduction from the bound proton was analyzed. Angular distributions of $d(\gamma, \eta)p(n)$ from coincident detection of η-mesons

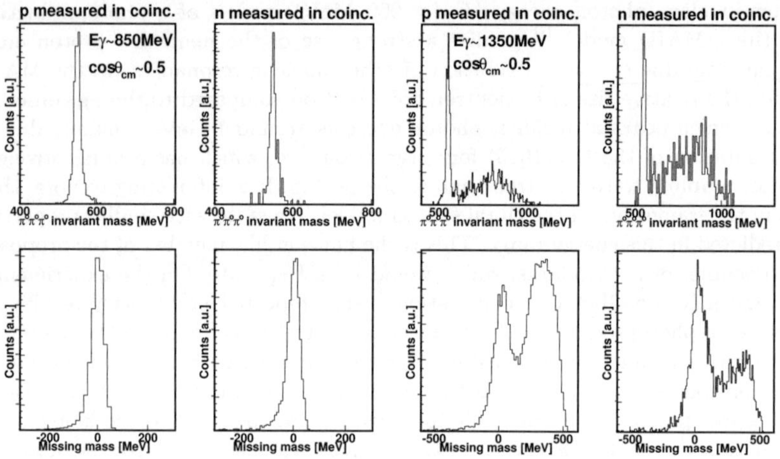

Fig. 1. Upper row: invariant mass spectra - Lower row: missing mass spectra after cut on invariant mass

and recoil protons are compared in fig. 2 to the reaction on the free proton [13] (the absolute scale of the quasifree data is arbitrary). The free and quasifree proton

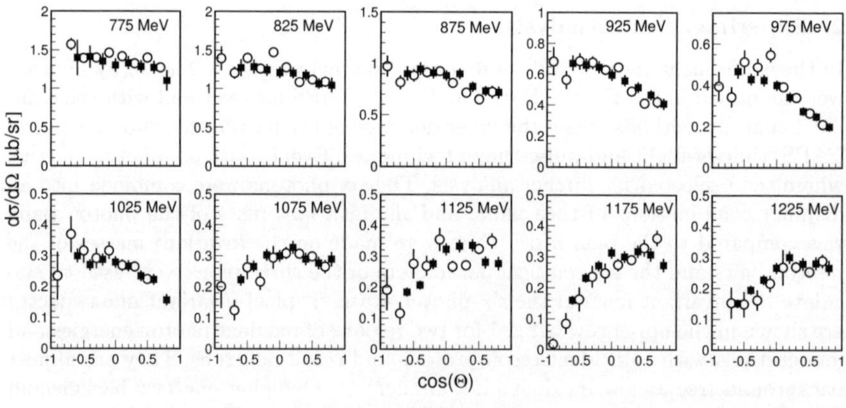

Fig. 2. Angular distributions for η production off the proton. Black squares: free proton [13], open circles: quasifree off protons bound in deuteron (arbitrary scaled)

data are in good agreement, which demonstrates that in this energy range, nuclear effects from the deuteron are only of minor importance. Angular distributions for the reaction off the bound neutron were extracted in an analogous way.

Excitation functions for η-photoproduction off deuterium with coincident detection of recoil protons and neutrons are shown in fig. 3 and compared to the MAID [10] results for free nucleons. The excitation function for the proton agrees nicely with the MAID parameterization and shows a smooth fall-off above the S_{11} region. However, the neutron data show a bump very similar to the GRAAL results [12] at W between 1.6 and 1.7 GeV. The ratio of the cross sections (see fig. 3) shows a behavior similar to the MAID prediction, where the enhancement for the neutron around 1.7 GeV is mainly due to the contribution of the $D_{15}(1675)$ resonance.

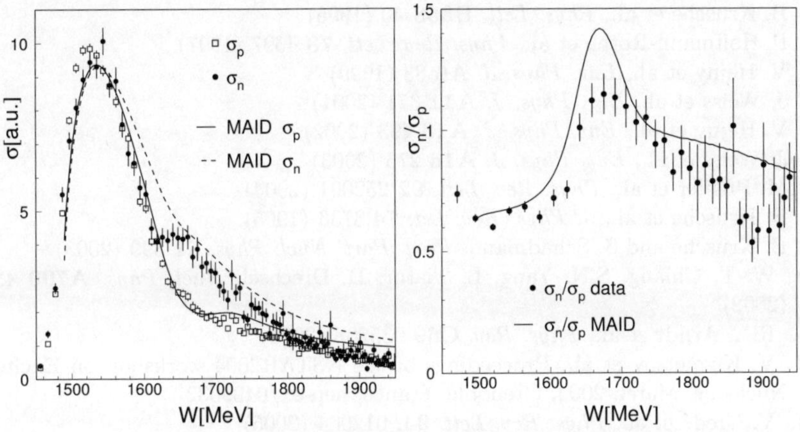

Fig. 3. Lefthand side: Total cross sections for quasifree η-production off protons and neutrons compared to the MAID predictions for free nucleons. All data and and calculations are arbitrary normalized to each other in the maximum. Right hand side: ratio of the quasifree cross sections compared to MAID prediction (all data very preliminary).

4. Conclusions and Outlook

Quasifree photoproduction of η-mesons from nucleons bound in the deuteron has been measured. The quasifree proton data are in good agreement with measurements off the free proton, indicating that nuclear corrections like final state interactions [etc.] are small. The reaction off the neutron shows a bump-like enhancement at incident photon energies around 1 GeV, which might be related to the excitation of the $D_{15}(1675)$ nucleon resonance, however other possibilities cannot yet be excluded. Angular distributions and photon beam asymmetries of this reaction, which are still under analysis, will help to restrict the possible quantum numbers of the resonant state. For the future, a conclusive test of possible distributions of the $D_{15}(1535)$ resonance can be achieved via the measurement of double polarization observables. In particular the oberservables E (circularly pol. beam on longitudinally pol. target) and G (linearly pol. beam on circularly pol. target) are very sensitive to this resonance and can basically map out the real and imaginary part of the corresponding amplitude.

Acknowledgments

The presented data are part of the experimental program of the CBELSA/TAPS collaboration. This work was supported by Schweizerischer Nationalfond and Deutsche Forschungsgemeinschaft.

References

1. B. Krusche et al., *Phys. Lett.* **B358** 40 (1995)
2. P. Hoffmann-Rothe et al., *Phys. Rev. Lett.* **78** 4697 (1997)
3. V. Hejny et al., *Eur. Phys. J.* **A6** 83 (1999)
4. J. Weiss et al., *Eur. Phys. J.* **A11** 371 (2001)
5. V. Hejny et al., *Eur. Phys. J.* **A13** 493 (2002)
6. J. Weiss et al., *Eur. Phys. J.* **A16** 275 (2003)
7. M. Pfeiffer et al., *Phys. Rev. Lett.* **92** 252001 (2004)
8. B. Krusche et al., *J. Phys. Rev. Let.* **74**,3736 (1995).
9. B. Krusche and S. Schadmand, *Prog. Part. Nucl. Phys.* **51** 399 (2003)
10. W.-T. Chiang, S.N. Yang, L. Tiator, D. Drechsel, *Nucl. Phys.* **A700** 429 (2002)
11. R.A. Arndt et al., *Phys. Rev.* **C69** 035208 (2004)
12. V. Kuznetsov et al., Proceedings of the NSTAR2004 workshop on Excited Nucleons, March 2004, Grenoble, France; hep-ex/0409032
13. V. Credé et al., *Phys. Rev. Lett.* **94**, 012004 (2005)

Λ AND Σ PHOTOPRODUCTION ON THE NEUTRON*

P. NADEL-TURONSKI AND B. L. BERMAN

The George Washington University, Washington, DC 20052

E-mail: turonski@jlab.org

The $\gamma n \to K_s^0 Y$ and $\gamma n \to K^+\Sigma^-(1385)$ channels are being analyzed using the CLAS g10 data set. As recent calculations show a large sensitivity of, *e.g.*, the proposed $D_{13}(1900)$ resonance to polarization observables, we hope to extend this study by making a new experiment with polarized real photon beams and an LD_2 target in CLAS, to measure all $\vec{\gamma} n \to K\vec{Y}$ reactions. N^* decays to low-lying KY^* and K^*Y states, as well as $\vec{Y}$-N final state interactions would also be investigated.

1. Introduction

Measuring strange N^* decays is important for learning more about known and missing resonances that do not couple strongly to pions [1,2], and to provide constraints for new coupled-channel analyses [3,4,5] that may resolving the ambiguities of older (isobaric) models [6,7,8]. For the proton, earlier kaon photoproduction data from SAPHIR [9,10] have now been supplemented by results from CLAS [11,12]. There are not yet any published data on the neutron, but two analyses of the $\gamma n \to K^+\Sigma^-$ cross section, using the CLAS g2 data set, are close to publication [13,14]. We are currently investigating, the $K^0\Lambda$, $K^0\Sigma^0$, and $K^+\Sigma^-(1385)$ channels using the significantly larger, but unpolarized g10 data set. The self-analyzing nature of the hyperons allows measurements of the recoil polarization, but access to further observables requires a polarized beam. A letter of intent has thus been submitted for a high-statistics kaon production experiment in CLAS with linearly and circularly polarized real photon beams [15].

Both the current analyses and the proposed experiment are important for resonances that have large neutron helicity amplitudes, and the $K^0\Lambda$ channel in particular is predicted to be sensitive to the much debated missing $D_{13}(1900)$ resonance [16]. The deuteron is also the best system for studies of final state interactions, which have to be taken into account when determining the quasi free processes. The Laget approach [17] has also made it possible to directly observe $\Sigma^-(1385)$ rescattering on the proton [18], and it is expected that the $\vec{Y}$-N interaction will be sensitive to the different predictions of modern hyperon-nucleon

*Supported by DoE grant DE-FG02-95ER40901.

potentials [19].

2. Strange N^* Decays

The high statistics (10 billion triggers) of the g10 experiment has allowed exclusive measurements of the $\gamma n \to K_s^0\Lambda$ and $\gamma n \to K_s^0\Sigma^0$ channels by detecting their charged decay into $\pi^+\pi^-\pi^-p$. The spectator proton can be reconstructed from kinematics. Since the Σ^0 decays to $\Lambda\gamma$, these two channels have to be analyzed simultaneously. Due to Fermi motion, they cannot be separated by the missing mass of the K_s^0 alone, but also require the missing mass of $K^0\Lambda$, as shown in Fig. 1. Quasi free events are selected using the momentum of the spectator protons, which have an almost isotropic angular distribution for $p < 0.3$ GeV/c. In the part of g10 that has a lower main torus field, there are about 30,000 exclusive $K_s^0\Lambda$ and 20,000 $K_s^0\Sigma^0$ events.

Among the N^* decays to excited kaons and hyperons, a prominent role is played by the $\Sigma(1385)$ [1]. As with the elementary reactions, we measure this process on the neutron, using the $\gamma n \to K^+\Sigma^-(1385) \to K^+\pi^-\pi^-p$ channel.

3. Polarization Observables

Exclusive measurements using unpolarized photon beams only allow the determination of the hyperon recoil polarization (P). We hope, however, soon to obtain polarized real photon data that will enable us to extract the polarization transfer observables ($C_{x'}, C_{z'}, O_{x'}, O_{z'}$), as well as the beam ($\Sigma$) and target ($T$) asymmetries. These will add important constraints for theoretical models. Eventually, a polarized target (HD or frozen-spin) may be used to obtain the remaining 8 observables, but a 40 cm long LD_2 target offers a higher rate, lower background, and one avoids the complication of deuteron tensor polarization. As demonstrated by the recent g8b experiment [20], the 6 GeV electron beam at Jefferson Lab offers a significant advantage for the degree of linear photon polarization, especially at $W = \sqrt{s} \sim 1.9$ GeV and above, as well as for the KY^* and K^*Y channels.

Preliminary calculations using the "Chiral-Symmetry-Inspired" model of Walyuo and Bennhold [5] indicate a large sensitivity of polarization observables ($O_{x'}, O_{z'}$) to the missing $D_{13}(1900)$ resonance in the $K^0\Lambda$ channel [15]. There are also known resonances, such as the $D_{15}(1675)$, which have neutron helicity amplitudes that are significantly larger than those for the proton [21]. The same can be true for some of the "missing" ones.

4. Final State Interactions

While initial and final state interactions in the deuteron make the study of quasi free processes on the neutron somewhat more complicated than for the free proton, they offer other opportunities. One topic that has prompted development of new rescattering models [17] is color transparency, which can be experienced by a hadron following a high-t reaction that cause the quarks to assume a point-like configuration with a very short relaxation time. But interesting things occur even

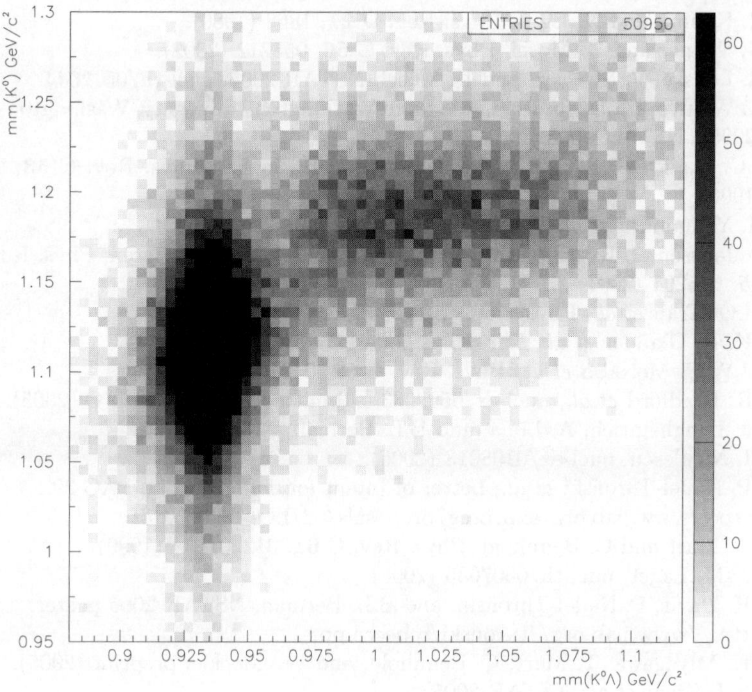

Fig. 1. For the $\gamma n \rightarrow K_s^0 Y$ channels, the Λ (left) is clearly separated from the Σ^0 (right) in a plot with the missing mass of the K_s^0 (hyperon) on the vertical and that of $K_s^0\Lambda$ (spectator) on the horizontal axis. The yields can be extracted from a fit to the projection on the $K_s^0\Lambda$-axis. To select quasi free events, a spectator momentum cut of $p < 0.3$ GeV/c has been applied. A maximum has also been set for the z-axis, saturating the Λ peak.

before the postulated onset of color transparency. The Laget approach has, for instance, recently led to the first direct observation of $\Sigma^-(1385)$ scattering on the proton [18]. Analyses of Λ-N and K-N channels are also in progress. More statistics would, however, be welcome.

Since photoproduced hyperons have a significant polarization, it possible to investigate the $\vec{Y} - N$ interaction in the final state. Recent calculations for circularly polarized photons find that $C_{z'}$ in particular is sensitive to most final state interaction effects, and polarization observables can be used to test the predictions of different modern Y-N interaction potentials [19]. The contributions for various partial waves can also be determined.

References

1. S. Capstick and W. Roberts, Phys. Rev. D **58**, 074011 (1998).
2. R. Koniuk and N. Isgur, Phys. Rev. D **21**, 1868 (1980).
3. G. Penner and U. Mosel, Phys. Rev. C **66**, 055212 (2002).
4. H. Lenske, V. Shklyar, and U. Mosel, NSTAR 2005, nucl-th/0512044.
5. A. Waluyo, Ph.D. thesis, The George Washington University, Washington, DC (2005); NSTAR 2005.
6. J.C. David, C. Fayard, G.H. Lamot, and B. Saghai, Phys. Rev. C **53**, 2613 (1996).
7. H. Yamamura *et al.*, Phys. Rev. C **61**, 014001 (1999).
8. S. Janssen, J. Ryckebusch, D. Debruyne, and T. Van Cauteren, Phys. Rev. C **65**, 015201 (2001).
9. M.Q. Tran *et al.*, Phys. Lett. **B445**, 20 (1998).
10. K.-H. Glander *et al.*, Eur. Phys. J. **A19**, 251 (2004).
11. J.W.C. McNabb *et al.*, Phys. Rev. C **69**, 042201(R) (2004).
12. R. Bradford *et al.*, nucl-ex/0509033, submitted to Phys. Rev. C (2005).
13. J. Langheinrich, A. Lima, and B.L. Berman, NSTAR 2005.
14. I. Niculescu, nucl-ex/0108013 (2005).
15. P. Nadel-Turoński *et al.*, Letter of Intent for Jefferson Lab PAC 29. http://www.jlab.org/exp_prog/proposals/06/LOI-06-005.pdf
16. T. Mart and C. Bennhold, Phys. Rev. C **61**, 012201(R) (1999)
17. J.-M. Laget, nucl-th/0507035 (2005).
18. R. Davis, P. Nadel-Turoński, and B.L. Berman, NSTAR 2005 poster. http://www.jlab.org/~turonski/rebecca.ppt
19. K. Miyagawa, T. Mart, C. Bennhold, and W. Glöckle, preprint (2005).
20. F.J. Klein *et al.*, NSTAR 2005.
21. S. Capstick, Phys. Rev. D **46**, 2864 (1992).

THE PROBLEM OF EXOTIC STATES: VIEW FROM COMPLEX ANGULAR MOMENTA

YA. I. AZIMOV

Petersburg Nuclear Physics Institute,
Gatchina, Leningrad area,
188300, Russia
E-mail: azimov@thd.pnpi.spb.ru

Having in mind present uncertainty of the experimental situation in respect to exotic hadrons, it is important to discuss any possible theoretical arguments, pro and contra. Up to now, there are no theoretical ideas which could forbid existence of the exotic states. Theoretical proofs for their existence are also absent. However, there are some indirect arguments for the latter case. It will be shown here, by using the complex angular momenta approach, that the standard assumptions of analyticity and unitarity for hadronic amplitudes lead to a non-trivial conclusion: the S-matrix has infinitely many poles in the energy plane (accounting for all its Riemann sheets). This is true for any arbitrary quantum numbers of the poles, exotic or non-exotic. Whether some of the poles may provide physical (stable or resonance) states, should be determined by some more detailed dynamics.

1. Exotic hadrons: brief overview

The problem of exotic states of hadrons has long history, nearly as long as quark themselves. After several unsuccessful experimental searches, it was formulated most clearly by Lipkin[1], as a problem for theorists: "Why are there no strongly bound exotic states ..., like those of two quarks and two antiquarks or four quarks and one antiquark?" This question could be eliminated by experimental observations of the first exotic baryon Θ^+, just consisting of at least four quarks (two u's and two d's) and one antiquark ($\bar{s}$). However, other experimental publications, with null results, cast doubts on its existence (recent history and present experimental status see in the talk [2]). Therefore, Lipkin's question may be still burning, having no answer. It seems reasonable in such a situation to analyze in detail all arguments, pro and contra exotic hadrons. Here we briefly discuss some of them.

The current experimental situation is rather uncertain and not decisive yet. It was summarized at this conference[2] by the words: "The narrow Θ^+ pentaquark is not in good health, but it is too early to pronounce it dead".

What about theoretical situation, the general postulates of Quantum Field Theory (QFT) do not provide any way to discriminate between exotic and non-exotic particles. Its particular case, Quantum Chromodynamics (QCD), believed to underlie the strong interactions and hadron spectroscopy, also cannot forbid exotic hadrons *vs.* non-exotic ones (at least, at the present level of understanding the structure and properties of QCD).

Moreover, in the framework of QCD, any hadron should be seen in some conditions (*e.g.*, for short time intervals) as a multi-quark system. Experiments on hard processes confirm this. But it is then difficult to understand why such multi-quark systems must be bound in all cases to have quantum numbers of a 3-quark (or quark-antiquark) system.

Attempts to calculate hadronic spectra in various approaches, which are assumed to be based on QCD (bag model, soliton-like models, sum rules, lattice calculations, and so on), as a rule, also demonstrate exotic states (though with their properties strongly dependent on the model used).

Rather unexpectedly, the method of complex angular momenta (CAM), usually related only to high-energy asymptotics of hadronic amplitudes, has also something to say in the problem of exotic states. It may suggest a new (indirect) argument for existence of exotic hadrons. This argument has been recently published [3]. Here it will be explained in more detail.

2. CAM and exotics

Let us begin with some necessary preliminaries, partly forgotten now. For simplicity, at first we neglect particle spins.

Consider a process 2-hadrons-into-2-hadrons. Its amplitude A is a function of two independent variables, for which one may choose the c.m. energy W and c.m. scattering angle θ. Another possibility is to use invariant variables. There are three of them (Mandelstam variables): one is the c.m. energy squared $s = W^2$; two others are squares of two c.m. momentum transfers t and u, between a particular initial hadron and one or the other final hadron. Usually, t is taken as proportional to $z = \cos\theta$, with u proportional to $-z$. Evidently, they are not independent. Moreover, the sum $s+t+u$ equals just to the sum of the squared masses for two initial and two final particles (being thus independent of both W and z).

The amplitude $A(s, z)$, as function of z, may be decomposed into partial waves. Every partial-wave amplitude $f_l(s)$ corresponds to a definite value l of the orbital momentum. For the physical amplitudes, l may only be equal to a non-negative integer number. In the case of purely elastic scattering, the physical amplitudes $f_l(s)$ satisfy the elastic unitarity condition

$$f_l(s) - f_l^*(s) = 2ik\, f_l(s)\, f_l^*(s)\,, \tag{1}$$

where k is the c.m. momentum.

Now we make two assumptions:

1) Amplitudes $f_l(s)$ admit unambiguous analytical continuation to non-integer, and even complex, values of l.

2) There are no massless hadrons and no massless hadron exchanges.

The first assumption is not trivial. Every function, defined on a set of discreet points, can be analytically continued, but the continuation is, generally, very ambiguous. Only in some cases there exists a preferred continuation, which may be clearly separated from all others. This is fulfilled, *e.g.*, if the amplitude $A(s,z)$ satisfies dispersion relations (DR) in momentum transfers t, u; the arising continuation is described by the integral Gribov-Froissart (GF) formula[4] (for more details see the monograph [5]). Such DR have never been formally proved, neither in general QFT, nor in QCD. Nevertheless, they (and their analogs) are widely and actively used in phenomenology of strong interactions. Up to now, they have not encountered any inconsistency. Note that the DR provide a sufficient condition for the unambiguous continuation; necessary conditions are essentially weaker.

The second assumption ensures a finite range of interactions, and also the threshold behavior $\sim k^{2l}$ for the elastic scattering amplitudes $f_l(s)$ with physical values of l, when $k \to 0$. The GF formula, where it is applicable, provides the same threshold behavior for the continued amplitudes $f_l(s)$. But the elastic unitarity condition for the continued amplitudes has somewhat modified form

$$f_l(s) - f_{l^*}^*(s) = 2ik\, f_l(s)\, f_{l^*}^*(s)\,, \tag{2}$$

which coincides with eq.(1) only at real l.

It is easy to see now that the continued unitarity relation (2) is not always consistent with the threshold behavior $f_l(s) \sim k^{2l}$. Indeed, near threshold, each of the left-hand side terms is $\sim k^{2\mathrm{Re}\, l}$, while the right-hand side is $\sim k^{4\mathrm{Re}\, l+1}$. Since the left-hand side terms may subtract each other, but not enhance, the two sides may be consistent with each other only at $\mathrm{Re}\, l > -1/2$. This problem was first discovered by Gribov and Pomeranchuk[6]. To solve it, they studied in more detail the small-k^2 region and showed that there are Regge poles, which condense near threshold to the point $l = -1/2$ and, thus, invalidate the k^{2l}-behavior of $f_l(s)$ for $\mathrm{Re}\, l < -1/2$. Near the threshold, these poles have trajectories

$$l_n(s) \approx -\frac{1}{2} + \frac{i\pi n}{\ln(R\sqrt{|k^2|})}\,, \tag{3}$$

with R being the effective interaction radius. The number n takes any positive and/or negative integer values, $n = \pm 1, \pm 2, \ldots$. Hence, there are infinitely many reggeons condensing to $l = -1/2$ at $k^2 \to 0$.

Till now, we have neglected particle spins. Taking them into account changes the orbital momentum l by the total angular momentum j. The reggeon condensation still exists at a two-particle threshold, though with the shifted limiting point [7]. For the spins σ_1 and σ_2 it is

$$j = -1/2 + \sigma_1 + \sigma_2\,, \tag{4}$$

instead of $j = -1/2$, without spins. The reason is simple: the condensation point, as before, corresponds to $l = -1/2$, and the highest value of j at fixed l equals $l + \sigma_1 + \sigma_2$. The movement of the condensing reggeons in the j-plane, near a threshold energy for two spinning particles, is also described by trajectories (3), but with the shifted limiting point (4).

The main conclusion of the above consideration is that the unitarity and the possibility of unambiguous analytical continuation in j (standard analytical properties, in particular), taken together, imply existence of the Gribov-Pomeranchuk (GP) threshold condensations of reggeons. They collect infinite number of reggeons and, therefore, imply that the total number of reggeons is always infinite.

The reggeon positions depend on energy and are determined by a relation of the form $F(s, j) = 0$. Each of its solution corresponds to an amplitude pole, which may be considered either as a pole in j, with position depending on the energy (on s), or as a pole in the energy (in s), with position depending on the total angular momentum j. This provides one-to-one correspondence between reggeons and spin-dependent poles in the energy plane. Therefore, the infinite number of reggeons corresponds to the infinite number of poles in the energy plane.

Note that no assumptions about quantum numbers of the poles have been used. Hence, the hadronic S-matrix should have infinite number of poles with any quantum numbers, exotic in particular.

The structure and properties of the reggeon condensations may be studied explicitly in the non-relativistic quantum mechanics with a final-range potential. It also generates the threshold behavior $\sim k^{2l}$ and satisfies elastic unitarity. Detailed investigation [8] for the Yukawa potential $V(r) = g\ \exp(-\mu r)/r$ confirms the general character of the GP condensations. It also shows, that poles related to bound states (or resonances) are "initially" members of the set of GP condensing poles, and "evaporate" from it, when attraction increases. The energy plane for the Yukawa potential has many Riemann sheets, and most of the energy-plane poles are "hidden" on remote sheets, while the poles related to bound or resonance states approach to the physical region. The infinite number of the Yukawa energy-plane poles can be "visualized" by taking the limit $\mu \to 0$, which transforms the Yukawa potential into Coulomb one and demonstrates the well-known accumulation of the infinite number of Coulomb bound states near the threshold (note that the double limiting transition $\mu \to 0$, $k \to 0$ is not equivalent here to the similar, but reversed limit $k \to 0$, $\mu \to 0$).

Existence of energy-plane poles with exotic quantum numbers is the necessary condition for existence of exotic hadrons. It appears satisfied under the familiar assumptions of unitarity and analyticity. But only more detailed dynamics may determine whether some of the poles emerge near the physical region, to provide indeed the stable or resonance exotic states.

Two more "technical" notes may be interesting:

1) Reggeons have been used above in a non-standard manner. Usually, to apply the CAM approach for obtaining results in the s-channel (the channel where the invariant s has the meaning of the squared c.m. energy), one begins from the crossed channel, where t and s are, respectively, the squared energy and squared momentum transfer (see Ref. [5]). Analytical continuation of partial-wave amplitudes in this t-channel allows, after returning into the s-channel, to study behavior of the invariant amplitude (and cross section) at the high energy s and at a fixed value of the momentum transfer t. To obtain the conclusion about energy-plane poles in the s-channel, we have used complex angular momenta in the same s-channel.

2) Gribov and Pomeranchuk suggested one more situation where infinitely many reggeons might exist near a fixed point in the l-plane, this time $l = -1$ [9]. However, accounting for existence of the moving branch points prevents

this reggeon accumulation from emerging. The threshold GP condensations of reggeons are not influenced by such branch points.

3. Summary and Conclusion

The presented results may be summarized as follows:

- Under the familiar assumptions of unitarity and analyticity, hadronic amplitudes have infinite number of energy-plane poles with any quantum numbers, both exotic and non-exotic. Thus, there is an infinite "reservoir" of poles, which satisfies the necessary condition for existence of exotics. Most of the poles are "hidden" on remote Riemann sheets of the energy plane.
- Real existence (or absence) of exotic hadrons, *i.e.*, of the S-matrix poles sufficiently near the physical region, may be guaranteed only by more detailed dynamics.

Meanwhile, the old wisdom[10], that "...either these states will be found by experimentalists or our confined, quark-gluon theory of hadrons is as yet lacking in some fundamental, dynamical ingredient...", is still alive.

Acknowledgments

I thank R.A. Arndt, K. Goeke, I.I. Strakovsky, and R.L. Workman, my coauthors in paper [3], for stimulating discussions. This talk is based on the work partly supported by the U.S. Department of Energy Grant DE–FG02–99ER41110, by the Jefferson Laboratory, by the Southeastern Universities Research Association under DOE Contract DE–AC05–84ER40150, by the Russian-German Collaboration (DFG, RFBR), by the COSY-Juelich-project, and by the Russian State grant RSGSS-1124.2003.2. My participation in NSTAR2005 was supported by the Organizing Committee, by the Jefferson Laboratory, and by the George Washington University. I thank Prof. Goeke for hospitality in the Theoretical Physics Institute II of the Ruhr-University Bochum during writing this text.

References

1. H. J. Lipkin, *Phys. Lett.* **45B**, 267 (1973).
2. V. Burkert, plenary talk at this conference.
3. Ya. I. Azimov, R. A. Arndt, I. I. Strakovsky, R. L. Workman, and K. Goeke, *Eur. Phys. J.* **A26**, 79 (2005); hep-ph/0504022.
4. V. N. Gribov, *Sov. Phys. JETP* **14**, 1395 (1962);
   M. Froissart, *Phys. Rev.* **123**, 1053 (1961).
5. P. D. B. Collins, *An Introduction to Regge Theory and High Energy Physics* (Cambridge Univ. Press, 1977)
6. V. N. Gribov and I. Ya. Pomeranchuk, *Phys. Rev. Lett.* **9**, 238 (1962).
7. Ya. I. Azimov, *Phys. Lett.* **3**, 195 (1963).
8. Ya. I. Azimov, A. A. Anselm, and V. M. Shekhter, *Sov. Phys. JETP* **17**, 246, 726 (1963).

9. V. N. Gribov and I. Ya. Pomeranchuk, *Phys. Lett.* **2**, 239 (1962).
10. R. L. Jaffe and K. Johnson, *Phys. Lett.* **60B**, 201 (1976).

SEARCH FOR Θ^+ AT CLAS IN $\gamma n \rightarrow \Theta^+ K^-$.

N. A. BALTZELL AND D. J. TEDESCHI*

Department of Physics and Astronomy
University of South Carolina
712 Main Street,
Columbia, SC 29208, USA
E-mail: baltzell@sc.edu

The existence of pentaquarks is being studied in recent experiments at Jefferson Lab. This analysis investigates the reaction $\gamma d \rightarrow \Theta^+ K^-(p)$ with the Θ^+ decaying to pK^0. Produced with a tagged photon beam of endpoint energy 3.6 GeV incident on a 24 cm liquid deuterium target, the $pK^0_s K^-(p)$ final state is measured exclusively. With well defined strangeness and no neutral meson background, this channel is an important place to look for the Θ^+. However, it contains large contributions from hyperons produced via $\gamma n \rightarrow Y^* K^0$, and the effects of mesons are also present in the $K^0 K^-$ system. The current focus is on understanding these backgrounds.

1. Introduction

The classification of nuclear states into baryons and mesons has been in place since the 1960's with the introduction of colored quarks. To this day, the spectra of almost all known hadrons can be described by such a constituent quark model. However, this naive model does not accommodate all the states predicted by QCD. Accordingly, searches for exotic states whose quantum numbers cannot be formed by color singlet three-quark and quark-antiquark states remains a method of testing our understanding.

The purported Θ^+ is an example of such an exotic state. With baryon number and strangeness of +1, its minimal quark content is $uudd\bar{s}$. After being predicted by the chiral soliton and diquark models, [1,2] the LEPS collaboration reported a positive experimental signal in 2003. [3] Since then, numerous searches for the Θ^+ have been performed, but the world's results to date are inconclusive. The highest statistics samples tend to yield negative results, but relating them to the positive sightings is difficult due to varying production mechanisms and energy regimes. Considering that not all groups report a cross section (or an upper limit on one), comparisons become even more difficult.

*This work supported by grant 0244982 of the National Science Foundation.

In 2004, the CLAS collaboration performed a photoproduction experiment on a deuterium target at Jefferson Lab to test the results in the $\Theta^+ \to nK^+$ channel from a previous data set. [4] The new, large statistics allow investigation of the same channel with a different decay mode, $\gamma n \to \Theta^+ K^- \to pK^0K^-$. This channel is advantageous in that no final state interaction is required, the strangeness of the pK^0 system is tagged, and the final state can be fully reconstructed.

2. The Experiment and Event Selection

The G10 experiment ran for two months in the Spring of 2004, accumulating almost 5 pb^{-1} of photoproduction data on a 24 cm liquid deuterium target. The CLAS detector is used to reconstruct multi-particle final states, and the photon tagging system measures the beam photon's energy, ranging from 0.8 to 3.7 GeV.

After standard reconstruction and particle identification algorithms, the detection of four final state particle is required, $p\pi^+\pi^-K^-$. Two undetected particles must be identified, the K_s^0 and a missing proton, so the signal regions of their effective mass distributions are selected for further analysis (see Fig. 1). It is also required that the momentum of the missing proton be consistent with that of a spectator nucleon inside the deuterium nucleus; a cut of $P_{miss} < 100$ MeV/c is used. Additionally, the reconstructed primary vertex must fall within the target cell.

To reduce background and conserve four-momentum, the reaction is kinematically fitted with two constraints, the masses of the K_s^0 and proton. Particle misidentification is considered by artificially switching a particle's mass and kinematically fitting this new system. The confidence levels from the fits are used to discriminate between real and fake $\gamma d \to pK_s^0K^-(p)$ events. After event selection, 22K events are kept with a S/B ratio of less than 10%.

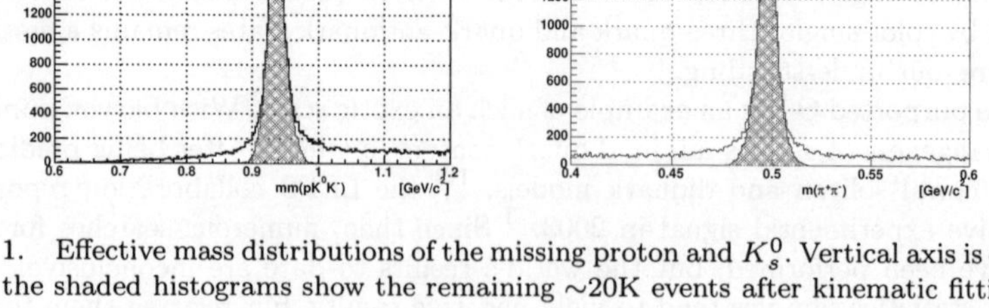

Fig. 1. Effective mass distributions of the missing proton and K_s^0. Vertical axis is counts, and the shaded histograms show the remaining ~20K events after kinematic fitting.

3. Background Processes

Identifying background processes and understanding their effects is important when searching for a rare, unconfirmed resonance. It has been suggested that kinematic reflections are the cause of reported Θ^+ signals. [5]

Negative strangeness baryons decaying to pK^- contribute significantly to background processes in this channel. About 2000 $\gamma n \rightarrow K_s^0 \Lambda(1520)$ events are seen. Being a narrow resonance and close to threshold, the $\Lambda(1520)$ can be removed with a cut on the invariant mass of the pK^- system. However, numerous four-star Λ and Σ states exist in the 1670-1820 MeV/c^2 mass region in this sample. With their large resonant widths, they cannot be efficiently removed.

Charged mesons decaying to $K^0 K^-$ also contribute, but none are narrow states. The shape of their effective mass distribution is highly dependent on energy. For example, the peak of the $a_2(1320)$ resonance is visible at large photon energies, and contribution from a $\rho(980)$ is suspected at low energy.

4. Simulation

A Monte Carlo model of the hyperon and meson resonances is being developed for a good understanding of efficiencies. Relativistic, complex Breit-Wigner amplitudes are assumed for each resonance, with relative weights dependent on center of mass energy and momentum transfer. Decay angular distributions are characterized by spherical harmonics whose quantum numbers are chosen based on the data when possible.

$$\mathit{Amplitude} = C(s) \cdot e^{-\alpha t + i\beta} \cdot \frac{m_o \Gamma(m)}{(m_0^2 - m^2) - i m_0 \Gamma(m)} \cdot Y_m^l(\theta, \phi)$$

Initial simulations were performed using 3-body phase space events smeared according to deuterium's Fermi momentum distribution. The resulting kinematic spectra do not sufficiently describe the data, but do provide a background shape with which to estimate the contributions from resonant processes. A few hyperons are chosen and relativistic Breit-Wigner shapes are fitted to the data on top of the phase space background in Fig. 2. Currently in the event generator, amplitudes are added coherently and fit parameters are tuned from the data using initial phase space fits, but the resulting kinematic spectra show good agreement with data without using a phase space background. [6]

5. Outlook

Hyperon and meson resonant amplitudes are being used to perform a maximum likelihood fitting technique to systematically model the data. With a good understanding of the backgrounds, accurate detector efficiencies can be extracted from simulation and confidence placed upon the results of the $\Theta^+ \rightarrow pK_s^0$ search.

References

1. D. Diakonov, V. Petrov, and M. Polyakov, *Z. Phys. A* **359**, 305 (1997).
2. M. Karliner and H. J. Lipkin, *Phys. Lett. B* **575**, 249 (2003);
   R. L. Jaffe and F. Wilczek, *Phys. Rev. Lett.* **91**, 232003 (2003).

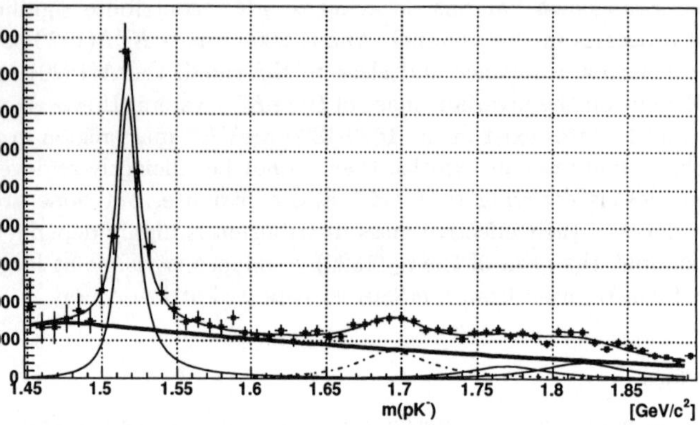

Fig. 2. The normalized pK^- invariant mass spectrum showing contributions from hyperon resonances above phase space background.

3. T. Nakano *et al.*, *Nucl. Phys. A* **721**, C112 (2003)
4. S. Stepanyan, *et al.*, *Phys. Rev. Lett.* **91**, 252001 (2003).
5. A. Dzerbia *et al.*, *Phys. Rev. D* **69**, 051901 (2004).
6. D. Tedeschi, Θ^+ *Search in CLAS from* $\gamma d \to pK^0K^-(p)$, http://conferences.jlab.org/pentaquark/talks/Tedeschi.pdf (2005).

$\Sigma^+(1189)$ PHOTOPRODUCTION OFF THE PROTON

M. NANOVA FOR CBELSA/TAPS COLLABORATION*

II. Physikalisches Institut
Justus Liebig Universitaet Giessen,
35392 Giessen, Germany
E-mail: M.Nanova@physik.uni-giessen.de

The exclusive measurement of the reactions $\gamma p \rightarrow K^{*0}\Sigma^+(1189)$ and $\gamma p \rightarrow K^0\pi^0\Sigma^+(1189)$ leading to the p $4\pi^0$ final state is presented. It is the first attempt to study nucleon resonance excitations in these channels. The experiment has been performed at the tagged photon facility of the ELSA accelerator (Bonn) in the beam energy range from threshold up to 2.6 GeV. The Crystal Barrel and the photon spectrometer TAPS were combined to a detector system, which provides an almost 4π coverage of the geometrical solid angle and which is very well suited to examine photoproduction with multi-photon final states. In addition the reaction $\gamma p \rightarrow p\pi^0\eta \rightarrow p4\pi^0$ has been studied. First results are presented and discussed.

1. Introduction

Photon-induced reactions on nucleons are very well suited to investigate baryonic resonances. Such experiments provide important information on many open questions in baryon structure. Model calculations predict more resonant states than have been oserved so far. Some of the so called *missing* resonances are predicted to decay not only into meson-nucleon but as well into meson-hyperon final states [1]. First theoretical predictions based on the quark model for K^* photoproduction suggested to study nucleon resonance excitations in the channels $\gamma p \rightarrow K^{*0}\Sigma^+$ and $\gamma p \rightarrow K^{*+}\Sigma^0$ [2]. The assumption of t-channel kaon exchange in this model showed a significant effect for K^{*0} photoproduction and strong forward peaking of the vector meson at higher energies (> 2 GeV).
The Σ^+ photoproduction has been studied in another channel $\gamma p \rightarrow \pi^0 K^0\Sigma^+$ in Ref. 3 using a chiral unitary approach for meson-baryon scattering in the energy range close to threshold. The $N^*(1535)$ resonance is dynamically generated and was cleary seen in the ηp invariant mass distribution by studying the $\gamma p \rightarrow p\pi^0\eta$ reaction. The same dynamical model was applied to calculate the cross section of $\gamma p \rightarrow \pi^0 K^0\Sigma^+$. [3]

*Work is supported by DFG (SFB/TR-16)

2. The Crystal Barrel and TAPS experiment at ELSA

Data were taken with the detector system Crystal Barrel [4] and TAPS [5,6] at the 3.3 GeV electron stretcher facility ELSA [7]. The setup is shown schematically in Figure 1. The extracted electrons from ELSA with energies E_0 hit a primary

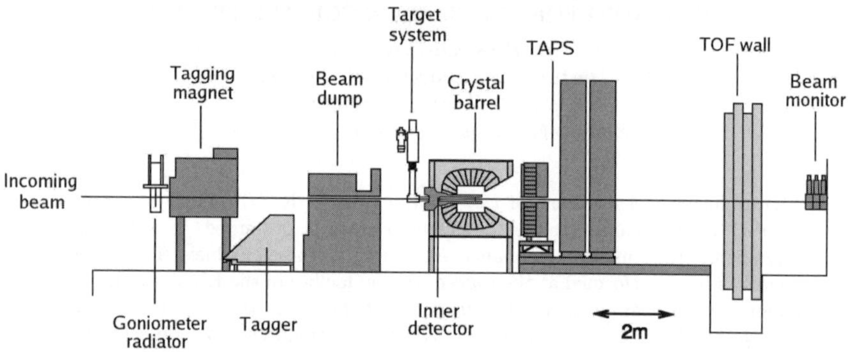

Fig. 1. Schematic overview of the beam line of the CBELSA/TAPS experiment.

radiation target and produce bremsstrahlung. The corresponding energy of the photons ($E_\gamma = E_0 - E_e^-$) is determined in a tagging system by the deflection of the scattered electrons in a magnetic field. Thereby a tagged beam is provided in the photon energy range from 0.8 GeV up to 3.1 GeV for an incoming electron energy of 3.3 GeV. The Crystal Barrel detector (CB), a photon calorimeter consisting of 1290 $CsI(Tl)$ crystals ($\approx$16 radiation lengths X_0), covers polar angles from 30^o to 160^o and the complete azimuthal angle range. The photoproduction target in the center of the CB (5x2.5 cm) is filled with liquid hydrogen and is surrounded by a scintillating fibre detector built to identify charged particles inside the CB. The forward angle (6^o- 30^o) region is covered by the TAPS calorimeter which consists of 528 hexagonal BaF_2 crystals ($\approx$12 X_0). In front of each BaF_2 module a 5 mm thick plastic scintillator is mounted for the identification of charged particle. The resulting combined system covers 99% of 4π geometrical solid angle and is very well suited to examine photoproduction with multi-photon final states.

3. Data Analysis

Because the reaction channel of interest has 8 photons and a proton in the final state, only those events have been selected which contain exactly 9 clusters. The 8 photons have been combined to reconstruct the 4 pions. The correct combination has been selected using a χ^2 test. It is required that the missing mass for the reaction $\gamma p \to 4\pi^0 p_x$ (proton is not used although measured) gives the proton mass. In addition the coplanarity of the $4\pi^0$ system and the proton is required. As the largest contribution to the events containing $4\pi^0$'s are from $\pi^0\eta$ - photoproduction, the next step is to remove these events from the data sample. Subsequently, the remaining $4\pi^0$ and the proton are combined to a $K^{*0}\Sigma^+$ pair, which can be done in four different combinations. The invariant mass spectrum is shown in Figure 2 for energies higher than 1840 MeV, which is the threshold for the reaction $\gamma p \to K^{*0}\Sigma^+$. After the applied cuts a considerable background is still left as can be seen from Figure 2(a). Monte Carlo studies have been done

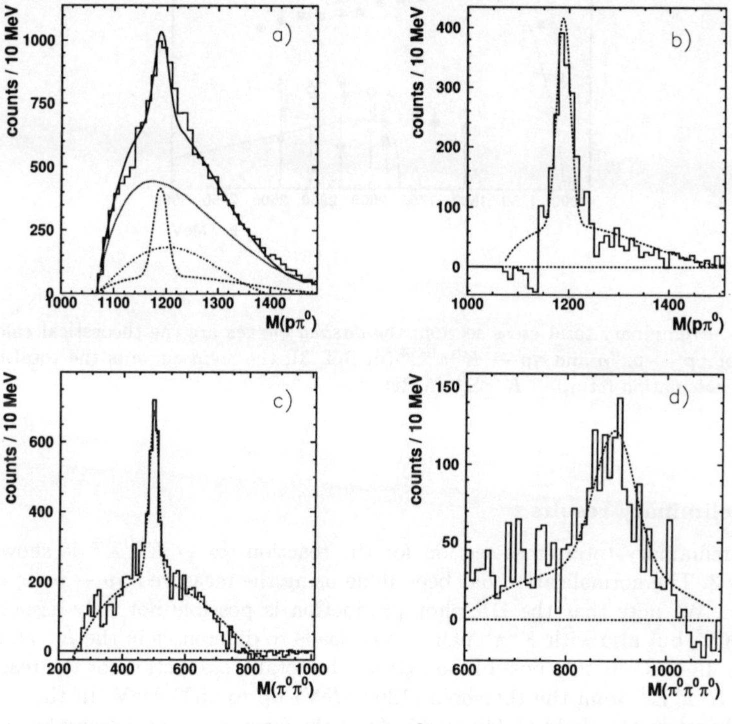

Fig. 2. (a)The invariant mass spectrum M(pπ^0): *dotted curve* is the simulated background p$4\pi^0$, *dash-dotted curve* is the simulated $\Delta 3\pi^0$ and the *dashed curve* is the simulated signal. Peak and combinatorial background after background subtraction (b) Σ^+ , (c) K^0 and (d) K^{*0} with the simulated signal respectively (*dashed curve*).

to understand the background. We assume that the main contribution to the background is from other channels such as $\gamma p \to p4\pi^0$ and $\gamma p \to \Delta 3\pi^0$ with the same final state. The total spectrum is fitted by a combination of the simulated background and a signal. After subtracting the background Σ^+,K^{*0} and K^0 signals are extracted. The agreement with the simulated signal and combinatorial background is rather good as can be seen in Figure 2.

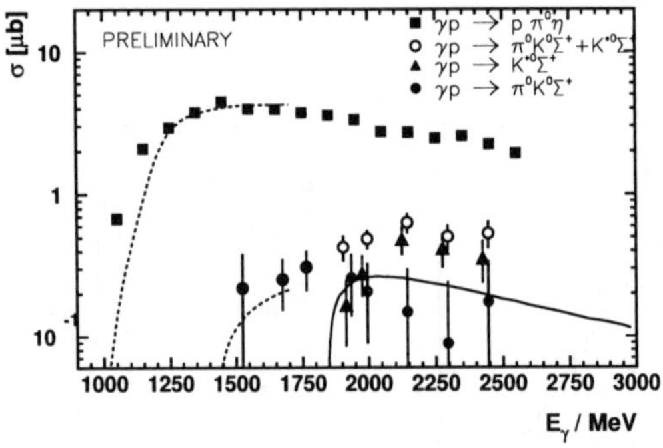

Fig. 3. Preliminary total cross section: the *dashed curves* are the theoretical calculations for $\gamma p \to p\pi^0\eta$ and $\gamma p \to K^0\pi^0\Sigma^+$(in Ref. 3), the *solid* curve is the total cross section calculation for $\gamma p \to K^{*0}\Sigma^+$ in Ref. 2 .

4. Preliminary results

The preliminary total cross section for the reaction $\gamma p \to K^{*0}\Sigma^+$ is shown in Figure 3. The normalization has been done using the measured $\gamma p \to p\pi^0\eta$ cross section. We note that the Σ^+ photoproduction is possible not only associated with K^{*0}, but also with $K^0\pi^0$ pairs which leads to differences in the K^* and Σ^+ counts. In this way it is possible to extract the total cross section for the reaction $\gamma p \to K^0\pi^0\Sigma^+$ from the threshold (1300 MeV) up to 2600 MeV. In the energy bins closest to threshold (1300 - 1500 MeV) the form of the background causes a problem and within the available statistics no signal can be extracted. The differential cross sections for both reactions have been derived, but the present accuracy does not allow to establish significant deviations from isotropy as predicted in the Ref. 2.

References

1. S. Capstick and W. Roberts, *Phys. Rev.* **D 57**, 4301 (1998).
2. Q. Zhao, J. S. Al-Khalili, C. Bennhold, *Phys. Rev.* **C 64**, 052201 (2001)

3. M. Doering, E. Oset, D. Strottman, *arXiv: nucl-th/ 0510015*
4. E. Aker et al., *Nucl. Instr. and Methods* **A 321**, 69 (1992).
5. R. Novotny et al., *IEEE Trans. Nucl. Sci.* **38**, 392 (1991).
6. A.R. Gabler et al., *Nucl. Instr. and Meth.* **A 346**, 168 (1994).
7. D. Husmann, W. J. Schwille, *Phys. Bl.* **44**, 40 (1988).

PHOTO-EXCITATION OF HYPERONS AND EXOTIC BARYONS IN $\gamma N \to K\overline{K}N$ *

Y. OH AND K. NAKAYAMA

Department of Physics, University of Georgia, Athens, Georgia 30602, U.S.A.

T.-S. H. LEE

Physics Division, Argonne National Laboratory, Argonne, Illinois 60439, U.S.A.

We investigate the reaction of $\gamma N \to K\overline{K}N$ focusing on the photoproduction of $\Lambda(1520)$ and putative exotic $\Theta(1540)$. We consider various background production mechanisms including the production of vector mesons, tensor mesons, and other Λ and Σ hyperons. We discuss the angular distribution of $\Lambda(1520)$ photoproduction cross section and the radiative decays of $\Lambda(1520)$. We also discuss what we expect for the invariant mass distributions if the $\Theta(1540)$ is formed in the reaction, with the parameters studied so far. We find that the peak in the KN invariant mass distribution, if confirmed, can hardly come from the kinematic reflections, especially, due to the tensor meson backgrounds.

Recent experimental activities searching for exotic pentaquark states[1,2] renewed the interests in double kaon photoproduction, i.e., $\gamma N \to K\overline{K}N$, motivating recent theoretical investigations on its production mechanisms[3–6]. Investigating various physical quantities of this reaction allows the study of hadron resonances that are formed during the reaction. For example, from its $K\overline{K}$ channel, we can study the photoproduction of mesons, which decay mostly into $K\overline{K}$, and the $\overline{K}N$ channel is directly related to the $S = -1$ hyperon production, where S is the strangeness. Furthermore, the KN channel of this reaction gives a tool of searching for the hypothetical $S = +1$ exotic baryons. In fact, this is the reaction used by LEPS Collaboration[1] to observe the exotic Θ, which caused a lot of theoretical and experimental investigations[6]. However, the signals for $\Theta(1540)$ could not be confirmed by other experiments, which lead to a doubt on its existence.

Another motivation of this work is the radiative decays of hyperon resonances. As stated above, double kaon photoproduction is used to study the production of hyperon resonances. Close inspection of the kinematic region of this reaction at low energies shows that the ϕ meson production is dominant in the $K\overline{K}$ channel and $\Lambda(1520)$ is in the $\overline{K}N$ channel. Therefore, this reaction may be used for study-

*Work supported by Forschungszentrum-Jülich, contract No. 41445282 (COSY-058) and U.S. DOE Nuclear Physics Division Contract No. W-31-109-ENG-38.

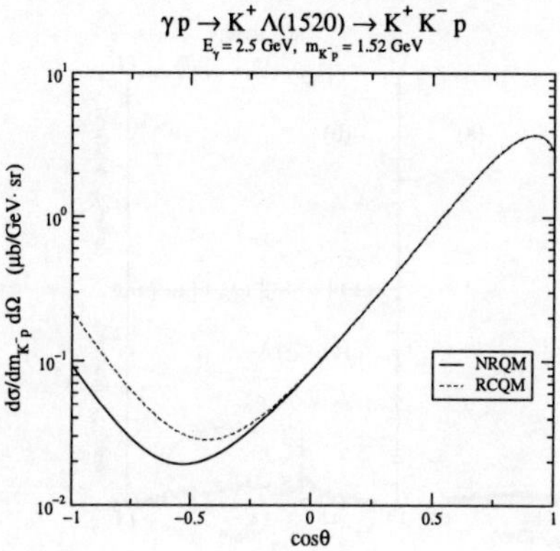

Fig. 1. Differential cross section for $\gamma p \to K^+\Lambda(1520) \to K^+K^-p$. The solid and dashed lines are obtained by using the nonrelativistic quark model[11] and relativistic quark model[12] predictions for the $\Lambda(1520) \to \Sigma\gamma$ decay.

ing the production mechanisms of $\Lambda(1520)$, which contain the radiative decays of $\Lambda(1520)$. Theoretical predictions on the radiative decays of $\Lambda(1520)$ are strongly model-dependent and their experimental data are very limited and uncertain[7]. There are recent measurements for $\Gamma(\Lambda(1520) \to \Lambda\gamma)$ by SPHINX Collaboration and CLAS Collaboration[8,9], of which results are consistent to each other. But the decay width of $\Lambda(1520) \to \Sigma\gamma$ has not been reported so far. (Note that the decay width of $\Lambda(1520) \to \Sigma\gamma$ in PDG[7] is not from a direct measurement, but from that of $\Lambda(1520) \to \Lambda\gamma$ by using some SU(3) relations.)

In this work, we consider the photoproduction of $\Lambda(1520)$ and that of $\Theta(1540)$. As background production mechanisms of the latter, we consider vector meson production, tensor meson production, and the t-channel (tree) Drell diagrams, as well as the production of other Σ and Λ hyperons. We also include the form factors and constrain the cutoff parameter by the total cross section data for $\gamma p \to K^+K^-p$. The details on the diagrams, effective Lagrangians used in this calculation, and the method to restore the charge conservation condition can be found in Ref. 6. The models for vector meson production can be found in Ref. 10.

Shown in Fig. 1 are the results for the double differential cross sections for $\gamma p \to K^+\Lambda(1520) \to K^+K^-p$, where θ is the angle of K^+ in the c.m. frame. In this calculation we use the value of CLAS Collaboration for the $\Lambda(1520) \to \Lambda\gamma$ decay[9], $\approx$ 167 keV. For the decay of $\Lambda(1520)$ into $\Sigma\gamma$ we take two model predictions: the nonrelativistic quark model prediction of Ref. 11, $\approx$ 55 keV, and that of the relativistic quark model of Ref. 12, $\approx$ 293 keV. The result shows that the double differential cross section at large scattering angles depends on the decay width of $\Lambda(1520) \to \Sigma\gamma$.

Next, we consider the production of exotic $\Theta(1540)$. For this calculation, we assume that the Θ belongs to $\overline{\mathbf{10}}$, and it has $J^P = \frac{1}{2}^+$ and a decay width

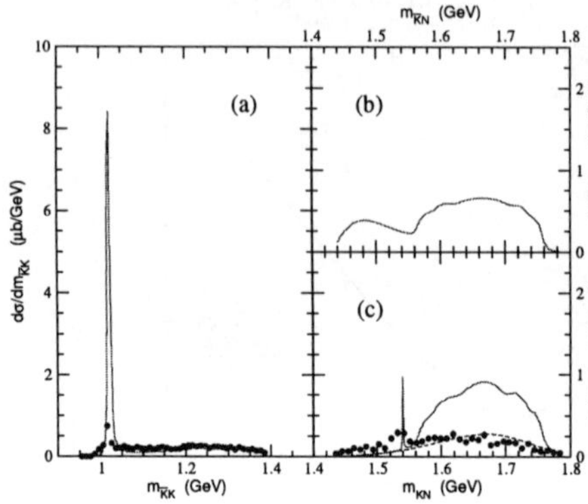

Fig. 2. (a) $K\overline{K}$, (b) $\overline{K}N$, and (c) KN invariant mass distributions for $\gamma n \to K^+K^-n$ at $E_\gamma = 2.3$ GeV. The data are from Ref. 2. The dashed line in (c) is obtained without the ϕ and Θ contributions.

of 1 MeV[7]. The SU(3) effective Lagrangian for the interaction of $\overline{\mathbf{10}}$ with the normal baryon and meson octet can be found, e.g., in Ref. 13. In Fig. 2, we present the results for the invariant mass distributions of various channels in the $\gamma n \to K^+K^-n$ reaction. Apparently, the peak coming from the $\Theta(1540)$ formation depends on the cross section of Θ photoproduction. In this work, we used the model of Ref. 14 for $\gamma N \to \overline{K}\Theta$. (See, e.g., Ref. 15 for the general properties of this reaction.)

One interesting question about this reaction is whether the peak in the KN channel could come from the other backgrounds, especially from tensor meson production[4]. This was motivated by an old experiment[16] on $\pi^-p \to K^-X$, where the observed peaks in the exotic channel were ascribed to the background, especially higher-spin meson, production mechanisms, since the peak positions moved by changing the beam energy. Therefore, it is crucial to test whether this can explain the peak in the KN channel of double kaon photoproduction. In this work, we include $a_2^0(1320)$ and $f_2(1275)$ tensor meson production as backgrounds. We found that (i) the tensor meson production alone gives a very broad peak not a sharp peak, although we confirmed that the peak position moved depending on the beam energy, and (ii) the tensor meson production part is suppressed very much compared with the other reaction mechanisms, and its contribution is hard to be seen even after removing the ϕ and $\Lambda(1520)$ production part from the backgrounds. Therefore, any sharp peak, if confirmed, can be hardly explained by kinematic reflections. In addition, a very sharp peak is expected in the KN invariant mass distribution if the exotic Θ is formed in the reaction. (We note that the most recent CLAS experiment[18] shows no evidence of $\Theta(1540)$, but its existence is still controversial[19,20].)

References

1. LEPS Collaboration, T. Nakano *et al.*, Phys. Rev. Lett. **91**, 012002 (2003).
2. CLAS Collaboration, S. Stepanyan *et al.*, Phys. Rev. Lett. **91**, 252001 (2003).
3. K. Nakayama and K. Tsushima, Phys. Lett. B **583**, 269 (2004).
4. A. R. Dzierba *et al.*, Phys. Rev. D **69**, 051901 (2004).
5. W. Roberts, Phys. Rev. C **70**, 065201 (2004); in these proceedings; A. I. Titov *et al.*, Phys. Rev. C **71**, 035203 (2005); A. Sibirtsev *et al.*, hep-ph/0509145.
6. Y. Oh, K. Nakayama, and T.-S. H. Lee, hep-ph/0412363, Phys. Rept. (in print).
7. Particle Data Group, S. Eidelman *et al.*, Phys. Lett. B **592**, 1 (2004).
8. SPHINX Collaboration, Yu. M. Antipov *et al.*, Phys. Lett. B **604**, 22 (2004).
9. CLAS Collaboration, S. Taylor *et al.*, Phys. Rev. C **71**, 054609 (2005).
10. Y. Oh *et al.*, nucl-th/0004055; Phys. Rev. C **63**, 025201 (2001); Y. Oh and T.-S.H. Lee, Phys. Rev. C **66**, 045201(2002); Phys. Rev. C **69**, 025201 (2004).
11. E. Kaxiras, E. J. Moniz, and M. Soyeur, Phys. Rev. D **32**, 695 (1985).
12. M. Warns, W. Pfeil, and H. Rollnik, Phys. Lett. B **258**, 431 (1991).
13. S. H. Lee, H. Kim, and Y. Oh, Phys. Rev. D **69**, 094009 (2004); J. Kor. Phys. Soc. **46**, 774 (2005); Y. Oh and H. Kim, Phys. Rev. D **70**, 094022 (2004).
14. Y. Oh, H. Kim, and S. H. Lee, Phys. Rev. D **69**, 014009 (2004); Nucl. Phys. A **745**, 129 (2004).
15. K. Nakayama, W. G. Love, Phys. Rev. C **70**, 012201 (2004).
16. E. W. Anderson *et al.*, Phys. Lett. **29B**, 136 (1969).
17. K. Hicks *et al.*, Phys. Rev. D **71**, 098501 (2005).
18. CLAS Collaboration, M. Battaglieri *et al.*, hep-ex/0510061.
19. H. Z. Huang (for the STAR Collaboration), hep-ex/0509037.
20. V. D. Burkert, hep-ph/0510309; in these proceedings.

ON THE NATURE OF THE $\Lambda(1405)$ AS A SUPERPOSITION OF TWO STATES

E. OSET

Departamento de Física Teórica and IFIC, Centro Mixto, Institutos de Investigación de Paterna - Universidad de Valencia-CSIC

V. K. MAGAS AND A. RAMOS

Departament d'Estructura i Constituents de la Matéria, Universitat de Barcelona, Diagonal 647, 08028 Barcelona, Spain

The $K^-p \to \pi^0\pi^0\Sigma^0$ reaction has an experimental $\pi^0\Sigma^0$ mass distribution of forming the $\Lambda(1405)$ with a peak at 1420 MeV and a relatively narrow width of $\Gamma = 38$ MeV. We use these data in combination with those of the $\pi^-p \to K^0\pi\Sigma$ reaction and elements of chiral unitary theory to prove that there are two $\Lambda(1405)$ states instead of one as so far assumed.

It has been long accepted that the $\Lambda(1405)$ is a dynamical resonance generated from the interaction of meson baryon components in coupled channels. In particular, the unitary extensions of chiral perturbation theory ($U\chi PT$) [1–5], using the lowest order chiral Lagrangian and unitarity in coupled channels generate the $\Lambda(1405)$ and lead to good agreement with the K^-p reactions. The surprise, however, came with the realization that there are two poles in the neighborhood of the $\Lambda(1405)$ both contributing to the final experimental invariant mass distribution [3–9]. The properties of these two states are quite different, one has a mass around 1390 MeV, a large width of about 130 MeV and couples mostly to $\pi\Sigma$, while the second one has a mass around 1425 MeV, a narrow width of about 30 MeV and couples mostly to $\bar{K}N$. The two states are populated with different weights in different reactions and, hence, their superposition can lead to different distribution shapes. Since the $\Lambda(1405)$ resonance is always seen from the invariant mass of its only strong decay channel, the $\pi\Sigma$, hopes to see the second pole are tied to having a reaction where the $\Lambda(1405)$ is formed from the $\bar{K}N$ channel. This is accomplished by the recently measured reaction $K^-p \to \pi^0\pi^0\Sigma^0$ [10] which allows us to test already the two-pole nature of the $\Lambda(1405)$.

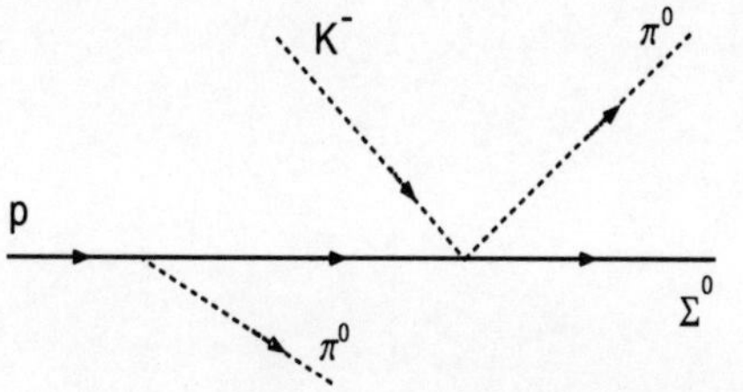

Fig. 1. Nucleon pole term for the $K^- p \to \pi^0 \pi^0 \Sigma$ reaction.

Our model for the reaction $K^- p \to \pi^0 \pi^0 \Sigma^0$ in the energy region of $p_{K^-} = 514$ to 750 MeV/c, as in the experiment [10], considers those mechanisms in which a π^0 loses the necessary energy to allow the remaining $\pi^0 \Sigma^0$ pair to be on top of the $\Lambda(1405)$ resonance. The first of such mechanisms is given by the diagram of Fig. 1. Other mechanisms that involve the meson meson interaction and baryon-baryon-three meson vertices were found negligible in the detailed study of [11].

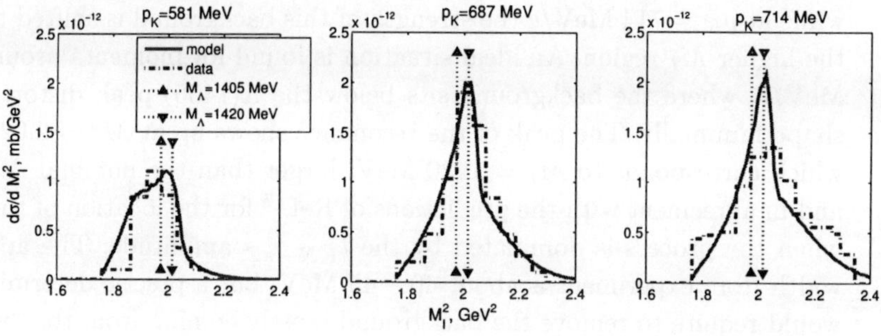

Fig. 2. The $(\pi^0\Sigma^0)$ invariant mass distribution for three different initial kaon momenta.

The indistinguishability of the two emitted pions requires the implementation of symmetrization. This is achieved by summing two amplitudes evaluated with the two pion momenta exchanged. In addition, a factor of 1/2 for indistinguishable particles is also included in the total cross section.

Our calculations show that the process is largely dominated by the nucleon pole term shown in Fig. 1. As a consequence, the $\Lambda(1405)$ thus obtained comes mainly from the $K^- p \to \pi^0 \Sigma^0$ amplitude which, as mentioned above, gives the largest possible weight to the second (narrower) state.

In Fig. 2 our results for the invariant mass distribution for three different energies of the incoming K^- are compared to the experimental data. Sym-

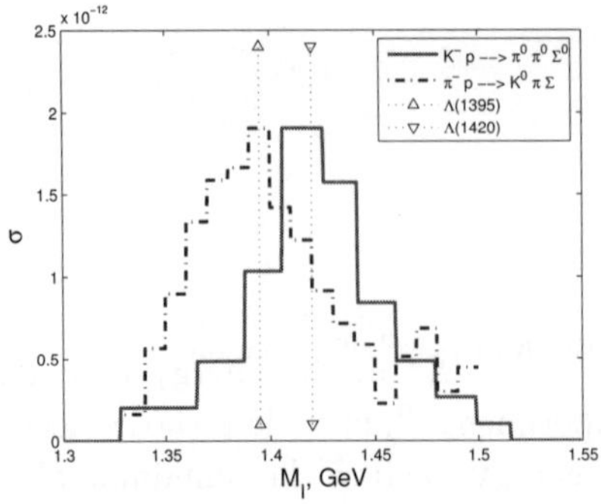

Fig. 3. Two experimental shapes of $\Lambda(1405)$ resonance. See text for more details.

metrization of the amplitudes produces a sizable amount of background. At a kaon laboratory momentum of $p_K = 581$ MeV/c this background distorts the $\Lambda(1405)$ shape producing cross section in the lower part of M_I, while at $p_K = 714$ MeV/c the strength of this background is shifted toward the higher M_I region. An ideal situation is found for momenta around 687 MeV/c, where the background sits below the $\Lambda(1405)$ peak distorting its shape minimally. The peak of the resonance shows up at $M_I^2 = 2.02$ GeV^2 which corresponds to $M_I = 1420$ MeV, larger than the nominal $\Lambda(1405)$, and in agreement with the predictions of Ref. [6] for the location of the peak when the process is dominated by the $t_{\bar{K}N\to\pi\Sigma}$ amplitude. The apparent width from experiment is about $40 - 45$ MeV, but a precise determination would require to remove the background mostly coming from the "wrong" $\pi^0\Sigma^0$ couples due to the indistinguishability of the two pions. A theoretical analysis permits extracting the pure resonant part by not symmetrizing the amplitude. This is done in Ref. [11]. The width of the resonant part is $\Gamma = 38$ MeV, which is smaller than the nominal $\Lambda(1405)$ width of 50 ± 2 MeV [13], obtained from the average of several experiments, and much narrower than the apparent width of about 60 MeV that one sees in the $\pi^- p \to K^0\pi\Sigma$ experiment [14], which also produces a distribution peaked at 1395 MeV. In order to illustrate the difference between the $\Lambda(1405)$ resonance seen in this latter reaction and in the present one, the two experimental distributions are compared in Fig. 3. We recall that the shape of the $\Lambda(1405)$ in the $\pi^- p \to K^0\pi\Sigma$ reaction was shown in Ref. [12] to be largely built from the $\pi\Sigma \to \pi\Sigma$ amplitude, which is dominated by the wider lower energy state.

The invariant mass distributions shown here are not normalized, as in experiment. But we can also compare our absolute values of the total cross sections with those in Ref. [10]. This is done in Ref. [11] and the agreement is good.

The study of the present reaction, complemental to the one of Ref. [12]

for the $\pi^- p \to K^0 \pi \Sigma$ reaction, has shown that the quite different shapes of the $\Lambda(1405)$ resonance seen in these experiments can be interpreted in favour of the existence of two poles with the characteristics predicted by the chiral theoretical calculations. Besides demonstrating once more the great predictive power of the chiral unitary theories, this combined study of the two reactions gives the first clear evidence of the two-pole nature of the $\Lambda(1405)$.

Acknowledgments: This work is partly supported by DGICYT contracts BFM2002-01868, BFM2003-00856, the Generalitat de Catalunya contract SGR2001-64, and the E.U. EURIDICE network contract HPRN-CT-2002-00311. This research is part of the EU Integrated Infrastructure Initiative Hadron Physics Project under contract number RII3-CT-2004-506078.

References

1. N. Kaiser, P. B. Siegel and W. Weise, Phys. Lett. B **362** (1995) 23
2. E. Oset and A. Ramos, Nucl. Phys. A **635** (1998) 99
3. J. A. Oller and U. G. Meissner, Phys. Lett. B **500** (2001) 263
4. D. Jido, A. Hosaka, J. C. Nacher, E. Oset and A. Ramos, Phys. Rev. C **66** (2002) 025203
5. C. Garcia-Recio, J. Nieves, E. Ruiz Arriola and M. J. Vicente Vacas, Phys. Rev. D **67** (2003) 076009 [arXiv:hep-ph/0210311].
6. D. Jido, J. A. Oller, E. Oset, A. Ramos and U. G. Meissner, Nucl. Phys. A **725** (2003) 181
7. C. Garcia-Recio, M. F. M. Lutz and J. Nieves, Phys. Lett. B **582** (2004) 49
8. T. Hyodo, S. I. Nam, D. Jido and A. Hosaka, Phys. Rev. C **68** (2003) 018201
9. S. I. Nam, H. C. Kim, T. Hyodo, D. Jido and A. Hosaka,
10. S. Prakhov *et al.* [Crystall Ball Collaboration], Phys. Rev. C **70** (2004) 034605.
11. V. K. Magas, E. Oset and A. Ramos, Phys. Rev. Lett. **95** (2005) 052301
12. T. Hyodo, A. Hosaka, E. Oset, A. Ramos and M. J. Vicente Vacas, Phys. Rev. C **68** (2003) 065203
13. K. Hagiwara *et al.* [Particle Data Group], Phys. Rev. D **66**, 010001 (2002).
14. D. W. Thomas, A. Engler, H. E. Fisk, and R. W. Kraemer, Nucl. Phys. B **56**, 15 (1973).

CHANNEL COUPLING EFFECTS IN PHOTO-INDUCED $\rho - N$ PRODUCTION

A. USOV AND O. SCHOLTEN

Kernfysisch Versneller Instituut
Zernikelaan 25
9745 CL, Groningen
The Netherlands
Email's: usov@kvi.nl and scholten@kvi.nl

We present an extension of our coupled channels calculation to include photo-induced $\rho - N$ production. We show that indirect contributions are large and can account for some of the typical discrepancies seen in a tree-level calculations.

1. Introduction

A number of the recent experiments on vector meson production[1–3] have greatly stimulated theoretical research on the subject. New and more accurate experimental data will hopefully allow to address the long-standing problem of missing resonances. However in order to make well-based predictions for the parameters of these new resonances a good understanding of the underlying non-resonant background is needed. Reviewing the recent publications[5] on the photo-induced ρ-meson production reveals a common problem – most of the tree-level calculations have difficulties describing differential cross-section in the high-t region. As a possible solution to this problem we propose a channel-coupled treatment, where the coupling to the pion production sector accounts for this discrepancy.

2. Model

Here we give a short overview of the model used to perform calculations and present some of the results. The more detailed description of the model will be published soon. In the following calculations we have used the same model as was used for kaon production[4]. The tree-level description of the $\gamma + N \to \rho + N$ channel was refined, taking the tree-level calculation (model

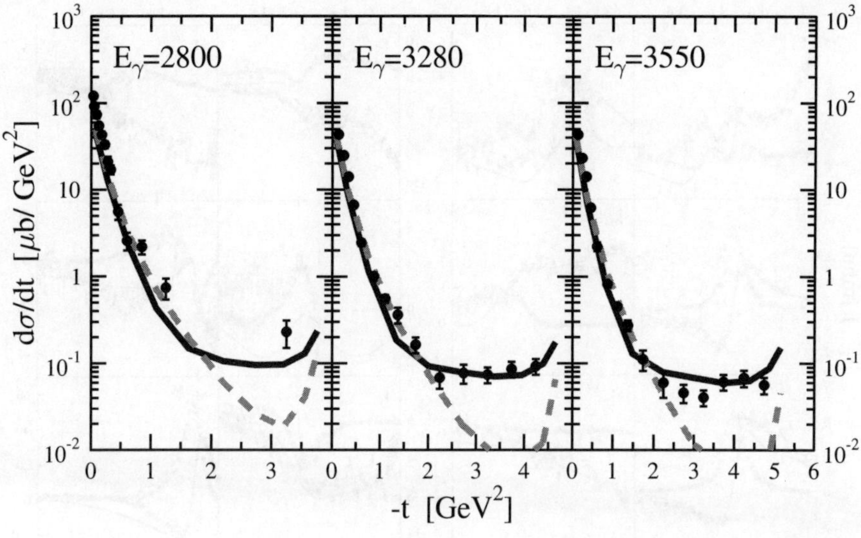

Fig. 1. Differential cross-section for $\gamma + N \to \rho + N$ channel. The solid line corresponds to the full calculation. The dashed line shows the tree-level calculation. The data is taken from ref.[2].

A) of Y. Oh and T.-S. H. Lee[5] as a starting point. However in order to make it compatible with our model a number of ingredients were reworked. The gauge invariance restoration scheme was changed to be compatible with a minimal substitution-based scheme used throughout the rest of the model. Additionally a number of parameters were changed to match their counterparts in other channels. Nevertheless the tree-level calculation is still compatible with ref.[5] and shares a common problem of the tree-level calculations – a dramatic under-prediction of the differential cross-section in the middle- and high-t region, as can be seen from Figure 1 (dashed line). The solid line in the same figure shows the full calculation, where the channel coupling effects have compensated this discrepancy.

As the dominant source of the indirect contributions we identify the following process $\gamma + N \to \pi + N \to \rho + N$. For the pion-production channel, as compared to our old calculation, we have introduced additional formfactors to regulate it's high-energy behavior. The resulting partial waves are shown in Figure 2. For the pion-induced ρ production channel there is unfortunately no direct experimental data available and we rely on the inelasticities in pion scattering. In the Figure 3 the phase shifts and inelas-

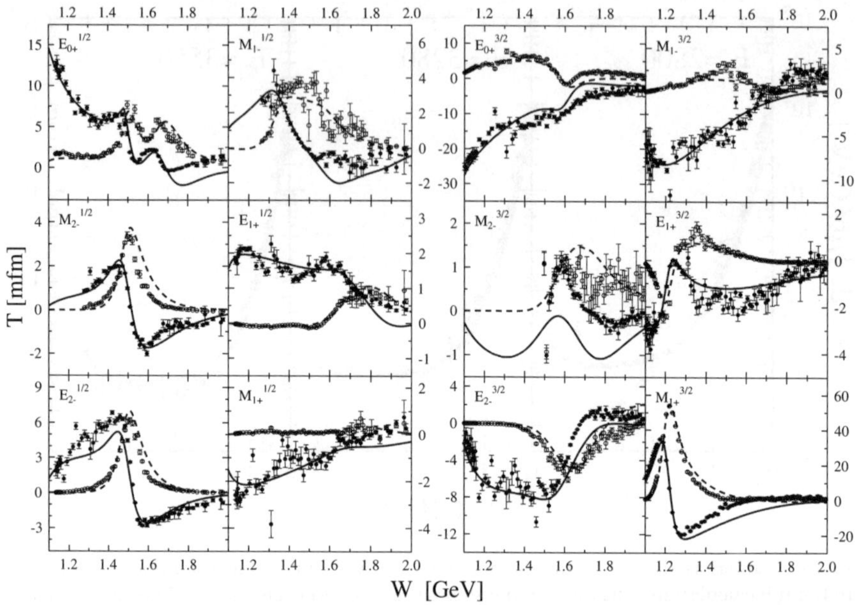

Fig. 2. Partial wave decomposition of the $\gamma + N \to \pi + N$ channel. The data is taken from the on-line database of the VPI group[6].

ticities for pion scattering are shown. The solid line denotes the full calculation and the dashed line shows the calculation where all ρ final states were excluded from the model space. While it's clear that the inelasticities generated due to the ρN channels fit the data, there is still an ambiguity in the $\pi + N \to \rho + N$ channel associated with the gauge invariance restoration procedure. As an example the dotted line of the Figure 3 shows a calculation where a different gauge invariance restoration prescription was used in the pion-induced ρ-production channel. While the resulting inelasticities show large differences, it's impossible to draw a definite conclusion. It's also worth to note that the above mentioned ambiguity results in a factor of approximately 2 difference for the strength of the coupled-channels effects in the photo-induced ρ production channel.

3. Conclusion

We have shown that channel coupling effects are important for the accurate description of the non-resonant background in vector meson production. This indirect contribution accounts for some of the discrepancies in the

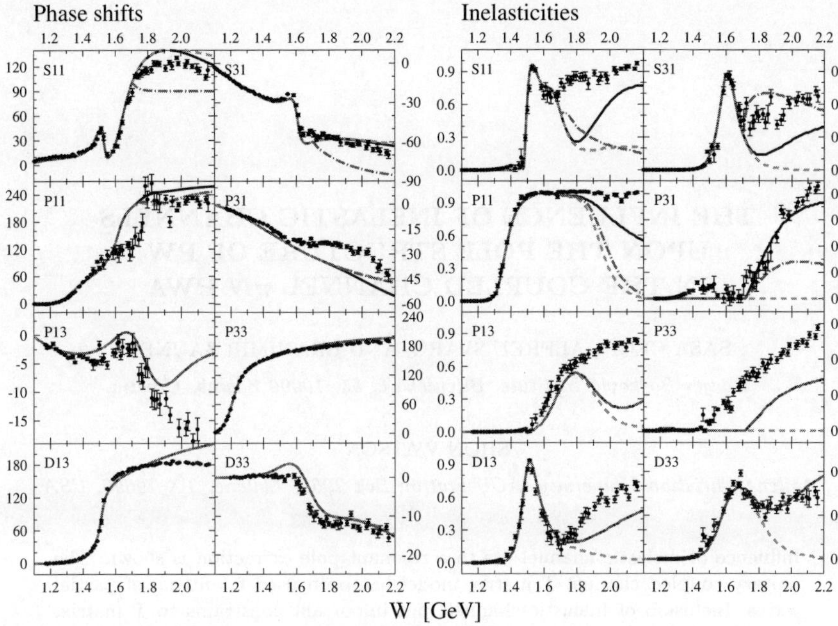

Fig. 3. Partial wave decomposition for the $\pi + N \to \pi + N$ channel. The solid line corresponds to the full calculation. The dashed line shows the calculation where all ρ final states were excluded. The dotted line denotes the calculation where a different gauge invariance restoration prescription was used in $\pi + N \to \rho + N$ channel. The data is taken from the on-line database of the VPI group[6].

description of experimental data which could otherwise be misinterpreted as being due to resonant contributions. However to determine accurately the strength of these "indirect" contributions experimental data for other channels is needed. In the particular case of ρ-meson production cross-sections for pion-induced ρ production would be of great help.

References

1. E. Anciant *et al.*, CLAS Collaboration, *Phys. Rev. Lett.* **85** 4682 (2000).
2. M. Battaglieri *et al.*, CLAS Collaboration, *Phys. Rev. Lett.* **87** 172002 (2001).
3. M. Battaglieri *et al.*, CLAS Collaboration, *Phys. Rev. Lett.* **90** 022002 (2003).
4. A. Usov and O. Scholten, *Phys. Rev.* **C72**, 025205 (2005).
5. Yongseok Oh and T.-S. H. Lee, *Phys. Rev.* **C69**, 025201 (2004).
6. R.A. Arndt, I.I. Strakovsky and R.L. Workman, *Phys. Rev.* **C53**, 430 (1996); *Phys. Rev.* **C66**, 055213 (2002); updates available on the web: http://gwdac.phys.gwu.edu.

THE INFLUENCE OF INELASTIC CHANNELS UPON THE POLE STRUCTURE OF PW IN THE COUPLED CHANNEL πN PWA

SAŠA CECI*, ALFRED ŠVARC, AND BRANIMIR ZAUNER
Rudjer Bošković Institute, Bijenička c. 54, 10000 Zagreb, Croatia.

SHON WATSON
Abilene Christian University, ACU Station Box 7963, Abilene, TX 79699, USA.

Influence of inelastic channel fits to a resonant pole extraction is shown. The unitary coupled channel T matrix model has been used for fitting of partial waves. Inclusion of inelastic channels put important constrains to T matrix. Poles not observed in elastic channels enter into picture through inclusion of the inelastic channels.

1. Introduction

We want to examine the influence of inelastic channel data to T matrix poles. For some time, there has been a controversy about number of resonant poles of T matrix in P_{11} partial wave. We will show that the number of poles can change after fitting of the additional channels. Arndt *et al.*[5] report that there is no N(1710). Mark Manly questions its existence at NSTAR 2004. meeting in Grenoble[6] because N(1710) does not show in the speed plot, but reports two widths 50 and 400 MeV. Batinić *et al.*[2] need four poles in P_{11} to give good description of experimental ηN data. Theoretical three quark models[7] offer at least six states in energy region bellow 3000 MeV which is known as the missing resonances problem.

We shall show that simultaneous fit of πN elastic single energy solutions for partial waves obtained by Arndt *et al.*[4], and $\pi N \to \eta N$ partial wave taken from Batinić *et al.*[2], would rise need for N(1710) resonance pole.

*Email: sasa.ceci@irb.hr

2. Coupled channel formalism

We use coupled channel model based on Cutkosky CMB [1–3]:

$$T = \sqrt{\mathrm{Im}\Phi}\,\gamma^T G\,\gamma\,\sqrt{\mathrm{Im}\Phi}$$

$$\mathrm{Im}\phi = \frac{q^{2L+1}}{\sqrt{s}\cdot(Q_1+\sqrt{Q_2^2+q^2})^{2L}}; \quad G = (s_B - s - \gamma\,\Phi\gamma^T)^{-1}$$

$$\phi(s) = \frac{s-s_o}{\pi}\int_{s_o}^{\infty}\frac{\mathrm{Im}\phi(s')}{(s'-s_o)(s'-s)}$$

Mandelstam s (square of center of mass energy) can be complex, and we look for resonant poles in the Riemann sheet closest to the physical sheet. For that purpose we use this mapping:

$$\phi(s) = \sum_{n=0}^{N} c_n\,Z^n(s); \quad Z(s) = \frac{\alpha+\sqrt{s_o-s}}{\alpha-\sqrt{s_o-s}},$$

where s_o is square of the threshold energy, while c_n and α are fit parameters.

3. Fits

When just elastic channel are fitted, we needed only one pole in Figure 2. If we add more poles, they would appear at energies higher than 2000 MeV. The second one is probably related to N(2100) P_{11} resonance - but there is not enough data close to this energy to be more specific.

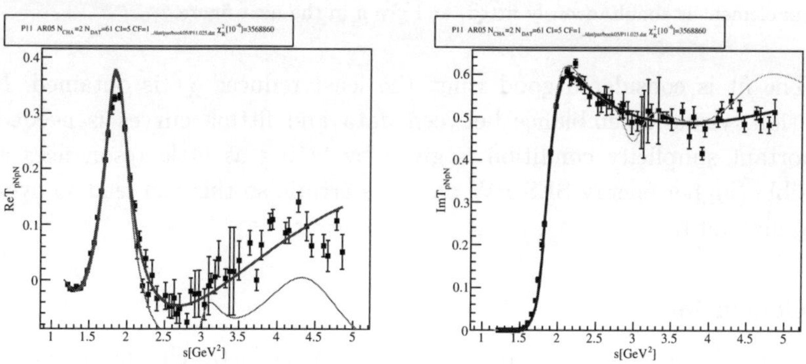

Fig. 1. Real and imaginary part of πN elastic fit.

Simultaneous fit of $\pi N \to \pi N$ and $\pi N \to \eta N$ partial waves produce pole close to 1700 MeV, that was invisible in elastic fits Figure 5.

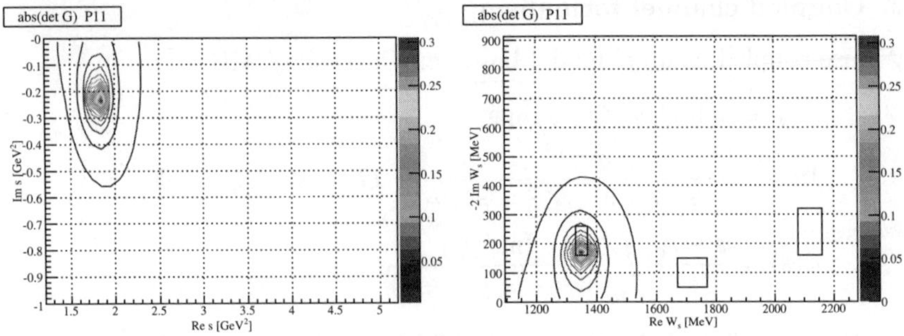

Fig. 2. Roper pole is clearly visible in πN elastic fit.

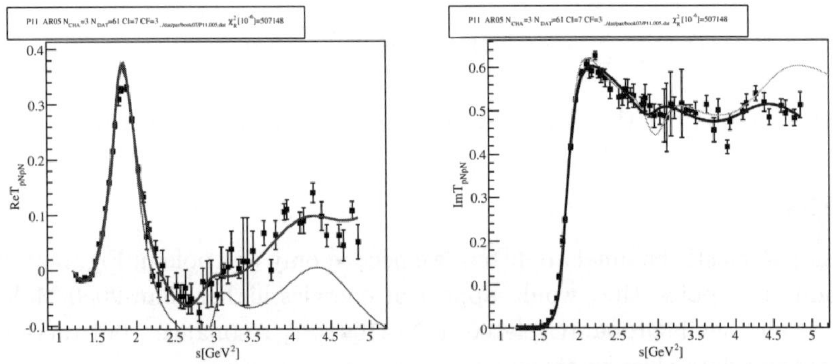

Fig. 3. Real and imaginary part of $\pi N \rightarrow \pi N$ T matrix element. The $\pi N \rightarrow \eta N$ T matrix element is simultaneously fitted, and given in the next figure.

The fit is considered good when the least reduced χ^2 is obtained. In addition, visual resemblance between data and fitting curves is needed. Important simplicity condition is given by fitting as little resonances as possible (higher energy SES PW data are erratic so this can lead to oversimplification).

4. Conclusions

We have shown that inclusion of $\pi N \rightarrow \eta N$ inelastic channel demands N(1710) resonance. Partial wave fits have been done to test the influence of inelastic channels to pole positions. Method and formalism are multichannel in nature, hence can use inelastic data (in addition to elastic) to "sharpen" resonant pole positions. Choice of inelastic channel is important - ηN is not the best choice for P11 in N(1710) energy region (KΛ looks promising).

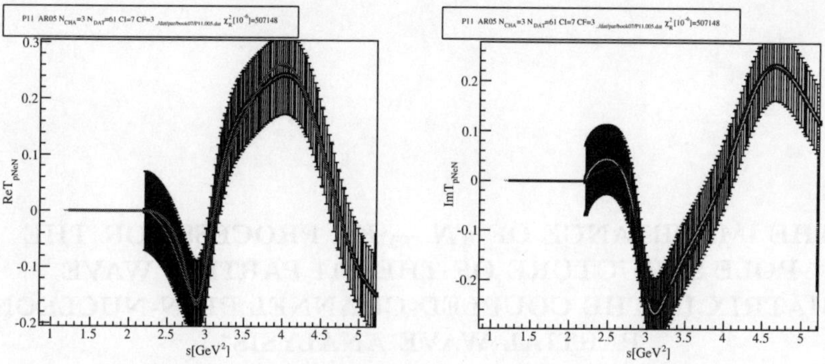

Fig. 4. Real and imaginary part of $\pi N \rightarrow \eta N$ T matrix element.

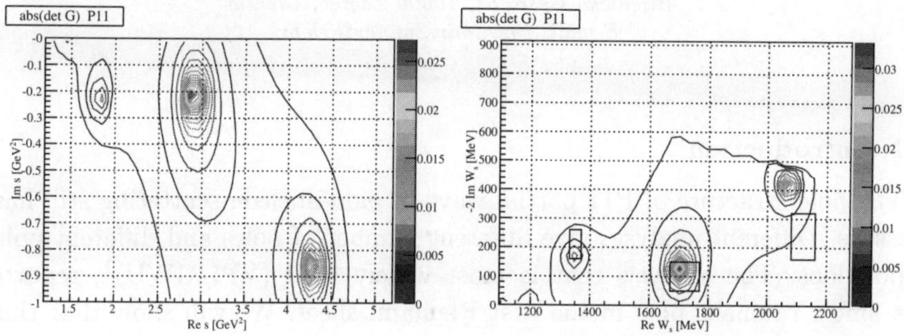

Fig. 5. Three poles are visible, very close to PDG values.

SES dataset is hard to fit, and fits do not seem reliable which is related to problems with continuum ambiguities.

References

1. R.E. Cutkosky *et al.*, Phys. Rev. **D20**, 2839 (1979).
2. M. Batinić, *et al.*, Physica Scripta **58**, 15 (1998).
3. T.P. Vrana, S.A. Dytman, and T.S.-H- Lee, Phys. Rep. **328**, 181 (2000).
4. see: http://gwdac.phys.gwu.edu/analysis/pin_analysis.html.
5. R. A. Arndt *et al.*, Phys. Rev. **C69**, 035213 (2004).
6. D. M. Manley, Proceedings of the Workshop on the Physics of Excited Nucleons, March 24-27, 2004. Grenoble, France., 382, (2004).
7. S. Capstick and W. Roberts, Prog.Part.Nucl.Phys. **45** S241 (2000)

THE IMPORTANCE OF $\pi N \to K\Lambda$ PROCESS FOR THE POLE STRUCTURE OF THE P11 PARTIAL WAVE T-MATRIX IN THE COUPLED-CHANNEL PION-NUCLEON PARTIAL WAVE ANALYSIS*

B. ZAUNER, S. CECI AND A. ŠVARC

Rudjer Boskovic Institute
Bijenička Cesta 54, 10000 Zagreb, Croatia
E-mail: Branimir.Zauner@irb.hr

1. Introduction

The pole structure of P11 partial wave in pion-nucleon scattering still has issues. Different analyses give different number of poles and different pole positions. The analysis that is most widely used (VPI/GWU)[1], reports a single resonant pole in the first Riemann sheet. We will show that this analysis does not exclude other poles, namely poles in W=1700 MeV region, and that they appear the moment we include the $\pi N \to K\Lambda$ inelastic channel.

2. Tools and data sets

Tools We are using a coupled-channel CMU-LBL type formalism. For the collection of formulae we refer the reader either to original paper by Cutkosky et.al[2] or to one of the more recent CC_PWAs; Zagreb[3] or Pittsburgh/ANL[4].

Data Sets For elastic channel, we used single energy solutions from SAID site[5]. For the $\pi N \to K\Lambda$ channel, we used T matrix element we obtained in a single channel partial wave analysis from experimental data. We fitted data to three partial waves - S11, P11 and P13. We had a single resonance per wave, except in P11, where we allowed for two. Error bars were put

*The complete set of transparencies can be found at http://hadron.physics.fsu.edu/NSTAR2005/TALKS/Friday/Parallel_C/Zauner.pdf

a posteriori, in order to make statistical weight of this partial wave data smaller.

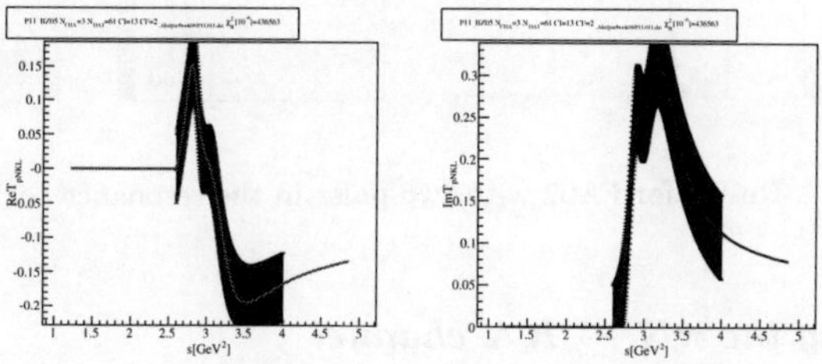

Fig. 1. The $\pi N \to K\Lambda$ P11 partial wave T matrix element.

3. Poles around 1700 MeV are a direct consequence of inelastic channels

3.1. *Fitting only FA02*

First, we shall show that poles around 1700 MeV are not artefacts of the CMU-LBL model. We fitted only FA02 SES (and had a free unitarizing channel to preserve unitarity) and found out that a single pole is sufficient to describe the data well. There is no sign of any pole structure in the 1700 MeV region.

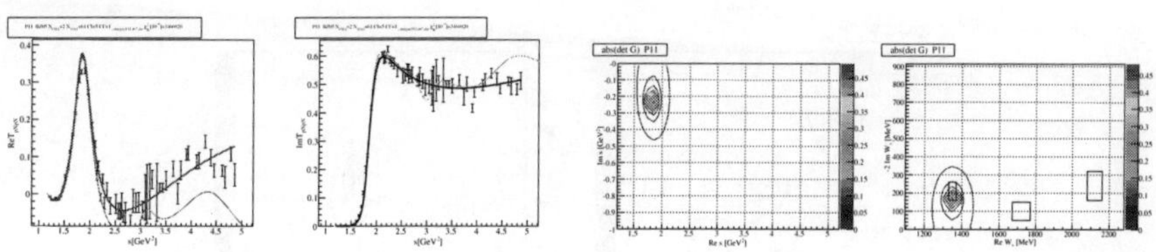

Fig. 2. Result for FA02 with a single pole in the resonance region. Two right graphs show pole positions, where one can see good agreement with Particle Data Group pole positions (black rectangles on the picture)

If we add another pole, agreement with the data is slightly improved in the high energy region, as seen in fig.3. In the right part of fig.3, one can see that Roper resonance did not move, and that we got another resonance in 2300 MeV region, far above the region we are interested in. Therefore we conclude that VPI/GWU SES do not need a pole in the 1700 MeV region. Consequently, there is no need for the N(1710)P11 resonance.

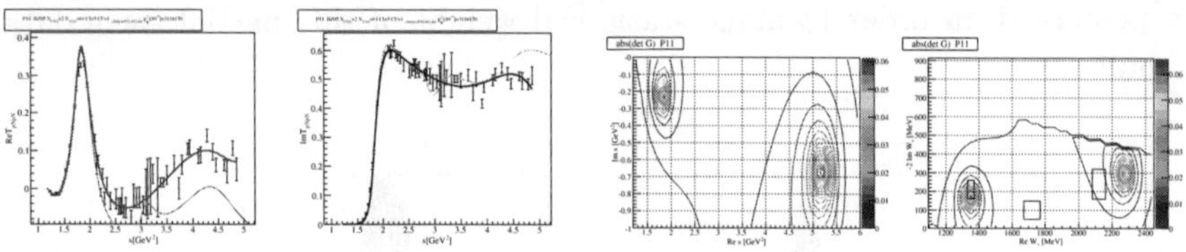

Fig. 3. Result for FA02 with two poles in the resonance region.

3.2. *Including the $\pi N \rightarrow K\Lambda$ channel*

When we include the $\pi N \rightarrow K\Lambda$ T matrix (now we have three channels to fit - elastic, inelastic and unitarizing), we see that a single pole in the resonance region miserably fails to reproduce the data. Moreover, in CMU-LBL model, we have two poles in the subthreshold region that are simulating background. Inclusion of the inelastic channel (when we put a single pole in the resonance region) forces them to simulate real resonances, as one can see in fig.4.

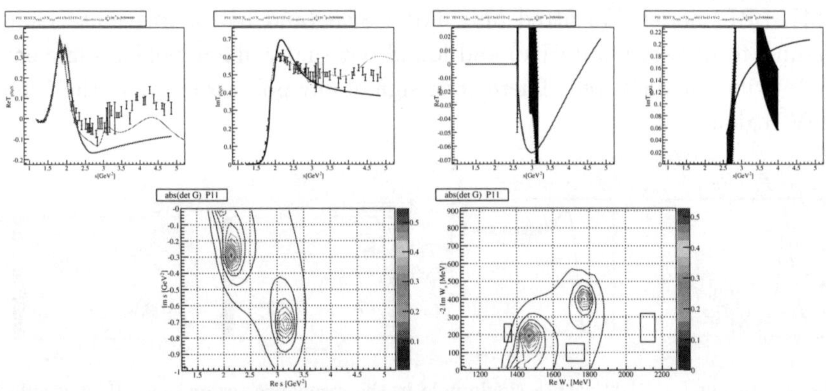

Fig. 4. Result for FA02 SES and $\pi N \rightarrow K\Lambda$ T matrix element. We used a single pole in resonant region.

After further fitting, and trying various number of resonances, first number of poles in the resonant region that reproduces both data sets well and does not force the background poles into resonant region is four. In the fig.4 we can see that we have pole structure at around 1700 MeV, i addition to the Roper resonance pole.

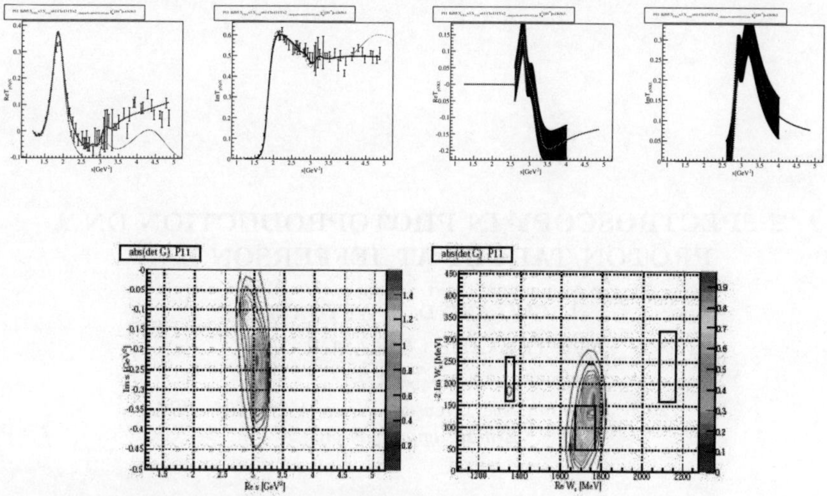

Fig. 5. Result for FA02 SES and $\pi N \to K\Lambda$ T matrix element. We used 4 poles in resonant position.

4. Conclusions

Procedure that was presented here points out that the lack of evidence for N(1710)P11 in many recent analyses is due to the single channel treatment. Introduction of inelastic channels (namely, $\pi N \to K\Lambda$) is crucial to the confirmation of the existence of N(1710)P11.

References

1. R. A. Arndt et. al., Phys. Rev. **C69**, 035213 (2004).
2. R.E. Cutkosky, C.P. Forsyth, R.E. Hendrick and R.L. Kelly, Phys. Rev. **D20**, 2839 (1979).
3. M. Batinić, I. Šlaus, A. Švarc and B.M.K. Nefkens, Phys. Rev. **C51**, 2310 (1995); M. Batinić, I. Šlaus, A. Švarc, B.M.K. Nefkens and T.S.-H. Lee, Physica Scripta **58**, 15 (1998).
4. T.P. Vrana, S.A. Dytman and T.S.-H- Lee, Phys. Rep. **328**, 181 (2000).
5. see: http://gwdac.phys.gwu.edu/
6. S. Eidelman et al., Physics Letters B592, 1 (2004)

Ξ SPECTROSCOPY IN PHOTOPRODUCTION ON A PROTON TARGET AT JEFFERSON LAB*

L. GUO AND D. P. WEYGAND
FOR THE CLAS COLLABORATION

Jefferson Lab,
12000 Jefferson Ave., Newport News, VA 23606, USA
E-mail: lguo@jlab.org

The CLAS Collaboration at Jefferson Lab conducted a photoproduction experiment on a proton target using a tagged photon beam with an energy range of 1.5-3.8 GeV during May-July 2004. With an integrated luminosity of about 75 pb^{-1}, this experiment provides the largest data set for photon-proton reactions ever collected. The reaction $\gamma p \rightarrow K^+K^+\Xi^-(1320)$ has been investigated and the preliminary results of the cross section measurement of $\Xi^-(1320)$ for the photon energy range of 2.7-3.8 GeV have been obtained. In a search for excited cascade states, the reaction of $\gamma p \rightarrow K^+K^+\pi^-(\Xi^0(1320))$ has also been explored.

1. Introduction and Photoproduction of cascade resonances

Compared with the non-strange baryons and $S = -1$ hyperon states, the Ξ resonances are in general not well studied. Only two cascade states, the Ξ(1320) and Ξ(1530), have four star status in the PDG. This is mainly due to small cross sections, and the fact that the cascade resonances cannot be produced through direct formation. It is important to view the investigation of the cascade resonances as an essential part of the baryon spectroscopy program. Flavor $SU(3)$ symmetry implies as many Ξ resonances as N^* and Δ^* states, meaning many more cascade resonances await to be discovered [1]. The facility at CEBAF offers the opportunity to study Ξ resonances in exclusive photoproduction. Kaon beam experiments that dominated the earlier Ξ studies suffer from low intensity, while hyperon beam experiments typically suffer from the high combinatoric background that complicates the analysis. It has been demonstrated that cascade production can be inves-

*This work is supported by DOE under contract DE-AC05-84ER40150 tigated through exclusive reactions such as $\gamma p \to K^+K^+(X)$ [2] in CLAS. However, low statistics hampered further investigation of the production mechanisms.

2. Ground state Ξ^- production in $\gamma p \to K^+K^+(X)$

The new CLAS photoproduction data set (g11) collected during summer 2004 used a tagged photon beam [3] [4] mainly in the range of 1.5-3.8 GeV (less than 5% data was collected with higher beam energy up to 4.75 GeV) incident on a proton target, and represents more than 75 pb^{-1} luminosity. The ground state Ξ^- is observed clearly in the K^+K^+ missing mass spectrum (Fig. 1, left), with the two kaons detected by the CLAS spectrometer. The first excited cascade resonance $\Xi^-(1530)$ is also visible. As discussed earlier, it is likely that the cascade resonances are produced through other intermediate resonances such as Y^* states. However, there has not been unambiguous evidence for such a production mechanism in photoproduction. The Ξ^-K^+ invariant mass spectrum (Fig. 1, right) seems to show interesting structures near 2 GeV. However, the two kaons being identical bosons and the uncertainty of the many overlapping hyperon states near 2 GeV makes it difficult to determine the relevant parameters. The Ξ^- cross section has been extracted by assuming a t-channel process producing hyperon states that decay to Ξ^-K^+. Various differential cross section results have been studied to improve the simulation, and the total cross sections intergrated from different methods are consistent with each other. It is worthwhile to note that the cross section[†] of Ξ^- seems to increase with the photon energy(Fig. 2). This would be possible if the hypothesis of $Y^* \to \Xi^-K^+$ is correct since higher photon energy naturally leads to more accessible Y^* states. The JLAB-MSU phenomenological approach [5] for exclusive reactions with three final particles is being developed to incorporate the $\Xi^-K^+K^+$ channel and possibly determine the Ξ^- photoproduction mechanism in the future.

3. Reaction $\gamma p \to K^+K^+\pi^-(\Xi^0(1320))$

Excited cascade resonances can be investigated through the reaction $\gamma p \to K^+K^+\pi^-(\Xi^0)$ using the new CLAS data set. The $\Xi(1530)$ decays to $\Xi\pi$ almost 100%. For higher mass cascade resonaces, both $\Xi(1620)$ (one star

[†]The cross sections are preliminary, and 20% systematic errors are expected

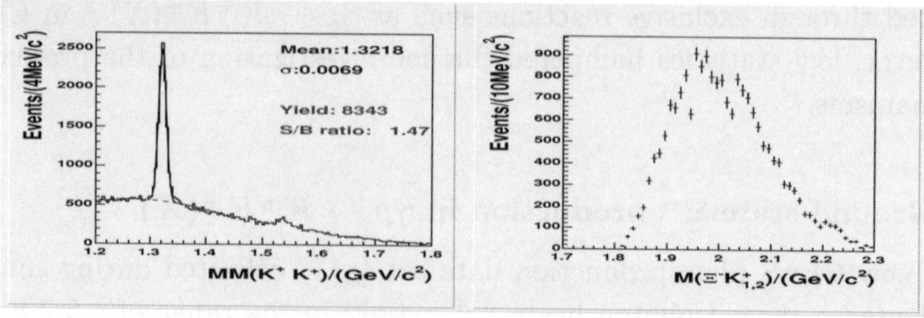

Fig. 1. Left: K^+K^+ missing mass spectrum with tighter timing cuts (less stingent cut yields more than 12 k Ξ^-); Right: $\Xi^- K^+$ invariant mass spectra after side band subtraction (Both K^+ are included).

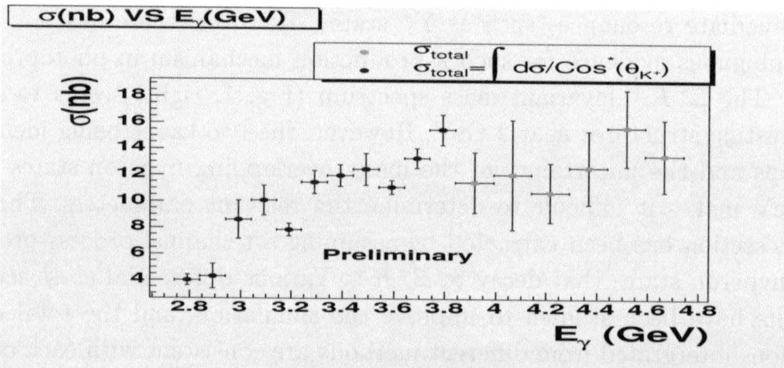

Fig. 2. Total cross sections(green) of $\Xi^-(1320)$. Results above 3.8 GeV are obtained without differential cross section measurement due to the very low statistics. Only statistical errors are shown.

state) and $\Xi(1690)$ have been observed in the $\Xi\pi$ channel, but further confirmation is needed. The three charged particles are identified by the CLAS spectrometer, while the Ξ^0 events are reconstructed from missing mass (Fig. 3, top right). Kinematic fitting procedures were applied by constraining the Ξ^0 mass. The $\Xi^-(1530)$ peak is seen to become narrower after the fitting (Fig. 3, bottom), while a small enhancement around the 1605 MeV region is also observed. Due to the low statistics, and the possible interference of $\gamma p \to K^+K^{*0}(\Xi^0)$, it is difficult to draw any definite conclusion regarding the origin of this small structure.

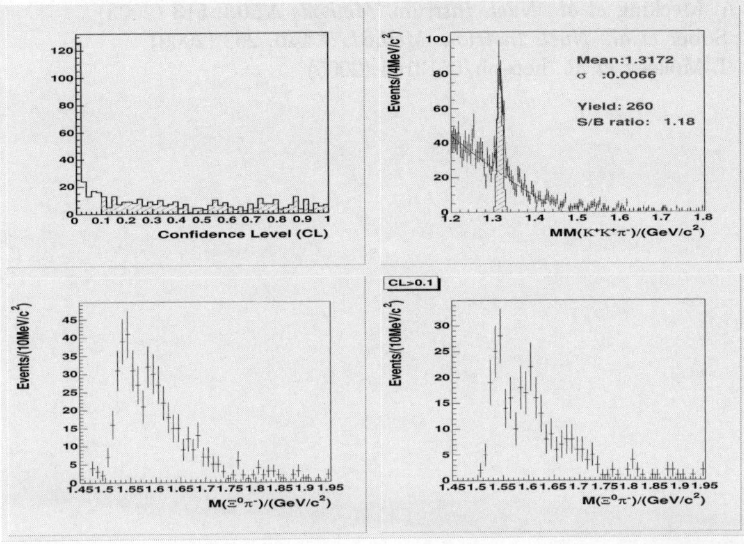

Fig. 3. Top Left: Confidence level (CL) distribution for events in the Ξ^0 peak (as shown on the right, shaded region indicates CL> 0.1); Top right: $(K^+K^+\pi^-)$ missing mass spectra(shaded region indicates CL> 0.1); Bottom left: $(\Xi^0\pi^-)$ invariant mass spectrum for events in the Ξ^0 peak; Bottom Right: $(\Xi^0\pi^-)$ invariant mass spectrum (using kinematic fitted four vectors) for events in the Ξ^0 peak and CL> 0.1;

4. Discussion

More than 10000 Ξ^- events has been observed in the reaction of $\gamma p \rightarrow K^+K^+(\Xi^-)$. The total cross section appears to increase with photon energy. Although the data suggests that the ground state Ξ^- could be a decay product of high mass hyperons, the current status of the analysis is not conclusive. The reaction of $\gamma p \rightarrow K^+K^+\pi^-(\Xi^0)$ has also been investigated. Although a small enhancement around 1605 MeV is observed near the $\Xi^-(1530)$ peak, the nature of this enhancement is not clear.

5. Acknowledgment

We wish to thank all of the CLAS collaborators, the extraordinary efforts of the CEBAF staff, and particularly the g11 running group.

References

1. B. M. K. Nefkens, *Baryon spectroscopy and chiral symmetry* (1995)
2. J. Price *et al.*, *Phys.Rev.* **C71**, 058201,(2005)

3. B.A. Mecking *et al.*, *Nucl. Instrum. Methods* **A503**, 513 (2003)
4. D. Sober *et al.*, *Nucl. Instrum. Methods* **A440**, 263 (2000)
5. V. I. Mokeev *et al.*, hep-ph/0512164 (2005)

STRUCTURE OF THE σ-MESON AND DIAMAGNETISM OF THE NUCLEON *

M. I. LEVCHUK

B. I. Stepanov Institute of Physics, BY-220072 Minsk, Belarus

A. I. L'VOV

P. N. Lebedev Physical Institute, RU-117924 Moscow, Russia

A. I. MILSTEIN

Budker Institute of Nuclear Physics, RU-630090 Novosibirsk, Russia

M. SCHUMACHER†

Zweites Physikalisches Institut der Universität, D-37077 Göttingen, Germany

The structure of the σ meson and the diamagnetism of the nucleon are shown to be topics which are closely related to each other. Arguments are found that the σ meson couples to two photons via its non-strange $q\bar{q}$ structure component. This ansatz leads to a quantitative explanation of the t-channel component of the difference of electromagnetic polarizabilities, $(\alpha-\beta)^t$, containing the diamagnetism of the nucleon. The prediction is $(\alpha-\beta)^t_{p,n} = (5\alpha_e g_{\pi NN})/(6\pi^2 m_\sigma^2 f_\pi) = 15.3$ in units of 10^{-4}fm^3 to be compared with the experimental values $(\alpha-\beta)^t_p = 15.1\pm1.3$ for the proton and $(\alpha-\beta)^t_n = 14.8\pm2.7$ for the neutron. The equivalent approach to exploit the $\pi\pi$ structure component of the σ meson via the BEFT sum rule leads to $(\alpha-\beta)^t_{p,n} = 14\pm2$, what also is in agreement with the experimental results.

1. Introduction

Diamagnetism is one of the dominant properties of the nucleon, contrasting with the fact that this phenomenon still is barely understood. A clearcut information is provided by dispersion theory[1–4] where Compton scattering is described by six invariant amplitudes $A_i(s,t)$. The amplitudes $A_i(s,t)$ are

*This work is supported by Deutsche Forschungsgemeinschaft through the projects SCHU222/25 and 436 RUS 113/510

†Corresponding author: e-mail mschuma3@gwdg.de

analytic functions in the two complex planes s and t and may be calculated from the singularities contained in these planes. The s-plane singularities are given by the meson-photoproduction cross sections contributing to the total photoabsorption cross section. The t-channel singularities were identified in the late 1950th and the beginning of the 1960th by Low, Jacob and Mathews (LJM[5]) and Hearn and Leader (HL[6]). These singularities reflect the degrees of freedom (d.o.f.) of the nucleon and we find it convenient to introduce the terms s-channel d.o.f and t-channel d.o.f. Translated into modern language, LJM[5] argued that instead of an excitation of the pion and constituent-quark structures of the nucleon, i.e. the s-channel d.o.f., the production of a π^0 meson in the intermediate state may lead to Compton scattering. This π^0-pole contribution is now generally accepted as part of the backward spin-polarizability γ_π. On the other hand the scalar-isoscalar $\pi\pi$ t-channel introduced by HL[6] making a large contribution to $(\alpha - \beta)$ is frequently ignored or insufficiently represented in theoretical approaches (see Ref.[4]).

2. Experimental status of electromagnetic polarizabilities

According to our recent analysis[7,4] the experimental polarizabilities may be summarized in the form given in Table 1. The quantities $\alpha_p, \beta_p, \alpha_n, \beta_n$

Table 1. Summary on electromagnetic polarizabilities in units of 10^{-4}fm^3

1		proton	neutron
2	BL sum rule	$(\alpha+\beta)_p = 13.9 \pm 0.3$	$(\alpha+\beta)_n = 15.2 \pm 0.5$
3	Compton scattering	$(\alpha-\beta)_p = 10.1 \pm 0.9$	$(\alpha-\beta)_n = 9.8 \pm 2.5$
4	BEFT sum rule	$(\alpha-\beta)^s_p = -5.0 \pm 1.0$	$(\alpha-\beta)^s_n = -5.0 \pm 1.0$
5	line 2 – line 3	$(\alpha-\beta)^t_p = 15.1 \pm 1.3$	$(\alpha-\beta)^t_n = 14.8 \pm 2.7$
6	experimental	$\alpha_p = 12.0 \pm 0.6$	$\alpha_n = 12.5 \pm 1.7$
7	s-channel only	$\alpha^s_p = 4.5 \pm 0.5$	$\alpha^s_n = 5.1 \pm 0.6$
8	t-channel only	$\alpha^t_p = 7.5 \pm 0.8$	$\alpha^t_n = 7.4 \pm 1.8$
9	experimental	$\beta_p = 1.9 \mp 0.6$	$\beta_n = 2.7 \mp 1.8$
10	s-channel only	$\beta^s_p = 9.5 \pm 0.5$	$\beta^s_n = 10.1 \pm 0.6$
11	t-channel only	$\beta^t_p = -7.6 \pm 0.8$	$\beta^t_n = -7.4 \pm 1.9$

are the experimental electric and magnetic polarizabilities for the proton and neutron, respectively. The quantities with an upper label s are the corresponding electric and magnetic polarizabilities where only the s-channel d.o.f. are included. These latter quantities have been obtained by making use of the fact that $(\alpha + \beta)$, when calculated from forward-angle dispersion theory as given by the Baldin or Baldin-Lapidus (BL) sum rule

$$(\alpha+\beta)=\frac{1}{2\pi^2}\int_{\omega_0}^{\infty}\frac{\sigma_{\rm tot}(\omega)}{\omega^2}d\omega, \tag{1}$$

has no t-channel contribution, i.e. $(\alpha+\beta)=(\alpha+\beta)^s$, and by using the estimate $(\alpha-\beta)^s_{p,n}=-5.0\pm1.0$ obtained form the s-channel part of the BEFT sum rule

$$(\alpha-\beta)^s=\frac{1}{2\pi^2}\int_{m_\pi+\frac{m^2\pi}{2M}}^{\infty}\sqrt{1+\frac{2\omega}{M}}\left[\sigma(\omega,E1,\cdots)-\sigma(\omega,M1,\cdots)\right]\frac{d\omega}{\omega^2} \tag{2}$$

both for the proton and the neutron (see Ref.[4]), where M is the nucleon mass. The numbers in line 5 of Table 1 are the t-channel contributions to $(\alpha-\beta)$ obtained from the experimental values $(\alpha-\beta)_p=10.1\pm0.9$ and $(\alpha-\beta)_n=9.8\pm2.5$ and the estimate for $(\alpha-\beta)^s_{p,n}$. We see that the experimental values for α are much larger than the s-channel contributions alone, whereas for the magnetic polarizabilities the opposite is true. For the magnetic polarizability it makes sense to identify the large difference between the experimental value and the s-channel contribution with the diamagnetic polarizability. This difference is filled up by β^t which, therefore, may be considered as the diamagnetic polarizability. It is important to notice that this definition of the diamagnetic polarizability has nothing in common with the "classical" diamagnetic term given e.g. in Eq. (12) of Ref.[4]. This latter term makes use of s-channel degrees of freedom only and, therefore, does not describe the physical origin of the diamagnetism.

3. The σ-pole and the BEFT sum rule

The σ meson having quantum numbers $I=0$ and $J=0$ may be considered as a quasi-stable $1/\sqrt{2}(|u\bar{u}\rangle+|d\bar{d}\rangle)$ $1^3\mathrm{P}_0$ state in a confining potential which is coupled to a di-pion state in the continuum $1/\sqrt{3}(|\pi^+\pi^-\rangle-|\pi^0\pi^0\rangle+|\pi^-\pi^+\rangle)$, showing up as a relative $S-$wave of the two pions. By exploiting the non-strange $q\bar{q}$ structure component we are led to the following relations for the σ-pole

$$F_{\sigma\gamma\gamma}=|M(\sigma\to2\gamma)|=\frac{\alpha_e}{\pi f_\pi}N_c\left[\left(\frac{2}{3}\right)^2+\left(-\frac{1}{3}\right)^2\right]=\frac{5}{3}\frac{\alpha_e}{\pi f_\pi} \tag{3}$$

$$(\alpha-\beta)^t_{p,n}=\frac{g_{\sigma NN}F_{\sigma\gamma\gamma}}{2\pi m_\sigma^2}=\frac{5\alpha_e g_{\pi NN}}{6\pi^2m_\sigma^2f_\pi}=15.3\times10^{-4}\mathrm{fm}^3 \tag{4}$$

where $\alpha_e=1/137.04$, $N_c=3$, $m_\sigma=665$ MeV, $f_\pi=(92.42\pm0.26)$ MeV and $g_{\sigma NN}=g_{\pi NN}=13.169\pm0.057$ Ref.[8]. The $\pi\pi$ structure component leads to the t-channel part of the BEFT sum rule in the following form

$$(\alpha-\beta)^t = \frac{1}{16\pi^2}\int_{4m_\pi^2}^{\infty}\frac{dt}{t^2}\frac{16}{4M^2-t}\left(\frac{t-4m_\pi^2}{t}\right)^{1/2} \times \left[f_+^0(t)F_0^{0*}(t) - \left(M^2-\frac{t}{4}\right)\left(\frac{t}{4}-m_\pi^2\right)f_+^2(t)F_0^{2*}(t)\right] \quad (5)$$

where $f_+^{J=0,2}(t)$ is the partial wave amplitude of the process $N\bar{N}\rightarrow\pi\pi$ and $F_{I=0}^{J=0,2}(t)$ the partial wave amplitude of the process $\pi\pi\rightarrow\gamma\gamma$. The predictions of the BEFT sum rule are $(\alpha-\beta)^t_{p,n} = +(14\pm 2)$ (Levchuk et al., see Ref.[4]), $(\alpha-\beta)^t_p = +16.5$ (Drechsel et al.[2]) with the arithmetic average $(\alpha-\beta)^t_{p,n} = +15.3\pm 1.3$.

Table 2 summarizes the results obtained for $(\alpha-\beta)^t_{p,n}$. It is remarkable that

Table 2. Difference of electromagnetic polarizabilities $(\alpha-\beta)^t_{p,n}$ in units of 10^{-4}fm^3

	$(\alpha-\beta)^t_p$	$(\alpha-\beta)^t_n$
experiment	15.1 ± 1.3	14.8 ± 2.7
σ-pole	15.3	15.3
BEFT sum rule	15.3 ± 1.3	15.3 ± 1.3

the experimental values are in agreement with the prediction of the σ-pole as well as the prediction of the BEFT sum rule. This leads to the tentative conclusion that both approaches are equivalent and to a confirmation of the expression given in (3) for $F_{\sigma\gamma\gamma}$, treating the two-photon coupling of the σ-meson analogous to the π^0-meson case.

References

1. A.I. L'vov, V.A. Petrun'kin, M. Schumacher, *Phys. Rev.* **C55**, 359 (1997)
2. D. Drechsel, B. Pasquini, M. Vanderhaeghen, *Phys. Rep.* **378**, 99 (2003)
3. F. Wissmann, *Springer Tracts in Modern Physics*, **200**, 1 (2004)
4. M. Schumacher, *Progress in Particle and Nuclear Physics* **55**, 567 (2005) [hep-ph/0501167]
5. F.E. Low, *Phys. Rev.* **120** 582 (1960) (and reference therein); M. Jacob, J. Mathews, *Phys. Rev.* **117**, 854 (1960)
6. A.C. Hearn, E. Leader, *Phys. Rev.* **126**, 789 (1962) ; R. Köberle, *Phys. Rev.* **166**, 1558 (1968)
7. M. Schumacher, *Proceedings: P.A. Cherenkov and Modern Physics, Moscow, June 22–25*, (2004) [nucl-ex/0411048]
8. M. Nagy, M.D. Scadron, G.E. Hite, *Acta Physica Slovaca* **54**, 427 (2004) [hep-ph/0406009]; M.D. Scadron et al., *Phys. Rev.* **D 69**, 014010 (2004)

EXCITED BARYONS IN THE $1/N_c$ EXPANSION

N. MATAGNE AND FL. STANCU
University of Liège, Physics Department,
Institute of Physics, B.5,
Sart Tilman, B-4000 Liège 1, Belgium
E-mail: nmatagne@ulg.ac.be, fstancu@ulg.ac.be

We review results for the mass spectrum of orbitally excited baryons obtained in the $1/N_c$ expansion. We show the dependence of various contributions to the mass operator as a function of the excitation energy.

1. Introduction

At low energies, typical for baryon spectroscopy, QCD does not admit a perturbative expansion in the strong coupling constant. About 30 years ago, 't Hooft suggested an alternative approach based on an $1/N_c$ expansion where N_c is the number of colors[1]. Witten described the counting rules for such an expansion[2]. The method works very well for the ground state baryons inasmuch as they display an SU($2N_f$) exact symmetry, where N_f is the number of flavors[3]. Although for excited states this symmetry is broken, in the last few years it has been realized that an $1/N_c$ expansion can as well be used to describe states belonging to various SU(6) multiplets. A particular attention has been paid to the $[\mathbf{70}, 1^-]$ multiplet[4–10].

Here we review recent work on the mass spectrum of baryons in the $N = 2$ and $N = 4$ bands. We are especially concerned with orbitally excited baryons belonging to the $[\mathbf{56}, 2^+]$, Ref. [14], $[\mathbf{70}, \ell^+]$ ($\ell = 0, 2$), Ref. [15] and $[\mathbf{56}, 4^+]$, Ref. [16], multiplets.

2. Mass operator

The mass operator up to $\mathcal{O}(1/N_c)$ has the general form

$$M = \sum c_i O_i + \sum b_i \bar{B}_i \tag{1}$$

where the operators O_i are SU(3)-flavor singlets and the operators $\bar{B}_i$, which are defined to have vanishing expectation values for non-strange states, break the flavor symmetry. The coefficients c_i and b_i that encode QCD dynamics are evaluated by a numerical fit to the available data. The operators O_i and $\bar{B}_i$ can be expressed as positive parity and rotationally invariant products of generators of SU(6) $\otimes$ O(3).

The analysis of symmetric $[\mathbf{56}, \ell^+]$ states is rather simple[14,16]. The total wave functions are obtained by coupling an orbital part $\sim Y_{\ell m}$ to spin-flavor symmetric states. But for mixed-symmetric representations, it is necessary to split the wave function into two parts : a symmetric core composed of $N_c - 1$ quarks and a excited quark. Generally, in the case of the mixed symmetric representations, one has both core generators ℓ^i_c, S^i_c, T^a_i and G^{ia}_c and excited quark generators ℓ^i_q, s^i, t^a and g^{ia}. The multiplet $[\mathbf{70}, 1^-]$ is a particular case with $\ell_c = 0$.

As an illustration, Table 1 gives the list of operators chosen for the study of the $[\mathbf{70}, \ell^+]$ multiplets[15]. O_1 is the SU(6) scalar operator of order N_c, O_2 and O_4 are the dominant parts of the spin-orbit and spin-spin operators respectively. Note that O_2 is $\mathcal{O}(N_c^0)$ for mixed-symmetric states, in contrast to the symmetric states case where it is $\mathcal{O}(N_c^{-1})$, see Refs. [15, 17]. $O_3 \sim \mathcal{O}(N_c^0)$ due to G^{ja}_c. Strange baryons are not included in this study. Table 1 contains the values of the coefficients c_i obtained by fitting the available experimental data. We present in Table 2 the masses of the resonances which we have interpreted as belonging to the $[\mathbf{70}, 0^+]$ or $[\mathbf{70}, 2^+]$ multiplets.

Table 1. List of operators and the coefficients resulting from the fit with $\chi^2_{\rm dof} \simeq 0.83$ for the $[\mathbf{70}, \ell^+]$ multiplets.

Operator	Fitted coef. (MeV)
$O_1 = N_c \mathbb{1}$	$c_1 = \;\; 555 \;\; \pm \;\; 11$
$O_2 = \ell^i_q s^i$	$c_2 = \;\; 47 \;\; \pm \;\; 100$
$O_3 = \frac{3}{N_c} \ell_q^{(2)ij} g^{ia} G_c^{ja}$	$c_3 = \;\; -191 \;\; \pm \;\; 132$
$O_4 = \frac{1}{N_c}(S^i_c S^i_c + s^i S^i_c)$	$c_4 = \;\; 261 \;\; \pm \;\; 47$

3. The dependence of the coefficients c_i on the excitation energy

It is interesting to see the change of c_i with the excitation energy. In Figure 1, we collect the presently know values of c_i $(i = 1, 2, 4)$ with error bars for the orbitally excited states studied so far in the large N_c expansion: $N = 1$,

Table 2. The partial contribution and the total mass (MeV) predicted for the $[\mathbf{70}, \ell^+]$ by the $1/N_c$ expansion as compared with the empirically known masses.

	$1/N_c$ expansion results						
	Partial contribution				Total	Empirical	Name, status
	c_1O_1	c_2O_2	c_3O_3	c_4O_4			
$^4N[\mathbf{70}, 2^+]\frac{7}{2}^+$	1665	31	42	217	1956 ± 95	2016 ± 104	$F_{17}(1990)$**
$^2N[\mathbf{70}, 2^+]\frac{5}{2}^+$	1665	10	0	43	1719 ± 34		
$^4N[\mathbf{70}, 2^+]\frac{5}{2}^+$	1665	-5	-106	217	1771 ± 88	1981 ± 200	$F_{15}(2000)$**
$^4N[\mathbf{70}, 0^+]\frac{3}{2}^+$	1665	0	0	217	1883 ± 17	1879 ± 17	$P_{13}(1900)$**
$^2N[\mathbf{70}, 2^+]\frac{3}{2}^+$	1665	-16	0	43	1693 ± 42		
$^4N[\mathbf{70}, 2^+]\frac{3}{2}^+$	1665	-31	0	217	1851 ± 69		
$^2N[\mathbf{70}, 0^+]\frac{1}{2}^+$	1665	0	0	43	1709 ± 25	1710 ± 30	$P_{11}(1710)$***
$^4N[\mathbf{70}, 2^+]\frac{1}{2}^+$	1665	-47	149	217	1985 ± 26	1986 ± 26	$P_{11}(2100)$*
$^2\Delta[\mathbf{70}, 2^+]\frac{5}{2}^+$	1665	-10	0	87	1742 ± 29	1976 ± 237	$P_{35}(2000)$**
$^2\Delta[\mathbf{70}, 2^+]\frac{3}{2}^+$	1665	16	0	87	1768 ± 38		
$^2\Delta[\mathbf{70}, 0^+]\frac{1}{2}^+$	1665	0	0	87	1752 ± 19	1744 ± 36	$P_{31}(1750)$*

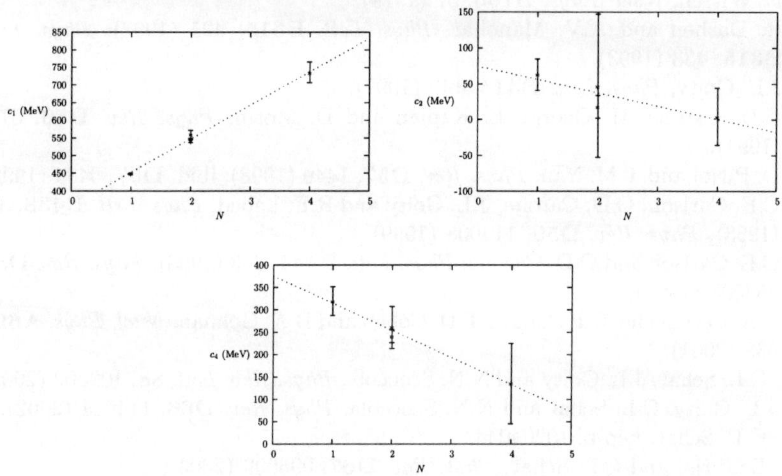

Fig. 1. Evolution of the coefficients c_i with the excitation energy corresponding to $N = 1, 2$ and 4. The straight lines are to guide the eye.

Ref. [10], $N = 2$ (lower values[14], upper values[15]) and $N = 4$, Ref. [16]. This behavior shows that at large energies the dominant contribution comes from c_1 and the contributions of the spin-dependent terms vanish. These results are consistent with the quark model picture where the linear term in N_c contains the free mass term, the kinetic and the confinement energy. An intuitive model based on the chiral symmetry restoration has already predicted that the spin dependent interactions vanish at high energies[18].

4. Conclusions

Our work is based on the assumption that there is no multiplet mixing and is restricted to non-strange baryons. Future work is devoted to strange baryons. To better understand the applications of the $1/N_c$ expansion more and better data is desirable.

Acknowledgments

The work of one of us (N. M.) was supported by the Institut Interuniversitaire des Sciences Nucléraires (Belgium).

References

1. G. 't Hooft, *Nucl. Phys.* **72**, 461 (1974).
2. E. Witten, *Nucl. Phys.* **B160**, 57 (1979).
3. R. Dashen and A.V. Manohar, *Phys. Lett.* **B315**, 425 (1993); *Phys. Lett.* **B315**, 438 (1993).
4. J.L. Goity, *Phys. Lett.* **B414**, 140 (1997).
5. C.D. Carone, H. Georgi, L. Kaplan and D. Morin, *Phys. Rev.* **D50**, 5793 (1994).
6. D. Pirjol and T.M. Yan, *Phys. Rev.* **D57**, 1449 (1998); ibid. **D57**, 5434 (1998).
7. C.E. Carlson, C.D. Carone, J.L. Goity and R.F. Lebed, *Phys. Lett.* **B438**, 327 (1998); *Phys. Rev.* **D59**, 114008 (1999).
8. C.E. Carlson and C.D. Carone, *Phys. Lett.* **B441**, 363 (1998); *Phys. Rev.* **D58**, 053005 (1998).
9. Z.A. Baccouche, C.K. Chow, T.D. Cohen and B.A. Gelman, *Nucl. Phys.* **A696**, 638 (2001).
10. C.L. Schat, J.L. Goity and N.N. Scoccola, *Phys. Rev. Lett.* **88**, 102002 (2002); J.L. Goity, C.L. Schat and N.N. Scoccola, *Phys. Rev.* **D66**, 114014 (2002).
11. C.L. Schat, hep-ph/0204044.
12. D. Pirjol and C.L. Schat, *Phys. Rev.* **D67**, 096009 (2003).
13. T.D. Cohen, D.C. Dakin, A. Nellore, R.F. Lebed, *Phys. Rev.* **D69**, 056001 (2004).
14. J.L. Goity, C.L. Schat and N.N. Scoccola, *Phys. Lett.* **B564**, 83 (2003).
15. N. Matagne and Fl. Stancu, *Phys. Lett.* **B631**, 7 (2005) [arXiv:hep-ph/0505118].
16. N. Matagne and Fl. Stancu, *Phys. Rev.* **D71**, 014010 (2005).
17. J.L. Goity, *Phys. Atom. Nucl.* **68**, 624-633 (2005); *Yad. Fiz.* **68**, 655-664 (2005); J.L. Goity, hep-ph/0504121.
18. L.Ya. Glozman, *Phys. Lett.* **B541**, 115 (2002).

SURPRISES IN $2\pi^\circ$ PRODUCTION BY π^- AND K^- AT INTERMEDIATE ENERGIES

B. M. K. NEFKENS
Department of Physics and Astronomy
University of California Los Angeles
Los Angeles, CA 90095-1547, USA

We report on our investigations of $2\pi^\circ$ production on protons by π^- and K^- at the AGS using the Crystal Ball multiphoton spectrometer. The incident beam momenta cover 0.5 to 0.75 GeV/c. The first surprise is the strong dependence of the density distribution of the $\pi^\circ\pi^\circ n$, $\pi^\circ\pi^\circ\Lambda$ and $\pi^\circ\pi^\circ\Sigma^\circ$ Dalitz Plots on the angle of the $2\pi^\circ$ system. The next surprise is the occurence of two peaks in the $2\pi^\circ$ invariant mass distribution of the $\pi^- p \to \pi^\circ\pi^\circ n$ and $K^- p \to \pi^\circ\pi^\circ\Lambda$ final states, but only one peak in $K^- p \to \pi^\circ\pi^\circ\Sigma$. The third surprise is the lack of direct evidence for the occurence of a $f_0(600)$ or σ meson. The final and most significant surprise is the applicability of (broken) SU(3) flavor symmetry when comparing $\pi^- p \to \pi^\circ\pi^\circ n$ to $K^- p \to \pi^\circ\pi^\circ\Lambda$ and $K^- p \to \pi^\circ\pi^\circ\Sigma$.

Our knowledge of the light scalar mesons is quite limited. There is even uncertainty over the name of the lightest member: The Review of Particle Physics[1] calls it the $f_0(600)$, the traditional symbol is σ. We will use $f_0(600)$ as the letter σ is needed for cross sections. The mass of the $f_0(600)$ is quoted[1] to be somewhere between 400 and 1200 MeV/c and the width between 600 and 1000 MeV. A good way for studying the properties of the $f_0(600)$ should be in $2\pi^\circ$ production by K^-, π^-, and γ on protons. The possible isospin values for two pion systems are I = 0,1,2. The favored production process is

$$K^- p \to \pi^\circ\pi^\circ\Lambda \tag{1}$$

as its final state has pure I = 0. There are five independent variables for a 3-body final state, such as the invariant masses of the $2\pi^\circ$ and $\pi^\circ\Lambda$, the $\theta_{\pi\pi}$ and ϕ angles and the overall CM energy W.

When the Λ of the $\pi^\circ\pi^\circ\Lambda$ final state decays into a π° and neutron there is a total of three π°'s resulting in six photons. Thus, the detector for Reaction (1) should have a high efficiency for the measurement of photons. This calls for a multiphoton spectrometer with a near–4π acceptance. The

Crystal Ball[2] is such a device. We have used it to make measurements[2,3] on $2\pi^\circ$ production by π^- and K^- using the AGS at the Brookhaven National Laboratory.

For the interpretation of our data we shall employ the isobar model that describes the production of a 3-body final state as a sequence of two-body processes. The first step in describing $K^- p \to \pi^\circ\pi^\circ\Lambda$ is the formation of intermediate-state hyperons, specifically, Λ^*'s. In our experiment which used eight incident K^- beams with momenta 0.52 to 0.75 GeV/c, or W from 1.57 to 1.68 GeV, the important intermediate states are the $\Lambda(1600)\frac{1}{2}^+$, $\Lambda(1670)\frac{1}{2}^-$, and $\Lambda(1690)\frac{3}{2}^-$, and they contribute with different strengths. The Λ^* decays occur likely by two parallel sequences. The first is $\Lambda^* \to \pi^\circ + \Sigma(1385)\frac{3}{2}^+$, followed by $\Sigma(1385) \to \pi^\circ + \Lambda$. The other is $\Lambda^* \to f_0(600) + \Lambda$, followed by $f_0(600) \to \pi^\circ + \pi^\circ$. Thus there are at least six important processes which must be considered when describing $K^- p \to \pi^\circ\pi^\circ\Lambda$.

A practical way for looking at the results of 3-body final state reactions is by Dalitz plots (DP). Some examples are shown in Fig. 1. For the axes we

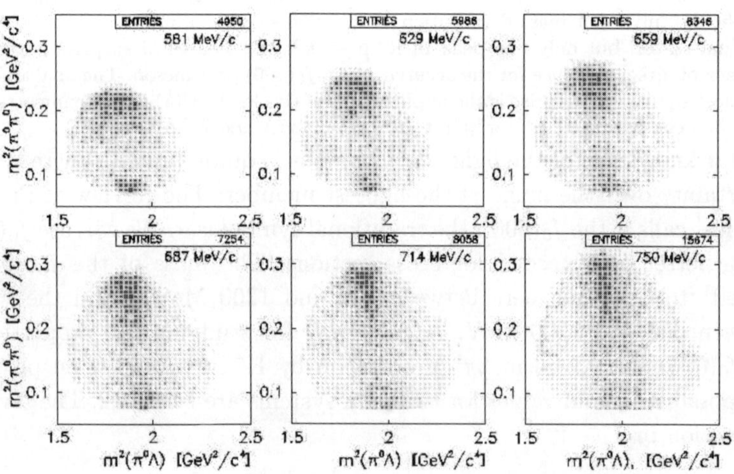

Fig. 1. Dependence of the Dalitz plot for $K^- p \to \pi^\circ\pi^\circ\Lambda$ on the incident beam momentum.

use the square of the invariant mass $m^2(x, y)$. This has the great virtue that any area in the DP is proportional to the Lorentz-invariant phase space. The horizontal axis is $m^2(\pi^\circ\Lambda)$ and the vertical $m^2(\pi^\circ\pi^\circ)$. The signature for a strong participation of the $\Sigma(1385)$ is a vertical band in the DP, this band is centered on the mass squared of the $\Sigma(1385)$ intermediate state and reflects the width of the $\Sigma(1385)$. Note that the $\Sigma(1385)$ reaction features two amplitudes for $\pi^\circ + \Lambda \to \Sigma(1385)$ due to the coupling of both π°'s in the final state to the Λ. The interference of these two amplitudes distorts the vertical band in the Dalitz plot giving rise to two separate high concentrations see Fig. 1. They are referred to as "islands" of high intensity separated by a region of low intensity.

The occurence of the $f_o(600)$ should manifest itself by a broad *uniform* horizontal band in the DP defined by the mass and width of the $f_0(600)$. Fig. 1 clearly shows that the sequence with the $\Sigma(1385)$ intermediate state is favored over the $f_0(600)$.

Next, we study the dependence of the DP on the dipion angle $\theta_{\pi\pi}$. This angle is the compliment of the angle of the recoil Λ. We have divided $\theta_{\pi\pi}$ into four quadrants 0–45°, 45°–90°, 90°–135° and 135°–180°. Shown in Fig. 2 are the four DP's for the events in the four quadrants. To

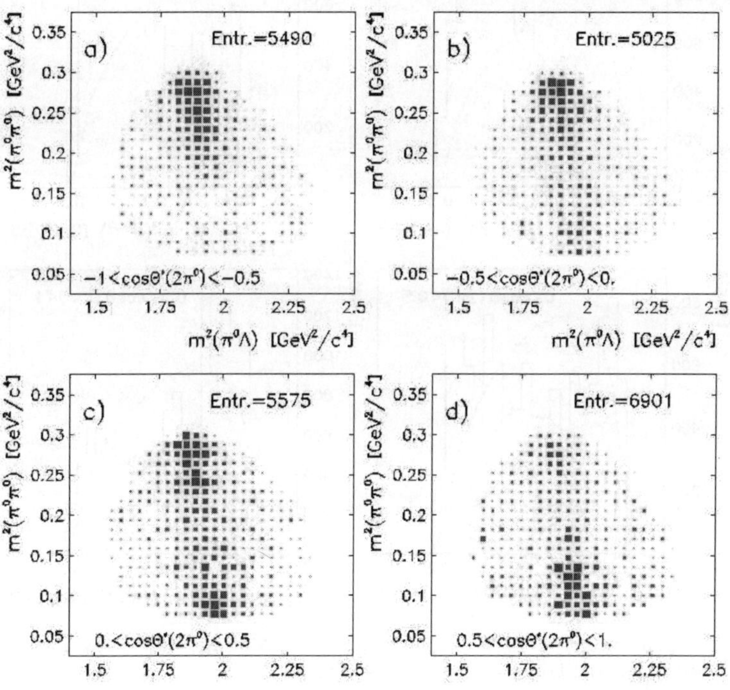

Fig. 2. Dependence of the Dalitz plot for $K^-p \to \pi^\circ\pi^\circ\Lambda$ at $p_K = 0.71$ GeV/c on $\cos\theta_{\pi\pi}$.

our surprise we found that the main features of the DP depend strongly on the $\theta_{\pi\pi}$ angle. The DP of the events in the backward quadrant, see Fig. 2a, is dominated by one island with many events covering a small area of phase space and located in the high $m^2(\pi^\circ\pi^\circ)$ region of the DP. In the next quadrant, see Fig. 2b, the DP has a band-like structure depleted in the middle. A somewhat similar description holds for the DP of events in the third quadrant in Fig. 2c. Finally, the DP for the forward region, see Fig. 2d, is dominated by an island in the low $m^2(\pi^\circ\pi^\circ)$ region and there is a very small island at large $m^2(\pi^\circ\pi^\circ)$. The salient features of a DP can be presented in quantitative form by means of the projections of the DP density on the invariant-mass-squared axes; the quadrant projections on the $m^2(\pi^\circ\pi^\circ)$ axes are shown in Fig. 3. All four projections are double

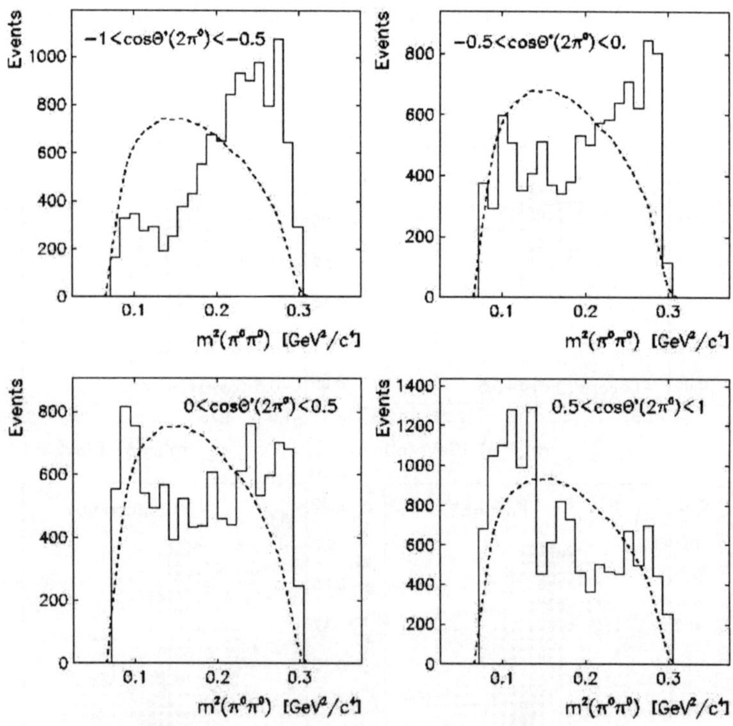

Fig. 3. Dependence of $m^2(\pi^\circ\pi^\circ)$ from $\mathrm{K^-p} \to \pi^\circ\pi^\circ\Lambda$ at $p_\mathrm{K} = 0.71$ GeV/c on $\cos\theta_{\pi\pi}$. The dashed line is the Lorentz-invariant phase space.

humped. In the backward section the large hump occurs close to maximum $m^2(\pi^\circ\pi^\circ)$ around $m^2 \simeq (0.53\ \mathrm{GeV}/c)^2$. The small hump is near the lowest

m^2 around $m^2 \simeq (0.31 \text{ GeV}/c)^2$. In the forward quadrant it is reversed; the large hump is near threshold and the little hump is near the maximum.

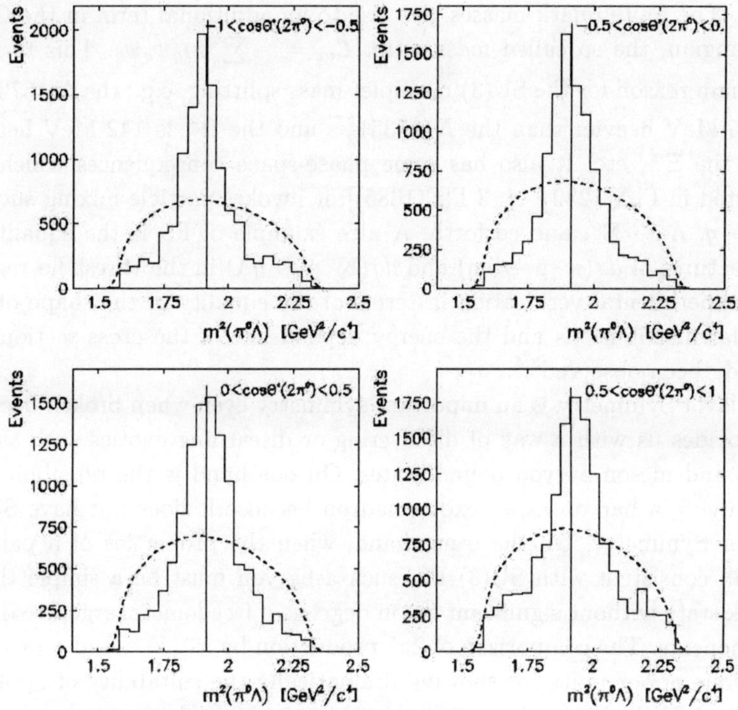

Fig. 4. Dependence of $m^2(\pi^\circ\Lambda)$ from $\text{K}^-\text{p} \rightarrow \pi^\circ\pi^\circ\Lambda$ at $p_\text{K} = 0.71$ GeV/c on $\cos\theta_{\pi\pi}$. The dashed line is the Lorentz-invariant phase space.

The DP density projection on the $m^2(\pi^\circ\Lambda)$ axis is shown in Fig. 4. All four figures are mono peaked; the maximum and the width of these peaks reflect the dominance of the $\pi^\circ\Lambda$ interaction via the $\Sigma(1385)\frac{3}{2}^+$ state. We estimate from Fig. 4 that the $f_0(600)$ is less than 25% of the $\Sigma(1385)$ contribution at $p_\text{K} = 0.7$ GeV/c.

It is of some interest to compare the features of $2\pi^\circ$ production in $\pi^\circ\text{p} \rightarrow \pi^\circ\pi^\circ\text{n}$ to those in $\text{K}^-\text{p} \rightarrow \pi^\circ\pi^\circ\Lambda$. For this we make use of (broken) SU(3) Flavor Symmetry, FS. In the theoretical limit of massless quarks the six different quark flavors are governed by the same QCD Lagrangian. Thus the interaction strength of all massless quarks is the same. This is the famous FS of QCD. If the 3 light quarks have SU(3) symmetry then SU(3) FS implies that all members of various SU(3) multiplets have the same mass and the features of FS–related reactions such as $\pi^-\mathrm{p} \to \eta\mathrm{n}$ and $\mathrm{K^-p} \to \eta\Lambda$ are the same, including angular distribution and energy dependence. In the real world the quarks have mass and FS is broken albeit not badly in most cases. The finite quark masses give rise to an additional term in the QCD Lagrangian, the so called mass term: $\mathcal{L}_m = -\sum_q m_q\bar{\psi}_q\psi_q$. This term is the main reason for the SU(3) multiplet mass splitting, e.g.: the $\Lambda(1670)\frac{1}{2}^-$ is 135 MeV heavier than the $N(1535)\frac{1}{2}^-$ and the Ω^- is 142 MeV heavier than the Ξ^-, etc. It also has some phase-space consequences which are reflected in $\Gamma[\Delta(1232)] \simeq 3\ \Gamma[\Sigma(1385)]$; it invokes particle mixing such as $\pi^\circ \leftrightarrow \eta$, $\Lambda \leftrightarrow \Sigma^\circ$, and so forth. A nice example of FS is the equality of the features of $d\sigma(\pi^-\mathrm{p} \to \eta\mathrm{n})$ and $d\sigma(\mathrm{K^-p} \to \eta\Lambda)$ in the threshold region. The experimental verification in terms of the equality of the shape of the angular distributions and the energy dependence of the cross section has already been observed[3,4].

Flavor Symmetry is an important symmetry even when broken because it provides us with a way of discovering or discarding exotics such as hybrids and meson-baryon bound states. On one hand is the possibility for identifying a baryon as an exotic hadron because it does not have SU(3) Flavor Symmetry. On the other hand, when the properties of a particle are all consistent with SU(3) FS, such a baryon must be a simple three-quark state without significant gluon degrees of freedom or large five quark components. The comparison of $2\pi^\circ$ production by π^-, K^-, and γ provides a unique opportunity for showing dramatically the suitability of applying (broken) SU(3) FS to the complex final state of 3-body processes.

Consider the following four reaction sequences:

$$\pi^-\mathrm{p} \to \mathrm{N}^* \to \pi^\circ\Delta(1232)\tfrac{3}{2}^+ \to \pi^\circ\pi^\circ\mathrm{n}, \tag{2}$$

$$\mathrm{K^-p} \to \Lambda^* \to \pi^\circ\Sigma(1385)\tfrac{3}{2}^+ \to \pi^\circ\pi^\circ\Lambda, \tag{3}$$

$$\mathrm{K^-p} \to \Sigma^* \to \pi^\circ\Lambda(1405)\tfrac{1}{2}^- \to \pi^\circ\pi^\circ\Sigma^\circ, \tag{4}$$

$$\gamma\mathrm{p} \to \mathrm{N}^* \to \pi^\circ\Delta(1232)\tfrac{3}{2}^+ \to \pi^\circ\pi^\circ\mathrm{n}. \tag{5}$$

It is clear that reaction (2) and (3) have the same SU(3) flavour structure. (3) and (4) have different intermediate states; the $\Sigma(1385)\frac{3}{2}^+$ is a member of a decuplet which has a totally symmetric flavor wave function while the $\Lambda(1405)\frac{1}{2}^-$ is a SU(3) signlet with a totally antisymmetric flavor wave function. Reactions (2) and (5) have the same $\pi\Delta$ intermediate state, however (5) is an electromagnetic process which is not subject to SU(3) Flavor Symmetry. Thus we come to the following prediction: $d^5\sigma(\pi^-\mathrm{p} \to \pi^\circ\pi^\circ\mathrm{n})$ should have the same features as $d^5\sigma(\mathrm{K^-p} \to \pi^\circ\pi^\circ\Lambda)$

but those of $d^5\sigma(\mathrm{K^-p} \to \pi^\circ\pi^\circ\Sigma)$ should be quite different. The similarity is perhaps best illustrated using the dependence of the DP on $\theta_{\pi\pi}$. Shown

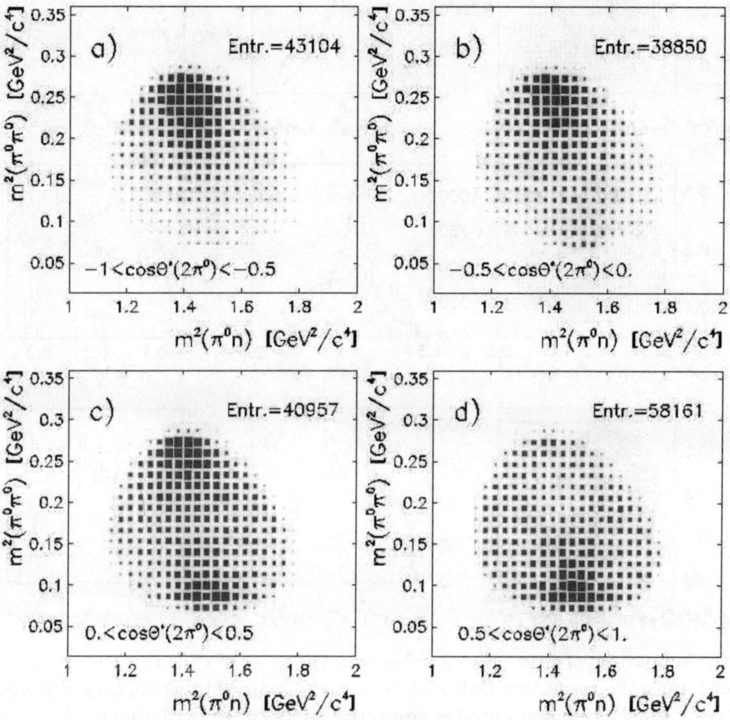

Fig. 5. Dependence of the Dalitz plot for $\pi^-\mathrm{p} \to \pi^\circ\pi^\circ\mathrm{n}$ at $p_\pi = 0.66$ GeV/c on $\cos\theta_{\pi\pi}$.

in Fig. 5 is the DP for the four quadrants of the reaction $\pi^-\mathrm{p} \to \pi^\circ\pi^\circ\mathrm{n}$. This is remarkably similar to that for $\mathrm{K^-p} \to \pi^\circ\pi^\circ\Lambda$, see Fig. 2. The main difference is in the width. The $\Sigma(1385)$ intermediate state is narrower than the $\Delta(1232)$ because there is less phase-space available.

A good overview of the features of $2\pi^\circ$ production by π^- and $\mathrm{K^-}$ is given in Fig. 6 where a–c represent $\pi^-\mathrm{p} \to \pi^\circ\pi^\circ\mathrm{n}$, d–f represent $\mathrm{K^-p} \to \pi^\circ\pi^\circ\Lambda$, and g–i represent $\mathrm{K^-p} \to \pi^\circ\pi^\circ\Sigma$. The DP for $\pi^-\mathrm{p} \to \pi^\circ\pi^\circ\mathrm{n}$ is very similar to that for $\mathrm{K^-p} \to \pi^\circ\pi^\circ\Lambda$, and both are dominated by a high density island. In contrast, the DP for $\mathrm{K^-p} \to \pi^\circ\pi^\circ\Sigma$ is dissimilar as it is rather uniformly populated. The $m^2(\mathrm{n}\pi^\circ)$ distribution for $\pi^-\mathrm{p} \to \pi^\circ\pi^\circ\mathrm{n}$ is dominated by a large peak with concave shaped sides, and has the features of the $\Delta(1232)$. The $m^2(\Lambda\pi^\circ)$ spectrum for $\mathrm{K^-p} \to \pi^\circ\pi^\circ\Lambda$ is also dominated by a large

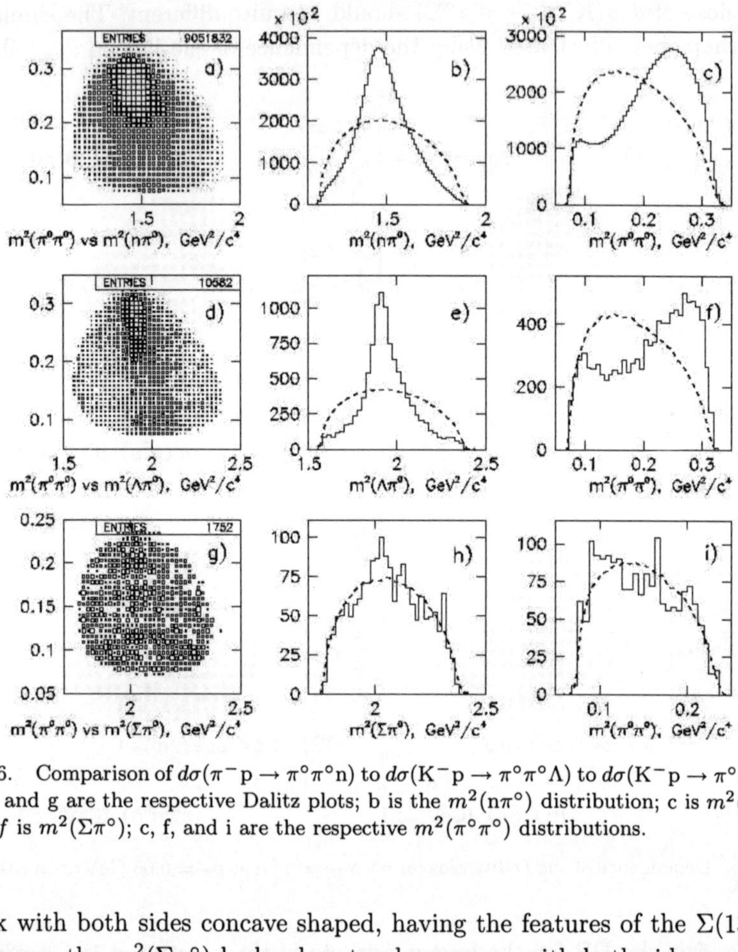

Fig. 6. Comparison of $d\sigma(\pi^-\mathrm{p} \to \pi^\circ\pi^\circ\mathrm{n})$ to $d\sigma(\mathrm{K}^-\mathrm{p} \to \pi^\circ\pi^\circ\Lambda)$ to $d\sigma(\mathrm{K}^-\mathrm{p} \to \pi^\circ\pi^\circ\Sigma)$. a, d, and g are the respective Dalitz plots; b is the $m^2(\mathrm{n}\pi^\circ)$ distribution; c is $m^2(\Lambda\pi^\circ)$ and f is $m^2(\Sigma\pi^\circ)$; c, f, and i are the respective $m^2(\pi^\circ\pi^\circ)$ distributions.

peak with both sides concave shaped, having the features of the $\Sigma(1385)$. However, the $m^2(\Sigma\pi^\circ)$ looks close to phase-space with both sides concave shaped.

The $m^2(\pi^\circ\pi^\circ)$ distribtuion for $\pi^-\mathrm{p} \to \pi^\circ\pi^\circ\mathrm{n}$ and $\mathrm{K}^-\mathrm{p} \to \pi^\circ\pi^\circ\Lambda$ are double peaked with similar shapes. The main peak occurs at high $m^2(\pi^\circ\pi^\circ)$, while $m^2(\pi^\circ\pi^\circ)$ for $\mathrm{K}^-\mathrm{p} \to \pi^\circ\pi^\circ\Sigma$ has only one meager peak at low $m^2(\pi^\circ\pi^\circ)$. Thus, the reactions $\pi^-\mathrm{p} \to \pi^\circ\pi^\circ\mathrm{n}$ and $\mathrm{K}^-\mathrm{p} \to \pi^\circ\pi^\circ\Lambda$ have very similar features. They are different than for $\mathrm{K}^-\mathrm{p} \to \pi^\circ\pi^\circ\Sigma$ as predicted by (broken) SU(3) Flavor Symmetry. This bodes well for searches for exotic baryonic matter based on the applicability of broken SU(3) Flavor Symmetry. Other interesting features are the major dependence of the Dalitz plots on the dipion angle $\theta_{\pi\pi}$. The $2\pi^\circ$ invariant mass spectra for $\pi^-\mathrm{p} \to \pi^\circ\pi^\circ\mathrm{n}$ and $\mathrm{K}^-\mathrm{p} \to \pi^\circ\pi^\circ\Lambda$ are characterized by two peaks while

$K^-p \rightarrow \pi^\circ\pi^\circ\Sigma$ has only one. There is no clear evidence for a substantial role of the $f_0(600)$.

Acknowledgments

This work is supported in part by US DOE.

References

1. S. Eidelman et al. (Particle Data Group), *Phys. Lett.* **B592**, 1 (2004).
2. S. Prakhov et al. (Crystal Ball Collab.), *Phys. Rev.* **C69**, 042202 (2004).
   ibid., *Phys. Rev.* **C69**, 045202 (2004).
   ibid., *Phys. Rev.* **C70**, 034605 (2004).
3. A. Starostin et al. (Crystal Ball Collab.), *Phys. Rev.* **C64**, 055205 (2001).
4. B.M.K. Nefkens et al. (NSTAR 2001), Mainz, Germany, *World Scientific*, 427 (2001).

SUMMARY OF THE BARYON RESONANCE ANALYSIS GROUP MEETING

S. CAPSTICK

Department of Physics,
Florida State University,
Tallahassee, FL 32306-4350, USA
E-mail: capstick@phy.fsu.edu

The Baryon Resonance Analysis Group (BRAG) met in Tallahassee the day before the NSTAR 2005 meeting, and discussed the current status and future of coupled-channel analysis, the publication of baryon resonance data, and how the BRAG could enhance the interaction of groups involved in coupled-channel analysis. A useful summary of the techniques used and groups currently involved in this effort was made, and is outlined here.

1. Coupled-channel analysis

The assumption that a resonance exists in analyses of elastic pion-nucleon data has often been used in the analysis of inelastic scattering data. It is agreed that progress in understanding baryon resonances depends on going beyond elastic pion-nucleon scattering. For example, the inclusion of both elastic and inelastic intermediate channels states is required to ensure unitarity of the pion-nucleon scattering amplitude. Similar amplitudes that involve very different baryon resonance parameters can result from analyses of elastic scattering data in isolation. However, those different parameters prescribe the general behavior in other channels, some of which may be measured accurately. This makes necessary coupled-channel analysis of all important data involving baryon resonances.

1.1. *Coupled-channel analysis approaches*

One important goal for the BRAG is to establish a single approach to coupled-channel analysis, the ingredients of which we agree on. For example, although there is more than one way to impose unitarity on two-body scattering amplitudes, one could ask if there is a best way. A more difficult issue is the imposition of unitarity constraints on three-body final states. Progress can be made by the quasi-two-body approach, which writes three-body states at a given energy as a superposition of an unstable particle and a stable particle, restricting the sum to those states important at that energy. It was pointed out that this approximation results in unphysical cuts in the scattering amplitude. The development of approaches which involve true three-body unitarity is important, but is difficult.

1.2. *Partial waves or observables*

An important issue is whether the goal of coupled-channel analysis should be to directly fit the entire database of scattering observables, such as differential and total cross sections and polarization asymmetries, instead of partial-wave amplitudes. In the case of (strong interaction stable) two-body final states, generally more useful at lower energy, it has been useful to first fit partial waves to the data using as little model input as possible, and then these partial waves are later fit with various models of the reaction. In the case of three-body final states, adopting this approach requires a quasi-two-body description of the final state, which has its limitations. Although initial progress may require extraction of and fitting to quasi-two-body partial waves, it was agreed that ultimately it is important for coupled-channel analysis to fit directly to scattering observables where three-body channels are open. It may be true that existing energy-dependent partial wave analyses, upon which many coupled-channel analyses are based, are poor.

1.3. *Including electromagnetic production amplitudes*

Although the inclusion of photoproduction amplitudes has a small effect on the description of hadron-production data, it was stressed that the inclusion of photoproduction data into coupled-channel analysis is crucial. This is because the fractional uncertainty in the data is what is important, and if a given set of baryon resonance parameters fits somewhat uncertain hadron production data well, but does not properly fit well measured photoproduction amplitudes, the result will be a large χ^2 and a poor overall fit. The precise photoproduction data resulting from recent experiments has a strong effect on the baryon resonance parameters resulting from coupled-channel analysis.

Recent electro-production data also has a high level of precision, and it was pointed out that in addition to information on baryon resonance masses, widths and decay branching fractions, this data contains important information on the momentum-transfer dependence of transitions amplitudes between baryons. Transition form factors have now been measured in this way for all four-star resonances whose parameters are reasonably well determined in hadronic scattering experiments. Coverage in the Δ region is the most extensive, with 16 observables measured in pion electroproduction experiments. Incorporation of this information into coupled-channel analysis will provide important constraints.

1.4. *Polarization and a complete set of experiments*

Recent and upcoming experiments will measure polarization (or spin) observables for photoproduction processes involving baryon resonances. It is clear from examining the SAID and MAID analyses that differences between the predictions for these polarization observables can be large, and so conversely they provide strong constraints on the analysis.

An interesting question is whether it is possible to design a complete set of experiments which will uniquely determine the scattering amplitude for a given process. Results from photoproduction experiments in the resonance region to measure single polarization observables are currently being analyzed, and experiments to measure double polarization observables will be carried out in the near future. For photoproduction of baryon ($J^P = 1/2^+$) pseudoscalar-meson final states, in principle eight experiments are required for a unique scattering amplitude to result. For example, in $K\Lambda$ and $K\Sigma$ photoproduction eight experiments are required to measure 30 observables. This set of experiments has the advantage of isospin selectivity and strong stability of the two-body final states. In the absence of such complete sets of measurements, the inclusion of data on polarization observables into coupled-channel analysis fits is likely to have strong effects on the extracted parameters for resonances at higher energies.

2. Publication of Baryon Resonance Data

It was agreed that a goal of the BRAG should be to publish a summary of baryon resonance parameters in a refereed journal, after the authors of the experimental papers upon which these results are based are consulted. The BRAG can have an important impact on the issue of the distiction between T-matrix pole positions and Breit-Wigner masses and widths for a resonance. Certainly Breit-Wigner fits need to be made simultaneously in all channels, but there also exist more than one possible Breit-Wigner pole parameterization. The BRAG should come up with a convention for the form of Breit-Wigner which is used to fit amplitudes, and insist that publications are explicit about the relationship between pole positions and energy dependence in the region of the pole, and the Breit-Wigner parameters. It was agreed that pole positions should be published only for four-star resonances, and that the position of their pion-nucleon elastic poles is most physically relevant.

A desire was expressed to establish a procedure for the BRAG to approve letters of intent and progress reports from its members. If a summary of recent work with preliminary results is communicated to the BRAG members, they can provide valuable input and suggestions, and this is likely to initiate collaborations among BRAG members.

3. Summary of groups, approaches and codes for coupled-channel partial-wave analysis

A detailed description of the theoretical approaches to the extraction of nucleon resonance parameters from meson-baryon reaction data is given by T.-S.H. Lee in these proceedings [1], which also describes how the various approaches are related to the underlying theory. During the BRAG meeting a brief description of these approaches and their state of development was made, and this is recorded here.

3.1. *Zagreb*

The Zagreb group [2] has performed a fit to the KH-80 πN elastic partial waves, plus total and differential cross sections for $\pi N \to K\Lambda$ and $\pi N \to \eta N$. Following the CMB-model approach, their coupled-channel T-matrices have two-body unitarity, and two-body analyticity results from the form of the channel propagators. Resonances are introduced into the s channel which couple to all channels, and a nonrelativistic (Lippmann-Schwinger) scattering equation is solved for the amplitudes. Nonresonant terms are represented by unphysical poles, two below threshold and one above the resonance region, in each channel. The model is fit to the partial waves and observables in the form of input T-matrix amplitudes, and the result is resonance pole positions.

3.2. *Pitt-Argonne-FSU*

This group uses the same approach as the Zagreb group, but fits the GWU (SAID) elastic πN amplitudes. The analysis of πN to the channels πN,

ηN, $K\Lambda$, ωN, and quasi-two-body $\pi\Delta$ and ρN channels is completed [3], and the work has been done to include photoproduction data in the fit [4]. A more physical treatment of non-resonant mechanisms is in development, by solving a scattering equation with driving terms given by simple t and u-channel exchanges. This approach is designed to sharply reduce the number of parameters required in the fit.

3.3. *Giessen/GWU*

The Giessen group [5] has a model of scattering from πN and γN to the two-body channels πN, ηN, $K\Lambda$, ωN using a K-matrix approach. As the real part of the channel propagator is dropped, the amplitudes do not satisfy constraints from analyticity, although this may not have a strong numerical impact. Baryon resonances with spin up to 5/2 are included and are described by bare poles which are dressed by the formalism, and non-resonant terms are provided by t-channel tree-level Born terms. The inclusion of the $\pi\pi N$ final state in a single effective channel is in development. The fit is to the SAID πN elastic and γN amplitudes, and differential cross sections in other channels.

Higher spin baryon resonances have been included using the approach of Pascalutsa [6] in a similar model developed by the GWU group [7].

3.4. *Kent State*

This S-matrix approach to analysis of strange meson production [8] introduces resonances via generalized multi-channel Breit-Wigner effects, and treats non-resonant terms by introducing unphysical poles. Data on strangeness photoproduction, and $\bar{K}N$ elastic and inelastic ($\pi\Lambda$, $\pi\Sigma$, $\pi\Sigma^*$, $\pi\Lambda^*$, K^*N, $\bar{K}\Delta$,...) channels are fit.

3.5. *Juelich/GWU/UGA*

The Juelich, George Washington University and University of Georgia group [9] has constructed a dynamical approach to the gauge-invariant treatment of photo- and electroproduction of pseudoscalar mesons off nucleons. The Juelich hadronic T-matrices are used as input to include final-state interactions. The model is fit to cross sections for the photoproduction of neutral and charged pions.

3.6. *Bonn-Gatchina*

The Bonn-Gatchina group uses a K-matrix approach to unitarity, Breit-Wigner resonances, and t-channel and u-channel Regge poles describing the non-resonant amplitudes. The model is applied to cross sections (two-body final states) and event-based analyses (three-body final states) for the photoproduction of πN, ηN, $\pi\Delta$, σN, $K\Lambda$ and $K\Sigma$, where σ is the broad S-wave enhancement in $\pi\pi$, and also $\pi^- p \to \pi^0\pi^0 n$.

Acknowledgments

This work is supported by the U.S. Department of Energy under contract DE-FG02-92ER40750.

References

1. T. S. H. Lee, these Proceedings; see also T. S. H. Lee, A. Matsuyama and T. Sato, Proceedings of Workshop on the Physics of Excited Nucleons (NSTAR 2004), Grenoble, France (2004), pp. 104, and arXiv:nucl-th/0406050.
2. A. Svarc, S. Ceci and B. Zauner, arXiv:hep-ph/0601033 and these Proceedings; B. Zauner, S. Ceci and A. Svarc, arXiv:hep-ph/0601035 and these Proceedings.
3. T. P. Vrana, S. A. Dytman and T. S. H. Lee, Phys. Rept. **328**, 181 (2000); A. Kiswandhi, S. Capstick and S. Dytman, Phys. Rev. C **69**, 025205 (2004).
4. S. A. Dytman, T. P. Vrana and T. S. H. Lee, PiN Newslett. **14**, 17 (1998).
5. V. Shklyar, H. Lenske, U. Mosel and G. Penner, Phys. Rev. C **71**, 055206 (2005) [Erratum-ibid. C **72**, 019903 (2005)].
6. V. Pascalutsa and R. Timmermans, Phys. Rev. C **60**, 042201 (1999).
7. A. B. Waluyo, S. C. Karppi, K. Foe and C. Bennhold, *Prepared for NSTAR 2002 Workshop on the Physics of Excited Nucleons, Pittsburgh, Pennsylvania, 9-12 Oct 2002.*
8. D. M. Manley, Nucl. Phys. A **754**, 221 (2005).
9. H. Haberzettl, K. Nakayama and S. Krewald, arXiv:nucl-th/0512072 and these Proceedings.
10. A. V. Sarantsev, V. A. Nikonov, A. V. Anisovich, E. Klempt and U. Thoma, Eur. Phys. J. A **25**, 441 (2005); A. V. Anisovich, A. Sarantsev, O. Bartholomy, E. Klempt, V. A. Nikonov and U. Thoma, Eur. Phys. J. A **25**, 427 (2005).

Scientific Program

Wednesday October 12

8:30 **Workshop Opening (Ballroom Rm 203)**
Kirby Kemper, FSU Vice President of Research

PLENARY SESSION A:

Focus Session on Coupled-Channel Analysis
(Chair: Simon Capstick)

8:45 **Overview of Models for Extracting N* Excitations in Meson Production Reactions**
T.S. Harry Lee

9:30 **MAID Analysis Technique**
Lothar Tiator

10:10 **Coffee Break**

10:45 **Meson Production on the Nucleon in Giessen K-matrix Approach**
Horst Lenske

11:25 **The Importance of Inelastic Channels in Eliminating Continuum Ambiguities in πN Partial Wave Analysis**
Alfred Svarc

12:05 **Lunch**

PLENARY SESSION A:

Focus Session on Coupled-Channel Analysis (continuation)

14:00 **Phenomenological Analysis of Recent CLAS Data on Double-Charged Pion, Photo, and Electro-Production off Proton**
Victor Mokeev

14:35 **Gauge-invariant Approach to Meson Photoproduction Including the FSI**
Helmut Haberzettl

15:10 **Focus Session Discussion**
Moderator: Simon Capstick

15:40 **Coffee Break**

16:00 - 18:00 PARALLEL SESSIONS P1

19:00 WELCOME RECEPTION (Holiday Inn Hotel)

Thursday October 13

PLENARY SESSION B:
Pentaquarks, Exotics, Recent Experimental Results, GDH Sum Rule

(Ballroom Rm 203, Chair: Carlo Schaerf)

8:30 **Status of Pentaquarks**
Volker Burkert

9:05 **BES Results from J/ψ Decays**
Zijin Guo

9:40 **Results from the GDH Experiment at MAMI and ELSA**
Alessandro Braghieri

10:15 **Coffee Break**

10:45 **Overview: Strangeness Production**
Dan Carman

11:20 **Recent Results from the Crystal-Barrel Experiment at ELSA**
Ulrike Thoma

11:55 **KΛ and KΣ Photoproduction in a Coupled-Channels Framework**
Olaf Scholten

12:30 **Lunch**

EXCURION to Wakulla Springs and River Boat Side

14:00 Bus Departs from Conference Center

BANQUET at Wakulla Springs Lodge

18:30 Dinner
20:30 Bus Departs from Wakulla Springs for Holiday Inn Hotel

Friday October 14

PLENARY SESSION C: Cascades

(Ballroom Rm 203, Chair: B.M.K. Nefkens)

8:30 **Cascade Physics: A New Window on Baryon Spectroscopy**
John Price

9:05 **Cascades Theory**
Simon Capstick

PLENARY SESSION D: Focus Session on Polarization

(Ballroom Rm 203, Chair: Mark Manley)

9:30 **Polarization Observables in the Photoproduction of Two Pseudoscalar Mesons**
Winston Roberts

10:15 **Coffee Break**

10:45 **The CB-ELSA Polarization Program**
Hartmut Schmieden

11:20 **Frozen-Spin Target and Polarized Photon Beams at CLAS**
Franz Klein

11:55 **Crystal Ball at MAMI**
Daniel Watts

12:30 **Lunch**

PLENARY SESSION D: Focus Session on Polarization
(Ballroom Rm 203, Chair: Winston Roberts)

14:00 **GRAAL: Recent Results**
Carlo Schaerf

14:35 **CLAS: Double-Pion Beam Asymmetry**
Steffen Strauch

15:10 **Focus Session Discussion**
Moderator: Winston Roberts

15:40 **Coffee Break**

16:00 - 18:00 *PARALLEL SESSIONS P2*

18:00 - 19:00 *WINE TASTING/POSTER SESSION*
(Room 201)

Saturday October 15

PLENARY SESSION E: Focus Session on Developments in Theoretical Descriptions of Baryon Spectrum, Including Lattice QCD and Coupled-Channel Initarized Chiral Models (Ballroom Rm 203, Chair: TBA)

8:30 **Dynamically Generated Resonances**
Matthias Lutz

9:05 **Chiral Symmetry Restoration in the Baryon Spectrum**
Thomas Cohen

9:40 **Large N_c Approach to the Description of the Baryon Spectrum**
Richard Lebed

10:15 **Coffee Break**

PLENARY SESSION E

(Ballroom Rm 203, Chair: Matthias Lutz)

10:45 **Towards a Determination of the Spectrum of QCD Using a Space-Time Lattice**
Colin Morningstar

11:20 **Dynamically Generated Spin 3/2 Resonances from Meson Octet-Baryon Decuplet Interaction**
Sourav Sarkar

11:55 **Focus Session Discussion**
Moderator: Matthias Lutz

Wednesday October 12

PARALLEL SESSION P1-A

(Room 203, Chair: Lothar Tiator)

16:00 **Coupled-Channel Fit to πN Elastic and η Production Data**
William Briscoe

16:20 **A PWA Program for Analyzing Photoproduction of Baryon Resonances**
Mike Williams

16:40 **Multichannel Partial-Wave Analysis of $\overline{\mathrm{K}}$N Scattering**
Hongyu. Zhang

17:00 **Helicity Amplitudes and Electromagnetic Decays of Strange Baryon Resonances**
Tim van Cauteren

17:20 **Progress Report for a New Karlsruhe-Helsinki Type πN PWA**
Timothy Watson

17:40 **Baryon Resonances through Meson Hadro- and Photoproduction in a Coupled-Channels Framework**
Agung Waluyo

PARALLEL SESSION P1-B

(Room 208, Chair: Dan Carman)

16:00 **Double-Polarization Observables in π Photoproduction from Polarized HD at LEGS**
Andrew Sandorfi

16:25 **Electroexcitation of Nucleon Resonances from CLAS Data**
Inna Aznauryan

16:50 **A Genetic-Algorithm Analysis of N* Resonances**
David Ireland

17:15 **Electroproduction of π^0 from $\Delta(1232)$ at High Q^2 with CLAS**
Maurizio Ungaro

17:35 **Measurement of Complete Angular Distributions of Exclusive nπ^+ Production From Protons for $Q^2 = 1.5 - 4$ GeV^2**
Kijun Park

17:55 **πN Charge Exchange in the N*(1440) Resonance Region**
Michael Sadler

PARALLEL SESSION P1-C

(Room 218, Chair: Eulogio Oset)

16:00 **CB-ELSA: N* Resonance Decay by the $K^0\Sigma^+$ Channel**
Ralph Castelijns

16:25 **Dynamical Coupled-Channel Study of the Process $\gamma p \to K^+\Lambda$**
T.-S. Harry Lee

16:45 **C_z, C_x for $K^+\Lambda$ and $K^+\Sigma^0$ Photoproduction**
Robert Bradford

17:05 **Light Hyperon Photoproduction on a Proton Target at JLab**
Lei Guo

17:25 **Strangeness Production via the Reaction $\gamma p \to K^{*0}\Sigma^+$**
Ishaq Hleiqawi

17:45 **Strangeness Produciton on the Neutron via the Reaction $\gamma n(p) \to K^+\Sigma^-$**
Jörn Langheinrich

Friday October 14

PARALLEL SESSION P2-A

(Room 203, Chair: Volker Credé)

16:00 **CLAS Results on Pseudoscalars**
Michael Dugger

16:25 **CB-ELSA: $\pi^0\pi^0$ Results**
Michael Fuchs

16:50 **Analyzing η' Photoproduction Data on the Proton at Energies of 1.5 - 2.3 GeV**
Kanzo Nakayama

17:15 **GRAAL Results on $n\eta$**
William Briscoe

17:35 **CB-ELSA: η and η' Photoproduction off the Deuteron**
Igal Jaegle

17:55 **Λ and Σ^0 Photoproduction on the Neutron**
Pawel Nadel-Turonski

PARALLEL SESSION P2-B

(Room 208, Chair: Paul Eugenio)

16:00 **The Problem of Exotic States: View From Complex Angular Momenta**
Yakov Azimov

16:25 Θ^+ **Search in CLAS with** $\gamma n \to pK^0_S K^-$
N.A. Baltzel

16:45 $\Sigma^+(1189)$ **Production in Photonuclear Reactions off the Proton**
Mariana Nanova

17:05 **Photoexcitation of Hyperons and Exotic Baryons in** $\gamma N \to K\overline{K}N$
Yongseok Oh

17:25 **On the Nature of the** $\Lambda(1405)$ **as a Superposition of Two States**
Eulogio Oset

17:45 **Channel-Coupling Effects in Photo-induced** ρN **Production**
Olaf Scholten

PARALLEL SESSION P2-C

(Room 218, Chair: T.-S. Harry Lee)

16:00 **The Influence of Inelastic Channels Upon the Pole Structure of Partial Wave T-matrices in the Coupled-Channel** πN **PWA**
Sasa Ceci

16:20 **The Importance of** $\pi N \to K\Lambda$ **Process for the Pole Structure of th the** P_{11} **Partial Wave T-matrix in the Coupled-Channel** πN **PWA**
Branimir Zauner

16:40 **Cascade Spectroscopy in Photoproduction on a Proton Target at JLab**
Lei Guo

17:00 **Structure of the** σ **Meson and Diamagnetism of the Nucleon**
Martin Schumacher

17:20 **Baryon Resonances in the** $1/N_c$ **Expansion**
Nicolas Matagne

17:40 **Major Surprises in** $\pi^0\pi^0$ **Production by** π^- **and** K^- **at Intermediate Energies**
Bernard Nefkens

Participants

Applegate, Douglas
Carnegie Mellon U
5526 Wilkins Ave
Pittsburgh, PA 15217 USA

Azimov, Yakov
Petersburg Nuclear Physics
Institute, PNPI
Orlova Rosha
188300 Gatchina, Russia

Aznauryan, Inna
Jefferson Lab
12000 Jefferson Avenue
Newport News, VA 23606 USA

Baltzell, Nathan
807 Wedgefield Road
Florence, SC 29501 USA

Blaszczyk, Lukasz
Florida State University
Department of Physics
102 Collins Research Laboratory
Tallahassee, FL 32306 USA

Bradford, Robert K.
University of Rochester
Department of Physics
500 Wilson Boulevard
Rochester, NY 14627-0171 USA

Braghieri, Alessandro
INFN - Pavia
Via Bassi 7
Pavia, Italy 27100

Briscoe, William J.
George Washington U
Department of Physics
725 21st Street, NE
Washington, DC USA

Burkert, Volker
Jefferson Lab
12000 Jefferson Ave
Newport News, VA 23606 USA

Capstick, Simon
Florida State University
Department of Physics
207 Keen Building
Tallahassee, FL 32306 USA

Carman, Daniel S.
Ohio University
Department of Physics
Athens, OH 45701 USA

Castelijns, Ralph
KVI, Zernikelaan 25
9747 AA Groningen
The Netherlands

Ceci, Sasa
Rudjer Boskovic Institute
Bijenicka c. 54
10000 Zagreb, Croatia

Chen, Shifeng
Florida State University
Department of Physics
102 Collins Research Laboratory
Tallahassee, FL 32306 USA

Cohen, Thomas D.
U of Maryland College Park
Department of Physics
082 Regents Drive
College Park, MD 20742 USA

Cook, Merritt S.
Florida International University
Physics Department
PO Box 940454
Miami, FL 33194 USA

Crawford, Ronald L.
University of Glasgow
D of Physics & Astronomy
Glasgow G12 8QQ
United Kingdom

Credé, Volker
Florida State University
Department of Physics
206 Keen Building
Tallahassee, FL 32306 USA

Davis, Rebecca M.
George Washington U
Physics Department
725 21st St, NE
Washington, DC USA

Dugger, Michael R.
Arizona State University
Physics Department
505 W. Baseline Rd.
Tempe, AZ 85287 USA

Eugenio, Paul M.
Florida State University
Department of Physics
205 Keen Building
Tallahassee, FL 32306 USA

Fuchs, Michael
Nanoscience Lab
Suwon
Jang-an Gu, Cheon cheon-dong
Sungkyunkwan University
Natural Science Bld. A, #31155
440 - 746, South Corea

Graham, Lewis P.
University of South Carolina
Physics Department
2121 Bee Ridge Rd
Columbia, SC 29223 USA

Guo, Lei
Jefferson Lab
12000 Jefferson Ave
Newport News, VA 23606 USA

Guo, Zijin
3400 North Charles Street
Bloomberg 366
Baltimore, MD 21218 USA

Hanretty, Charles M.
Florida State University
Department of Physics
102 Collins Research Laboratory
Tallahassee, FL 32306 USA

Haberzettl, Helmut
George Washington U
Department of Physics
Washington, DC 20052 USA

Helminen, Christina I.
University of Helsinki
Department of Physical Sciences
Helsinki, Finland 00014

Hleiqawi, Ishaq
403 N. Second St.
Hampton, VA 23664 USA

Igal, Jaegle
Universität Basel
Institut für Physik
Klingelbergstrasse 82
Basel, Switzerland CH-4056

Ireland, Dave
Dept of Physics & Astronomy
University Avenue
Glasgow, G12 8QQ
United Kingdom

Kiswandhi, Alvin
Florida State University
Department of Physics
202 Keen Building
Tallahassee, FL 32306 USA

Klein, Franz
CUA
Physics Department
210 Hannan Hall
Washington, DC 20064 USA

Krusche, Bernd
University of Basel
Institut für Physik
Klingelbergstr. 82
Basel, BL 4056 Switzerland

Langheinrich, Jörn H.
University of South Carolina
734 1/2 N 1st Street; Apt #B
Hampton, VA 23664 USA

Lebed, Richard F.
Arizona State University
Dept of Physics & Astronomy
Tempe, AZ 85287 USA

Lee, T. S. Harry
Argone National Laboratory
Physics Division
Argonne, IL 60439 USA

Lenske, Horst
Universität Giessen
Institut für Theoretische Physik
Heinrich-Buff-Ring 16
35392 Giessen, Germany

Lu, Haiyun
USC
3534 Lee Hills Dr.
Columbia, SC 29209 USA

Lutz, Matthias F.M.
GSI
Planck Str. 1
64291 Darmstadt, Germany

Lyczek-Way, Mica
Florida State University
Department of Physics
Tallahassee, FL 32306 USA

Manley, Mark
Kent State University
Department of Physics
105 Smith Hall
Kent, OH 44242 USA

Matagne, Nicolas
University of Liege
BAT.B5
Physique Theorique Fondamentale
Allee du 6-aout 17
Liege, Belgium 4000

Mokeyev, Victor I.
Jefferson Lab
12000 Jefferson Avenue
Newport News, VA 23606 USA

Morningstar, Colin
Carnegie Mellon U
Department of Physics
5000 Forbes Avenue
Pittsburgh, PA 15213 USA

Nadel-Turonski, Pawel A.
Jefferson Lab
12000 Jefferson Avenue
Newport News, VA 23606 USA

Nakayama, Kanzo
University of Georgia
Dept of Physics & Astronomy
Athens, GA 30602 USA

Nanova, Mariana
Giessen University
II. Physikalisches Institut
Heinrich Buff Ring 16
35392 Giessen, Germany

Nefkens, Bernard M.
UCLA
Dept of Physics & Astronomy
Box 951547
Los Angeles, CA 90095-1547 USA

Oh, Yongseok
University of Georgia
Dept of Physics & Astronomy
Athens, GA 30602 USA

Oset, Eulogio
IFIC
Edificio de Institutos de
Investigacion de Paterna
Aptdo. Correos 22085
46071 Valencia, SPAIN

Paris, Mark W.
Jefferson Lab
12000 Jefferson Avenue
Newport News, VA 23606 USA

Park, Kijun
2490 Fish Hatchery Rd., Unit #I-4
W. Columbia, SC 29172 USA

Pervin, Muslema
Florida State University
Department of Physics
Tallahassee, FL 32306 USA

Price, John W.
California State U
Dominguez Hills
CSUDH Dept of Physics
1000 E. Victoria St.
Carson, CA 90747 USA

Roberts, Winston
Florida State University
Department of Physics
211 Keen Building
Tallahassee, FL 32312 USA

Sadler, Michael
Abilene Christian University
ACU Box 27963
320B Foster Science Building
Abilene, TX 79699 USA

Sandorfi, Andrew M.
Brookhaven National Lab
Physics Department
Building 510
Upton, NY 11973 USA

Santoro, Joseph P.
Catholic University of America
273 Eastwood Dr.
Newport News VA 23602 USA

Sarkar, Sourav
University of Valencia
Nuclear Theory Group
Apartado Correos 22985
Valencia, Spain 46071

Schaerf, Carlo
University of Rome "Tor Vergata"
Dipartimento di Fisica
Via della Ricerca Scientifica
00133 Roma, Italy

Schmieden, Hartmut
Universität Bonn
Physikalisches Institut
Nussallee 12
53115 Bonn, Germany

Scholten, Olaf
KVI. Zernikelaan 25
Groningen, 9747 AA
The Netherlands

Schumacher, Martin
University of Goettingen
Zweites Physikalisches Institut
Friedrich-Hund-Platz 1
37077 Goettingen, Germany

Strauch, Steffen
University of South Carolina
Dept. of Physics & Astronomy
712 Main Street
Columbia, SC 29208 USA

Svarc, Alfred
Rudjer Boskovic Institute
Division of Experimental Physics
P.P. 180
Bijenicka c. 54
10000 Zagreb, Croatia

Taruna, Jutri
Florida State University
Department of Physics
FSA Box 61353
Tallahassee, FL 32306 USA

Tedeschi, David J.
University of South Carolina
Dept. of Physics & Astronomy
Columbia, SC 29208 USA

Thoma, Ulrike
Bonn University
HISKP, Universitaet Bonn
Nussallee 14-16
53115 Bonn, Germany

Tiator, Lothar
Institut für Kernphysik
Universität Mainz
Becher-Weg 45
55099 Mainz, Germany

Ungaro, Maurizio
UConn/JLab
563 Logan Plaace, Apt 13
Newport News, VA 23601 USA

Usov, Alexander
KVI, Zernikelaan 25
Groningen, 9747 AA
The Netherlands

Van Cauteren, Tim
Ghent University
Inst. v. Nucl. Wetenschappen
Proeftuinstraat 86
Ghent, Belgium 9000

Volya, Alexander
Florida State University
Department of Physics
208 Keen Building
Tallahassee, FL 32306 USA

Waluyo, Agung B.
George Washington U
12 West Deer Park Rd - Apt 203
Gaithersburg, MD 20877 USA

Watts, Daniel P.
University of Edinburgh
Room 8209 JCMB
Kings Buildings, Mayfield Rd
Edinburgh, EH9 3JZ
United Kingdom

Williams, Mike
Carnegie Mellon U
Department of Physics
5000 Forbes Ave
Pittsburgh, PA 15213 USA

Zauner, Branimir
Rudjer Boskovic Institute
Department of Exp. Physics
Bijenicka c. 54
Hanamanova 3
10000 Zagreb, Croatia

Zhang, Hongyu
Kent State University
Department of Physics
105 Smith Hall
Kent, OH 44242 USA

Zhao, Zhiwen
USC
Physics Department
712 Main Street
West Columbia, SC 29208 USA

AUTHOR INDEX